COMBUSTION and GASIFICATION in FLUIDIZED BEDS

Related Titles

Simeon Oka, Fluidized Bed Combustion   0824746996

Hiroshi Tsuji, Ashwani K. Gupta, Toshiaki Hasegawa, Masashi Katsuki, Ken Kishimoto, Mitsunobu Morita, High Temperature Air Combustion: From Energy Conservation to Pollution Reduction   0849310369

Marcio L. de Souza-Santos, Solid Fuels Combustion and Gasification: Modeling, Simulation, and Equipment, Operation   0824709713

Charles E. Baukal, Jr., Industrial Burners Handbook   0849313864

Charles E. Baukal, Jr., The John Zink Combustion Handbook   0849323371

COMBUSTION and GASIFICATION in FLUIDIZED BEDS

Prabir Basu

Taylor & Francis Group
Boca Raton London New York

A CRC title, part of the Taylor & Francis imprint, a member of the
Taylor & Francis Group, the academic division of T&F Informa plc.

Chapter 3 on "Fluidized Bed Gasification" is co-authored by Dr. S.C. Bhattacharya and Dr. P. Basu.

FIRST INDIAN REPRINT 2009

Published in 2006 by
CRC Press
Taylor & Francis Group
6000 Broken Sound Parkway NW, Suite 300
Boca Raton, FL 33487-2742

© 2006 by Taylor & Francis Group, LLC
CRC Press is an imprint of Taylor & Francis Group

Printed and bound in India by Replika Press Pvt. Ltd.

International Standard Book Number-10: 0-8493-3396-2 (Hard cover)
International Standard Book Number-13: 978-0-8493-3396-5 (Hard cover)
Library of Congress Card Number 2005053843

This book contains information obtained from authentic and highly regarded sources. Reprinted material is quoted with permission, and sources are indicated. A wide variety of references are listed. Reasonable efforts have been made to publish reliable data and information, but the author and the publisher cannot assume responsibility for the validity of all materials or for the consequences of their use.

No part of this book may be reprinted, reproduced, transmitted, or utilized in any form by any electronic, mechanical, or other means, now known of hereafter invented, including photocopying, microfilming, and recording, or in any information storage or retrieval system, without written permission from the publishers.

For permission to photocopy or use material electronically from this work, please access www.copyright.com (http://www.copyright.com/) or contact the Copyright Clearance Center, Inc. (CCC) 222 Rosewood Drive, Danvers, MA 01923, 978-750-8400. CCC is a not-for-profit organization that provides licenses and registration for a variety of users. For organizations that have been granted a photocopy license by the CCC, a separate system of payment has been arranged.

Trademark Notice: Product or corporate names may be trademarks or registered trademarks, and are used only for identification and explanation without intent to infringe.

Library of Congress Cataloging-in-Publication Data

Basu, Prabir,1946-
    Combustion and gasification in fluidized beds / Prabir Basu.
       p. cm.
    Includes bibliographical references and index.
    International Standard Book Number 10: 0-8493-3396-2
    1. Steam boilers. 2. Fluidized bed combustion. I. Title.

TJ288.B36 2005
621.1′83--dc22
                                          2005053843

Visit the Taylor & Francis Web site at
http://www.taylorandfrancis.com

and the CRC Press Web site at
http://www.crcpress.com

FOR SALE IN SOUTH ASIA ONLY

Preface

Fritz Winkler and Douglas Elliott, two pioneers of fluidized bed boilers and gasifiers, never thought the application of fluidized beds for energy production would be as extensive as it is today. Concerns about climate change and economic forces have made the need for clean energy from fossil and biomass fuels more important than ever. Several thousands of fluidized bed boilers and gasifiers are in operation around the world with capacities approaching 500–600 MWe. Use of fluidized beds for energy production as well as waste utilization is increasing steadily. Yet, it was difficult to find a comprehensive book on the basic design and working principles of fluidized bed gasifiers and boilers. The author felt this need especially while conducting professional development seminars, teaching fluidized beds courses in the university and working on fluidized bed power plants.

This monograph, designed to meet the above need, will provide engineers, scientists and students with a comprehensive understanding of the working of fluidized bed boilers and gasifiers. Prior knowledge of the process of fluidization is not essential, as the book presents both basics and advanced materials on the subject. Engineers, operators and technicians involved in energy industries can understand how fluidized bed equipment works, while students can learn how the basic principles of thermodynamics and fluid mechanics are used in the design of fluidized bed boilers and gasifiers. Experienced fluidization researchers can use this book as a good reference to identify areas of incomplete comprehension of the subject. Project engineers will find it useful in planning new projects or understanding the working of or trouble-shooting existing boiler and gasification plants. The planners will gain a better appreciation of the process, including its capabilities and limitations.

The book is comprised of 13 chapters and 4 appendices, including 7 tables that are useful for the design of fluidized bed equipment. The first chapter introduces readers to fluidized bed boilers and gasifiers while comparing other options for power generation from fossil or biomass fuels. Chapter 2 covers the basics of fluidized bed hydrodynamics. Chapter 3 discusses the fundamentals of gasification and its application to the design of fluidized bed gasifiers. Basics of combustion of solid fuels and its application in different types of fluidized beds are presented in Chapter 4. Chapter 5 concentrates on the pollution aspects of fluidized bed plants including issues related to climate change mitigation. Heat transfer in fluidized beds is discussed in Chapter 6. Chapter 7 pulls together information from earlier chapters to explain how bubbling fluidized bed boilers work and how these are designed. A comprehensive treatment of circulating fluidized bed boilers is presented in Chapter 8. The relevance of design and feed stock parameters to the operation of such boilers is also discussed in this chapter. Chapter 9, which discusses issues related to construction and maintenance of fluidized bed equipment, covers the construction materials exposed to special service conditions in fluidized bed gasifiers and boilers.

A fluidized bed boiler or gasification plant also includes several mechanical components, such as feeding systems, air distribution grids, gas–solid separators, and solid recycle systems, which are discussed in Chapters 10–13, respectively. Appendix 1 discusses relevant physical characteristics of granular solids typically used in fluidized beds. Calculations needed for the heat and mass balance of combustion reactions are presented in Appendix 2. A simplified model for computation of sorbent required for sulfur capture is given in Appendix 3. Useful data for design and other calculations related to fluidized bed equipment are presented in a series of tables in Appendix 4.

Information needed for the design, evaluation, and analysis of the performance of fluidized bed gasifiers and boilers was found scattered in many research papers and textbooks in related fields. Efforts were made to collect and research them, and then to organize them into a coherent design sequence. My special thanks go to Dr. Sribas C. Bhattacherya and Rui Zhang based on whose writings Chapters 3 and 7 were prepared. Dr. Murat Koksal of Dalhousie University and Dr. Animesh Dutta of Greenfield Research Inc. went through several chapters of this book and gave many helpful suggestions for improvements. Sami Sakalla and Subrata Roy drew all the diagrams. Several students, including Mathew Bligh, greatly helped the editing of the manuscript. The author is grateful to several publishers who permitted free use of materials for this book. Final thanks go to my wife Rama, who guided me through the project, and to my children Atreyee and Atreya for their encouragement and support.

Prabir Basu
Halifax, Nova Scotia

Author

Professor Prabir Basu is a noted international expert in fluidized beds. He has been working on fluidized bed energy conversion systems for more than 30 years. His notable contributions include the world's first commercial biomass fired fluidized bed boiler and the first English and Chinese language text books on circulating fluidized bed boilers. Professor Basu also founded the prestigious International Circulating Fluidized Bed Conference Series. He is known for his mission of spreading the fluidized bed-based clean energy option around the world. His close interaction with power industries made his original research work very much practice oriented. He has authored more than 160 research papers and 3 books and owns several patents. Presently, Dr. Basu heads the circulating fluidized bed research laboratory at the Dalhousie University in Canada.

Table of Contents

Chapter 1   Introduction 1
1.1 Background 1
1.2 Fluidized Bed Equipment 4
    1.2.1   Applications 4
    1.2.2   Fluidized Bed Gasifiers 5
    1.2.3   Fluidized Bed Boilers 6
    1.2.4   Application in Cement Industries 9
1.3 Features of Fluidized Beds 10
    1.3.1   Advantages of Fluidized Bed Boilers 10
    1.3.2   Features Specific to CFB Boiler 12
1.4 Fluidized Bed and Other Technology Options 13
    1.4.1   Technological Options for Generation of Power from Solid Fuels 13
    1.4.2   Improved Firing 15
    1.4.3   Combined-Cycle Plants 16
    1.4.4   Cost Implications 18
References 19

Chapter 2   Hydrodynamics 21
2.1 Regimes of Fluidization 21
    2.1.1   Packed Beds 25
    2.1.2   Bubbling Fluidized Beds 25
    2.1.3   Slugging 29
    2.1.4   Turbulent Beds 29
    2.1.5   Bubbling to Turbulent Transition 29
    2.1.6   Terminal Velocity of a Particle 31
    2.1.7   Freeboard and Furnace Heights 35
2.2 Fast Fluidized Bed 35
    2.2.1   Characteristics of Fast Beds 35
    2.2.2   Transition to Fast Fluidization 37
    2.2.3   Dense Suspension Up-Flow 42
2.3 Hydrodynamic Regimes in a CFB 42
    2.3.1   Difference in Operating Regimes of CFB Boilers and Reactors 43
2.4 Hydrodynamic Structure of Fast Beds 43
    2.4.1   Axial Voidage Profile 43
    2.4.2   Lateral Distribution of Voidage in a Fast Bed 48
2.5 Gas–Solid Mixing 52
    2.5.1   Gas–Solid Slip Velocity 52
    2.5.2   Dispersion 52
2.6 Scale-Up 53
    2.6.1   Circulating Fluidized Bed 53
    2.6.2   Bubbling Fluidized Bed 54
    2.6.3   Cyclone 54

Nomenclature 55
    Greek Symbols 56
    Dimensionless Numbers 56
References 57

Chapter 3   Fluidized Bed Gasification 59
3.1   Types of Gasifiers 60
    3.1.1   Entrained Bed 61
    3.1.2   Fluidized Bed 62
    3.1.3   Spouted Bed 62
    3.1.4   Fixed/Moving Bed 62
3.2   Theory 63
    3.2.1   Pyrolysis or Devolatilization 64
    3.2.2   Combustion 66
    3.2.3   Gasification 66
    3.2.4   Composition of Gas Yield 67
3.3   Effect of Operating Parameters on Gasification 69
    3.3.1   Boudouard Reaction 69
    3.3.2   Water–Gas 69
    3.3.3   Methanation 69
3.4   Effect of Feed Properties on Gasification 71
    3.4.1   Fuel Reactivity 71
    3.4.2   Volatile Matter 72
    3.4.3   Ash 73
    3.4.4   Moisture 73
3.5   Fluidized Bed Gasification 74
    3.5.1   Comparison of Bubbling and Circulating Fluidized Bed Gasifiers 75
    3.5.2   Types of Fluidized Bed Gasifiers 75
3.6   Examples of Some Fluidized Bed Gasifiers 77
    3.6.1   Winkler Gasifier 77
    3.6.2   KRW Gasifier 79
    3.6.3   Spouted Bed Partial Gasifier 79
    3.6.4   Foster Wheeler Atmospheric CFB Gasifier 80
    3.6.5   Battelle/Columbus Indirect Gasifier 81
    3.6.6   Halliburton KBR Transport Reactor 82
    3.6.7   Carbon Dioxide Acceptor Process 82
3.7   Gas Cleaning 83
    3.7.1   Sulfur and Other Chemical Contaminants Removal 83
    3.7.2   Tar Removal 84
    3.7.3   Particulate Removal 85
3.8   Design Considerations 85
    3.8.1   Gasifier Efficiency 85
    3.8.2   Equivalence Ratio 86
    3.8.3   Gas Composition 88
    3.8.4   Energy and Mass Balance 89
    3.8.5   Bed Materials 89
    3.8.6   Gasifier Sizing 90
    3.8.7   Alternative Design Approach 95
3.9   Modeling of Fluidized Bed Gasification 96
Appendix 3A 97
    Estimating Equilibrium Gas Composition 97

Appendix 3B 99
    An Example of a Fluidized Bed Biomass Gasifier 99
Nomenclature 99
    Greek Symbols 100
References 100

Chapter 4   Combustion 103

4.1 Stages of Combustion 103
    4.1.1 Heating and Drying 103
    4.1.2 Devolatilization 104
    4.1.3 Volatile Combustion 106
    4.1.4 Char Combustion 107
    4.1.5 Burning Rate of Char in Fluidized Beds 109
    4.1.6 Communition Phenomena during Combustion 115
4.2 Factors Affecting Combustion Efficiency 117
    4.2.1 Feed Stock 117
    4.2.2 Operating Conditions 119
4.3 Combustion in Bubbling Fluidized Bed Boilers 120
    4.3.1 Recirculation of Fly Ash 120
    4.3.2 Effect of Design Parameters on Combustion Efficiency 120
4.4 Combustion in Circulating Fluidized Bed Boilers 122
    4.4.1 Combustion of Char Particles 123
    4.4.2 Effect of Cyclone Design on Combustion Efficiency 125
    4.4.3 Some Combustion Design Considerations 125
    4.4.4 Performance Modeling 127
4.5 Biomass Combustion 127
    4.5.1 Agglomeration 129
    4.5.2 Fouling 130
    4.5.3 Corrosion Potential in Biomass Firing 130
Nomenclature 130
    Greek Symbols 131
    Dimensionless Numbers 132
References 132

Chapter 5   Emissions 135

5.1 Air Pollution 135
    5.1.1 Pollutants 136
    5.1.2 Emission Standards 141
5.2 Sulfur Dioxide Emission 141
    5.2.1 Chemical Reactions 142
    5.2.2 Reactions on Single Sorbent Particles 146
    5.2.3 Reactivity of Sorbents 148
    5.2.4 Sulfur Capture in Fluidized Beds 150
    5.2.5 Simple Model for Sulfur Capture in CFB Boilers 154
    5.2.6 Selection of Sorbent 157
5.3 Nitrogen Oxide Emission 159
    5.3.1 Sources of NO_x 159
    5.3.2 Methods of Reduction of NO_x Emissions 159
    5.3.3 NO_x Emission from CFB 161

5.4	Nitrous Oxide Emission	162
	5.4.1 Mechanism of Formation of N_2O in CFB	162
	5.4.2 Effects of Operating Parameters on N_2O	163
	5.4.3 Reduction of N_2O	164
5.5	Mercury Emission	164
	5.5.1 Sorbent Injection	164
	5.5.2 Electro-Catalytic Oxidation	164
	5.5.3 Precombustion Technologies	165
5.6	Carbon Monoxide Emission	165
5.7	Carbon Dioxide Emission	165
5.8	Emission of Trace Organics	166
5.9	Particulate Emission	166
	5.9.1 Ash Characteristics	166
Nomenclature		168
	Greek Symbols	169
References		169

Chapter 6 Heat Transfer 173

6.1	Gas-to-Particle Heat Transfer	173
	6.1.1 Gas-to-Particle Heat Transfer Equations	173
	6.1.2 Heating of Gas and Solids in the Fast Bed	176
6.2	Heat Transfer in Circulating Fluidized Beds	178
	6.2.1 Mechanism of Heat Transfer	178
	6.2.2 Theory	179
	6.2.3 Experimental Observations in a CFB Boiler	184
	6.2.4 Effect of Vertical Fins on the Walls	187
6.3	Heat Transfer in Bubbling Fluidized Bed	189
	6.3.1 Mechanistic Model	190
	6.3.2 Experimental Observations in Bubbling Fluidized Beds	195
	6.3.3 Heat-Transfer Correlation in Bubbling Beds	197
	6.3.4 Use of Design Graphs	197
	6.3.5 Calculation Based on Tables	200
6.4	Freeboard Heat Transfer in Bubbling Fluidized Beds	201
6.5	Heat Transfer in Commercial-Size CFB Boilers	202
	6.5.1 Heat Transfer to the Walls of Commercial CFB Boilers	202
	6.5.2 Wing Walls	202
	6.5.3 Heat-Transfer Variation along the Furnace Height	203
	6.5.4 Circumferential Distribution of Heat-Transfer Coefficient	203
Nomenclature		205
	Greek Symbols	207
	Dimensionless Numbers	207
References		207

Chapter 7 Bubbling Fluidized Bed Boiler 211

7.1	Description of a BFB Boiler	211
	7.1.1 Feedstock Preparation and Feeding	211
	7.1.2 Air and Gas Handling	213
	7.1.3 Ash and Emission Systems	213
	7.1.4 Steam Generation System	215
	7.1.5 Combustion Chamber	215

7.2	Features of BFB Boilers	215
	7.2.1 Advantages	215
	7.2.2 Limitations	216
7.3	Thermal Design of Bubbling Fluidized Bed Boilers	216
	7.3.1 Overall Heat Balance of the Boiler	216
	7.3.2 Boiler Efficiency	219
	7.3.3 Energy Balance in Dense Bubbling Bed	221
	7.3.4 Heat Balance of Boiler Components	223
	7.3.5 Mass Flow Rates of Fuel and Air	226
	7.3.6 Gas-Side Heat Balance	226
7.4	Combustion in Bubbling Fluidized Bed	228
	7.4.1 Coal Combustion in BFB Boilers	228
	7.4.2 Minimization of Combustible Losses in BFB Combustors	230
7.5	Furnace Design	234
	7.5.1 Basic Design of Furnace	235
	7.5.2 Part-Load Operations	238
	7.5.3 Other Operating Conditions	240
7.6	Start-Up Procedure and Control	242
	7.6.1 Start-Up Procedure	243
	7.6.2 Control of BFB Boiler	243
7.7	Some Operational Problems and Remedies	246
	7.7.1 Erosion of In-Bed Components	246
7.8	Design of Feed Systems for BFB Bed Boiler	248
Nomenclature		249
	Greek Symbols	251
References		251

Chapter 8 Circulating Fluidized Bed Boiler 253

8.1	General Arrangement	253
	8.1.1 Air System	253
	8.1.2 Flue Gas Stream	255
	8.1.3 Solid Stream	255
	8.1.4 Water–Steam Circuit	255
8.2	Types of CFB Boilers	257
	8.2.1 Boilers without Bubbling Bed Heat Exchangers	257
	8.2.2 Boilers with Bubbling Fluidized Bed Heat Exchanger	258
	8.2.3 Boilers with Inertial or Impact Separators	259
	8.2.4 Boilers with Vertical, Noncircular Cyclones	259
	8.2.5 Other Types	259
8.3	Non-CFB Solid Circulation Boilers	259
	8.3.1 Circo-Fluid Boilers	259
	8.3.2 Low-Solid Circulation Boiler	260
	8.3.3 Circulating Moving Bed Boiler	260
8.4	Combustion in a Circulating Fluidized Bed Furnace	263
8.5	Design of CFB Boilers	264
	8.5.1 Combustion Calculations	264
	8.5.2 Heat Balance	264
	8.5.3 Mass Balance	266
	8.5.4 Flow Split	266
	8.5.5 Control of Bed Inventory	267
	8.5.6 Some Operating Issues	270

8.6	Furnace Design	271
	8.6.1 Furnace Cross-Section	273
	8.6.2 Width and Breadth Ratio	273
	8.6.3 Furnace Openings	276
8.7	Design of Heating Surfaces	277
	8.7.1 Effect of Fuel Type on Furnace Heat Load	278
	8.7.2 Heat Absorption in the External Heat Exchanger	280
	8.7.3 Heat Absorption in the Furnace and Rest of the Boiler	280
	8.7.4 Heat Balance around a CFB Loop	285
8.8	Auxiliary Power Consumption	287
8.9	Control of CFB Boilers	287
	8.9.1 Part-Load Operation	287
	8.9.2 Load Control Options	289
8.10	Supercritical CFB Boiler	291
	8.10.1 Supercritical Operation	292
	8.10.2 Vertical vs. Wrapped Tube Arrangement	292
	8.10.3 Low Heat Flux	293
	8.10.4 Cost Advantage	293
Nomenclature		293
	Greek Symbols	296
References		296

Chapter 9 Material Issues 299

9.1	Material Selection Criteria	299
	9.1.1 Structural Requirements	299
	9.1.2 Heat-Transfer Duty	300
9.2	Erosion Potential	301
	9.2.1 Basics of Erosion	301
	9.2.2 Types of Erosion	304
	9.2.3 Erosion in Bubbling Fluidized Beds	304
	9.2.4 Options for Reduction of Erosion in Bubbling Beds	306
	9.2.5 Erosion in CFB Boilers	307
9.3	Corrosion Potentials	311
	9.3.1 Corrosion Mechanisms	312
	9.3.2 Examples of Fire-Side Corrosion	312
	9.3.3 Chemistry of Fire-Side Corrosion of Heating Surfaces	313
	9.3.4 Chlorine Corrosion	315
	9.3.5 Vanadium Corrosion	316
	9.3.6 Prevention of High-Temperature Corrosion	316
	9.3.7 Erosion–Corrosion	317
	9.3.8 Fouling and Deposit Formation	318
9.4	Steels Used in Fluidized Bed Boilers	318
	9.4.1 Carbon and Alloy Steels	318
9.5	Refractory and Insulations	320
	9.5.1 Importance of Lining	320
	9.5.2 Properties of Refractory	320
	9.5.3 Types of Refractory	321
	9.5.4 Types of Insulation	322
	9.5.5 Anchors	324
	9.5.6 Design Considerations of Lining	325
	9.5.7 Areas of Refractory Use in Fluidized Bed Boilers	325

		9.5.8	Failure Analysis	329

         9.5.8   Failure Analysis 329
         9.5.9   Refractory Maintenance 331
  9.6   Expansion Joints 333
  Nomenclature 334
         Greek Symbols 334
  References 334

Chapter 10   Solid Handling Systems for Fluidized Beds 337

  10.1   Solid Handling Systems 337
         10.1.1   Fossil Fuel Plants 337
         10.1.2   Municipal Solid Waste 339
  10.2   Biomass Handling Systems 341
         10.2.1   Feeding of Harvested Fuels 341
         10.2.2   Feeding of Nonharvested Fuels 341
  10.3   Feed System for Bubbling Beds 342
         10.3.1   Types of Feeders 342
         10.3.2   Mode of Feeding in Bubbling Beds 347
         10.3.3   Feed Point Allocation 350
  10.4   Feed System for Circulating Fluidized Beds 352
         10.4.1   Fuel Feed Ports 352
         10.4.2   Limestone Feed System 353
  10.5   Design of Pneumatic Transport Lines for Solids 353
         10.5.1   Pressure Drop in a Pneumatic Transport Line 353
         10.5.2   Design of a Pneumatic Injection System 355
         10.5.3   Fuel Splitter 355
  Nomenclature 357
         Greek Symbols 357
  References 357

Chapter 11   Air Distribution Grate 359

  11.1   Distributor Plates 359
         11.1.1   Plate-Type Distributor 360
         11.1.2   Nozzle-Type Distributor 362
         11.1.3   Sparge Pipe Distributor 363
         11.1.4   Distributor Grids for CFB 364
         11.1.5   Distributor Grids for Bubbling Beds 365
  11.2   Operation of Distributors 366
         11.2.1   What Causes Nonuniform Fluidization 366
  11.3   Design Methods 368
         11.3.1   Design Procedure for Bubbling Fluidized Beds 368
         11.3.2   Design Procedure for Circulating Fluidized Beds 370
  11.4   Practical Considerations 371
         11.4.1   Plenum or Air Box 372
         11.4.2   Sealing of Distributor 373
         11.4.3   Attrition 374
         11.4.4   Back-Flow of Solid 375
         11.4.5   Opening and Closing of Nozzles 377
         11.4.6   Erosion and Corrosion of Nozzles 378
  Nomenclature 379
         Greek Symbols 379
  References 380

Chapter 12   Gas–Solid Separators 381

12.1   Cyclones 381
    12.1.1   Types of Cyclone 382
    12.1.2   Theory 385
    12.1.3   Critical Size of Particles 385
    12.1.4   Working of Cyclones for CFB Combustors or Gasifiers 390
    12.1.5   Pressure Drop through a Cyclone 392
    12.1.6   Cyclone Collection Efficiency 394
    12.1.7   Reentrainment of Solids 399
    12.1.8   Cyclone Geometry 400
    12.1.9   Effect of the Vortex Finder 400
    12.1.10 Practical Considerations for CFB Cyclones 402
12.2   Impact Separator 405
    12.2.1   Features and Types 405
    12.2.2   Separation through Impaction 406
    12.2.3   Separation through Interception 407
    12.2.4   Grade Efficiency of Impingement Collectors 409
    12.2.5   Some Types of Impact Separators 409
12.3   Inertial Separators 410
    12.3.1   Separation through Gravity Settling 411
    12.3.2   Cavity Separator 411
    12.3.3   Design Steps 411
Nomenclature 413
    Greek Symbols 415
References 415

Chapter 13   Solid Recycle Systems 417

13.1   Types of Nonmechanical Valves 417
    13.1.1   Loop-Seal 418
    13.1.2   L-Valve 420
    13.1.3   V-Valve 420
13.2   Principles of Operation 421
    13.2.1   Solid Flow from Standpipe to Recycle Chamber 422
    13.2.2   Flow of Solids through Recycle Chamber 424
13.3   Design of Loop-Seal 426
    13.3.1   Pressure Balance 426
    13.3.2   Size of Loop-Seal 427
    13.3.3   Air-Flow through Loop-Seal 428
13.4   Design of L-Valve 431
    13.4.1   Maximum Solid Flow Rate 432
    13.4.2   Additional Design Considerations 433
13.5   Practical Considerations 433
    13.5.1   Plugging of Loop-Seal 433
    13.5.2   Pressure Surge 434
    13.5.3   Avoiding Loop-Seal Plugging 434
Nomenclature 435
    Greek Symbols 436
References 436

Appendix 1   Characteristics of Solid Particles 439

A1.1   Solid Particles 439
    A1.1.1   Equivalent Diameters 439
    A1.1.2   Sphericity (ϕ) 440
    A1.1.3   Mean Particle Size and Its Measurement 440
A1.2   Packing Characteristics 442
A1.3   Particle Classification 442
    A1.3.1   Group C 443
    A1.3.2   Group A 443
    A1.3.3   Group B 444
    A1.3.4   Group D 444
Nomenclature 444
References 444

Appendix 2   Stoichiometric Calculations 445

A2.1   Chemical Reactions 445
A2.2   Air Required 446
A2.3   Sorbent Requirement 446
A2.4   Solid Waste Produced 446
A2.5   Gaseous Waste Products 447
    A2.5.1   Carbon Dioxide 447
    A2.5.2   Water Vapor 448
    A2.5.3   Nitrogen 448
    A2.5.4   Oxygen 448
    A2.5.5   Sulfur Dioxide 448
    A2.5.6   Fly Ash 448
A2.6   Heating Value of Fuels 448
Nomenclature 450
Reference 451

Appendix 3   Simplified Model for Sulfur Capture 453

Nomenclature 454

Appendix 4   Tables of Design Data 455

Index 461

1 Introduction

On December 16, 1921 a new chapter opened in the history of the energy and power industries. Fritz Winkler of Germany introduced gaseous products of combustion into the bottom of a crucible containing coke particles, creating the first demonstration of gasification of coal in a fluidized bed. Winkler saw the mass of particles lifted by the drag of the gas to look like a boiling liquid (Squires, 1983). This experiment initiated a new process called *fluidization*, the art of making granular solids behave like a liquid.

Though some would argue that many others observed the phenomenon of fluidized beds in the past, the credit for the invention of the *bubbling fluidized bed* (BFB) process, which we use for scores of processes including combustion and gasification, should go to Winkler. He observed the process, took measurements, filed patents, and built commercial fluidized bed gasifiers 12 m^2 in cross section — very large even by today's standards.

1.1 BACKGROUND

The idea of burning instead of gasifying coal in a bubbling fluidized bed has been pursued and promoted most vigorously by Douglas Elliott since the early 1960s (Howard, 1983). He recognized the merit of burning coal in fluidized beds to generate steam by immersing boiler tubes in it. He advocated the use of fluidized beds for steam generation with the British Coal Utilization Research Association and the National Coal Board of the U.K. An active program for development of fluidized bed combustion started at the Central Electricity Generation Laboratory, Marchwood, shortly after Elliott's exploratory work.

Simultaneous developments in bubbling fluidized bed (BFB) boilers continued in the U.S.A. and China, but the lack of a recorded history of the development of the fluidized bed boiler in these two countries does not permit those developments to be included here. Steam generation from biomass-fired BFB boilers began perhaps in 1982 with the commissioning of a 10 t/h rice-husk-fired BFB boiler in India, designed by the author. Since then, many types of bubbling fluidized bed boilers, burning a wide variety of fuels, have been developed and commercialized around the world. The BFB boiler has rightfully taken over from the old stoker-fired boilers of the past.

The *pulverized coal* (PC) fired boiler (Figure 1.1), which first appeared in 1911, was the workhorse of the utility industries until BFB boilers (Figure 1.2) appeared in the early sixties. The main difference between PC and BFB boilers lies in the design of their furnaces. The convective section, which accommodates the economizer and superheater elements, is very similar in these boilers.

The *circulating fluidized bed* (CFB) (Figure 1.3) had a curious beginning. Warren Lewis and Edwin Gilliland conceived a new gas–solid process at the Massachusetts Institute of Technology in 1938 when they were trying to find an appropriate gas–solid contacting process for fluid catalytic cracking. Interestingly, they invented the *fast fluidized bed* process while unaware of the invention of the other form of essentially the same fluidized bed process invented by Winkler for gasification at least 17 years earlier (Squires, 1986).

Though the circulating fluidized bed process was used extensively in the petrochemical industries especially for fluid catalytic cracking in the post war period, it did not have a direct entry

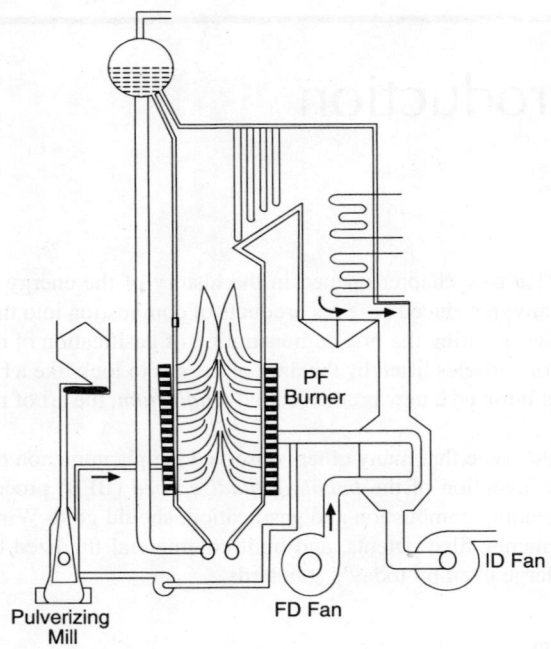

FIGURE 1.1 A pulverized coal-fired boiler.

FIGURE 1.2 Bubbling fluidized bed boiler.

Introduction

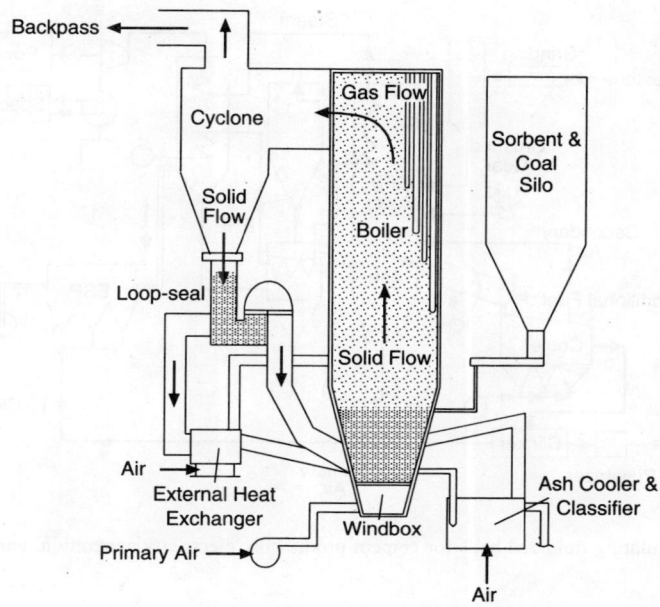

FIGURE 1.3 A circulating fluidized bed boiler.

into the field of coal combustion or gasification. Amongst a number of groups working independently was Lurgi, who found in CFB an excellent technique for carrying out operations with fine solids at very high velocities. Based on laboratory-scale work in their Metallgesellschaft laboratories, Lurgi developed an aluminium calcining process. Calcination being an endothermic process, gas or oil had to be burnt. The combustion heat was recovered in a multistage bubbling fluidized bed cooler, where waste gases exchanged heat with feed materials. Use of the circulating fluidized bed process allowed uniform control of the calcining temperature within required limits. As a result of this attractive feature, a large number of CFB calciners were soon put into commercial operation (Reh, 1986).

For the precalcining stage of the cement clinkering process, also endothermic, several companies used high-ash coal or shale to generate heat for the calcination of limestone (Figure 1.4), thus demonstrating the efficacy of CFB combustion for utilization of low-grade coal in cement production (Kuhle, 1984).

The first CFB boiler, designed exclusively for the supply of steam and heat, was built in the Vereingte Aluminum Werke at Luenen, Germany in 1982. This plant generated 84 MW total (9 MW electricity, 31 MW process steam, 44 MW molten salt melt) by burning low-grade coal washery residues in the presence of limestone.

The Ahlstrom group in Finland, on the other hand, started out in the late 1960s with the development of BFB boilers. In an effort to improve the performance of their bubbling fluidized bed sludge incinerator, Ahlstrom experimented with a hot cyclone in a bed operating at a velocity of 3 m/sec to capture fine solids, which left the bed, and recycle them back to it. Following a series of experiments in their Hans Ahlstrom Laboratory, they built the first commercial CFB boiler in Pihlava, Finland. It was a 15 MW (thermal output) boiler retrofit to an existing oil-fired boiler to replace expensive oil with peat. Initially the circulating fluidized bed boilers built by Ahlstrom were primarily for burning multifuel or low-grade fuels including bark, peat, and wood waste. Later boilers were designed exclusively for coal.

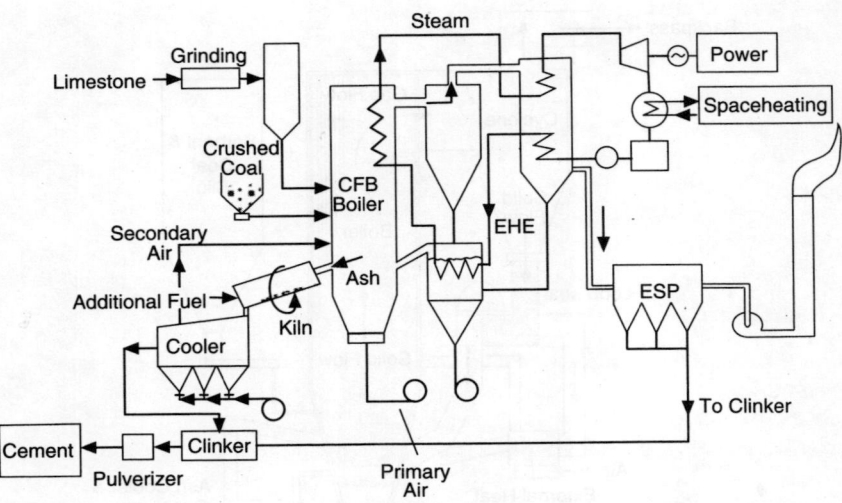

FIGURE 1.4 Circulating fluidized beds for cement production, electricity generation, and space heating.

1.2 FLUIDIZED BED EQUIPMENT

The principal application of gas–solid fluidization is in the following industries:

1. Energy conversion
2. Petro-chemical processing
3. Mineral processing
4. Chemical and pharmaceutical
5. Physical processing

The following section presents a brief description of some applications of fluidized beds.

1.2.1 APPLICATIONS

Some major applications in each of the above areas are given below:

1. Energy conversion
   - Steam generation
   - Gasification
   - Incineration
2. Petro-chemical processes
   - Fluid catalytic cracking (FCC) units
   - Fischer–Tropsch Synthesis
3. Mineral Processing
   - Calcination of alumina
   - Roasting of ores
   - Prereduction of iron ores
   - Precalcination for cement manufacture
4. Chemical and Pharmaceutical
   - Phthalic anhydride from naphthalene
   - Decomposition of sulfate, chloride

Introduction

- Butane oxidation to maleic anhydrite
- Methanol to olefins conversion
5. Physical Processing
    - Drying
    - Coating of particles
    - Heat exchanger and flue gas cleaning
    - Heat treatment

This book deals only with energy conversion.

1.2.2 Fluidized Bed Gasifiers

Gasification is perhaps the earliest commercial application of fluidized beds (Schimmoller, 2005). The major motivation for use of this type of gas–solid contacting process is its excellent solid mixing capability. A typical gasifier is shown in Figure 1.5. Crushed coal is fed into a bubbling fluidized bed of hot solids at 950°C. Steam, the major gasifying medium, is fed into the base of the fluidized bed through a *sparge pipe*-type of distributor. This fluidizes the raw coal along hot solids in the bed. Gasification products leave the bed from the top. The gas is cleaned and used.

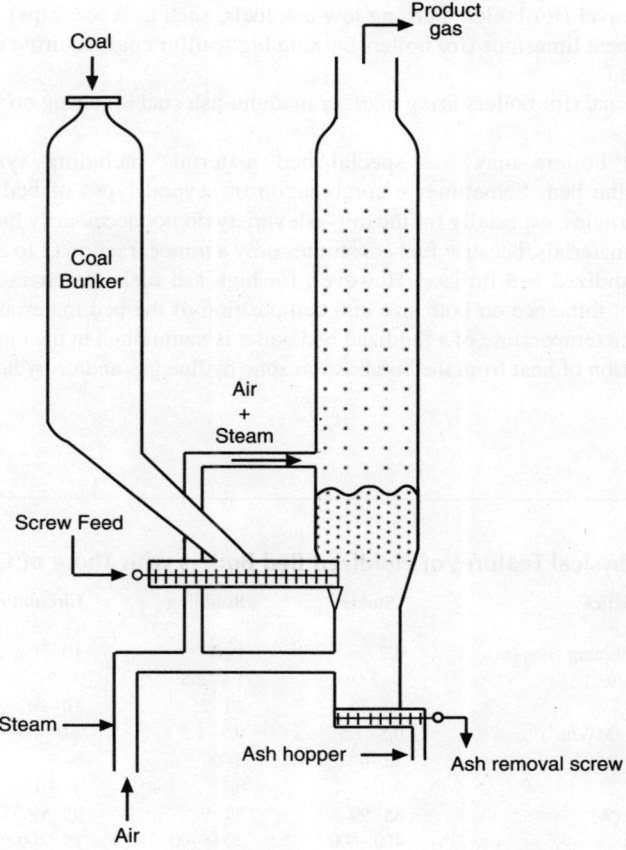

FIGURE 1.5 Schematic of a high-temperature fluidized bed gasifier.

The burning of the char in the lower section of the bed provides the heat required for endothermic gasification reactions.

1.2.3 FLUIDIZED BED BOILERS

A boiler is essentially a combined combustor and heat exchanger where fuel is burnt in a particular mode of gas–solid contacting. Each type of boiler uses a distinct type of gas–solid process. For example, pulverized coal boilers use transport mode, bubbling bed boilers use a bubbling fluidized mode, circulating fluidized bed boilers use the *fast fluidized bed* (see Chapter 2) mode of gas–solid contacting. Figure 2.3 shows the hydrodynamic regimes of gas–solid contacting for different types of boilers. Table 1.1 compares the physical features of different types of boilers.

1.2.3.1 Bed Materials

A fluidized bed boiler is a type of steam generator where fuel is burnt in a fluidized state (Figure 1.2). The furnace of a fluidized bed boiler contains a mass of granular solids, generally in the size range of 0.1 to 0.3 mm or 0.25 to 1.0 mm depending on the type of fluidized boiler it is. These solids are called *bed materials*, which could be made of the following:

- Sand or gravel (for boilers burning low-ash fuels, such as woodchips)
- Fresh or spent limestone (for boilers burning high-sulfur coal requiring control of sulfur emissions)
- Ash from coal (for boilers firing high- or medium-ash coal requiring no sulfur retention)

Biomass-fired boilers may use special bed materials including synthetics to avoid agglomeration in the bed. Sometimes a combination of several types of bed materials is used. The size of fuel particles, especially for the low-ash variety do not necessarily have a major bearing on the size of bed materials, because fuel constitutes only a minor fraction (1 to 3%) of the total bed materials in the fluidized bed furnace. However, for high-ash fuels the characteristic of the fuel exerts an important influence on both size and composition of the bed materials.

The combustion temperature of a fluidized bed boiler is maintained in the range of 800 to 900°C through the extraction of heat from the combustion zone by flue gas and/or by heat-absorbing tubes buried in it.

TABLE 1.1
Comparison of Physical Features of Fluidized Bed Boilers with Those of Other Types

Characteristics	Stoker	Bubbling	Circulating	Pulverized
Height of bed or fuel burning zone (m)	0.2	1–2	10–30	27–45
Superficial velocity (m/sec)	1–2	1.5–2.5	3–5	4–6
Excess air (%)	20–30	20–25	10–20	15–30
Grate heat release rate (MW/m^2)	0.5–1.5	0.5–1.5	3.0–4.5	4–6
Coal size (mm)	32–6	6–0	6–0	<0.1
Turn down ratio	4:1	3:1	3–4:1	3:1
Combustion efficiency (%)	85–90	90–96	95–99.5	99–99.5
Nitrogen oxides (Ppm)	400–600	300–400	50–200	400–600
Sulfur dioxide capture in furnace (%)	None	80–90	80–90	None

Introduction

There are two principal types of fluidized bed boilers:

1. Bubbling fluidized bed (BFB)
2. Circulating fluidized bed (CFB)

Both types could operate under either atmospheric pressure for steam generation or under high pressure for combined-cycle applications. The pressurized bubbling fluidized bed is popularly called *PFBC* while the pressurized circulating fluidized bed is known as a *PCFB*.

1.2.3.2 Bubbling Fluidized Bed Boiler

A bubbling fluidized bed boiler (Figure 1.2) comprises a fluidizing grate through which primary combustion air passes and a containing vessel, which is either made of (lined with) refractory or heat-absorbing tubes. The vessel would generally hold bed materials with or without heat-absorbing tubes buried in it. The open space above this bed, known as *freeboard*, is enclosed by heat-absorbing tubes. The secondary combustion air is injected into this section.

The boiler can be divided into three sections:

1. Bed
2. Freeboard
3. Back-pass or convective section.

The back-pass accommodates remaining heat transfer surfaces including the air heater. Fuel is fed either from the top, as shown in Figure 1.2, or through the bottom of the grate, while the ash generated in the bed is drained from it. A more detailed description is given in Chapter 7.

1.2.3.3 Circulating Fluidized Bed Boiler

A circulating fluidized bed boiler is one where fuel is burnt in a *fast fluidized bed regime*. A definition and details of this regime are in given in Chapter 2.

In a CFB boiler furnace the gas velocity is sufficiently high to blow all the solids out of the furnace. The majority of the solids leaving the furnace is captured by a gas–solid separator, and is recirculated back to the base of the furnace at a rate sufficiently high to cause a minimum degree of vertical mixing of solids in the furnace.

A CFB boiler is shown schematically in Figure 1.3. The primary combustion air (usually substoichiometric in amount) is injected through the floor or grate of the furnace and the secondary air is injected from the sides at a certain height above the furnace floor. Fuel is fed into the lower section of the furnace, where it burns to generate heat. A fraction of the combustion heat is absorbed by water- or steam-cooled surfaces located in the furnace, and the rest is absorbed in the convective section located further downstream, known as the *back-pass*.

The creation of a special hydrodynamic condition, known as *fast-bed* is key to the CFB process as it produces a high slip velocity between the gas and the solids (Figure 1.6). A unique combination of gas velocity, recirculation rate, solids characteristics, volume of solids, and geometry of the system gives rise to this special hydrodynamic condition. More detail of this special motion of gas and solid is provided in Chapter 2.

A circulating fluidized bed boiler can be divided into two sections (see Figure 6.1). The first section is the CFB loop, while the second is the back-pass, comprised of the reheater, superheater, economizer, and air-preheater surfaces. The back-pass of a CFB boiler is similar to that of a PC or BFB boiler.

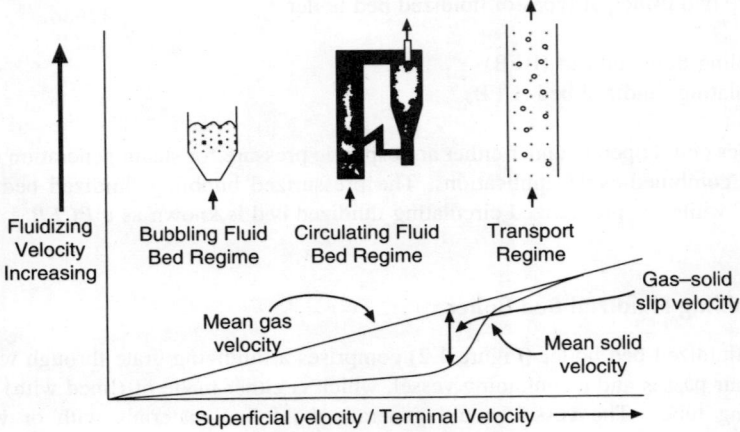

FIGURE 1.6 Motion of gas and solids resulting in a high slip velocity between particles.

The CFB loop of a CFB boiler includes the followings elements:

1. Furnace or fast fluidized bed
2. Gas–solid separator (cyclone or impact separator)
3. Solid recycle device (loop seal, seal pot, or L valve)
4. External heat exchanger (optional).

These components form a solid-circulating loop in which fuel is burned. The furnace enclosure of a CFB boiler is generally made of water tubes (Figure 1.3), which absorb a part of the generated heat. Additional, but less crucial components attached to a CFB boiler are the bed drain and solid classifier.

The lower part of the furnace is smaller and is often tapered in cross section. This helps maintain good fluidization, even with size-segregated particles. Walls of the lower section, which also serve as a thermal storage, are lined with refractory up to the level of secondary air entry. Above this level the furnace is uniform in cross section. The gas–solid separator (Cyclone) and the nonmechanical (Loop-seal) valve for solid recycle are located outside the furnace (Figure 1.3). These are also lined with refractory.

In some designs part of the hot solids recycling between the cyclone and the furnace are diverted through an *external heat exchanger* (Figure 1.3), which absorbs an additional fraction of the combustion heat. This heat exchanger is a bubbling fluidized bed in which heat-transfer surfaces are immersed. Very little combustion takes place in the external heat exchanger.

Fuel is generally injected into the lower section of the furnace, either directly or through the loop-seal. Limestone is fed into the bed in a similar manner. The fuel burns while it is mixed with the hot bed solids.

The primary combustion air enters the furnace through an air distributor or grate located at the furnace floor, while the secondary air is injected at some height above the grate to complete the combustion. Bed solids are well mixed throughout the height of the furnace. Thus, the bed temperature is nearly uniform in the range of 800 to 900°C, although heat is extracted along its height. Relatively coarse particles of sorbent and unburned char are captured in the gas–solid separator and recycled back near the base of the furnace. Finer solid residues (ash or spent sorbents) generated during combustion and desulfurization leave the furnace, escaping through the gas–solid separators, but are finally collected by a bag-house or electrostatic precipitator located further downstream.

Introduction

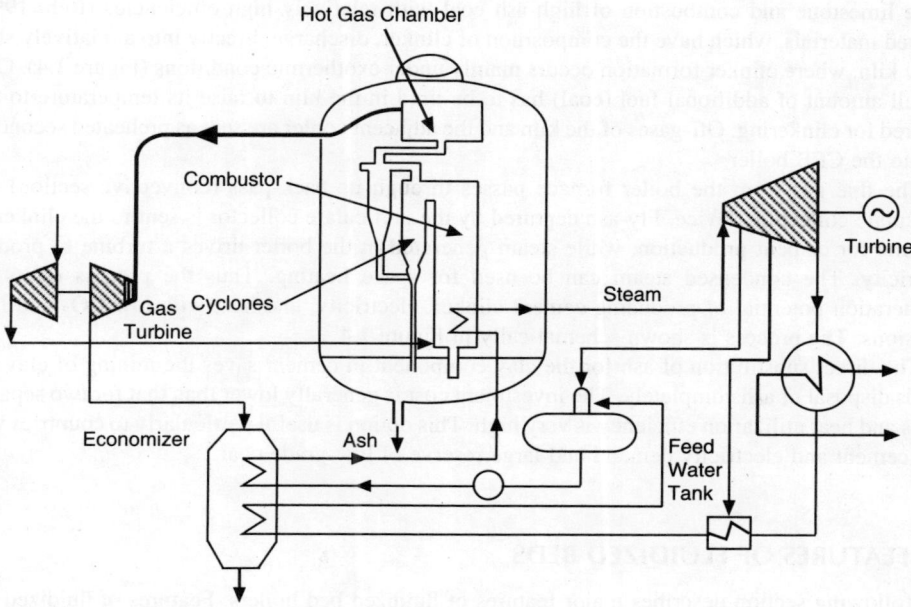

FIGURE 1.7 A combined-cycle plant using pressurized fluidized bed combustion.

1.2.3.4 Supercritical and Pressurized Fluidized Bed Boilers

Conventional thermal power plants typically operate with an overall efficiency of 33 to 36%. Such plants, though well proven, are inefficient and emit significant amounts of harmful gases per unit of power generated. As such, there is a great demand for more efficient power plants. High-efficiency thermal power plants can be of three types:

1. Rankin-cycle plant with supercritical boiler
2. Combined-cycle plants with subcritical boiler
3. Combined-cycle plants with supercritical boiler

Circulating fluidized bed firing is especially suitable for supercritical boiler applications as it is free of a number of limitations from which supercritical pulverized or gas-/oil-fired boilers suffer. Details of these are given in Section 8.10.

The pressurized fluidized bed (PFBC) was one of the first commercial applications of the coal-fired combined cycle (Figure 1.7). Here, coal is burnt in a pressurized bubbling fluidized bed boiler. The steam is expanded in a steam turbine. The high-pressure hot gas (800 to 850°C) from the pressurized fluidized bed is cleaned and expanded in a gas turbine. Thus, power is generated in a gas turbine as well as in a steam turbine. The efficiency of this plant is around 38 to 40% (Cuenca and Anthony, 1995).

1.2.4 APPLICATION IN CEMENT INDUSTRIES

The use of fluidized beds in cement industries is a promising area of application, especially in the context of greenhouse gas reduction. Two raw materials, limestone and high-ash coal, are prepared separately by grinding and then fed to a CFB boiler at around 850°C for simultaneous precalcining of the limestone and combustion of high-ash coal with relatively high efficiencies (Reh, 1999). Hot bed materials, which have the composition of clinker, discharge directly into a relatively short rotary kiln, where clinker formation occurs mainly under exothermic conditions (Figure 1.4). Only a small amount of additional fuel (coal) has to be fired in the kiln to raise its temperature to that required for clinkering. Off-gases of the kiln and the adjacent cooler are sent as preheated secondary air into the CFB boiler.

The flue gas from the boiler furnace passes through its back-pass (convective section) and particulate collection device. Fly-ash captured by the particulate collector is sent to the clinkering machine for cement production, while steam generated in the boiler drives a turbine to produce electricity. The condensed steam can be used for space heating. Thus the process offers the trigeneration potential of producing cement clinker, electricity, and heat with low CO_2 and NO_x emissions. The process is shown schematically in Figure 1.4.

The direct substitution of ash for the clay component in cement saves the mining of clay and avoids disposal of ash completely. The investment cost is generally lower than that for two separate plants and heat utilization efficiency is very high. This option is useful particularly to countries with high cement and electricity demand and large reserve of low-grade coal.

1.3 FEATURES OF FLUIDIZED BEDS

The following section describes major features of fluidized bed boilers. Features of fluidized bed gasifiers are detailed in Chapter 3.

1.3.1 ADVANTAGES OF FLUIDIZED BED BOILERS

Fluidized bed (FB) boilers have a number of unique features that make them more attractive than other types of solid fuel-fired boilers. These features are described below:

1.3.1.1 Fuel Flexibility

This is one of the major attractive features of FB boilers. They can burn a wide variety of fuels without a major penalty in performance. For example, if a PC boiler, designed for 45% ash coal, is fed with 10% ash coal it could cause a serious problem and vice versa, but that does not happen in a fluidized bed boiler. Table 1.2 lists different types of fuels a fluidized bed boiler can fire.

TABLE 1.2
Some Fuels a Fluidized Bed Boiler Can Fire

Coal	Coal Residue	Wood Residue	Sludge	Municipal Waste	Petroleum Product	Gas	Agricultural Waste
Anthracite	Bituminous gob	Bark	Paper mill	Refuse derived fuel	Oil	Off gas	Straw
Bituminous	Anthracite culm	Wood chips	De-inking	Garbage	Delayed coke	Natural	Olive waste
Sub-bituminous	Coal slurry	Saw dust	Municipal	Waste paper	Fluid coke	Other gases	Husk
Lignite	Mill rejects Washery waste	Forest residue Demolition waste	Gasifier fines	Shredded tires	Oil shale		

Fluidized bed gasifiers also share this fuel-flexibility feature of fluidized beds. Unlike fixed bed gasifiers, a fluidized bed gasifier is not restricted to one type or size of fuel.

Fuel particles constitute less than 1 to 3% by weight of all solids in the bed of a typical FB boiler. The rest of the solids are noncombustible, such as sorbents, fuel-ash or sand. The special hydrodynamic condition in the FB furnace allows for excellent gas–solid and solid–solid mixing. Thus, fuel particles fed to the furnace are quickly dispersed into the large mass of bed solids, which rapidly heat the fuel particles above their ignition temperature without any significant drop in the temperature of the bed solids. This feature of a FB furnace would ideally allow it to burn any fuel without the support of a superior fuel, provided its heating value is sufficient to raise the combustion air and the fuel itself above its ignition temperature. Thus, a wide range of fuels can be burned in the specific boiler without any major change in its hardware. Practical considerations, like the capacity of auxiliary equipment, heat balance etc., may limit the range of fuel that can be economically burnt in a given boiler.

To maintain the combustion temperature within an optimum range, it is necessary to absorb a certain fraction of the generated heat. This fraction varies from one fuel to another. For this reason, when the fuel is changed, a bubbling fluidized bed (BFB) boiler needs to adjust the furnace heat absorption by adjusting the amount of bed tubes immersed in the bed. This stipulation provides a limited range of fuel flexibility to the BFB boiler. A CFB boiler, on the other hand, can adjust itself to different types of fuels by controlling the heat absorption in the furnace through an adjustment of its operating parameters instead of the hardware modification as required by BFB boilers.

1.3.1.2 High Combustion Efficiency

The combustion efficiency of a CFB boiler is generally in the range of 97.5 to 99.5%, while that for a bubbling bed is lower, in the range of 90 to 98%. The following features contribute to the high combustion efficiency of circulating fluidized bed combustors:

- Better gas–solid mixing
- Higher burning rate (especially for coarser particles)

Fresh coal often contains a large amount of fine carbon particles. In addition, a considerable amount of carbon fines are generated during combustion through attrition. In a bubbling fluidized bed combustor, these fines are easily entrained out of the fluidized bed, which is usually 0.5 to 1.5 m deep. The freeboard above the bubbling bed, where particles are ejected, is not conductive to efficient combustion because of its poor gas–solid mixing and relatively low temperature. Thus an appreciable amount of carbon fines escaping into the freeboard of a BFB combustor leaves the boiler unburnt. The combustion efficiency of a BFB boiler can be improved to a certain extent by recycling the unburned carbon particles back to the furnace, although carbon particles, once cooled, do not burn as effectively as when hot.

In a CFB boiler, the combustion zone extends up to the top of the furnace (as much as 40 m in large utility boilers) and beyond into the hot cyclone. Fines collected by the hot cyclone are recycled back to the base of the furnace without cooling. Thus, carbon fines generated in the furnace have a longer time to burn during their travel through the height of the furnace and then through the rest of the circulating loop. The only combustible loss is due to the escape of carbon fines from the CFB loop. In some boilers, reinjection of these fines from downstream sections (economizer hopper, precipitator, for example) of the cyclone is also used to minimize the carbon loss.

Unlike bubbling beds, CFB boilers retain their efficiency over a wide range of operating conditions, even when firing fuels with a considerable amount of fines.

1.3.1.3 *In Situ* Sulfur Removal

Unlike other boilers, fluidized bed boilers can absorb the sulfur dioxide generated during combustion in the furnace itself. Low combustion temperature (800 to 900°C) allows a fluidized bed boiler, fed with limestone, to absorb the sulfur as solid calcium sulfate without having to use any add-on equipment as does a PC boiler.

A bubbling fluidized bed boiler may require 2.5 to 3.5 times the stoichiometric amount of limestone for 90% sulfur capture while a typical CFB boiler needs only about 2.0 times the stoichiometric amount of sorbent for this capture. The superior sulfur capture capability of the CFB boiler is due to the larger specific surface area of the sorbents and the longer residence time offered by its furnace.

1.3.1.4 Nitrogen Oxide Emissions

Low emission of nitric oxide (NO_2) is a major attractive feature of both BFB and CFB boilers. The combustion temperature in a fluidized bed boiler (800 to 900°C) is too low for the nitrogen in air to be oxidized into nitric oxide. Data collected in commercial CFB boilers suggest NO_2 emission in the range of 50 to 150 ppm (Hiltunen & Tang, 1988) or 20 to 150 mg/MJ (Kullendorff et al., 1988). The emission from bubbling fluidized bed boilers is slightly larger than this figure. In a CFB boiler the nitric oxide emission is especially low because of its staged addition of combustion air.

It may however be noted that the emission of nitrous oxide (N_2O) from fluidized bed boilers is much higher (100 to 200 ppm) compared to that from pulverized coal-fired boilers, where the nitrous oxide emission is considerably lower due to the latter's high combustion temperature (1200 to 1500°C).

1.3.2 FEATURES SPECIFIC TO CFB BOILER

The above-mentioned features are common to both bubbling and circulating fluidized bed boilers, but the following section lists a few features exclusive to CFB boilers.

1.3.2.1 Smaller Grate Area

Grate heat-release rate is an important design characteristic of a firing system. It gives the amount of combustion heat the furnace can release per unit cross-sectional area. Figure 1.9 compares the heat-release rate of a number of firing systems. The CFB firing has a grate heat-release rate in the range of 3.0 to 4.5 MW/m^2, which is close to that of PC. For both pressurized furnaces (PFBC and PCFB), permissible heat-release rates are higher, and are directly proportional to the combustor pressure. The grate heat-release rate of a BFB boiler is much smaller, and as such, for a given thermal output, it would require a furnace grate area 2 to 3 times larger than that of a CFB boiler.

The high heat-release rate of CFB boilers arises from the high superficial gas velocity (3.5 to 6 m/sec) used in them. The intense gas–solid mixing promotes a high rate of heat liberation through rapid generation and dispersion of heat in the bed. The smaller grate area makes CFB firing suitable for adoption into pulverized coal-fired or oil-fired boilers.

It has been shown that this unique feature allows renovation of old power plants by converting existing pulverized coal- or oil-fired boilers to CFB coal firing (Basu and Talukdar 2001). A major attraction of this option is that it could even increase the generation of the boiler, which may have declined considerably due to aging and firing of inferior grade coal, back to its original design value.

Introduction

1.3.2.2 Fewer Feed Points

The fuel feed system is much simpler in a CFB boiler due to its relatively small number of feed points, and less grate area for a given thermal output. Furthermore, good mixing and the extended combustion zone allow one feed point to serve a grate area much larger than that in a bubbling bed. Table 10.3 and Table 10.4 in Chapter 10 show that the bed area served by a feed point and the thermal input per feed point in a CFB boiler are much larger than those in bubbling fluidized bed boilers using under-bed feeding.

1.3.2.3 Good Turndown and Load-Following Capability

A high fluidizing velocity and easy control of heat absorption allows quick response of CFB boilers to varying loads. Some commercial units (Belin et al., 1999) report a load-following capability of 4 to 6% per minute. The initial success of CFB boilers in paper mills where load fluctuates rapidly attests to the load-following capability of CFB boilers.

A turndown ratio of 3 or 4:1 has been reported in commercial plants without oil support (Belin et al., 1999). Cold start-up of the furnace without any consideration of the turbine or refractory can be achieved in 6 to 7 h while that for a PC boiler is 9 to 10 h. A CFB boiler can start up in 2 h after a 12 h shutdown, while PC boiler would take about 5 to 6 h only after 8 h shutdown.

1.4 FLUIDIZED BED AND OTHER TECHNOLOGY OPTIONS

The technology choice for a new coal-based energy project involves a number of considerations. The availability of fuel, its price, and the local environmental regulations are a few major ones. The following section discusses some of the options available for power generation from fossil or biomass fuels.

1.4.1 TECHNOLOGICAL OPTIONS FOR GENERATION OF POWER FROM SOLID FUELS

There are several options for a clean coal-based power plant. These options can be broadly classified in the following groups:

1. Selection or pretreatment of fuel (switching to low-sulfur coal or beneficiation of coal)
2. Combustion modification (low NO_x burner, staged combustion, sorbent injection)
3. Postcombustion treatment (flue gas scrubber, NO_x reducer, or particulate control equipment)
4. Improved firing methods (pulverized coal firing, fluidized bed firing)
5. System modification (supercritical plant, combined-cycle plant)

Table 1.3 compares the overall plant efficiencies of different options and their respective carbon dioxide emissions. The latter parameter is increasingly becoming important in the wake of concerns about global warming.

1.4.1.1 Fuel Switch

If sulfur emission is a major concern, the option of switching to low-sulfur coal is one of the simplest options, but it may be unacceptable due to the nonavailability of low-sulfur coal at the particular site for economic or political reasons.

The removal of sulfur and ash prior to combustion through beneficiation is an option that needs to be carefully evaluated by using an economic model. However, besides removing pyritic sulfur, the beneficiation process also adds moisture and discards some useful heat through the washery

TABLE 1.3
Comparison of Technologies for Power Generation from a 600 MWe Coal-Based Plant

Technology			Net Plant Efficiency (%) LHV	CO_2 Emission Factor (gC/kWh) for Coal
Conventional system	Pulverized coal	Reference plant	36	252
		Subcritical steam	39	253
		Supercritical steam	42–45	202–216
	AFBC	Subcritical steam	39	233
Combined cycle	IGCC	Subcritical steam	38–43	211–239
	PFBC	Subcritical steam	44	206
	Hybrid[a]	Supercritical steam	52	174
	Fuel cells	Subcritical steam	47–60[b]	151–193
Combined heat and power	Pulverized coal	Subcritical steam	85[b]	107
		Supercritical steam	91–92	99–100
Fuel blending	Pulverized coal	Subcritical steam, coal/natural gas (85/15)	37	231
	Topping GT/PC	Subcritical, natural gas/coal (33/67)	41	121

[a] Gasifier and PFBC.
[b] Converted from HHV using a conversion factor of 1.04. Emission factor for reference coal: 25.2 gC/MJ (LHV); natural gas: 15.3 gC/MJ; and oil: 20.0 gC/MJ.

tailings. Beneficiation thus converts the high-sulfur, high-ash coal into low-sulfur low-ash coal and a stream of washery tailings.

Fuel switch and beneficiation do not contribute to the reduction in greenhouse gas or NO_x emissions, but decrease the particulate emission and reduce the solid waste disposal problem.

1.4.1.2 Combustion Modifications

An existing plant can be made more environmental friendly by improving the operation of its furnace. Such modifications can be done in a relatively short time and for less cost. Following options are available:

1. Use of over-fire air to reduce NO_x
2. Replace existing burners with low NO_x burner for NO_x control
3. Injection of sorbents and ammonia in the furnace for reductions in SO_2 and NO_x, respectively

Sorbent injection and staged combustion both have good promise for the control of SO_2 and NO_x. The staged addition of combustion air in a low NO_x burner reduces the NO_x emission but only to an extent below most national standards.

1.4.1.3 Postcombustion Scrubbing

The postcombustion cleaning of the flue gas is a popular and proven option. Flue gas produced by the boiler is cleaned by one of the various scrubbing units. Units belonging to this group are:

1. Flue gas desulfurizing scrubbers for reduction of SO_2 emission
2. Selective catalytic reducer for control of NO_x
3. Selective noncatalytic scrubber for NO_x

Introduction

4. Mercury and/or particulate capture in electrostatic precipitator or bag-house
5. Electrostatic precipitator or fabric filter for reduction in particulate emissions

Of late, some combined scrubbers are available which remove multiple pollutants from one unit. High capital and operating costs of flue gas desulfurization (FGD) and catalytic conversion should be carefully weighed against the overall economics of the plant. The plant availability factor could reduce because of the addition of an in-series unit.

1.4.1.4 Firing and System Modifications

Pollution from thermal power plants can be reduced by using either improved firing technique or by improving the overall efficiency of the plant. This gives a much wider range of options, and as such it is discussed in a separate section below.

1.4.2 IMPROVED FIRING

Here we discuss firing options available for use in the traditional Rankin-cycle steam power plant. Steam boiler-based power plants (Rankine-cycle) are broadly divided into two groups, subcritical and supercritical, depending on whether the pressure of the boiler is below or above the critical point of 22.09 MPa or 220.9 bar

1. Subcritical plant
2. Supercritical plant

Each of these two types of plants could use one of the following two types of firing systems:

1. Pulverized coal firing
2. Fluidized bed firing

1.4.2.1 Supercritical Boiler

The typical efficiency of subcritical Rankine-cycle steam power plants is around 30 to 35%, whereas that for supercritical plants could be as high as 40%. Supercritical boilers have been used in several Rankine-cycle plants resulting in improved overall efficiency and therefore lower CO_2 emissions. However, the construction of supercritical boilers for PC firing is not as convenient as it is for CFB firing. For this reason, new supercritical boilers with CFB firing are being built (Lundqvist, 2003).

Supercritical steam cycles also have a high efficiency compared to that in integrated gasification combined-cycle systems as shown in Table 1.3. This relatively simple option can compete favorably with IGCC.

1.4.2.2 Pulverized Coal Firing

Pulverized coal-fired boilers have been in use for more than 50 years for subcritical or supercritical boiler plants. They can use low NO_x burners to reduce nitric oxide emissions, which again come with their own penalty of combustible loss. The nitric oxide emission from PC boilers even with low NO_x burners is higher than that from fluidized bed boilers. However, PC firing produces significantly lower emissions of the greenhouse gas, nitrous oxide than that by fluidized bed firing. Combustion efficiency of PC firing is slightly higher that that of fluidized bed firing and its combustion air fans consume less power. Presently, the capacity of PC-fired boilers (~ 660 MWe) is larger than that for fluidized bed firing (< 460 MWe).

1.4.2.3 Fluidized Bed Firing

Fluidized bed combustion (bubbling and circulating), though much younger than PC firing, has demonstrated itself on a commercial scale. The intrinsic properties of this process allow it to restrict the emission of both NO_x and SO_2 below the limits set by most environment regulatory bodies. Also, it does not require any pretreatment of the fuel. No additional equipment (catalytic converters or special burners) is required to reduce NO_x emissions. However, limestone is required for absorption of SO_2. It requires between 3.5 (in bubbling) and 2 (in circulating) times the stoichiometric amount of limestone for the capture of sulfur dioxide. Thus it produces a proportionately larger quantity of solid wastes and carbon dioxide. The solid waste, being dry, can be disposed of with relative ease and inexpensively.

If fuel-flexibility and air-pollution stipulations prompt the choice of fluidized bed combustion, the planners are then required to decide between bubbling and circulating fluidized bed firing. CFB firing has a number of advantages over BFB firing as outlined in Section 1.3.2. In older designs with hot cyclones etc., the advantage of CFB firing was not apparent in small (<20 to 30 MWe) capacity boilers, but modern subcompact CFB boilers can be effective down to very small sizes (Basu et al., 2005). A CFB boiler could be a judicious choice if the following conditions are met:

- Capacity of the boiler is medium to large
- Sulfur retention and NO_x control are important
- The boiler must fire a low-grade fuel or highly fluctuating fuel quality
- Two-shift operation or short start-up time are required
- The plant requires high availability

1.4.3 COMBINED-CYCLE PLANTS

The use of combined-cycle plants for the generation of power from coal is also an attractive option. It offers high overall efficiency as well as low emissions of all pollutants. The following types of combined-cycle plants are available for coal:

1. Pressurized fluidized (bubbling or circulating) bed combined-cycle
2. Integrated gasification combined-cycle
3. Partial gasification pressurized fluidized bed combined-cycle

1.4.3.1 Pressurized Fluidized Bed Combined-Cycle

The application of *pressurized bubbling or circulating fluidized bed combustion* to combined-cycle power generation (Figure 1.7) has proved to be a very attractive alternative due to its high overall efficiency and superior environmental performance. Here, the combustion of coal takes place under high pressure. The pressurized hot gas is cleaned and expanded through a gas turbine. The waste heat is used to generate steam in a boiler. The steam generated, partially in the pressurized CFB combustor and partially in the waste heat boiler, produces electricity by means of a steam turbine, while this process is as effective in sulfur capture and NO_x-emission reduction as other fluidized bed boilers. Since the combined-cycle is more efficient than the steam-cycle alone, this option will produce a smaller amount of solid and gaseous wastes per unit of electricity produced.

1.4.3.2 Integrated Gasification Combined-Cycle

The integrated gasification combined-cycle (IGCC) process involves complete gasification of the fuel under pressure. The raw gas, produced being fairly hot, is cooled (Figure 1.10) in a heat recovery steam generator (HRSG). The cooled, medium-heating-value gas is then desulfurized and burnt in the combustor of a gas turbine to produce electricity. Waste flue gas from the turbine is used to heat the water for the boiler. Steam from the boiler produces additional electricity in a steam turbine. The combined efficiency of this process is in the range of 42 to 45%. The waste gas leaves the gas turbine at a relatively low temperature, thus it cannot produce high-temperature steam, which a modern turbine would require. A hybrid system that will burn coal fines in the gas turbine exhaust to raise the combustion temperature may help to increase the cycle efficiency.

This option generates very low levels of air and solid pollutants, however, the cost-effectiveness and overall plant availability of this system have to be evaluated carefully against other options. The power generation cost is not yet competitive with conventional plants, but government supports and emission credits might make this option viable.

1.4.3.3 Partial Gasification Combined-Cycle

This advanced version of the combined-cycle plant involves partial gasification of coal in a pressurized pyrolyzer instead of its complete gasification as done in an IGCC. As shown schematically in Figure 1.8, char produced in the pyrolyzer or partial gasifier burns in a pressurized CFB boiler to generate steam at supercritical conditions and high-pressure flue gas with excess oxygen. The pyrolysis gas is cleaned and burnt in the lean-combustion burner of a gas turbine by using the excess oxygen in the flue gas from the supercritical boiler. This can raise the temperature of the flue gas to the maximum permissible limit set by the turbine blades (~1250°C). The hot gas then expands through a gas turbine to produce power.

The high-pressure, high-temperature steam produces additional power in a steam turbine. Waste gas from the gas turbine preheats the water going into the boiler. An important difference between IGCC and this system is that gas turbine and steam turbine cycles are to a great extent independent of each other. Thus, both the steam turbine and gas turbine cycles may be operated at their peak efficiency, resulting in a very high (50%) overall efficiency of power generation (Davidson et al., 1991). The second stage combustion at 1000 to 1250°C may reduce the nitrous oxide generated in the CFB combustor at 850°C.

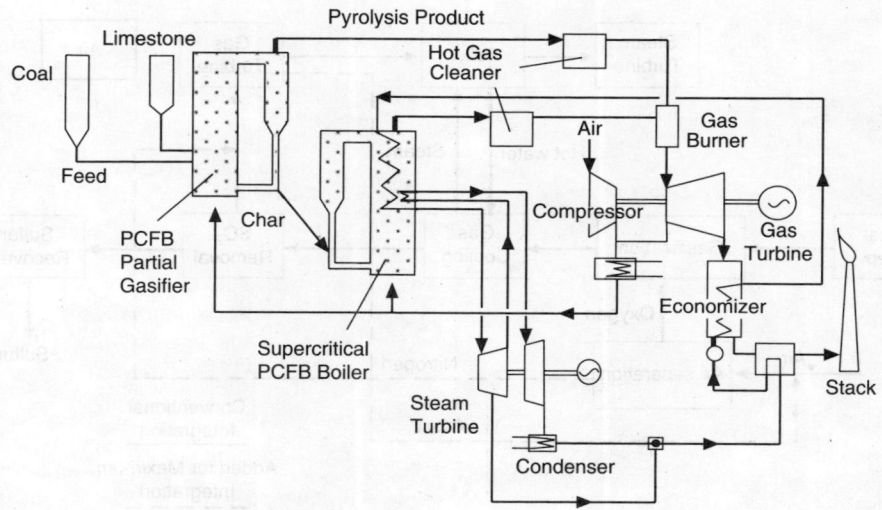

FIGURE 1.8 Advanced combined-cycle plant using partial gasification of fuel.

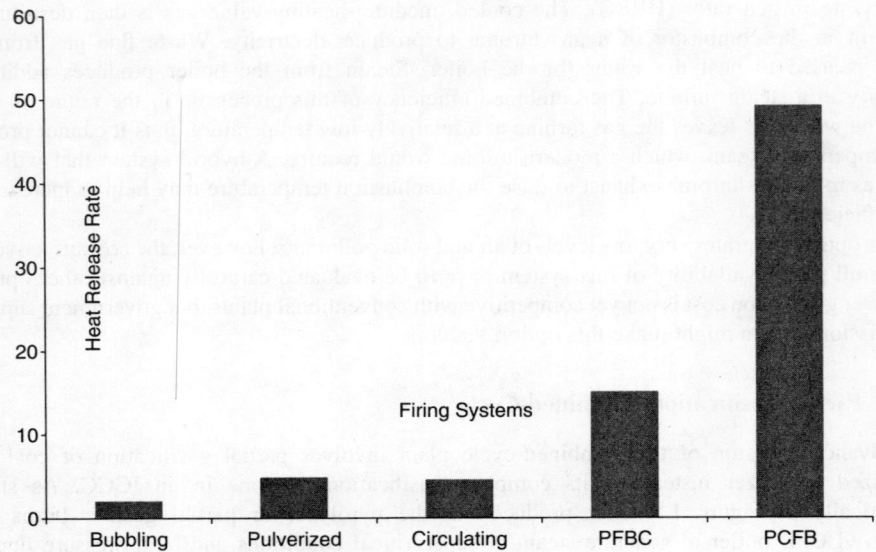

FIGURE 1.9 Comparison of heat release-rates of different firing systems.

1.4.4 Cost Implications

Choice of technology is governed to a great extent by the capital and operating cost of the plant.

Before selecting the appropriate technology for a specific site, a comprehensive study of different systems and the projected generation cost of each of them are to be worked out. Table 1.4 presents an example of such a study, which compared the *levelized costs* of several technological options for a specific site prepared using a commercial software (PlantCost©), This gives a clearer

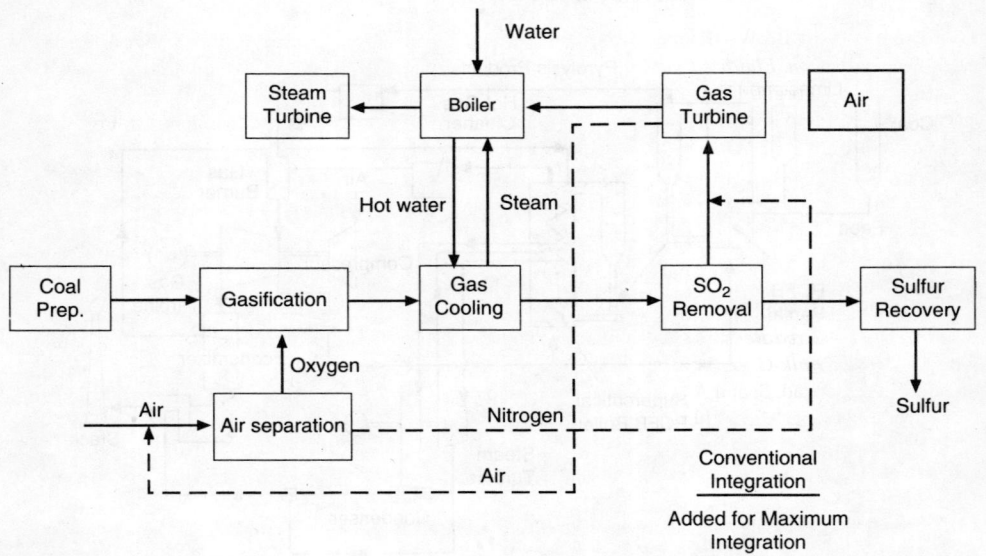

FIGURE 1.10 Integrated gasification combined-cycle.

TABLE 1.4
Comparison of Technologies for Generation from Coal from a 250 MWe Plant

Technology	Levelized Cost in USD ($/KWh)
Pulverized coal-fired (PC) basic	0.0592
Circulating fluidized bed-fired (CFB) basic	0.0620
PC + FGD	0.0649
PC + FGD + SCR	0.0667
PC + LNB	0.0595
PC + ESP	0.0609
CFB + ESP	0.0667
Combined-cycle with gas	0.0690

picture as to which technology might be most cost-effective over the lifetime of a plant for a given set of constraints. This calculation does not include the carbon tax or any credit for reductions in carbon dioxide emissions. One can see from Table 1.3 that the carbon intensity varies from one technology to other.

REFERENCES

Basu, P. and Talukdar, J., Revamping of a 120 MWe pulverized coal fired boiler with circulating fluidized bed firing, In *Proceedings of the 16th International Conference on Fluidized Bed Combustion*, Geiling, D. W., Ed., ASME, Reno, NV, May 13–16, 2001.

Basu, P., Dutta, A., Ghosh, A., and Chakraborty, P., An innovative solution to the problem of mill rejects in thermal power plants, In *Proceedings of the 18th International Conference on Fluidized Bed Combustion*, Li, J., Ed., ASME, Toronto, May 13–16, 2005.

Belin, E., Kavidass, S., Maryamchik, M., Walker, D. J., Mondal, A. K., and Price, C. E., Update of operating experience of B&W–IR circulating fluidized bed boilers, In *Proceedings of the 15th International Conference on Fluidized Bed Combustion*, Reuther, Robert B., Ed., 1999, 3 May 16–19, 1999, Savannah Georgia, USA, Paper FBC99-082.

Cuenca, A. M. and Anthony, E. J., *Pressurized Fluidized Bed Combustion*, Blackie Academic & Professional, London, p. 371, 1995.

Davidson, J. E., Cross, P. J. I., and Topper, J. M., Application of CFBC to power generation in the UK, In *Circulating Fluidized Bed Technology III*, Basu, P., Hasatani, M. and Horio, M., Eds., Pergamon Press, Oxford, 1991.

Hiltunen, M. and Tang, J. T., NO_x abatement in Ahlstrom pyropower circulating fluidized bed boilers, In *Circulating Fluidized Bed Technology II*, Basu, P. and Large, J. F., Eds., Pergamon Press, Oxford, pp. 429–436, 1988.

Howard, J. R., *Fluidized Beds — Combustion and Applications*, p. 131, 1983.

Kuhle, K., *Zement-Kalk-Gips*, 34, 219–225, 1984.

Kullendorff, A., Herstad, S. and Andersson, C., Emission control by combustion in circulating fluidized bed — operating experiences, In *Circulating Fluidized Bed Technology II*, Basu, P. and Large, J. F., Eds., Pergamon Press, Oxford, pp. 445–456, 1988.

Lundqvist, R. G., Designing large-scale circulating fluidised bed boilers, *VGB PowerTech*, 83(10), 41–47, 2003.

PlantCost — A Tool for Financial Analysis of Power Projects, Greenfield Research Inc., 2003 (www.plantcost.com).

Reh, L., The circulating fluidized bed reactor — the key to efficient gas–solid contacting process, In *Circulating Fluidized Bed Technology-I*, Basu, P., Ed., Pergamon Press, Toronto, pp. 105–118, 1986.

Reh, L., Challenges of circulating fluid bed reactors in energy and raw materials industries, In *Circulating Fluidized Bed Technology-VI*, Werther, J., Ed., Dechema, e.v, Frankfurt, pp. 10–11, 1999.

Schimmoller, B. K., "Coal gasification" *Power Engineering*, March, p. 36, 2005.

Squires, A. M., *Fluidized Bed Combustion and Applications*, Howard, J. R., Ed., Applied Science Publishers, Barking, p. 278, 1983.

Squires, A. M., The story of fluid catalytic cracking, In *Circulating Fluidized Bed Technology — I*, Basu, P., Ed., Pergamon Press, Toronto, 1986.

2 Hydrodynamics

The basic difference between a fluidized combustor or gasifier and a pulverized or stoker-fired boiler is in the hydrodynamics of the gas–solid motion in their furnaces or reactors. Interestingly, most of their environmental and operating characteristics are a direct result of the hydrodynamics of gas–solid interaction. For this reason, the performance of a fluidized bed boiler or gasifier drops significantly when the operation of its furnace deviates from the designed fluidization regime. Thus, a good understanding of the gas–solid motion in the furnace or reactor of a fluidized bed unit is very important. The present chapter presents a brief summary of the hydrodynamics of bubbling and circulating fluidized beds.

2.1 REGIMES OF FLUIDIZATION

Definition

Fluidization is defined as the operation through which fine solids are transformed into a fluid-like state through contact with a gas or liquid.

In a fluidized bed, the gravitational pull on fluidized particles is offset by the upward fluid drag of the gas. This keeps the particles in a semisuspended condition. A fluidized bed displays the following characteristics similar to those of a liquid, as one can observe in Figure 2.1:

1. The static pressure at any height is approximately equal to the weight of the bed solids per unit of cross-sectional area above that level
2. An object denser than the bulk density of the bed will sink, while one lighter will float. Thus, a steel ball sinks in the bed, while a light badminton cock floats on the surface (Figure 2.1a)
3. The solids from the bed may be drained like a liquid through an orifice at the bottom or on the side of the container (Figure 2.1b). The solids' flow-stream is similar to a water jet from a vessel
4. The bed surface maintains a horizontal level, independent of how the bed is tilted. Also, the bed assumes the shape of the vessel (Figure 2.1c)
5. Particles are well mixed, and the bed maintains a nearly uniform temperature throughout its body when heated

To understand how a fluidized bed is formed imagine a gas moving up through a bed of granular solids resting on the porous bottom of a column as shown in Figure 2.2. As the gas velocity through the solid particles increases, a series of changes in the motion of the particles could occur. For example, at a very low velocity the particles remain stationary on the floor (Figure 2.3a), but at a sufficiently high velocity they are transported out of the vessel (Figure 2.2d). With changes in gas

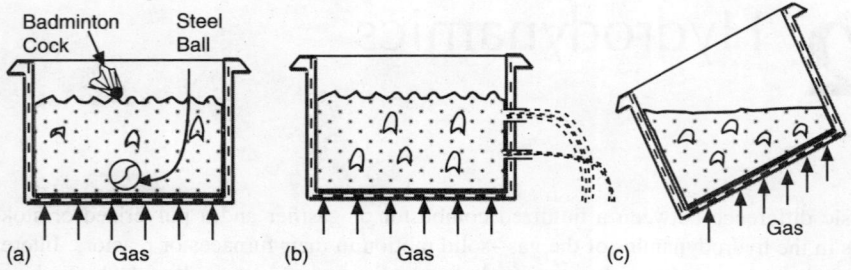

FIGURE 2.1 A fluidized bed demonstrates all the characteristics of a fluid.

velocity, the solids move from one state or regime to another. These regimes, arranged in order of increasing velocities, are:

- Packed bed (fixed) (Figure 2.2a)
- Bubbling bed (Figure 2.2b)
- Turbulent bed (Figure 2.2c)
- Fast bed (used in circulating fluidized beds)
- Transport bed (pneumatic or entrained bed) (Figure 2.2d)

Table 2.1 compares the features of the above gas–solids contacting processes.

TABLE 2.1
Comparison of Principal Gas–Solid Contacting Processes

Property	Packed Bed	Fluidized Bed	Fast Bed	Pneumatic Transport
Application in boilers	Stoker fired	Bubbling fluidized	Circulating fluidized	Pulverized coal fired
Mean particle diameter (mm)	<300	0.03–3	0.05–0.5	0.02–0.08
Gas velocity through combustor zone (m/sec)	1–3	0.5–2.5	4–6	15–30
Typical U/U_t	0.01	0.3	2	40
Gas motion	Up	Up	Up	Up
Gas mixing	Near plug flow	Complex two phases	Dispersed plug flow	Near plug flow
Solids motion	Static	Up and down	Mostly up, some down	Up
Solid–solid mixing	Negligible	Usually near perfect	Near perfect	small
Overall voidage	0.4–0.5	0.5–0.85	0.85–0.99	0.98–0.998
Temperature gradient	Large	Very small	Small	Maybe significant
Typical bed-to-surface. Heat transfer coefficient, (W/m^2 K)	50–150	200–550	100–200	50–100

Source: Modified from Grace, J.R., *Can. J. Chem. Eng.*, 64, 353–363, 1986.

Note:

The term *bed* has been used loosely in Table 2.1 and elsewhere in the text, referring to a body of gas–solid in one of the above contacting modes.

Hydrodynamics

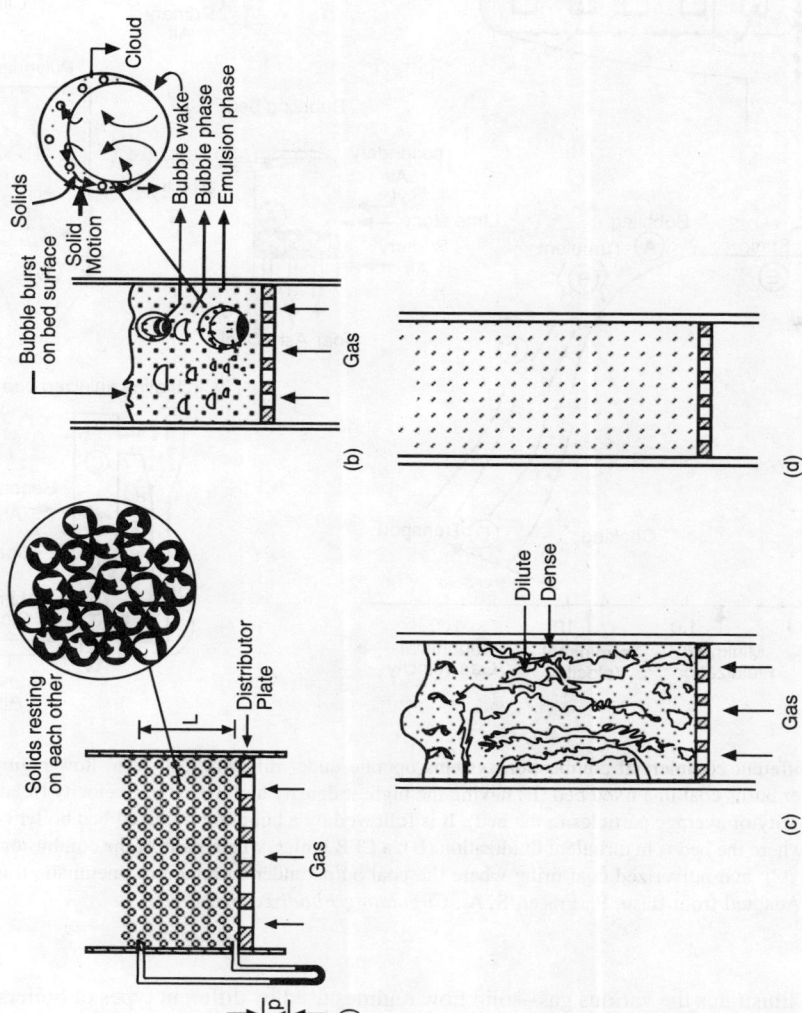

FIGURE 2.2 (a) A fixed bed of particles through which gas is flowing. (b) A bubbling fluidized bed showing gas circulation around bubbles. (c) Turbulent fluidized bed. (d) A transport or entrained bed.

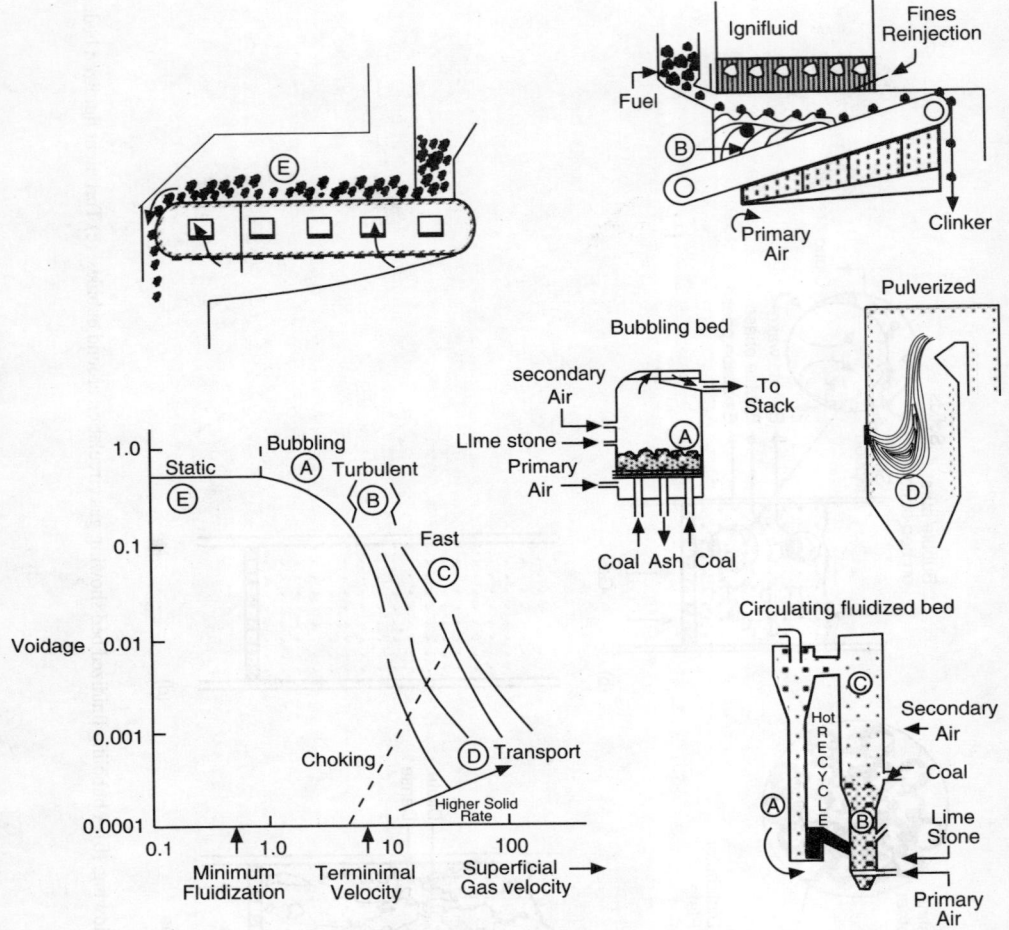

FIGURE 2.3 Different commercial combustion systems operate under different gas–solid flow regimes. A stoker-fired boiler burns coal in a fixed bed (E) having the highest density and lowest gas velocity (relative to the terminal velocity of average particles in the bed). It is followed by a bubbling fluidized bed boiler (A); an Ignifluid boiler where the bed is in turbulent fluidization (B); a CFB boiler, where parts of the combustor are in fast fluidization (C); and pulverized coal firing where the coal burns under entrained or pneumatic transport condition (D). (Adapted from Basu, P., Fraser, S. A., *Circulating Fluidized Bed Boiler*.)

Figure 2.3 illustrates the various gas–solid flow regimes used by different types of boilers. The diagram plots the volume fraction of solids in the furnace or combustion zone against the superficial gas velocity through this zone. It is apparent that the solid volume fraction decreases with increasing superficial velocity. Stoker-fired boilers use the fixed-bed regime, and have the densest combustion zone, while pulverized boilers, which operate in the pneumatic transport regime, have the leanest combustion zone. The furnace of a bubbling (BFB) or a circulating (CFB) fluidized bed boiler lies between these two extremes.

Each of the above flow regimes has distinct characteristics. Table 2.1 presents a comparison of some characteristic features of different gas–solid processes used in various types of boilers. The following sections discuss features of some of these gas–solid flow regimes.

Hydrodynamics

2.1.1 Packed Beds

A *fixed* or *packed* bed refers to a bed of stationary particles residing on a perforated grid, through which a gas flows (Figure 2.2a). In *moving packed* beds, the solids move with respect to the walls of the column, like those in the return leg of a CFB. In either case, the particles do not move relative to each other. When the gas flows through a packed bed of solids it exerts a drag force on the particles, causing a pressure drop across the bed. The pressure drop per unit height of a packed bed, $\Delta P/L$ of uniformly sized particles, d_p is correlated as (Ergun, 1952):

$$\frac{\Delta P}{L} = 150 \frac{(1-\varepsilon)^2}{\varepsilon^3} \frac{\mu U}{(\phi d_p)^2} + 1.75 \frac{(1-\varepsilon)}{\varepsilon^3} \frac{\rho_g U^2}{\phi d_p} \qquad (2.1)$$

where, ε is the void fraction in the bed, ϕ is the sphericity (see Appendix 1) of bed solids, μ is the dynamic viscosity, and ρ_g is the density of the gas. The *superficial gas velocity*, U is defined as the gas flow rate per unit cross section of the bed.

2.1.2 Bubbling Fluidized Beds

If the gas-flow rate through the fixed bed is increased, the pressure drop due to the fluid drag continues to rise as per Equation 2.1, until the superficial gas velocity reaches a critical value known as the *minimum fluidization velocity*, U_{mf}. At the velocity where the fluid drag is equal to a particle's weight less its buoyancy, the fixed bed transforms into an *incipiently fluidized bed*. In this state the body of solids behaves like a liquid, as illustrated in Figure 2.1. Since the pressure drop across the bed equals the weight of the bed, the fluid drag F_D is written as:

$$F_D = \Delta P A = AL(1-\varepsilon)(\rho_p - \rho_g)g \qquad (2.2)$$

where A and L are the cross-sectional area and height of the bed, respectively.

The minimum fluid velocity at which the bed just becomes fluidized (U_{mf}) may be obtained by solving Equation 2.1 and Equation 2.2 simultaneously to obtain:

$$\mathrm{Re}_{mf} = \frac{U_{mf} d_p \rho_g}{\mu} = [C_1^2 + C_2 \mathrm{Ar}]^{0.5} - C_1 \qquad (2.3)$$

where Archimedes number,

$$\mathrm{Ar} = \frac{\rho_g(\rho_p - \rho_g)g d_p^3}{\mu^2}$$

and d_p is the surface-volume mean diameter of particles (see Appendix 1).

The values of the empirical constants C_1 and C_2 as taken from experiments are 27.2 and 0.0408, respectively (Grace, 1982).

At minimum fluidization, the bed behaves as a pseudoliquid. A typical bubbling bed is shown in Figure 2.2. For Group B and D particles (see Appendix 1), a further increase in gas flow can cause the excess gas to flow in the form of *bubbles*. The gas–solid suspension around the bubbles and elsewhere in the bed is called the *emulsion phase* (Figure 2.2b). The superficial gas velocity through this phase is of the order of U_{mf} and it has a characteristic voidage ε_{mf}.

A fluidized bed of Group A (see Appendix 1) particles does not start bubbling as soon as the superficial velocity exceeds U_{mf}, but instead the bed starts expanding. The bubbles start appearing when the superficial gas velocity exceeds another characteristic value called *minimum bubbling velocity*, or U_{mb}. The minimum bubbling velocity for Group A particles is given by

Abrahamsen and Geldart (1980) as:

$$U_{mb} = 2.07 \exp(0.716F) d_p \left[\frac{\rho_g^{0.06}}{\mu^{0.347}} \right] \quad (2.4)$$

where F is the mass fraction of particles less than 45 μm, d_p is the mean surface-volume diameter of particles in m, ρ_g is the density of gas in kg/m³, and μ is the viscosity of gas in kg/m sec.

2.1.2.1 Bubbles

Bubbles are essentially gas voids with little or no solids. Due to the buoyancy force, bubbles would rise through the emulsion phase, by-passing the particles (Figure 2.4). The bubble size increases with particle diameter, d_p, excess gas velocity, $(U - U_{mf})$, and with its distance above the distributor or grid of the bed.

The bubble can grow only to a maximum size, beyond which it collapses. The stable maximum bubble size, D_{bmax} is given by:

$$D_{bmax} = \frac{2(U_t^*)^2}{g} \quad (2.5)$$

where, U_t^* is the *terminal velocity* (see Appendix 1) of particles having a diameter 2.7 times that of averaged size of the bed solids (Grace, 1982).

As the bubbles rise in the bed, gas would ordinarily enter from the bottom and leave from the top of the bubble. This happens because, at a given height, the static pressure inside the bubble is constant, while that outside in the emulsion phase decreases linearly from the bottom to the top of the bubbling bed. For this reason the pressure at the bottom of the bubble is lower than that in its neighborhood and at the top of the bubble the pressure is greater than around it.

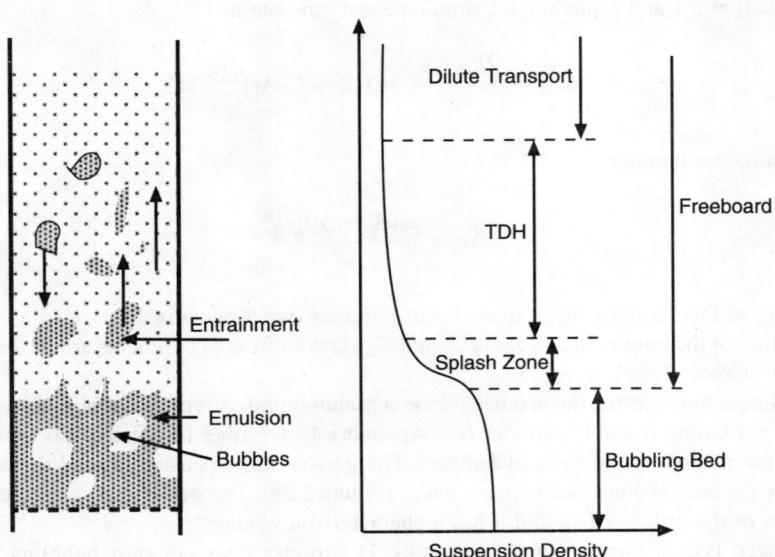

FIGURE 2.4 The bubbles erupt on the surface of a bubbling bed ejecting solids which travel up through the freeboard. The solid concentration reduces along the height but remains unchanged beyond the TDH.

Hydrodynamics

Depending upon the velocity of the bubble relative to the gas in the emulsion phase, the gas flow in and out of the bubble could change. For example, when the bubble velocity is much higher than the gas velocity, the gas circulates entirely inside the bubble instead of moving out of it. On the other hand, when the bubble rises at the same velocity as the gas in the emulsion phase, all gases from the emulsion phase will enter its bottom and leave from its top. Further details are available in Kunii and Levenspiel (1991).

2.1.2.2 Practical Implication

Most bubbling fluidized bed boilers use Group B or D particles, which results in large bubbles and hence high bubble rise velocity. From the above analysis we note that such bubbles are likely to have a limited exchange of gas with the emulsion phase. Burning fuel particles, requiring oxygen, generally reside in the emulsion phase. The gas flow through the emulsion phase is of the order of $[AU_{mf}]$, which is, for most fluidized bed boilers, a small fraction of that passing through the bubbles $[A(U - U_{mf})]$. In the absence of efficient gas exchange between the two phases, there is major by-passing of oxygen while the fuel particles are kept oxygen-starved in the emulsion phase. Although bubbles, by definition, are free of solids, some fine char particles and volatiles are carried in them. These fuels burn efficiently, utilizing part of the bubble-phase oxygen.

Some mixing of the oxygen from the bubble and emulsion phases occurs in the splash zone immediately above the bed surface. Fine particles could receive the oxygen here and continue to burn in the freeboard.

2.1.2.3 Properties of Bubbles

In BFB boilers and gasifiers, the amount of gas passing through the emulsion phase is equal to that required for minimum fluidization, i.e., AU_{mf}. Therefore, the remaining gas, $A(U - U_{mf})$ passes through the bubble phase. The equivalent volume diameter d_b, of a bubble at a height Z, above the distributor is a function of the nozzle area of the distributor A_0, and the superficial gas velocity through the bubble phase $(U - U_{mf})$. It can be estimated using the expression given by Darton et al. (1977) as:

$$d_b = 0.54(U - U_{mf})^{0.4}\left(Z + 4\sqrt{A_0}\right)^{0.8} g^{-0.2} \qquad (2.6)$$

where g is the acceleration due to gravity (m/sec^2).

Mori and Wen (1975) suggested the following correlation for Group B and D particles:

$$\frac{d_{bm} - d_b}{d_{bm} - d_{b0}} = \exp\left(-\frac{0.3Z}{D}\right) \qquad (2.7)$$

where D is the diameter of the bed. The limiting bubble size, d_{bm} is correlated as:

$$d_{bm} = 0.65\left[\frac{\pi}{4}D^2(U - U_{mf})\right]^{0.4} \quad \text{(cm)} \qquad (2.8)$$

Note: Equation 2.8 is dimensional, U and U_{mf} are in cm/sec, and D is in cm.

The initial bubble size d_{b0}, near the bottom of the bed is (Kunii and Levenspiel, 1991):

$$d_{b0} = \frac{2.78(U - U_{mf})^2}{g} \qquad (2.9)$$

A single bubble rises through the emulsion phase with a velocity (Kunii and Levenspiel, 1991):

$$u_{bs} = 0.711\sqrt{gd_b} \qquad (2.10)$$

The above velocity of bubbles is not the absolute velocity; it is with respect to the emulsion phase. When a new bubble enters the bottom of the bed, it pushes up the emulsion phase by a volume corresponding to its own. Thus, the emulsion phase itself rises up with an absolute velocity $(U - U_{mf})$ up to the bed surface. We do not see the emulsion phase rising above the bed height because its upper surface drops back as soon as a bubble bursts. The absolute velocity of the bubble, U_b is:

$$U_b = U - U_{mf} + u_{bs} \qquad (2.11)$$

2.1.2.4 Bed Expansion

A bed is at the minimum fluidized condition when the superficial gas velocity is equal to U_{mf}. In this state, there are no bubbles, and only emulsion phase remains. For Group B and D particles, addition of any gas in excess of this goes to the formation of bubbles, which push their way into the emulsion phase of solids, resulting in bed expansion. Davidson and Harrison (1963) derived the following expression for bed expansion, assuming the bubble velocity to be constant at a mean value of U_{bm}:

$$\frac{H - H_{mf}}{H_{mf}} = \frac{U - U_{mf}}{U_{bm}} \qquad (2.12)$$

2.1.2.5 Freeboard

A bubbling fluidized bed boiler would normally have an open space above its bed surface, known as the *freeboard*. Each bubble carries some solids upwards in its wake. Bubbles erupt at the surface of the bed, throwing or entraining particles into the freeboard above. Due to their momentum and local gas drag, the entrained particles travel upwards through the freeboard. The particles may travel upwards either in groups or individually as shown in Figure 2.4. Since the fluid drag is not necessarily greater than the weight of all particles or groups of particles, some of the particles may disengage from the gas in the freeboard, returning to the dense bed due to gravitational force. This process of disengagement reduces the upward flux of particles exponentially along the height of the freeboard.

Beyond a certain height, only a negligible amount of particles disengage from the gas to return to the dense bed. This height is known as *transport disengaging height* (TDH). Beyond this height, only particles whose weight is sufficiently small to be balanced by the fluid drag are carried up. More specifically, particles with terminal velocities (see Section 2.1.6) lower than the superficial velocity in the freeboard are carried away. The flux of particles carried away beyond the TDH is known as the *elutriation rate*.

2.1.3 SLUGGING

Commercial fluidized bed boilers or gasifiers are generally too large in diameter to face the problem of slugging. However, this topic is discussed below to complete discussion of all flow regimes of gas–solid flows. From Equation 2.6 and Equation 2.7 we note that the size of bubble increases as the fluidizing velocity U, or the bed height Z, increases. If the bed is narrow and deep, the bubble may increase to a size comparable to the width of the bed. In this case the bubble passes through the bed as a slug. This phenomenon is known as *slugging*. A necessary condition for the formation of slugs is that the maximum stable bubble size D_{bmax}, must be greater than 0.6 times the diameter of the bed D (Geldart, 1986). Slugging naturally could not occur in all beds under normal operating conditions. The criterion for slug formation at choking is given by (Yang, 1976):

$$\frac{U_t^2}{gD} \geq 0.123 \qquad (2.13)$$

where U_t is the terminal velocity of the average-sized bed solid. The minimum slugging velocity U_{sl}, is given by Stewart and Davidson (1967):

$$U_{sl} = U_{mf} + 0.07(gD)^{0.5} \qquad (2.14)$$

where D is the diameter of the bed. For a rectangular bed or a bed of any other cross-sectional, D may be taken as 4 × area/perimeter.

2.1.4 TURBULENT BEDS

When the velocity of a gas through a bubbling fluidized bed is increased above the minimum bubbling velocity, the bed starts expanding. A continued increase in the velocity may eventually show a change in the pattern of bed expansion, indicating a transition into a new regime called a turbulent bed. The transition from bubbling to turbulent may be due to an increase in the bubble fraction, an expansion of the emulsion phase (Nakajima et al., 1991) and/or thinning of the emulsion walls separating the bubbles. In the turbulent bed regime, the bubble phase loses its identity due to rapid coalescence and breakup of bubbles. This results in a violently active and highly expanded bed with a change in the pattern of bed expansion throwing particles into the freeboard above the bed. The bed will have a surface but it is considerably diffused. Such beds are referred to as *turbulent beds* (Figure 2.2c).

Werther and Wein (1994) presented data comparing turbulent to bubbling bed regimes for 200-μm ash particles. They showed that both axial and radial voidage profiles of turbulent fluidized beds are much different from those of bubbling beds. The lower section of most CFB boilers, characterized by violent coalescing of bubbles, operates in the turbulent regime.

2.1.5 BUBBLING TO TURBULENT TRANSITION

The pressure drop across a turbulent bed fluctuates rapidly. As the velocity is increased, the amplitude of fluctuation increases, reaching a peak at the velocity u_c. It then reduces to a steady value as the fluidizing velocity is increased further to the velocity u_k (Figure 2.5). The transition from a bubbling to turbulent bed does not take place sharply at one velocity. The onset of this transition begins at the velocity u_c and is completed at the velocity u_k. The transition appears to start at the upper surface of the bed and move downward. There is no well-established correlation for calculating the velocity of transition from the bubbling to turbulent bed. However, some

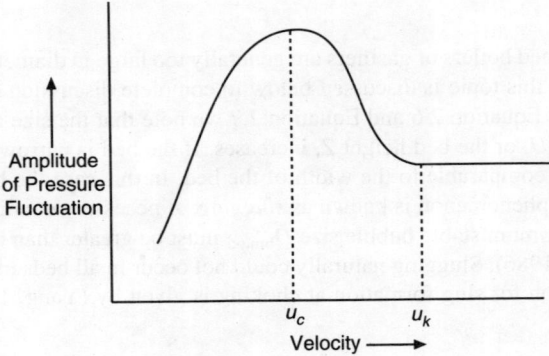

FIGURE 2.5 The amplitude of pressure fluctuation across the bed increases as the bed approaches turbulent fluidization.

correlations (Grace, 1982) based on small-diameter beds are given below:

$$u_c = 3.0\sqrt{\rho_p d_p} - 0.17 \quad \text{m/sec} \quad (2.15)$$

$$u_k = 7.0\sqrt{\rho_p d_p} - 0.77 \quad \text{m/sec} \quad (2.16)$$

where $(\rho_p d_p)$ is in the range 0.05 to 0.7 kg/m². Equation 2.15 and Equation 2.16 are dimensional. Horio and Morishita (1988) presented an alternative set of equations for this transition:

$$Re_c = \frac{u_c d_p \rho_g}{\mu} = 0.936 Ar^{0.472} \quad (2.17)$$

$$Re_k = \frac{u_k d_p \rho_g}{\mu} = 1.46 Ar^{0.472} \quad (Ar < 10^4)$$
$$= 1.41 Ar^{0.56} \quad (Ar > 10^4) \quad (2.18)$$

TABLE 2.2
Some Experimental Values of Velocities of Transition from Bubbling to Turbulent Fluidization

Solid	Density (kg/m³)	Size (μm)	u_c (m/sec)	u_k (m/sec)	Diameter of Bed (mm)	Reference
Sand	2665	69	1.1	2.58	76	Rhodes and Geldart, 1986a, 1986b
Sand	2648	134	0.9	2.94	76	Rhodes and Geldart, 1986a, 1986b
Sand	2600	270	1.8	3.84	76	Rhodes and Geldart, 1986a, 1986b
Sand	2600	606	2.0	5.36	76	Rhodes and Geldart, 1986a, 1986b
Sand	2660	1090	2.4	6.68	76	Rhodes and Geldart, 1986a, 1986b
Sand	2640	250–400	1.3	3.3	92	Perales et al., 1991
Catalyst	1800	64	0.4		800	Sun and Chen, 1989

Hydrodynamics

Table 2.2 presents some experimental values of velocities of transition to turbulent fluidization for some solids. The transition from bubbling to turbulent fluidization occurs at a lower velocity in larger-diameter vessels (Sun and Chen, 1989). An absence of data for large beds prevents defining the actual extent of the effect of bed diameter on large commercial-sized units.

Fine particles enter turbulent fluidization at a velocity sufficiently above their terminal velocity, whereas coarser particles may enter turbulent fluidization at a velocity below their terminal velocity.

The gas–solid contact in this regime is good and the reactor performance approaches that of an ideal back-mix reactor. Heat transfer coefficients in turbulent regimes are lower than those in the bubbling bed regime, but that is due to the lower solid concentration in a turbulent bed. If one takes this difference in suspension density into account, heat transfer in a turbulent bed would be higher than that in a bubbling bed confirming a higher degree of solid mixing in the former regime.

2.1.6 Terminal Velocity of a Particle

In the regimes discussed so far, solids are generally retained within a certain height above the grid. Except for some entrainment, there is no large-scale migration of particles with the gas, so, these regimes (fixed, bubbling and turbulent) are in the *captive* stage. Subsequent discussions will concentrate on *transport* stage which involves large-scale migration of particles out of the vessel.

Consider a particle falling freely from rest and accelerating under gravity in an infinite and stationary medium. The buoyancy force and the fluid drag oppose the effect of gravity. The particle accelerates until it reaches an equilibrium velocity called the *terminal velocity*.

When a fluid flows over a stationary particle or travels at a velocity higher than that of an upward-moving particle, the particle experiences an upward fluid drag. It is also subjected to a buoyancy force, and downward gravitational force, as shown in Figure 2.6. The *drag force* on the particle is related to the kinetic energy of the fluid and the projected area of the particle, and is

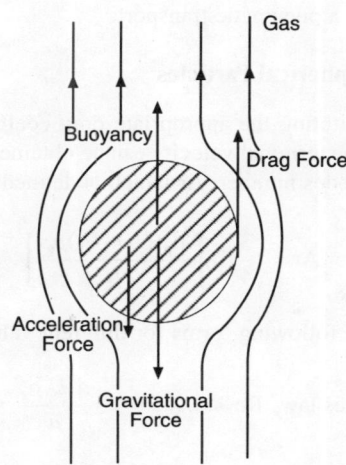

FIGURE 2.6 Force balance on a particle moving in an upward gas stream.

defined as:

$$F_D = C_D \frac{\pi d_p^2}{4}\left(\frac{\rho_g U^2}{2}\right) \quad (2.19)$$

where, U is the free stream gas velocity, and C_D is the drag coefficient, which is a function of the Reynolds number, $\text{Re} = (d_p U \rho_g / \mu)$

The characteristic relation between C_D and Re for particles of different shape factors is expressed as:

$$C_D = \frac{a_1}{\text{Re}^{b_1}} \quad (2.20)$$

where the constants a_1 and b_1 can be approximated as below (Howard, 1989).

Range of Re	Region	a_1	b_1
$0 < \text{Re} < 0.4$	Stoke's law	24	1.0
$0.4 < \text{Re} < 500$	Intermediate law	10	0.5
$500 < \text{Re}$	Newton's law	0.43	0.0

If both the gas and particle move upwards with velocities U and U_s, respectively, the particle will experience a fluid drag due to the relative velocity $(U - U_s)$ resisting its fall. Thus, the force balance under steady state can be written as:

Gravitational force = Buoyancy force + Drag force

$$m_p g = m_p \frac{\rho_g g}{\rho_p} + C_D \frac{\pi (U - U_s)^2 \rho_g}{8} d_p^2 \quad (2.21)$$

The limiting relative velocity $(U - U_s)$, or the velocity of the particle when the gas velocity U, is zero, is known as the *terminal velocity*, U_t. The terminal velocity is essentially the slip velocity between the fluid and particle in a pneumatic transport.

2.1.6.1 Terminal Velocity of Spherical Particles

For spherical particles, by substituting the appropriate drag coefficient values in Equation 2.21, the Reynolds number based on the terminal velocity can be obtained. To simplify it we arrange the equation in terms of the Archimedes number, Ar which is defined as $[g d_p^3 \rho_g (\rho_p - \rho_g)/\mu^2]$:

$$\text{Ar} = \frac{3}{4} C_D \left[\frac{d_p (U - U_s) \rho_g}{\mu}\right]^2 \quad (2.22)$$

Equation 2.22 reduces to the following forms for different values of the drag coefficient:

$$\text{Stokes law, Re} < 0.4 : \quad \frac{d_p U_t \rho_g}{\mu} = \frac{\text{Ar}}{18} \quad (2.23)$$

$$\text{Intermediate law, } 0.4 < \text{Re} < 500 : \quad \frac{d_p U_t \rho_g}{\mu} = \left[\frac{\text{Ar}}{7.5}\right]^{0.666} \quad (2.24)$$

Hydrodynamics

Newton's law, $500 < Re$: $\quad \dfrac{d_p U_t \rho_g}{\mu} = \left[\dfrac{Ar}{0.33}\right]^{0.5}$ (2.25)

2.1.6.2 Terminal Velocity of Nonspherical Particles

The above equations are only valid for spherical particles. The drag on nonspherical particles could be significantly different from that on spherical ones, though their equivalent particle diameter may be the same. This is illustrated by the difference in free-fall velocity of a piece of popcorn and a sphere of an equivalent diameter. An approximation is to multiply the terminal velocity of the spherical particle by a factor, K_t (Pettyjohn and Christiansen, 1948):

$$U_t = K_t U_t \text{ (spherical)} \quad (2.26)$$

where, sphericity, ϕ lies between 0.67 and 0.996.

The correction factor, K_t is obtained by:

For $Re < 0.2$: $\quad K_t = 0.843 \log_{10}\left[\dfrac{\phi}{0.065}\right]$

For $Re > 1000$: $\quad K_t = \left[\dfrac{4(\rho_p - \rho_g)g d_v}{3\rho_g(5.31 - 4.88\phi)}\right]^{0.5}$ (2.27)

where d_v is the volume mean diameter of the particle.

For Reynolds number (Re) between 0.2 and 1000, K_t can be obtained by interpolating the values of K_t calculated for the above two ranges of Reynolds numbers.

Example 2.1

Find the minimum fluidization, minimum bubbling, terminal velocities, and velocity for the onset of transition to turbulent fluidization for 300-μm sand ($\rho_p = 2500$ kg/m^3) in a 0.203 m × 0.203 m bed operating under the following conditions:

	A	B
Bed temperature (°C)	825	27
Gas density (kg/m^3)	0.316	1.16
Gas viscosity (N sec/m^2)	4.49×10^{-5}	1.84×10^{-5}

Solution

1. *First, check to which group the particle belongs*:

$$\rho_p = 2500 \text{ kg/m}^3 = 2.5 \text{ g/cm}^3$$

$$\rho_p - \rho_g = 2.5 - 0.00116 = 2.499 \text{ g/cm}^3 \quad (27°C)$$

$$= 2.5 - 0.000316 = 2.499 \text{ g/cm}^3 \quad (825°C)$$

From Figure A1.2 of Appendix 1 we find that a 300-μm particle will belong to Group B.

	A	B
	27°C	825°C
$Re_c = 0.936(2267)^{0.472}$	35.9	8.36
$u_c = 35.9 \times 1.84 \times 10^{-5}/(0.0003 \times 1.16)$	1.897 m/sec	3.96 m/sec
$Re_k = 1.46(2267)^{0.472}$	56.0	3.05
$u_k = 56 \times 1.84 \times 10^{-5}/(0.000003 \times 1.16)$	2.959 m/sec	6.18 m/sec

2. *Minimum fluidization velocity*: From Equation 2.3

$$Re_{mf} = [27.2^2 + 0.0408 Ar]^{0.5} - 27.2$$

For 27°C:

$$Ar = \frac{(1.16)(2500 - 1.16)(9.81)(0.0003)^3}{(0.0000184)^2} = 2268$$

$$Re_{mf} = [27.2^2 + 0.0408 \times 2268]^{0.5} - 27.2 = 1.65$$

From above:

$$U_{mf} = \frac{(1.65)(0.0000184)}{(0.0003)(1.16)} = 0.087 \text{ m/sec}$$

Similarly for 825°C:

$$Ar = 103.7 \qquad Re_{mf} = 0.0777$$

$$U_{mf} = 0.0368 \text{ m/sec}$$

3. *Minimum bubbling velocity*: One could use Equation 2.4 to find U_{mb}, but since it is a Group B particle, there is no U_{mb}. The bed starts bubbling immediately after reaching U_{mf}

$$U_{mb} = U_{mf} = 0.087 \text{ m/sec at } 27°C$$

Transition to turbulent fluidization is found using either Equation 2.15 and Equation 2.16 or Equation 2.17 and Equation 2.18

Here, $\rho_p d_p = 2500 \times 0.0003 = 0.75$, but Equation 2.15 and Equation 2.16 are valid only for $0.05 < \rho_p d_p < 0.7$.

Therefore, we use Equation 2.17 and Equation 2.18 to find the transition velocities: We use the values of Archimedes number calculated above:

$$Ar|27°C = 2268$$

$$Ar|825°C = 103.7$$

4. *Terminal velocity*: At 27°C, Ar = 2268. Using Equation 2.24,

$$U_t = \frac{1.84 \times 10^{-5}}{0.0003 \times 1.16} \left(\frac{2268}{7.5}\right)^{0.666} = 2.37 \text{ m/sec}$$

$$Re = \frac{2.37 \times 0.0003 \times 1.16}{1.84 \times 10^{-5}} = 44.8$$

Hydrodynamics

This verifies that the Reynolds number is within the intermediate law region (0.4 < Re < 500), which justifies the use of Equation 2.24.

Similarly, at 825°C, Ar = 103.7. Hence, we use Equation 2.24 to get:

$$U_t = \frac{4.49 \times 10^{-5}}{0.0003 \times 0.316} \left(\frac{103.7}{7.5}\right)^{0.666} = 2.72 \text{ m/sec}$$

2.1.7 Freeboard and Furnace Heights

The freeboard height is an important design parameter, especially in a bubbling fluidized bed boiler. In a boiler with water-walls but without in-bed tubes, the freeboard height is generally dictated by the heating-surface area requirement of the evaporator. When in-bed tubes are used, the freeboard height may be chosen for other considerations. An important consideration is to minimize the entrainment of unburnt carbon by choosing a freeboard exceeding or close to the *transport disengaging height* (TDH).

2.1.7.1 Transport Disengaging Height

On the basis of their experiments, Zenz and Weil (1958) presented a graph of TDH against $(U - U_{mb})$ for different bed diameters up to 3 ft. Several empirical correlations are available, but most of them are for fine particles and none of them have been tested with coarse particles or large-diameter beds. Horio et al. (1980) presented the following equation for TDH:

$$\text{TDH} = 4.47 d_{eq,s}^{1/2} \text{ m} \qquad (2.28)$$

$d_{eq,s}$ is the equivalent volume diameter of a bubble at the bed surface (m)

2.2 FAST FLUIDIZED BED

In the context of its use in Circulating fluidized bed (CFB) boilers, the *fast fluidized bed* may be defined as:

> A high velocity gas–solid suspension where particles, elutriated by the fluidizing gas above the terminal velocity of individual particles, are recovered and returned to the base of the furnace at a rate sufficiently high to cause a degree of solid refluxing that will ensure a minimum level of temperature uniformity in the furnace.

2.2.1 Characteristics of Fast Beds

The furnace of a typical circulating fluidized bed operates in this regime. The term, *fast bed* was coined by Yerushalmi et al. (1976). It was described as a regime lying between turbulent fluidization and pneumatic transport. In a typical fast fluidized bed, one observes a nonuniform suspension of slender particle agglomerates or clusters moving up and down in a dilute, upwardly flowing gas–solid continuum (Figure 2.7).

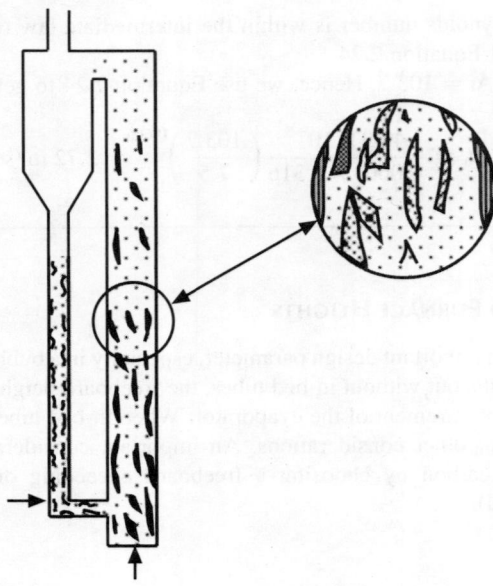

FIGURE 2.7 A fast fluidized bed is comprised of solid agglomerates moving up and down in a very dilute dispersion of solids.

High slip velocity between the gas and solids, formation and disintegration of particle agglomerates and excellent mixing are major characteristics of this regime. Typical axial and radial variation of suspension density is another physical characteristic of the fast bed.

The formation of solid agglomerates is not an adequate condition for the fast fluidized bed, but is an important and necessary feature of this regime.

A qualitative description of the phenomenon leading to the formation of clusters in a pneumatic transport column is presented in Figure 2.8. In this imaginary experiment, a solid is

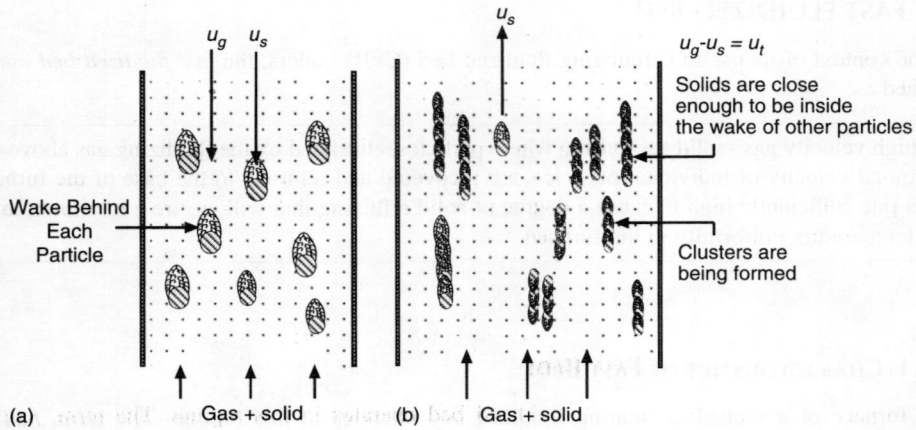

FIGURE 2.8 (a) Solids moving up in a gas stream may have a small wake behind each particle. (b) Transition from pneumatic transport to fast fluidization may occur when the rate of solid circulation is continually increased.

Hydrodynamics

continuously fed to the rising gas stream in the column. At a very low feed rate, the particles will be uniformly dispersed in the gas stream, each particle traveling in isolation. The relative velocity between the gas and the solid will form a small wake behind each particle (Figure 2.8a).

For a given gas velocity, the solid feed rate may be increased to a level where the solid concentration will be so high that one particle will enter the wake of the other. When that happens, the fluid drag on the leading particle will decrease, and it will fall under gravity until it drops on to the trailing particle (Figure 2.8b). The effective surface area of the pair just formed is lower, so the fluid drag will be lower than their combined weight, causing the pair to fall further and collide with other particles. Thus, an increasing number of particles combine together to form particle agglomerates known as clusters. However, these clusters are not permanent, being continuously torn apart by the up-flowing gas. In this way, the cycle of cluster formation and disintegration continues.

2.2.2 Transition to Fast Fluidization

A clear picture of the transition to and from fast fluidization is lacking at the moment. Thus, a description of the process, like that presented in Figure 2.8, is only tentative.

Figure 2.9 presents another description of the transition. Imagine that gas is flowing upwards through a vertical column to which a solid is fed at a given rate W_1, so the suspension is initially in pneumatic transport. If the superficial gas velocity through the column is decreased (Figure 2.9) without changing the solid feed rate, the pressure drop per unit height of the column will decrease due to the reduced fluid friction on the wall (C–D). However, the suspension will become increasingly denser with decreasing gas velocities. Thus, the gas–solid drag will dominate the pressure drop across the column. There would be a pressure drop across the column due to both wall–gas friction and gas–solid drag (friction). Under steady-state conditions the solid–gas friction is equal to the weight of the solids. With a continued decrease in the superficial velocity (Figure 2.9), the bed becomes denser and the pressure drop begins to increase (D–E). The point of reversal (D) marks the onset of fast-bed (Reddy-Karri and Knowlton, 1991) from pneumatic transport.

If the gas velocity is decreased further, the solid concentration in the column increases up to a point where the column is saturated with solids, the gas being unable to carry any more. The solids start accumulating, filling up the column, and causing a steep rise in the pressure. This condition (E) is called *choking*. In smaller diameter columns, the bed starts slugging, while in larger ones it

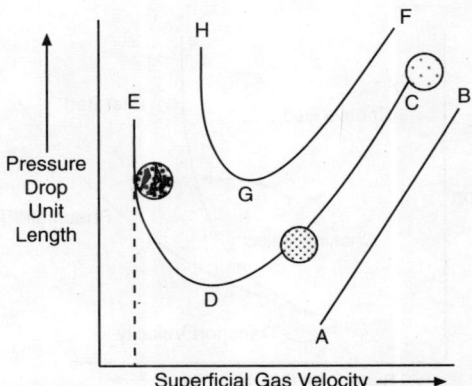

FIGURE 2.9 Schematic representation of different regimes of gas–solid flow through a vertical column.

undergoes transformation into a nonslugging dense-phase fluidized bed, such as a turbulent bed. The gas–solid regimes below this velocity have the generic name of *captive* state. The captive state may include turbulent, slugging, bubbling, and fixed beds.

2.2.2.1 Transition from Bubbling to Fast Bed

The discussion above suggests that the choking velocity, U_{ch} may be used to mark the transition from captive state to fast fluidization. Yang's (1983) correlation, developed for Group A particles at room temperatures using small-diameter (<0.3 m) pneumatic transport columns, may be used for a first approximation of the choking velocity in fluidized bed boilers for Group A particles. However, one must recognize its limitations.

$$\frac{U_{ch}}{\varepsilon_c} = U_t + \sqrt{\frac{2gD(\varepsilon_c^{-4.7} - 1)\rho_p^{2.2}}{6.81 \times 10^5 \rho_g^{2.2}}} \quad (2.29)$$

where $D < 0.3$ m

$$G_s = (U_{ch} - U_t)(1 - \varepsilon_c)\rho_p \quad (2.30)$$

where G_s is the solid circulation rate, ε_c is the voidage at choking, and U_t is the terminal velocity of single particles. For given values circulation rates U_t, ρ_p and ρ_g, one can calculate U_{ch} through the simultaneous solution of the above two equations.

This correlation suggests a square root dependence of bed diameter, while experimental data (Knowlton, 1990b) show that the dependence is on the fifth root of the diameter and the bed diameter may not have any effect on the transition velocity for beds larger than 0.5 m diameter (Knowlton, 1990a). Furthermore, the validity of this in high-temperature CFB boilers or gasifiers has yet to be verified. Thus, Equation 2.29 and Equation 2.30 should be used with caution.

A qualitative flow regime diagram, developed by Reddy-Karri and Knowlton (1991), is presented in Figure 2.10. The regimes are shown on a plot of solid circulation rate and fluidizing or superficial velocity. The line A–B (locus of choking velocity) marks the boundary between the captive and fast-bed conditions. Figure 2.10 further shows that at higher circulation rates, the transition to fast fluidization occurs at higher velocities.

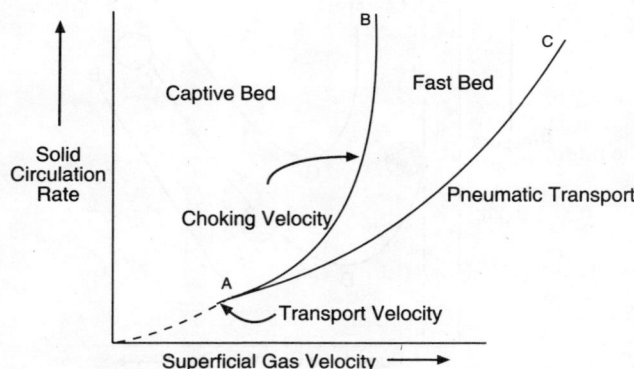

FIGURE 2.10 The fast fluidization is bounded by two velocities that depend on the circulation rate.

Hydrodynamics

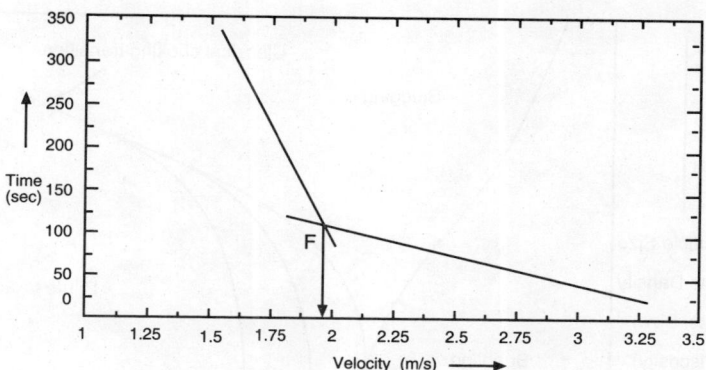

FIGURE 2.11 The time of emptying a column decreases with increasing gas velocity through the column. The transport velocity is marked by the intersection (F) of slopes of the curves. (Adapted from Perales, J.F. et al., in *Circulating fluidized Bed Technology III*, Basu, P., Hasatani, M., Hortio, M., Eds., Pergamon Press, Oxford, 1991.)

2.2.2.2 Transport Velocity

There is a minimum velocity below which fast fluidization cannot occur, irrespective of the circulation rate. This velocity, known as *transport velocity*, U_{tr}, is marked by point (A) in Figure 2.10. The critical velocity may be described as follows. If a bed is fluidized above the terminal velocity described 2.1.1 of the individual bed particles, all solids will be entrained out of the column unless they are replaced simultaneously. However, it takes a certain amount of time to empty the column. As the velocity in excess of the terminal velocity increases, the time needed to empty the vessel decreases gradually until a critical velocity is reached, above which there is a sudden drop in the time for emptying the vessel (Figure 2.11). An empirical relation for the transport velocity is given by Perales et al. (1991) based on their experiments in a 92-mm diameter column:

$$U_{tr} = 1.45 \times \frac{\mu}{(\rho_g d_p)} Ar^{0.484} \tag{2.31}$$

where $20 < Ar < 50{,}000$.

Thus, a fast bed may be defined more rigorously as below:

> Fast fluidization is a regime of gas–solid fluidization above the transport velocity. For a given solid circulation rate, it is bound by the choking velocity on the lower side and the velocity corresponding to the minimum pressure drop on the gas–solid contacting phase diagram on the higher side.

2.2.2.3 Transition from Pneumatic Transport to Fast Bed

It is apparent from Figure 2.10 that for each solid circulation rate there are two limiting velocities for the fast bed: the choking velocity on the lower side and another velocity on the higher side that marks the transition from fast fluidization to pneumatic transport. This is also the velocity at which, for a given circulation rate, the average pressure drop across the height of the column reaches a minimum value (point D, Figure 2.9).

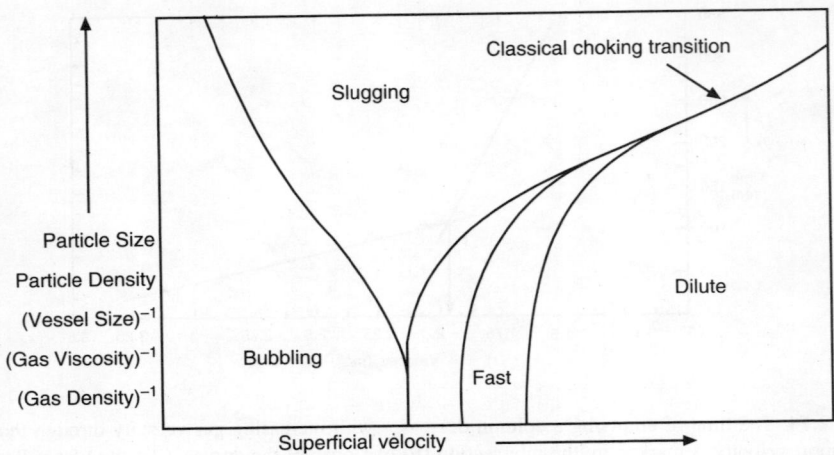

FIGURE 2.12 The transition from one regime to another depends on design parameters, such as particle size, particle density, column size, gas density, and gas viscosity.

It is difficult to detect this transition point experimentally. An indirect experimental method can be used (Biswas and Leung, 1987) to find this velocity by exploiting the axial nonuniformity feature of the fast fluidized bed. If the pressure drop across two small but equal sections in the upper and the lower section of a column carrying solids in pneumatic transport, is measured the values will be similar. Now, if the gas velocity is decreased at a given circulation rate, the pressure drop (hence, suspension densities) at those two points will increase by identical degrees until the transition velocity to fast fluidization is reached, at which point the column becomes axially nonuniform. Thus, the suspension density near the lower section of the column will start exceeding that at the top. No comprehensive correlation is available at the moment to estimate this velocity. Only experimental data can provide the necessary guide to this transition.

2.2.2.4 Flow Regime Diagram

The transition to fast fluidization is affected by a number of parameters, such as particle size (Drahos et al., 1988), particle density, gas viscosity, gas density, and tube size. Figure 2.12 is a qualitative representation of transition values of those parameters as a function of superficial velocity. It was used by Shingles and Dry (1986) to describe the regime transition.

Figure 2.12 does not show the transition for variation in operating variables such as solid circulation rate, which was shown in Figure 2.10.

These diagrams are presently used for qualitative assessment only. An important point, shown in Figure 2.12 is that the operating range of fast beds reduces for coarser particles. This is why it is more difficult to maintain fast fluidization for Group B particles than for Group A particles.

Example 2.2

Find the minimum velocity for fast fluidization for 300-μm sand particles at 27 and 825°C for the following conditions. The desired solid circulation rate in the fast regime is 30 kg/m^2 sec. The cross section of the bed is 0.203 m × 0.203 m. The density of the particles is 2500 kg/m^3.

Hydrodynamics

Temperature (°C)	825	27
Gas density (kg/m^3)	0.316	1.16
Gas viscosity (N sec/m^2)	4.49×10^{-5}	1.84×10^{-5}
Terminal velocity (m/sec)	2.72	2.27

Solution

1. *We will first use Equation 2.31 for the transport velocity. For 27°C:*

$$\text{Ar} = \frac{1.16 \times (2500 - 1.16) \times 9.81 \times 0.0003^3}{(1.84 \times 10^{-5})^2} = 2268$$

So, $20 < \text{Ar} < 50{,}000$, so:

$$U_{tr} = 1.45 \times 1.84 \times 10^{-5} \times 2268^{0.484}/(1.16 \times 0.0003) = 3.227 \text{ m/sec}$$

Similarly at 825°C:

$$U_{tr} = 6.49 \text{ m/sec}$$

2. *We will now use Equation 2.29 and Equation 2.30 to find the choking velocity and the circulation rate*:

$$\frac{U_{ch}}{\varepsilon_c} = U_t + \left| \frac{2gD(\varepsilon_c^{-4.7} - 1)\rho_p^{2.2}}{6.81 \times 10^5 \rho_g} \right|^{0.5}$$

$$G_s = (U_{ch} - U_t)(1 - \varepsilon_c)\rho_p$$

For an equivalent bed diameter:

$$D = 4(0.203 \times 0.203)/(0.203 + 0.203 + 0.203 + 0.203) = 0.203 \text{ m}$$

The values of U_t are taken from Example 2.1:

$$U_{t,27°C} = 2.37 \text{ m/sec} \quad \text{and} \quad U_{t,825°C} = 2.72 \text{ m/sec}$$

By substituting values at 27°C into Equation 2.29, we get:

$$U_{ch}/\varepsilon_c = 2.37 + 11.229(\varepsilon_c^{-4.7} - 1)^{0.5}$$

$$30 = (U_{ch} - 2.37)(1 - \varepsilon_c)2500$$

For 825°C the above equation takes the form:

$$U_{ch}/\varepsilon_c = 2.722 + 46.9(\varepsilon_c^{-4.7} - 1)^{0.5}$$

$$30 = (U_{ch} - 2.722)(1 - \varepsilon_c)2500$$

Iterative solutions of above equations give U_{ch} and ε_c. These values can be compared with the transport velocities, which also seem to mark the transition between turbulent and fast beds.

Temperature (°C)	U_{ch} (m/sec)	ε_c	U_{tr} (m/sec)
27	4.19	0.9937	3.22
825	7.70	0.9976	6.49

The restricted validity of the correlation for choking velocity makes it imperative that the transition velocity computed from Equation 2.25 and Equation 2.26 is used with caution. Experiments should be the best guide.

2.2.3 DENSE SUSPENSION UP-FLOW

Circulating fluidized bed boilers and gasifiers operate in the fast-bed regime, while most CFB fluid catalytic crackers (FCC) operate in a similar but different regime known as *dense suspension up-flow*. Table 2.3 lists the difference between these two flow regimes. Both regimes belong to the circulating fluidized bed family and are characterized by a high level of solid circulation around the CFB loop. FCC reactors need high solid inventory and for that reason, this regime operates with a suspension density one order of magnitude higher than that used by a CFB boiler operating in the fast-bed regime. The difference is elaborated further in Section 2.3.1.

2.3 HYDRODYNAMIC REGIMES IN A CFB

In a circulating fluidized bed (CFB) boiler, hot solids circulate around a loop carrying heat from burning coal to heat-absorbing surfaces, and finally leave the furnace with the flue gas. Here, solids move through a number of hydrodynamic regimes while passing through different sections of the boiler. In Table 2.4, we note that in a typical CFB boiler or gasifier the solids are in the turbulent fluidized bed regime in the lower furnace, fast fluidized bed regime in the upper furnace, swirling flow in the cyclone, moving bed in the standpipe, bubbling bed in the loop-seal and pneumatic transport in the back-pass.

Most of the combustion and sulfur-capture reactions take place in the upper furnace above the secondary air level, which operates under a special hydrodynamic condition termed *fast fluidization*. The attractive novel features of CFB boilers, such as fuel flexibility, low NO_x emissions, high combustion efficiency, good limestone utilization for sulfur capture and fewer feed points are direct results of this special mode of gas–solid motion in their furnaces. The following section will discuss the nature of this regime in some details.

TABLE 2.3
Difference between CFB Boilers and CFB Reactors

Application	Boilers	Reactors
Regime	Fast	Dense phase suspension up-flow
Geldart particle groups	B or D	A, AC, C, or B
Particle diameter (μm)	200	70
Suspension density in upper region (kg/m³)	1–10	10–100
External circulation rate (kg/m² s)	10–50	500–1100
Solid residence time (sec)	30–600	3–15
Velocity (m/sec)	<6	<25
Temperature (°C)	850	550

Hydrodynamics

TABLE 2.4
Flow Regimes in Different Components of a Circulating Fluidized Bed

Location	Regime
Furnace (below secondary air level)	Turbulent or bubbling fluidized bed
Furnace (above secondary air level)	Fast fluidized bed
Cyclone	Swirl flow
Return leg (standpipe)	Moving packed bed
Loop seal/external heat exchanger	Bubbling fluidized bed
Back-pass	Pneumatic transport

2.3.1 DIFFERENCE IN OPERATING REGIMES OF CFB BOILERS AND REACTORS

The two most popular uses of circulating fluidized bed process are boilers and FCC reactors. Hundreds of FCC reactors are in commercial operation in refineries around the world. There is a basic difference between these two prominent applications of the CFB process — it is in the flow regime of operation of their risers.

Auxiliary power consumption being a major concern, boiler furnaces require minimum pressure drop across the riser and maximum internal solid circulation. For this reason, CFB boilers use a low (1 to 10 kg/m² sec) external solid circulation rate and a relatively dilute suspension density of about 1 to 5 kg/m³ in the upper part of the riser. In CFB catalytic reactors, on the other hand, internal back-mixing is not desirable as it decreases the yield. Therefore, they use large external solid circulation rates (500 to 1100 kg/m² sec), yielding high suspension densities in the range of 10 to 100 kg/m³. This regime is more appropriately called *dense suspension up-flow* instead of *fast-bed* as in CFB boilers.

Additionally, CFB boilers use group A, B, or even D particles, while FCC reactors use mostly Group AC, A particles. Some important differences are listed in Table 2.3.

2.4 HYDRODYNAMIC STRUCTURE OF FAST BEDS

The hydrodynamic condition influences parameters like auxiliary power consumption, heat absorption, temperature distribution, combustion condition, bed inventory, and erosion. So, a good understanding of the hydrodynamics of the flow regime of the furnace is important for a rational design of the boiler. The following section presents a short description of the hydrodynamic structure of the fast bed, as well as some empirical equations to predict the distribution of local suspension densities in the radial and axial directions.

2.4.1 AXIAL VOIDAGE PROFILE

A fast fluidized bed is nonuniform in both the axial and lateral directions. The axial distribution of the cross-sectional averaged voidage of FCC solids in a CFB reactor is shown in Figure 2.13. The axial profile shown in Figure 2.13a, is for a fluid catalytic cracking (FCC) particle and is representative of CFB risers used in FCC units. There is a change in profile, giving a point of inflexion at a certain height. Kwauk et al. (1986) explained this on the basis of cluster diffusion in the axial direction and concluded that the axial distribution of the voidage is S-shaped, as in Figure 2.13a. This profile is for Group A and weakly Group B particles, while that in Figure 2.13b is for Group B particles.

The profile shown in Figure 2.13a is clearly different from that seen in Figure 2.13b for a CFB boiler, where the secondary air is fed at a certain height above the bottom of the riser. The profiles

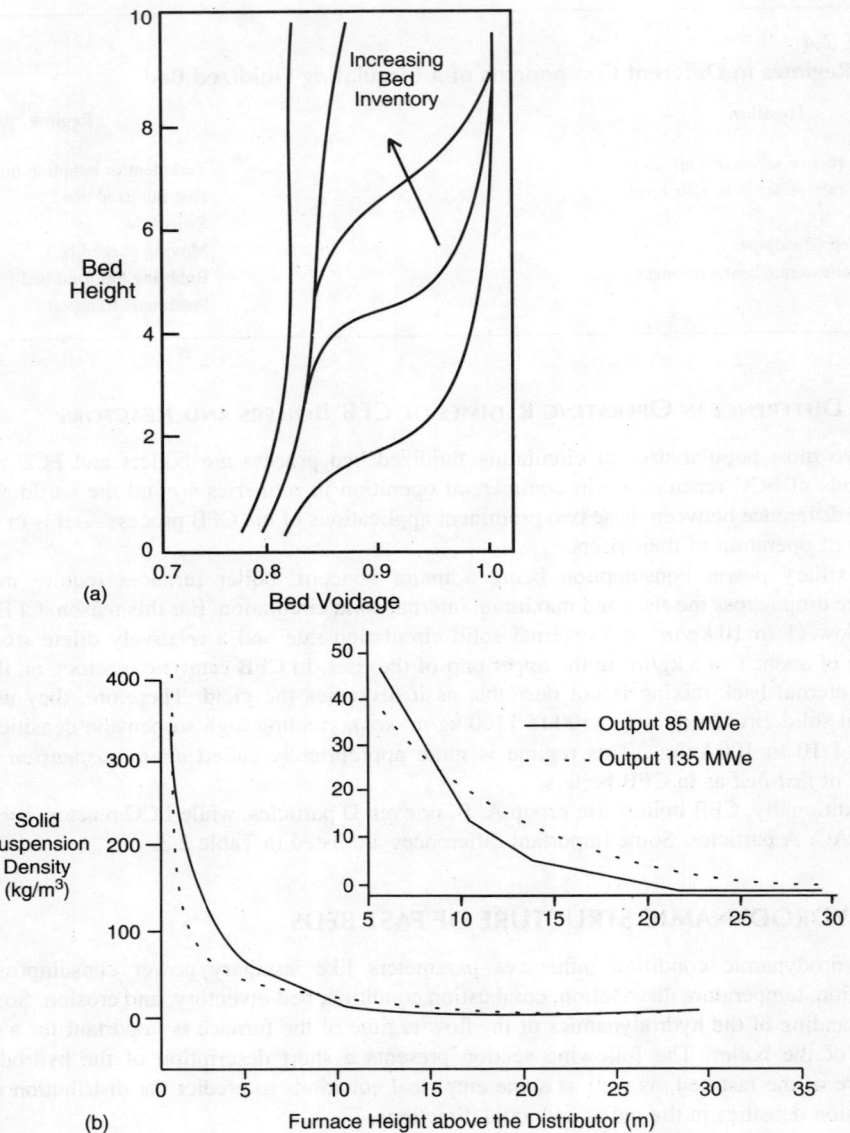

FIGURE 2.13 (a) Axial profile of cross section average bed voidage for FCC particles for different bed inventories. (b) Bed density profile drawn using data from a 135 MWe CFB boiler (Zhang et al., 2005) for two loads (85 and 135 MWe). For better resolution of the variation in suspension density, values above 5 m is replotted with an expanded scale.

of Figure 2.13a are typical for fast fluidized beds where the entire fluidizing gas is fed from the bottom of the riser, which is generally the case for CFB reactors, which are considerably denser than those in CFB boilers (Table 2.3).

Figure 2.13b shows the suspension density profile of a typical large CFB boiler. The figure shows that the lower section of the fast bed, especially below the secondary air-injection level, is dense while the upper section is relatively dilute. Beyond that height, there is an exponential decay of the suspension density. This decay clearly does not show an increase in the suspension density near the riser exit, observed in bench-scale units. An increase in suspension density is generally not expected in large commercial units unless the boiler designer decides to use a constricted furnace exit or place some tube panels restricting the gas–solid flow at the furnace exit. This would cause the solids to decelerate, leading to an increased suspension density.

An interesting point to observe in Figure 2.13b is that the suspension density profile changes with boiler load. For example, at higher load the density of the lower bed (below the secondary air) reduces and that of the lean upper bed increases compared to that observed in a lower load. The opposite is true at lower loads. In commercial CFB boilers, the ratio of primary to secondary air is greater at higher boiler load. As a result, solids from the dense lower bed are transported to the upper part of the riser to make it denser.

Based on their experimental data, Zhang et al. (2005) developed the following empirical correlation for suspension density profiles in the risers of large commercial CFB boilers:

$$\rho_{sus} = \frac{1}{g}\left[A_1 \exp\left(-\frac{h}{t_1}\right) + A_2 \exp\left(-\frac{h}{t_2}\right)\right] \qquad (2.32)$$

where ρ_{sus} is the bulk solid suspension density; g is gravity, h is height in the furnace, A_1, A_2, t_1 and t_2 are fitting parameters for each pressure drop profile, which depends on the boiler load.

2.4.1.1 Effects of Circulation Rate on Voidage Profile

Axial profiles of voidage or suspension density, shown in Figure 2.13a and Figure 2.14, show a change in the gradient. This point where the change takes place (point of inflexion) and separates the lean and dense regions of a fast bed is a function of the solid circulation rate and the amount of solids in the system. This effect arises from the pressure balance around the circulating fluidized bed loop, which is explained with the help of Figure 2.14.

The lower section of the fast bed is denser and therefore it results in higher pressure drops per unit height of the bed relative to the upper section, which is leaner. For a given amount of bed inventory, the solids are distributed between the bed and the return leg in such a way that

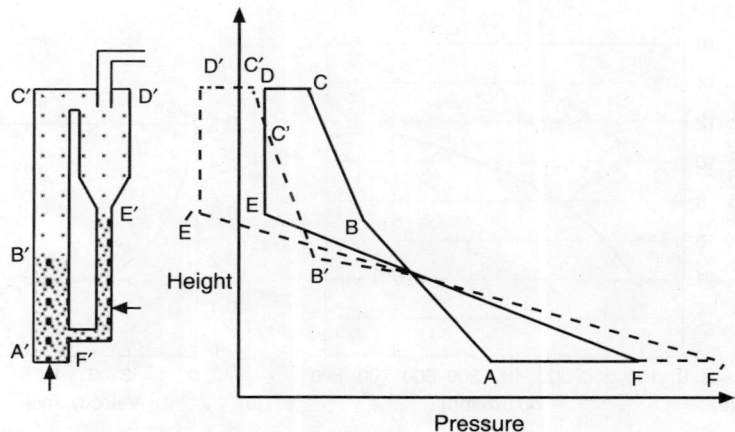

FIGURE 2.14 The voidage profile is governed by the pressure balance around the CFB loop. The dotted lines indicate the pressure profile at a higher circulation rate.

the pressure drops across two legs of the loop (Figure 2.14) balance each other out. Thus, we see in Figure 2.14 a pressure profile closed around the CFB loop.

The pressure drop across the L-valve controlling the solid flow is proportional to its flow rate. So, when the solid flow-rate is increased by increasing the aeration air, the pressure at point F (with reference to A) increases, and the pressure drop across the moving packed bed in the return leg increases from (EF) to (E'F'). The pressure drop across the cyclone also changes from CD to C'D'.

For stable operations the pressure balance around the loop may be written as:

$$\Delta P_{F-A} + \Delta P_{A-B} + \Delta P_{B-C} + \Delta P_{C-D} = \Delta P_{D-E} + \Delta P_{E-F} \quad (2.33)$$

This pressure balance depends on the different operating parameters, so, the response of the bed to a variation in an operating parameter can be predicted from the pressure balance.

An alternative approach to the prediction of axial distribution of voidage in a fast bed is based on the entrainment model (Rhodes and Geldart, 1986a, 1986b; Bolton and Davidson, 1988; Kunii and Levenspiel, 1990). The equation of axial voidage distribution developed by Kunii and Levenspiel (1991) is:

$$\frac{\varepsilon_d - \varepsilon}{\varepsilon_d - \varepsilon_a} = \exp[-a(h - h_i)] \quad (2.34)$$

where $h > h_i$ and $aU = 2$ to 5 sec^{-1} for $d_p < 70$ μm, and $aU = 4$ to 12 sec^{-1} for $d_p > 88$ μm, and h_i is the height of the point of inflexion.

Alternatively, for fast beds, the density decay constant, a, can be correlated with superficial velocities using experimental data, as shown in Figure 2.15. The asymptotic voidage in the denser section ε_a, for Group A particles is typically:

$\varepsilon_a = 0.88$ to 0.78 for fast beds
$\quad = 0.60$ to 0.78 for turbulent beds
$\quad = 0.45$ to 0.60 for bubbling beds (Kunii and Levenspiel, 1990).

The voidage ε_a, would be even higher for bed materials typically used in commercial CFB boilers (coal ash, sorbent, sand). In fast beds with secondary air injections, the point of inflexion h_i, generally occurs at or below the secondary air-injection level. The asymptotic voidage in the

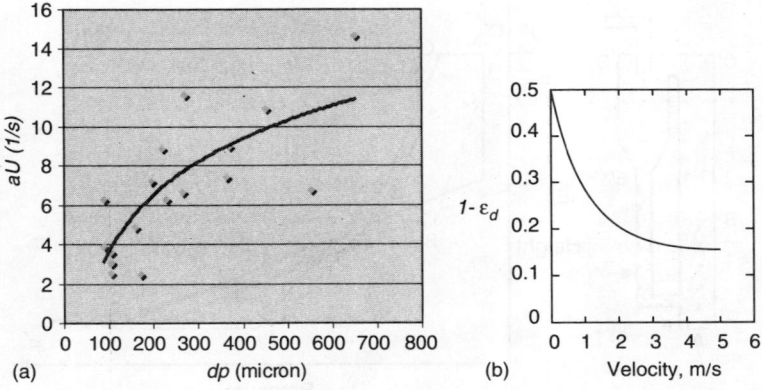

FIGURE 2.15 Decay constant 'a' in axial bed density profile as used in Equation 2.34. Each of the curves represents the product of velocity and the decay constant. (Adapted from Kunii, D. and Levenspial, O., in *Circulating Fluidized Beds III*, Basu, P. et al., Eds., Pergamon Press, Oxford, 1991.)

Hydrodynamics

dilute phase (upper section) ε_d, may be linked with the voidage below which cluster formation starts. Thus, one can approximate ε_d as the choking voidage ε_c, which can be found from Equation 2.29.

The voidage at the furnace exit at height H can be found by substituting H for h in Equation 2.34 as:

$$\varepsilon_e = \varepsilon_d - (\varepsilon_d - \varepsilon_a)\exp[-a(H - h_i)] \qquad (2.35)$$

The mean bed voidage, ε_s above the point of inflexion (or above the secondary air level in CFB boilers) can be found by integrating Equation 2.34 between H and h_i or the height of the secondary air level.

$$\varepsilon_s = \varepsilon_d - \frac{\varepsilon_a - \varepsilon_e}{a(H - h_i)} \qquad (2.36)$$

It follows that the amount of solids in the furnace, W can be found as:

$$W = A\rho_p[h_i(1 - \varepsilon_a) + (H - h_i)(1 - \varepsilon_s)] \qquad (2.37)$$

In a CFB boiler, one can take h_i to be equal to the height of the secondary air level for a first approximation.

If one assumes that the voidage at the top of the furnace is so high that all particles are completely dispersed and there is no refluxing, the voidage of the gas–solid suspension leaving the bed may be approximated as $[G_s/\rho_p U]$. If there is appreciable refluxing at that height, the voidage should be estimated from the following equation:

$$(1 - \varepsilon_e)\rho_p = \left[\frac{G_u}{U_s} + \frac{G_d}{U_d}\right] \qquad (2.38)$$

where G_u is the solid flux moving upward, G_d is the downward solid flux near the exit, U_s is the linear velocity of upward-moving solids, and U_d is the velocity of downward-moving solids.

Instead of computing it from Equation 2.34, the voidage profile above the secondary air level is sometimes fitted by a simpler equation of the following form:

$$\frac{c}{(h - h_i)^n},$$

where c and n are fitted constants and h is the height measured from the grate.

2.4.1.2 Effect of Particle Size on Suspension Density Profile

The particle size has an important practical influence on the suspension density profile in the furnace. Of late, large numbers of CFB boilers are operating at temperatures much higher than those for which they are designed. It has been observed that once the particle size of the circulating solids reduces, the bed temperature is reduced due to increased heat transfer because of higher suspension density. This positive development happens often without any penalty to fan power consumption. When a boiler does not give the desired output, the operators often feed finer sizes of coal into the furnace. This helps increase the pressure drop across the bed and consequently increases the furnace heat transfer. This is an important practical means of solving the problem of excessive furnace temperatures.

When the particle size is finer, more particles will be transported to the upper riser to make the upper heat-absorbing section denser. If the particle size distribution moves towards larger sizes, more of the particles will remain in the lower furnace instead of rising into the upper bed. This decreases the heat transfer coefficient in the upper bed.

TABLE 2.5
Response of Dependent Variables to Changes in Operating Parameters

Situations	1	2	3	4	5	6
Operating parameters						
Gas velocity	−	0	0	0	−	−
Valve opening	0	−	0	−	0	+
System inventory	0	0	−	+	−	0
Response to						
Solid circulation rate	−	−	−	0	0	0
Inflexion point	?	−	−	0	+	+
Dilute phase density	+	−	−	0	+	+
Dense phase density	+	−	−	0	+	+
Return leg inventory	−	+	−	+	+	−

Symbols used: −, decrease; +, increase; 0, unchanged.
Source: Adapted from Matsen, J.M., in *Circulating Fluidized Bed Technology II*, Basu, P. and Large, J.F., Eds., Pergamon Press Plc., 1988. (With permission).

2.4.1.3 Effect of Bed Inventory on Suspension Density Profile

In operating CFB it is very difficult to change either the gas velocity or the circulation rate forcing operators to look for some other parameter such as bed inventory. Yue et al. (2005) suggested that by changing the bed inventory one can influence the suspension density. Higher solids in the loop-seal leg could help higher solids flow through the loop seal which in turn may increase the suspension density in the furnace.

Matsen (1988) prepared a table of response of a typical circulating fluidized bed to the variation in three operating parameters, fluidization velocity, inventory, and resistance coefficient in the loop seal (value opening). Effects of these parameters on operating characteristics are listed in Table 2.5. Some of the parameters may not be changed in industrial units, but this table gives a qualitative picture of the response of a CFB to changes in system parameters.

2.4.2 LATERAL DISTRIBUTION OF VOIDAGE IN A FAST BED

Macroscopically, the fluidizing gas (with thinly dispersed solids) moves upwards in a plug flow. Detailed measurements show that the gas velocity near the wall is considerably lower than that in the core of the bed. Sometimes it could even be downwards. The majority of solids are back-mixed, meaning that they move up and down about the bed in the form of agglomerates of fine solids (known as clusters) that move together in the fast bed as a single body for a brief period of time and then dissolve. The formation of these clusters is explained in Figure 2.7. The clusters tend to assume shapes with the least drag, resulting in slender shapes. Also, the concentration of clusters is higher near the wall than that at the axis of the furnace. Clusters are less likely to appear in very dilute beds of coarse particles.

2.4.2.1 Core-Annulus Model

A typical furnace (riser) of a CFB boiler may be split into two vertical regions: *core and annulus*, as shown in Figure 2.16. The velocity of the gas in the annulus is low to negative, while that in the core is well above the superficial velocity through the riser. Solids move upwards through the core in a dilute suspension, with the occasional presence of clusters. When the clusters drift sideways due to

Hydrodynamics

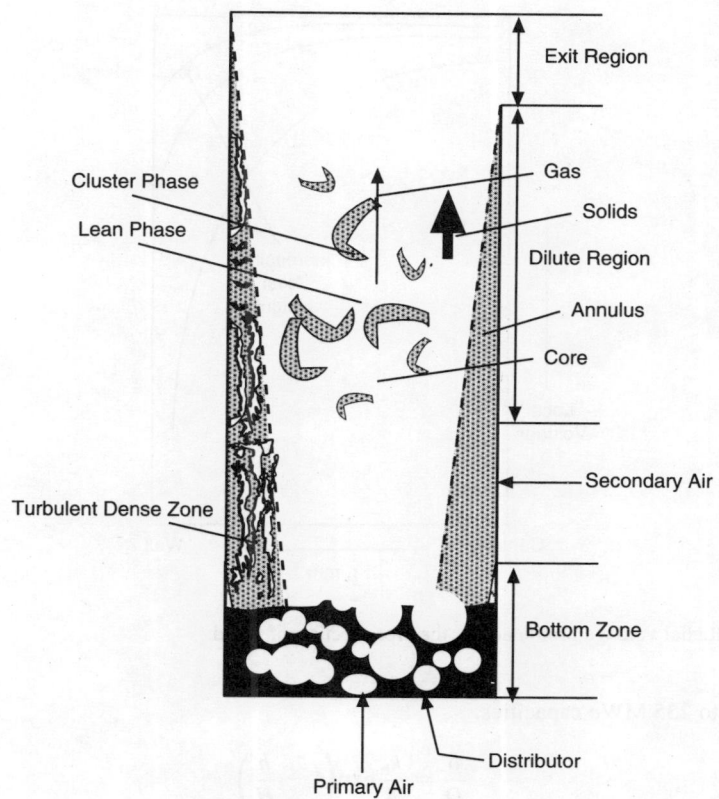

FIGURE 2.16 Core-annulus structure of CFB.

hydrodynamic interactions they are exposed to gas velocities that are too low or even negative to carry the cluster upwards. As a result the clusters start to fall in the low-velocity annular region on the wall. The downward velocity of the clusters is between 1 and 2 m/sec in the laboratory units, while it could be in the range of 2 to 8 m/sec (Figure 2.18) in large commercial CFB boilers (Werther, 2005).

Occasionally, an upward-moving cluster from the core that is suddenly swept to the wall may travel a short distance with its upward momentum before falling. The solids falling near the wall occasionally drift towards the core to be picked up again in the up-flowing gas, initiating their upward journey. This results in two lateral fluxes: one towards the wall and another away from the wall.

The up-and-down movement of solids in the core and annulus sets up an internal circulation in the bed, in addition to the external circulation, where solids captured by the cyclone are returned to the bed. Experimental measurement (Horio and Morishita, 1988) has shown the internal circulation rate to be many times the external circulation rate. The temperature uniformity of the bed is a direct result of this internal solid circulation.

The thickness of the annular zone increases from the top of the riser to its bottom. The average thickness of the annulus may vary from a few millimeters in laboratory-scale units (Horio and Morishita, 1988; Knowlton, 1990a) to several tens of centimeters in commercial units. Experimental results on wall-thickness measured in several large commercial CFB boilers show that it could increase from 50 to 300 mm over a height of 20 m. Johansson et al. (2005) found the following empirical correlation to have the best fit of wall-layer thickness, δ with experimental data

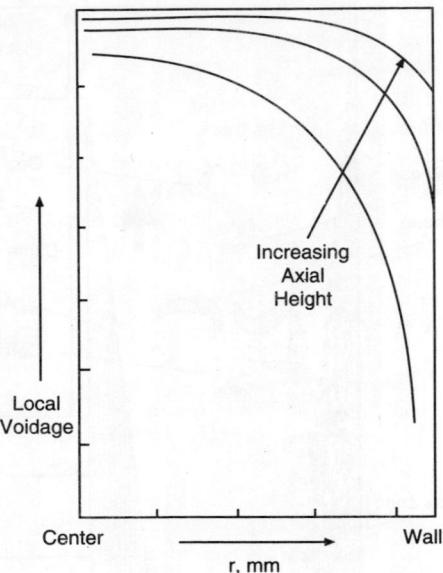

FIGURE 2.17 Radial voidage profile across the cross section of a bed.

from units up to 235 MWe capacities.

$$\frac{\delta}{D} = \frac{k_b}{4}\exp\left(1 - \frac{h}{H}\right) \tag{2.39}$$

where h is the height above distributor, D is the equivalent diameter of the bed, H is the total bed height, and the back-flow ratio k_b is $0.154 \, H/D$.

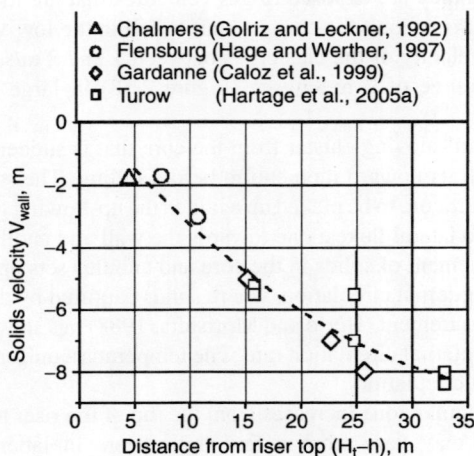

FIGURE 2.18 The velocity of solids sliding down the walls as measured in several large CFB boilers. (From Werther, J., in *Circulating Fluidized Beds VIII*, Kefa C., Ed., International Academic Publishers, Beijing, 2005. With permission.)

Hydrodynamics

The local voidage and the gas and solid velocities change continuously from the axis to the wall (Hartge et al., 1988; Horio and Morishita, 1988). The voidage is highest along the axis of the column and lowest on the wall (Arena et al., 1988; Tung et al., 1988). The radial voidage distribution is much flatter in the upper section of the bed, as well as at lower circulation rates. A typical radial voidage profile in a fast bed is shown in Figure 2.17. Measurements in laboratory-scale units suggest that the local voidage $\varepsilon(r)$, is a function of only the cross-sectional average voidage ε_{av}, and the nondimensional radial distance (r/R), from the axis of the bed. One such correlation prepared for an FCC using hollow glass beads and alumina particles (Zhang et al., 1994) is given below:

$$1 - \varepsilon(r) = (1 - \varepsilon_{av})\left[0.191 + \left(\frac{r}{R}\right)^{2.5} + 3\left(\frac{r}{R}\right)^{11}\right] \quad (2.40)$$

where $1 > (r/R) > 0.75$.

Large commercial units show similar distributions of voidage (Schaub et al., 1989), but a lack of adequate data in commercial units does not allow verification of the above empirical equation. In commercial CFB boilers, the suspension density determined from static pressures measured on the wall often shows values as low as 1 to 5 kg/m³, while indirect assessments suggest a much higher inventory of solids in the furnace. This suggests a very large volume of solids in downward motion near the wall. Since they are not fully supported by the drag of upward-moving gas, the static pressure drop may not necessarily detect these solids. This effect is, however, less pronounced in small-diameter, bench-scale CFB columns.

The actual mechanism of solid movement and mixing in the bed is more complex than that portrayed above.

Example 2.3

For a CFB furnace operating at 825°C, estimate the bed inventory and the bed voidage at 4 m above a fast bed that is 20 m tall. Also, find the voidage at the wall at this height using the empirical equation of Tung et al. (1988), given $\rho_p = 2500$ kg/m³, $U = 8$ m/sec, and $d_p = 300$ μm. The secondary air is injected at a level of 3 m. The bed cross section is 2.5 m × 10 m below and 5 m × 10 m above this level.

Solution

We know $h_i = 3$ m.
The asymptotic voidage is estimated from Figure 2.13b as $\varepsilon_a = 0.85$. For details on this choice, refer to Kunii and Levenspiel (1991).
The choking voidage in the dilute phase was calculated in example 2.2 as 0.9976. We assume the asymptotic voidage in the dilute phase to be equal to this. So, $\varepsilon_d = 0.9976$.
The decay constant a, for $U = 8$ m/sec is taken from Figure 2.15 as $a = 1.025$ m^{-1}. In the absence of adequate data, the use of Figure 2.15 for a is somewhat arbitrary. Further research is needed for a more precise value.
The voidage at the furnace exit, ε_e can be found from Equation 2.34:

$$\varepsilon_e = \varepsilon_d - (\varepsilon_d - \varepsilon_a)\exp[-a(H - h_i)]$$
$$= 0.9976 - (0.9976 - 0.85)/\exp[1.025(20 - 3)]$$
$$\approx 0.998$$

The axial mean bed voidage, ε_s above the secondary air level is found from Equation 2.36:

$$\varepsilon_s = \varepsilon_e - \frac{\varepsilon_e - \varepsilon_a}{a(H - h_i)} = 0.99760 - \frac{0.9976 - 0.85}{1.025(20 - 3)} = 0.989$$

The bed cross-sectional area below the secondary air A_i, is (2.5 m × 10 m) and that above it A_s, is (5 m × 10 m). Thus, Equation 2.37 is modified to find the solid inventory W:

$$W = \rho_p[h_i A_i(1 - \varepsilon_a) + (H - h_i)A_s(1 - \varepsilon_s)]$$
$$= 2500[3 \times 2.5 \times 10(1 - 0.85) + (20 - 3)5 \times 10(1 - 0.989)]$$
$$= 51500 \text{ kg}$$

The voidage at a height of 4 m is found from Equation 2.34:

$$\varepsilon = 0.9976 - (0.9976 - 0.85)/\exp[1.025(4 - 3)] = 0.9082$$

The voidage on the wall at this level is calculated from Equation 2.40, where $r/R = 1$

$$\varepsilon(r) \text{ at wall} = 1 - (0.191 + 1 + 3)(1 - 0.989) = 0.954$$

2.5 GAS–SOLID MIXING

The mixing of gas and solids is an important aspect of the design of any circulating fluidized bed that involves reactions, whether gasification or combustion. This section presents a brief account of some of the important information related to this.

2.5.1 GAS–SOLID SLIP VELOCITY

A major attraction of CFB reactors is their relatively high gas–solid slip velocity. Considering the bulk motion of gas and solids in a fast bed, the slip velocity U_{slip}, between them is written as:

$$U_{\text{slip}} = \frac{U}{\varepsilon} - \frac{G_s}{(1 - \varepsilon)\rho_p} \qquad (2.41)$$

However, we know that solids generally travel upwards in a dispersed phase through the core of the bed, where the slip velocity is of the order of the terminal velocity of individual particles. Near the wall, the gas velocity is much lower and may even be downwards in some cases. Thus, when the solids descend along the wall, they do not necessarily come across a very high gas–solid slip velocity. This description, however, ignores the presence of clusters, which travel at a very different velocity through the core, which may account for the high mass transfer rate (an indication of slip velocity) observed in many CFB processes.

Direct measurement of local velocities of gas and particles shows that the slip velocity increases towards the wall.

2.5.2 DISPERSION

The gas in a fast fluidized bed is generally assumed to be in plug flow. However, there is some dispersion in both the radial and axial directions.

2.5.2.1 Gas Dispersion

The gas dispersion from a point can be written as:

$$D_r \frac{1}{r}\frac{d}{dr}\left[r\frac{dC}{dr}\right] + D_a \frac{d^2C}{dz^2} = U\frac{dC}{dz} \qquad (2.42)$$

where C is the concentration of a gas species.

The radial dispersion is much lower than the axial dispersion. The axial dispersion coefficient decreases with velocity and voidage. Based on their experiments with dispersion in a bed of 58 μm FCC (1575 kg/m^3), Li and Wu (1991) found that the axial dispersion coefficient for turbulent, fast, and entrained beds can all be expressed by a simple empirical relation as a function of voidage:

$$D_g = 0.1953 \varepsilon^{-4.1197} \text{ m}^2/\text{sec} \qquad (2.43)$$

Schugerl (1967) expressed the effective gas dissipation coefficient D_g, by combining the contributions of axial D_{ga}, as well as lateral dispersion D_{gr}

$$D_g = D_{ga} + \beta U^2 \frac{D^2}{D_{gr}} \qquad (2.44)$$

where β for the turbulent single phase is 0.00005.

2.5.2.2 Solids Dispersion

A simple way to interpret data on solid dispersion distribution is the dispersion equation which includes contribution of convection and diffusion.

$$\frac{\partial C}{\partial t} = D_{ax}\frac{\partial^2 C}{\partial x^2} - \frac{\partial(v_p C)}{\partial x} \qquad (2.45)$$

Rhodes et al. (1991) found the values of axial dispersion coefficient D_{ax}, for solids to be in the range of 3 to 19 in the operating range of $G_s = 5$ to 80 kg/m^2 sec and velocity $U = 3$ to 5 m/sec.

2.6 SCALE-UP

For successful scale-up of a fluidized bed, the hydrodynamic scaling law is a powerful tool. A good scale-up would involve a balance among hydrodynamic, combustion and heat transfer processes, which is very difficult to obtain.

2.6.1 CIRCULATING FLUIDIZED BED

For macroscopic similarity of the flow structures of two CFB units one could use (Horio, 1997) the following scale-up relations using the ratio of gravity and viscous force constant:

$$\frac{U}{U^0} = \frac{V_s}{V_s^0} = \sqrt{m} \qquad (2.46)$$

where U is the superficial gas velocity, and the superscript (^{0}) stands for the values of the prototype. The superficial solid velocity V_s, is given by (G_s/ρ_p). The geometric scale m is based on the characteristic size of the combustor. For example, if a riser of diameter D is being scaled up to a diameter D^0, the scale parameter m, will be:

$$m = \frac{D}{D^0} \tag{2.47}$$

Two other conditions, which fix the particle size between model and prototypes, must be satisfied for proper scaling. They are:

$$U_t = \sqrt{m}U_t^0 \qquad \varepsilon_{mf} \leq \varepsilon < 1 \tag{2.48}$$

$$U_{mf} = \sqrt{m}U_{mf}^0 \qquad \varepsilon_{mf} \leq \varepsilon < 1 \tag{2.49}$$

Xu et al. (2005) suggested an alternative approach to the modeling of fast-bed risers using the equality of the Froud number, which is the ratio of inertia and gravity forces.

2.6.2 Bubbling Fluidized Bed

For a bubbling bed one could use the following relations (Horio et al., 1986) derived from the bubble coalescence/splitting model.

2.6.2.1 Group B Solids

For similar bubble-fraction, bubble-size distribution, solids-circulation and mixing, the following relation is satisfied:

$$U - U_{mf} = \sqrt{m}(U^0 - U_{mf}^0) \tag{2.50}$$

2.6.2.2 Group A Solids

Both bubble distribution and interstitial gas flow can be made similar if Equation 2.50 as well as the equation, $U_{mf} = \sqrt{m}U_{mf}^0$, are satisfied

It may be noted that if particle shapes are too different, the scaling law may not be very accurate.

Glicksman et al. (1993) suggested an alternative approach, which keeps the following nondimensional parameter unchanged between model and prototype, in addition to keeping the particle size distribution similar.

$$\frac{U^2}{gL}, \frac{U}{U_{mf}}, \frac{L_1}{L_2}, \frac{G_s}{\rho_p U} \tag{2.51}$$

2.6.3 Cyclone

Solid particles in a cyclone are subjected to drag and centrifugal forces as opposed to gravity and drag forces in the riser. For this reason Tagashira et al. (2005) used a Reynolds number and an Archimedes number by replacing gravity with centrifugal force.

$$\text{Re} = \frac{\rho_g d_p u_i}{\mu}$$

$$\text{Ar}_{centrifugal} = \frac{\rho_g(\rho_p - \rho_g)d_p^3 u_i^2}{\mu^2 r_c} \tag{2.52}$$

Equivalence of these two non-dimensional numbers proved effective in Scale-up of cyclones.

NOMENCLATURE

a:	decay constant in axial voidage profile in Equation 2.34
a_1:	constant in Equation 2.20
A:	cross-sectional area of bed (m^2)
A_0:	area of the distributor per orifice (m^2)
A_1, A_2:	fitting parameters in Equation 2.32
b:	constant in Equation 2.35
b_1:	exponent in Equation 2.20
c:	fitted constant in Equation 2.38
C_1:	constants in Equation 2.3
C_2:	constants in Equation 2.3
C_D:	drag coefficient
d_{b0}:	initial bubble size, units (m)
d_b:	equivalent volume diameter of a bubble (m)
d_{bm}:	limiting bubble size (cm)
$d_{eq,s}$:	equivalent volume diameter of a bubble at the bed surface (m)
d_p:	surface volume mean diameter of bed particles (m)
d_v:	volume mean diameter of particle (m)
D:	diameter of bed (m)
D^0:	scale-up diameter (m)
D_{ax}:	axial dispersion coefficient of solids (m^2/sec)
D_{bmax}:	maximum stable bubble size (m)
D_g:	dispersion coefficient (m^2/sec)
D_{ga}:	axial dispersion coefficient of gas (m^2/sec)
D_{gr}:	radial dispersion coefficient of gas (m^2/sec)
D_r:	lateral or radial dispersion coefficient (m^2/sec)
F:	mass fraction of particles less than 45 μm
F_D:	drag force on particles (N)
G_d:	downward solid flux (kg/m^2 sec)
G_s:	net solid circulation rate (kg/m^2 sec)
G_u:	upwards solid flux (kg/m^2 sec)
g:	acceleration due to gravity (9.81 m/sec^2)
L:	height of a section of bed (m)
L_1, L_2:	linear dimension of model and prototype used in scale-up (m)
H:	total height of the furnace measured above the distributor (m)
H_{mf}:	minimum fluidization velocity (m/sec)
h:	height above the grate in the bed (m)
h_i:	height of location of point of inflexion (m)
h_0:	characteristics height in Equation 2.15 (m)
k_b:	back-flow ratio (down-flow/net circulation rate)
K_t:	factor in Equation 2.26 and Equation 2.27
m:	geometric scale
m_p:	mass of particle (kg)
n:	fitted constant in Equation 2.38
P:	pressure (N/m^2)
r:	radial distance from the axis of the bed (m)
r_c:	radius of the cyclone (m)
R:	radius of the bed (m)
t_1, t_2:	fitting parameters of Equation 2.32
TDH:	transport disengaging height (m)

u_{bs}: velocity of bubble through emulsion phase (m/sec)
u_i: velocity at cyclone inlet (m/sec)
u_c: onset velocity for turbulent fluidization (m/sec)
u_k: velocity for completion of transition to turbulent fluidization (m/sec)
U: superficial gas velocity (m/sec)
U^0: prototype superficial gas velocity (m/sec)
U_b: absolute velocity of bubble (m/sec)
U_{bm}: mean bubble velocity (m/sec)
U_{bs}: bubble rise velocity (m/sec)
U_{ch}: choking velocity (m/sec)
U_d: downward velocity of solids (m/sec)
U_{mb}: minimum bubbling velocity (m/sec)
U_{mf}: minimum fluidization velocity (m/sec)
U_{mf}^0: minimum fluidizing velocity of prototype (m/sec)
U_s: upward velocity of solids (m/sec)
U_{sl}: minimum slugging velocity (m/sec)
U_{slip}: slip velocity between gas and solids (m/sec)
U_t: terminal velocity of particle of size d_p (m/sec)
U_T^0: prototype terminal velocity (m/sec)
U_t^*: terminal velocity of particle of size $2.7 d_p$ (m/sec)
U_{tr}: transport velocity (m/sec)
V_s: superficial velocity of solids (m/sec)
V_s^0: prototype superficial solid velocity (m/sec)
W: solid inventory (kg)
Z: height above distributor (m)

Greek Symbols

β: constant in Equation 2.44
ε: void fraction in the bed
ε_a: asymptotic voidage for dense phase below the point of inflexion
ε_{av}: cross-sectional average voidage at given height of the bed
ε_c: voidage at choking
ε_d: asymptotic voidage in the dilute section of the fast bed
ε_s: mean voidage in the entire bed
$\varepsilon(r)$: voidage at a radial distance r from the axis
ε_{mf}: voidage at minimum fluidization
δ: thickness of the wall or annulus zone (m)
μ: dynamic viscosity of gas (kg/m sec)
ϕ: sphericity of particle
ρ_{sus}: cross-sectional average suspension density of bed (kg/m^3)
ρ_g: density of gas (kg/m^3)
ρ_p: density of bed particle (kg/m^3)
ΔP: pressure drop across bed (N/m^2)

Dimensionless Numbers

Ar: Archimedes number ($\rho_g(\rho_p - \rho_g)g d_p^3 / \mu^2$)
Ar$_{centrifugal}$: Archimedes number based on centrifugal force ($\rho_g(\rho_p - \rho_g) u_i^2 d_p^3 / r \mu^2$)
Re$_{mf}$: Reynolds number at minimum fluidized condition ($U_{mf} d_p \rho_g / \mu$)

Re_c: $u_c d_p \rho_g / \mu$
Re_k: $u_k d_p \rho_g / \mu$

REFERENCES

Abrahamsen, A. R. and Geldart, D., *Powder Technol.*, 26, 35, 1980.
Arena, U., Cammarota, A., Massimilla, L., and Pirozzi, P., The hydrodynamics behavior of two circulating fluidized bed units of different sizes, In *Circulating Fluidized Bed Technology II*, Basu, P. and Large, J. F., Eds., Pergamon Press, Oxford, pp. 223–229, 1988.
Biswas, J. and Leung, L. E., Applicability of choking correlations for fast-fluid bed operation, *Powder Technol.*, 51, 179–180, 1987.
Bolton, L. W. and Davidson, J. F., Recirculation of particles in fast fluidized risers, In *Circulating Fluidized Bed Technology II*, Basu, P. and Large, J. F., Eds., Pergamon Press, Oxford, pp. 139–146, 1988.
Darton, R. C., LaNauze, R. D., Davidson, J. F., and Harrison, D., *Trans. Inst. Chem. Eng.*, 55, 274, 1977.
Davidson, J. F. and Harrison, D. H., *Fluidised Particles*, Cambridge University Press, Cambridge, MA, 1963.
Drahos, J., Cermak, J., Guardani, R., and Schugerl, K., Characterization of flow regime transitions in a circulating fluidized bed, *Powder Technol.*, 56, 41–48, 1988.
Ergun, S., Fluid flow through packed columns, *Chem. Eng. Prog.*, 48, 89–94, 1952.
Geldart, D., *Gas fluidization technology*, Wiley, Chichester, p. 88, 1986.
Glicksman, L. R., Hyre, M., and Woloshun, K., Simplified scaling relationships for fluidized beds, *Powder Technol.*, 77, 177–199, 1993.
Grace, J. R., Fluidized bed hydrodynamics, In *Handbook of Multiphase Systems*, Hestroni, G., Ed., Hemisphere, Washington, DC, chap. 8.1, 1982.
Hartge, E. U., Rensner, D., and Werther, J., Solids concentration and velocity patterns in circulating fluidized beds, In *Circulating Fluidized Bed Technology II*, Basu, P. and Large, J. F., Eds., Pergamon Press, Oxford, pp. 165–180, 1988.
Horio, M., Hydrodynamics of circulating fluidized beds, In *Circulating Fluidized Bed Technology IV*, Avidan, A. A., Grace, J. R., and Knowlton, T. M., Eds., Blackie A&P, London, pp. 65–72, 1997.
Horio, M. and Morishita, K., Flow regimes of high velocity fluidization, *Jpn. J. Multiphase Flow*, 2(2), 117–136, 1988.
Horio, M., Notaka, A., Sawa, Y., and Muchi, I., A similarity rule of fluidized bed for scale-up, *AICE J.*, 32, 1466–1488, 1986.
Horio, M., Taki, A., Hsieh, Y. S., and Muchi, I., *Fluidization*, Grace, J. R. and Matsen, J. M., Eds., Plenum Press, New York, p. 509, 1980.
Howard, J. R., *Fluidized Bed Technology — Principles and Applications*, Adam Hilger, Bristol, UK, 1989.
Johansson, A., Johansson, F., Leckner, B., and Gadowski, J., Solids back-mixing in CFB risers, In *Circulating Fluidized Bed Technology-VIII*, Kefa, C., Ed., International Academic Publishers, Beijing, pp. 151–158, 2005.
Knowlton, T. M., Hydrodynamics and non-mechanical solid recycle and discharge system in circulating fluidized bed systems, In *Proceedings of the Workshop on Materials Issue in Circulating Fluidized Bed*, EPRI, GS-6747, February, pp. 11–21, 1990a.
Knowlton, T. M., Private Communication, 1990b.
Kunii, D. and Levenspiel, O., Entrainment of solids from fluidized beds ii operation of fast fluidized bed, *Powder Technol.*, 61, 193–206, 1990.
Kunii, D. and Levenspiel, O., Flow modeling of fast fluidized beds, In *Circulating Fluidized Bed Technology III*, Basu, P., Hasatani, M., and Horio, M., Eds., Pergamon Press, Oxford, pp. 91–98, 1991.
Kwauk, M., Ningde, W., Youchu, L., Bingyu, C., and Zhiyuan, S., Fast fluidization at ICM, In *Circulating Fluidized Bed Technology — I*, Basu, P., Ed., Pergamon Press, Toronto, pp. 33–62, 1986.
Li, Y. and Wu, P., Axial gas mixing in circulating fluidized bed, In *Circulating Fluidized Bed Technology III*, Basu, P., Hasatani, M., and Horio, M., Eds., Pergamon Press, Oxford, pp. 581–586, 1991.
Matsen, J. M., Rise and fall of recurrent particles, In *Circulating Fluidized Bed Technology II*, Basu, P. and Large, J. F., Eds., Pergamon Press, Oxford, pp. 3–12, 1988.
Mori, S. and Wen, C. Y., *AICE J.*, 21, 109, 1975.

Nakajima, M., Harada, M., Asai, M., Yamazaki, R., and Jimbo, G., Bubble fraction and voidage in an emulsion phase in the transition to a turbulent fluidized, bed, In *Circulating Fluidized Bed Technology III*, Basu, P., Hasatani, M., and Horio, M., Eds., Pergamon Press, Oxford, pp. 79–85, 1991.

Perales, J. F., Coll, T., Llop, M. F., Puigjaner, L., Arnaldos, J., and Casal, J., On the transition from bubbling to fast fluidization regimes, In *Circulating Fluidized Bed Technology III*, Basu, P., Hasatani, M., and Horio, M., Eds., Pergamon Press, Oxford, pp. 73–78, 1991.

Pettyjohn, E. A. and Christiansen, E. B., *Chem. Eng. Prog.*, 44, 157, 1948.

Reddy-Karri, S. B. and Knowlton, T., A practical definition of fast fluidized bed, In *Circulating Fluidized Bed Technology III*, Basu, P., Hasatani, M., and Horio, M., Eds., Pergamon Press, Oxford, pp. 67–72, 1991.

Rhodes, M. J. and Geldart, D., Transition to turbulence, In *Fluidization V*, Ostergaard, K. and Sorensen, A., Eds., Engineering Foundation, New York, pp. 281–288, 1986a.

Rhodes, M. J. and Geldart, D., The hydrodynamics of recirculating fluidized beds, In *Circulating Fluidized Bed Technology*, Basu, P., Ed., Pergamon Press, Toronto, pp. 193–200, 1986b.

Rhodes, M., Zhou, S., Hirama, T., and Cheng, H., Effects of operating conditions on operating condition on longitudinal solids mixing in a CFB riser, *AICE J.*, 37, 1450–1458, 1991.

Schaub, G., Reimert, R., and Albrecht, J., Investigation of emission rates from large scale CFB combustion plants, In *Proceedings of the 10th International Conference on Fluidized Bed Combustion*, Manaker, A., Ed., ASME, New York, pp. 685–691, 1989.

Schugerl, K., Experimental comparison of mixing in a three and two phase fluidized beds, In *Proceedings of the International Symposium on Fluidization*, Drinkenburg, A. A. H., Ed., Netherlands University Press, Amsterdam, pp. 782–794, 1967.

Shingles, T. and Dry, R., Circulating fluidized beds, *Encyclopedia of Fluid Mechanics*, Vol. 4, Cherimisinoff, N. P., Ed., Gulf Publishing, p. 1067, 1986, chap. 33.

Stewart, P. S. B. and Davidson, J. F., Slug flow in fluidized beds, *Powder Technol.*, 1, 61, 1967.

Sun, G. and Chen, G., Transition to turbulent fluidization and its prediction, In *Fluidization VI*, Grace, J. R., Shemilt, L. W., and Bergougnou, M. A., Eds., Engineering Foundation, New York, pp. 33–40, 1989.

Tagashira, T., Toril, I., Myouyou, K., and Horio, M., Performance analysis of CFB cyclone based on Ar–Re scaling, In *Circulating Fluidized Bed Technology III*, Cen, K., Ed., International Academic Publishers, Beijing, pp. 915–922, 2005.

Tung, Y., Li, J., and Kwauk, M., Radial voidage profiles in a fast fluidized bed, In *Fluidization'88*, Kwauk, M. and Kunii, D., Eds., Science Press, Beijing, pp. 139–145, 1988.

Werther, J., Fluid mechanics of large-scale CFB units, In *Circulating Fluidized Bed Technology IV*, Avidan, A. A., Ed., AICE, New York, pp. 1–14, 1993.

Werther, J., Fluid dynamics, temperature and concentration fields in large-scale CFB combustors, In *Circulating Fluidized Bed Technology-VIII*, Kefa, C., Ed., International Academic Publishers, Beijing, pp. 1–25, 2005.

Xu, H., Cao, Z., Li, J., and Reh, L., Similarity analysis of CFB with two different riser heights, In *Circulating Fluidized Bed Technology-VIII*, Kefa, C., Ed., International Academic Publishers, Beijing, pp. 402–408, 2005.

Yang, W.C., In *Proceedings of the Pneumotransport 3*, BHRA Fluid Engineering, Bedford, E5–49, 1976.

Yang, W. C., Criteria for choking in vertical pneumatic conveying lines, *Powder Technol.*, 35, 143–150, 1983.

Yerushalmi, J., Turner, D. H., and Squires, A. M., The fast fluidized bed, *Industr. Eng. Chem., Process Design and Development*, 15, 47–53, 1976.

Yue, G., Lu, J., Zhang, H., Yong, H., Zhang, J., and Liu, Q., Design theory of circulating fluidized bed boilers, In *Proceedings of the 18th International Conference on Fluidized Bed Combustion*, Jia, L., Ed., ASME, New York, 2005, paper: FBC 78134.

Zhang, H., Lu, J., Yang, H., Yong, J., Wang, Y., Xiao, X., Zhao, X., and Yue, G., Heat transfer measurements and predictions inside the furnace of 135 MWe CFB boiler, In *Circulating Fluidized Bed Technology III*, Cen, K., Ed., International Academic Publishers, Beijing, pp. 254–260, 2005.

Zhang, W., Johnsson, F., and Leckrun, B., Characteristics of the lateral particle distribution in circulating fluidized bed boilers, In *Circulating Fluidized Bed Technology IV*, Avidan, A. A., Ed., AichE, New York, pp. 266–273, 1994.

Zenz, F. A. and Weil, N. A., A theoretical–empirical approach to mechanism of particle entrainment from fluidized beds, *AICE J.*, 4, 472, 1958.

3 Fluidized Bed Gasification

Gasification is a relatively old technology. Coal gasification was invented in 1792 and was extensively used to produce town gas in the 19th century, but was quite expensive. For this reason, people in North America started using whale oil for illumination, driving the whale population nearly to extinction. In 1846, Dr. Abraham Gesner, a native of Nova Scotia, Canada, invented a process for extracting lighting oil from solid fuels using pyrolysis. He called this oil *kerosene* (from the Greek word for wax and oil). In 1849, he developed a process for extracting kerosene from oil, just discovered in the U.S.A. and Canada, saving the whales from extinction.

Definition

> Gasification refers to a group of processes that converts solid or liquid fuels into a combustible gas with or without contact with a gasification medium.

Gasification is generally carried out by reacting fuel such as coal, biomass, petroleum coke, or heavy oil, with a restricted amount of oxygen and often in combination with steam. Heat evolved from the exothermic reaction of oxygen with the fuel serves to maintain the gasifier at the operating temperature and drives certain endothermic reactions taking place inside it. Steam can be the sole gasification medium if an external source can provide the heat necessary for the endothermic gasification reactions.

Gasification of coal offers certain important advantages over direct combustion. For example, for a given throughput of fuel processed, the volume of gas obtained from gasification is much less compared to that obtained from a combustion system. The reduced volume of gas needs smaller equipment, and hence results in lower overall costs. For small capacity power packs, a unit comprised of a gasifier and a compression–ignition engine is less expensive than one comprising a boiler, condenser, steam engine, etc. Thus, gasification provides an attractive option for remote locations.

There is, however, one important shortcoming of the gasifier: the carbon conversion efficiency is rarely 100%. As a result, a useful part of the fuel energy remains in the char. Furthermore, for cold gas applications, the sensible heat of the gas could be lost unless a heat recovery system is used.

Clean gas obtained from a gasifier can fuel a large combined cycle system for electricity generation, where the gasified fuel is first burnt in a combustion turbine–generator unit, and then the hot exhaust gas from its gas turbine is used for generating steam to produce further power in a steam turbine–generator unit. The combination of a gasifier and a combined cycle is called the integrated gasification combined cycle (IGCC). Such a system results in a greater reduction in emissions from coal-based energy systems compared to that with direct combustion of the coal for power generation. An IGCC system offers a generating efficiency ($\sim 40\%$), higher than that for a conventional direct combustion pulverized coal fired plant ($\sim 34\%$) (Schimmoller, 2005). The target efficiency of gasification-based combined-cycle electricity generation of the Department of Energy of the U.S.A. is 52% on HHV basis (DOE, 2001), which is even higher than the efficiency of modern supercritical pulverized coal-fired power plants ($\sim 40\%$).

Another emerging application of gas from solid fuels is high-efficiency production of electricity using fuel cells. Coal can be gasified to produce syngas — a gas containing CO and H_2. This gas, after a high degree of purification can be substituted for natural gas in fuel cells, which could help enhance the power generation efficiency to above 50%.

Coal gas can be further converted to transportable liquid fuels, or used in chemicals and fertilizer. Thus, as petroleum and natural gas reserves are depleted, coal gasification may emerge as a vital process for developing alternatives to these fuels.

This chapter complements topics covered in the other chapters and provides readers with an insight into the thermo-chemical processes occurring during gasification and the design of different types of fluidized bed gasifiers.

3.1 TYPES OF GASIFIERS

Depending upon the gasification medium, gasifiers can be broadly classified into two groups:

1. Air-blown, where air is the gasification medium
2. Oxygen-blown, where pure oxygen is the gasification medium

Air gasification produces a low heating value (5000 to 6000 kJ/kg or 3 to 6 MJ/m^3, LHV) gas, which contains about 50% nitrogen and can fuel engines and furnaces.

Oxygen blowing is free from diluents like nitrogen. As a result it produces higher (15,000 kJ/kg or 10 to 12 MJ/m^3, LHV) heating value gas, which is, however, still leaner than natural gas typically having a heating value of 50,000 kJ/kg (40 MJ/m^3). Table 3.1 compares the heating values and composition of gasification products with those of other gaseous fuels.

Depending upon how the gas and fuel contact each other, gasifiers can be further divided into following four types:

1. Entrained bed
2. Fluidized bed (Bubbling or Circulating)

TABLE 3.1
Comparison of Gasification Products with Other Gaseous Fuels

Fuels	Composition by Volume (%)						Higher Heating Value kJ/kg
	CH_4	C_mH_n	H_2	CO	CO_2	Others	
Gasifier							
Air-blown	1		25		7	49	5000
Oxygen-blown	5		60		3	2	15,000
Natural gas	97.3	2.4				0.74	55,000
Synthetic fuels							
Water gas	15.5	7	34	32	4.3	6.5	23,000
Coke oven gas	24.9	3.2	50.2	9	3.2	9.5	35,000
Producer gas			14	27	4.5	50.9	5,500
Blast furnace gas	3	$O_2 = 0.2$ $H_2O = 3.4$	2.4	23.4	14.4	56.4	2400
Commercial gases	C_4H_{10}	C_3H_8	C_3H_6	C_2H_6	C_2H_4	C_4H_8	
Propane	2	65.5	30	2	0.5	21.8	50,000
Butane	68.6	6.4			3.2	21.8	49,000

Fluidized Bed Gasification

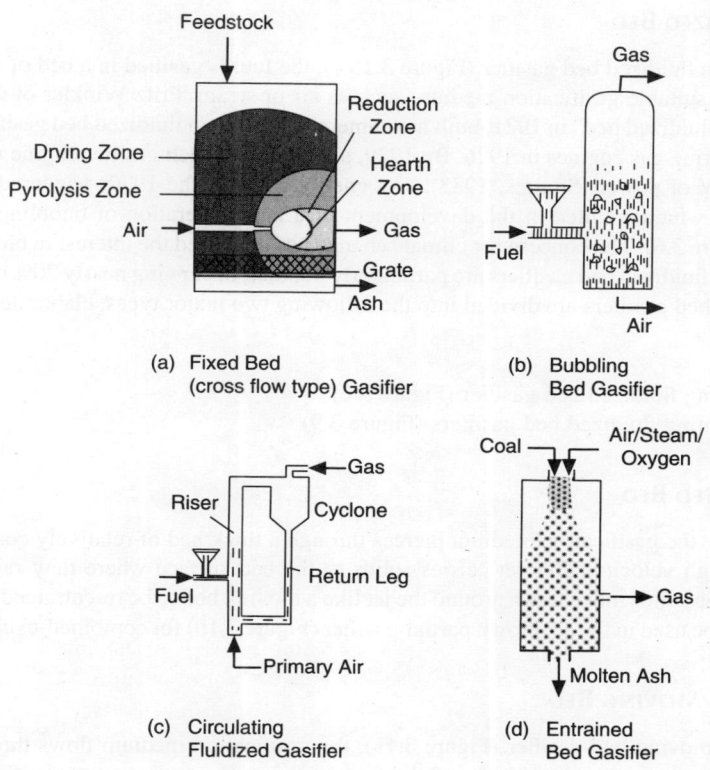

FIGURE 3.1 Gasification can be carried out in different gas–solid contacting processes.

3. Spouted bed
4. Fixed or moving bed

Figure 3.1 shows schematics of different types of gasifiers. Spouted bed, a relatively new and less developed, is not shown here. It is shown later in Figure 3.10 as a part of a combined-cycle plant.

The following section provides a brief description of each of the above types of gasifiers.

3.1.1 Entrained Bed

Entrained-flow systems (Figure 3.1d) gasify pulverized fuel particles suspended in a stream of oxygen (or air) and steam. Ash in the coal melts at the high operating temperature of the gasifier and is removed as liquid slag. A number of manufacturers offer commercial entrained-bed gasifiers for large-scale applications, such as Texaco, Shell, and Koppers–Totzek. These gasifiers typically operate at pressures up to 35 bar and use oxygen as the gasifier medium (William et al., 2000). Many IGCC plants utilize entrained bed gasifiers.

Entrained bed gasifiers are available in much larger capacities (> 100 MWe) than other types, but these are more commonly used for fossil fuels like coal, refinery wastes, etc. Their use for biomass gasification is rather limited, as it requires the fuel particles to be very fine (in the order of 80 to 100 μm).

3.1.2 Fluidized Bed

In the case of a fluidized bed gasifier (Figure 3.1b,c), the fuel is gasified in a bed of small particles fluidized by a suitable gasification medium such as air or steam. Fritz Winkler of Germany, who invented the "fluidized bed" in 1921, built a commercial air-blown fluidized bed gasifier, which was used for powering gas engines in 1926. By 1929, a total of five such gasifier engine units produced about 100 MW of power (Squires, 1983). The energy crisis of the 1970s triggered an interest in fluidized bed, which resulted in the development of a new generation of bubbling fluidized bed gasifiers (Figure 3.6). The concern for climate change has increased the interest in biomass gasification for which fluidized bed gasifiers are particularly popular, occupying nearly 20% of their market.

Fluidized bed gasifiers are divided into the following two major types, elaborated upon later in Section 3.5:

1. Bubbling fluidized bed gasifier (Figure 3.6)
2. Circulating fluidized bed gasifiers (Figure 3.9)

3.1.3 Spouted Bed

In this system, the gasification medium pierces through a thick bed of relatively coarse (Group D) particles at high velocity. This jet carries solids to the bed surface where they rain down like a fountain. These solids move down around the jet like a moving bed to be re-entrained by the jet once again. It may be used in the air-blown partial-gasifier (Figure 3.10) for combined-cycle applications.

3.1.4 Fixed/Moving Bed

In a fixed or moving bed gasifier (Figure 3.1a), the gasification medium flows through, and thus comes into contact with, a fixed bed of solid fuel particles. Depending upon the flow direction of the gasifying medium through the bed of fuel, this type of gasifier can be of three types:

1. Updraft (medium flows upwards) (Figure 3.2)
2. Downdraft (medium flows downward)
3. Sidedraft (fuel is fed from the top and gas flows sideways through it) (Figure 3.1a)

The best-known and most widely used large, moving bed system is the Lurgi gasifier, which has been used in South Africa for indirect liquefaction of coal since 1955. The technology for building small moving bed gasifiers is more than a century old and was well established by 1900. During the World War II, more than one million such gasifiers were in use in different parts of the world for operating trucks, buses, taxis, boats, trains and other vehicles.

Downdraft gasifiers are very popular, especially for biomass gasification. They are typically used for small capacities (<1.5 MWth) and occupy more than 75% of the biomass gasification market (Maniatis, 2002)

A fixed or moving bed gasifier would consist of two zones;

1. Combustion
2. Gas–solid back-mixed gasification

Figure 3.2 shows a schematic of a fixed bed gasifier. Here, we assume that coal is devolatilized in a separate zone and only pure char is gasified here. The only reaction in the lower zone is carbon combustion to produce carbon dioxide and the heat necessary for the endothermic gasification reaction occurring up above. In this zone, char burns to ash, which is drained as shown in Figure 3.2. Combustion gases and steam go up to the next zone, which provides char to

Fluidized Bed Gasification

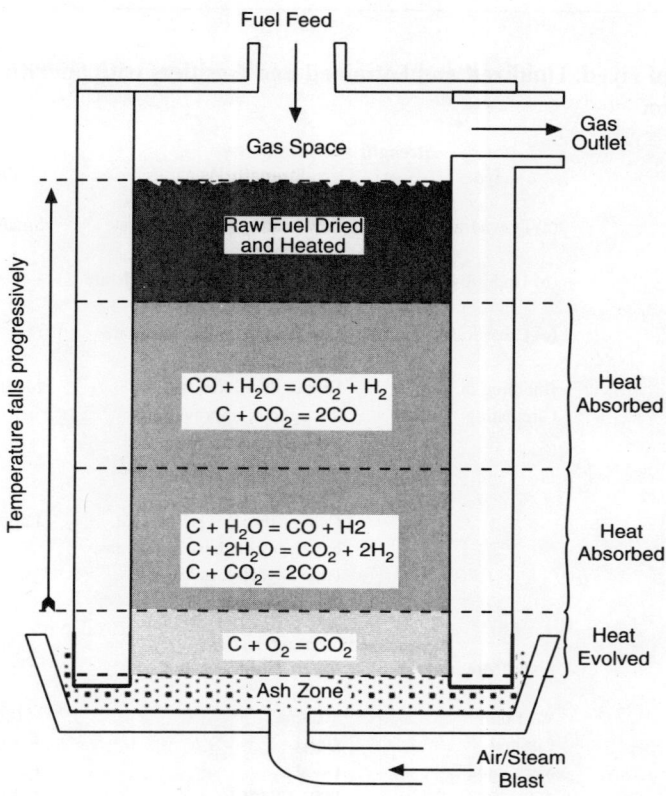

FIGURE 3.2 Updraft gasifier.

the lower combustion zone. All rate-controlling steam–carbon and hydrogen–carbon reactions occur in this zone.

Table 3.2 compares relative strengths and weaknesses of different types of gasifiers. More detailed comparison of their performance is presented in Table 3.3.

3.2 THEORY

In typical gasifiers the following physicochemical processes take place at temperatures indicated within brackets

1.	Drying	($>150°C$)
2.	Pyrolysis (devolatilization)	($150–700°C$)
3.	Combustion	($700–1500°C$)
4.	Reduction	($800–1100°C$)

Processes 1, 2 and 4 absorb heat provided by the exothermic combustion process. In the drying process, the moisture in the solid fuel evaporates. The pyrolysis process separates the water vapor, organic liquids and noncondensable gases from the char or solid carbon of the fuel. The combustion process oxidizes fuel constituents in an exothermic reaction, while the gasification process reduces them to combustible gases in an endothermic reaction.

TABLE 3.2
A Comparison of Fixed, Fluidized and Entrained Bed Gasifiers with Specific Reference to Coal Gasification

Strength and Weakness

Class	Types	Strength/Weakness	Power Production
Fixed bed	(a) Downdraft	Low heating value, moderate dust, low tar	Small to medium scale
	(b) Updraft	Higher heating value, moderate dust, high tar	
	(c) Crossdraft	Low heating value, moderate dust, high tar	
Fluidized bed	Bubbling or Circulating	Higher than fixed bed throughput, improved mass and heat transfer from fuel, higher heating value, higher efficiency	Medium scale
Entrained flow		Can gasify all types of coal, large sensible heat in flue gas, large capacity, involves slagging of ash	Large scale

Operational characteristics

Parameters	Fixed/Moving bed	Fluidized bed	Entrained bed
Feed size	<51 mm	<6 mm	<0.15 mm
Tolerance for fines	Limited	Good	Excellent
Tolerance for coarse	Very good	Good	Poor
Exit gas temperature	450–650°C	800–1000°C	>1260°C
Feed stock tolerance	Low-rank coal	Low-rank coal and excellent for biomass	Any coal including caking but unsuitable for biomass
Oxidant requirements	Low	Moderate	High
Reaction zone temperature	1090°C	800–1000°C	>1990°C
Steam requirement	High	Moderate	Low
Nature of ash produced	Dry	Dry	Slagging
Cold gas efficiency	80%	89.2%	80%
Application	Small capacities	Medium-size units	Large capacities
Problem areas	Tar production and utilization of fines	Carbon conversion	Raw gas cooling

Source: Developed with data from Stultz and Kitto, 1992.

Some authors (DOE, 1978) view gasification as a three-stage process:

1. Devolatilization
2. Rapid-rate methane formation
3. Low-rate char gasification

3.2.1 Pyrolysis or Devolatilization

Pyrolysis (also called *partial gasification*) was first observed in the 17th century and was later used by Murdoch in 1797 to produce town gas for street lighting, and then by Gesner in 1846 to produce clean, transportable oil for homes and other uses. Presently, there is a renewed interest in

TABLE 3.3
Comparison of Basic Characteristics of Different Types of Gasifiers

	Fixed Beds Types			Fluidized Beds	
	Updraft	Downdraft	Side-Draft	Bubbling	Circulating
Sensitivity to:					
Fuel specification	Moderate	Specific	Moderate	Flexible	Flexible
Fuel size	Very good	Good	Good	Fair	Fair
Moisture content	Very good	Fair	Good	Good	Good
Ash content	Poor	Poor	Poor	Very good	Very good
Reaction temperature	1000°C	1000°C	900°C	850°C	850°C
Fuel mixing	Poor	Poor	Poor	Very good	Excellent
Gas exit temperature	250°C	800°C	900°C	800°C	850°C
Tar in gas	Very high	Very low	Very high	Moderate	Low
Dust in gas	Good	Moderate	High	Very high	Very high
Turndown ratio	Good	Fair	Good	Very good	Good
Scale-up potential	Good	Poor	Poor	Good	Very good
Start-up facility	Poor	Poor	Poor	Good	Good
Control facility	Fair	Fair	Fair	Very good	Very good
Carbon conversion	Very good	Very good	Poor	Fair	Very good
Thermal efficiency	Excellent	Very good	Good	Good	Very good
LHV of gas	Poor	Poor	Poor	Poor	Fair

partial gasification for advanced combined-cycle power generation from coal (see Section 1.4 in Chapter 1).

A series of complex physical and chemical processes occur during the devolatilization or pyrolysis processes, which start slowly at less than 350°C, accelerating to an almost instantaneous rate above 700°C. The composition of the evolved products is a function of the temperature, pressure, and gas composition during devolatilization. The pyrolysis process is initiated at around 230°C, when the thermally unstable components, such as lignin in biomass, and volatiles in coal, are broken down and evaporate with other volatile components. This process can be represented by the following general reaction:

$$\text{Coal (or biomass)} + \text{Heat} \rightarrow \text{Char} + \text{Gases} + \text{Vapors or Liquid} \quad (3.1)$$

(Coke Ash)

(CO, CH_4, H_2O, CO_2)

Liquid (PAH, Tar)

The vaporized liquid product contains tar and polyaromatic hydrocarbons (PAH,). The tar, being sticky, represents a great challenge to downstream machines like filters, engines, etc. If the pyrolysis product could be made to pass through a high temperature (1100 to 1200°C) zone, a large fraction of the tar would break down to smaller hydrocarbons. The heating value of the gas produced in pyrolysis is low (3.5 to 9 MJ/m^3).

Pyrolysis generally produces the following three products:

1. Light gases such as H_2, CO, CO_2, H_2O, CH_4
2. *Tar*, a black, viscous and corrosive liquid composed of heavy organic and inorganic molecules

3. Char, a solid residue mainly containing carbon

Composition of the pyrolysis product depends on several factors including the temperature and rate of heating. The tar and gas content of coal pyrolysis product increases with temperature up to 900 to 1000°C.

3.2.2 COMBUSTION

The oxidation or combustion of char is one of the most important chemical reactions taking place inside a gasifier, providing practically all the thermal energy needed for the endothermic reactions. Oxygen supplied to the gasifier reacts with the combustible substances present, resulting in the formation of CO_2 and H_2O, which subsequently undergo reduction upon contact with the char produced from pyrolysis.

$$C + O_2 = CO_2 + 393.77 \text{ kJ/mol carbon} \qquad (3.2)$$

The other combustion reaction is the oxidation of hydrogen in fuel to produce steam.

$$H_2 + \tfrac{1}{2} O_2 = H_2O + 742 \text{ kJ/mol } H_2 \qquad (3.3)$$

3.2.3 GASIFICATION

Gasification involves a series of endothermic reactions supported by the heat produced from the combustion reaction described above. Gasification yields combustible gases such as hydrogen, carbon monoxide, and methane through a series of reactions. The following are four major gasification reactions:

1. Water–gas reaction
2. Boudouard reaction
3. Shift conversion
4. Methanation

Brief descriptions of these reactions are given below.

3.2.3.1 Water–Gas Reaction

Water–gas reaction is the partial oxidation of carbon by steam, which could come from a host of different sources, such as water vapor associated with the incoming air, vapor produced from the evaporation of water, and pyrolysis of the solid fuel. Steam reacts with the hot carbon according to the heterogeneous water–gas reaction:

$$C + H_2O = H_2 + CO - 131,38 \text{ kJ/kg mol carbon} \qquad (3.4)$$

In some gasifiers, steam is supplied as the gasification medium with or without air or oxygen.

3.2.3.2 Boudouard Reaction

The carbon dioxide present in the gasifier reacts with char to produce CO according to the following endothermic reaction, which is known as the *Boudouard reaction*:

$$CO_2 + C = 2CO - 172,58 \text{ kJ/mol carbon} \qquad (3.5)$$

3.2.3.3 Shift Conversion

The heating value of hydrogen is higher than that of carbon monoxide. Therefore, the reduction of steam by carbon monoxide to produce hydrogen is a highly desirable reaction.

$$CO + H_2O = CO_2 + H_2 - 41,98 \text{ kJ/mol} \tag{3.6}$$

This endothermic reaction, known as *water–gas shift*, results in an increase in the ratio of hydrogen to carbon monoxide in the gas, and is employed in the manufacture of synthesis gas.

3.2.3.4 Methanation

Methane could also form in the gasifier through the following overall reaction:

$$C + 2H_2 = CH_4 + 74,90 \text{ kJ/mol carbon} \tag{3.7}$$

This reaction can be accelerated by nickel-based catalysts at 1100°C and 6 to 8 bar. Methane formation is preferred especially when the gasification products are to be used as a feedstock for other chemical process. It is also preferred in IGCC applications due to methane's high heating value.

3.2.4 COMPOSITION OF GAS YIELD

The composition of the gas obtained from a gasifier depends on a number of parameters such as:

1. Fuel composition
2. Gasifying medium
3. Operating pressure
4. Temperature
5. Moisture content of the fuels
6. Mode of bringing the reactants into contact inside the gasifier, etc.

It is very difficult to predict the exact composition of the gas from a gasifier. Consideration of chemical equilibrium of the components of the gas often provides a useful insight into understanding of the performance of a gasifier through experimental research. However, Li et al. (2004) found that the gasifier products do vary from their equilibrium values. In any case, the chemical equilibrium gives good starting values.

Chemical equilibrium requires that at each temperature, there is an equilibrium constant for each reaction. Given sufficient time the concentration of these gases will reach their equilibrium concentrations.

For example, in the case of the water–gas shift reaction (Equation 3.6), the equilibrium equation is written as:

$$K_{ps} = \frac{[CO_2][H_2]}{[CO][H_2O]}$$

where $[CO_2]$, $[H_2O]$, $[CO]$ and $[H_2]$ are equilibrium concentrations of CO_2, H_2, CO and H_2O expressed in respective partial pressures.

In the case of gasification, the Boudouard reaction (Equation 3.5) and the water–gas reaction (Equation 3.4) are related by the water–gas shift reaction (Equation 3.6). Thus, only two of these reactions and the methane formation reaction (Equation 3.7) need to be considered for estimating the equilibrium gas composition. Table 3.4 shows the values of equilibrium constants K_{pw}, K_{pb}, and K_{pm} for the water–gas reaction (Equation 3.4), Boudouard reaction (Equation 3.5) and Methanation reaction (Equation 3.7), respectively. Appendix 3A shows a simple case of estimation of the equilibrium concentration of the constituents of the product gas of gasification.

TABLE 3.4
Equilibrium Constants for Water Gas, Boudouard and Methane Formation Reactions (JANAF Thermochemical Tables, 1985)

Temperature (K)	K_{pw} (Equation 3.4) $C + H_2O = H_2 + CO$	K_{pb} (Equation 3.5) $CO_2 + C = 2CO$	K_{pm} (Equation 3.7) (Li, 2005) $C + 2H_2 = CH_4$
400	7.709×10^{-11}	5.225×10^{-14}	2.989×10^{5}
600	5.058×10^{-5}	1.870×10^{-6}	9.235×10^{1}
800	4.406×10^{-2}	1.090×10^{-2}	1.339×10^{0}
1000	2.617×10^{0}	1.900×10^{0}	9.632×10^{-02}
1500	6.081×10^{2}	1.622×10^{3}	2.505×10^{-03}

Source: From Li, Xuantian, personal communication, May 2005. With permission.

Example 3.1

> Calculate the equilibrium gas composition at 1 atm and 1500 K for the water–gas reaction:
>
> $$C + H_2O = CO + H_2$$

Solution

At equilibrium, there will be some amount of both reactants and products in the mixture. Let

V_{CO} = Volume fraction of CO in the product mixture
V_{H_2} = Volume fraction of H_2 in the product mixture
V_{H_2O} = Volume fraction of H_2O in the product mixture

From Equation 3.4 for the water–gas reaction, we can write

$$\frac{V_{H_2} \times V_{CO} P}{V_{H_2O}} = K_{pw} \quad \text{(i)}$$

The sum of fractional volume of gases must be equal to 1.0.

$$V_{H_2O} + V_{CO} + V_{H_2} = 1 \quad \text{(ii)}$$

For every mole of hydrogen, 1 mole of carbon monoxide is produced in the reaction. Since an equal number of moles of CO and H_2 are produced, we have:

$$V_{CO} = V_{H_2} \quad \text{(iii)}$$

Thus,

$$V_{H_2O} = 1 - 2V_{CO} \quad \text{(iv)}$$

From Equation i,

$$K_{pw} = \frac{V_{H_2} V_{CO} P}{V_{H_2O}} = \frac{(V_{CO})^2 P}{(1 - 2V_{CO})} \quad \text{(v)}$$

The value of the equilibrium constant for the water–gas reaction (K_{pw}) at 1500 K is taken from Table 3.4 to be 6.081×10^2. The pressure P is 1 atm.

$$608.1(1 - 2V_{CO}) = (V_{CO})^2$$

On solving for V_{CO}

$$V_{CO} = -608.1 + [608.1^2 + 608.1]^{0.5} = -608.1 + 608.5 = 0.4998$$

Since $V_{H_2} = V_{CO} = 0.4998$

Also $V_{H_2O} = 1 - 2V_{CO} = 0.0004$

The volumetric composition of the gas is therefore: CO: 49.98%; H_2: 49.98% and H_2O: 0.04%.

3.3 EFFECT OF OPERATING PARAMETERS ON GASIFICATION

The composition of the product gas of gasification, which is a mixture of carbon monoxide, methane, hydrogen, nitrogen, carbon dioxide, etc., depends on its operating parameters as well as on the characteristics of the feedstock. Among the operating parameters, the temperature and the pressure of the gasifier have the greatest effect on the product composition. Li et al. (2004) found in their CFB gasifier that the HHV of the produced gas increased by 10% for an increase in the temperature from 700 to 800°C. The following section describes effect of temperature and pressure on the gasification reactions described above.

3.3.1 BOUDOUARD REACTION

Figure 3.3a shows the variation of volumetric carbon monoxide concentration with temperature for the Boudouard reaction ($CO_2 + C = 2CO$) at a pressure of 1 atm. It can be seen from the figure that high temperature favors carbon monoxide formation. Pressure has the opposite effect. Figure 3.2b shows that at a given temperature, carbon monoxide formation is favored by low pressure. Thus, these figures suggest that if high carbon monoxide content gas is desired, the gasifier should be designed to operate at a high temperature and low pressure.

3.3.2 WATER–GAS

Figure 3.4a and b show the variation of carbon monoxide and hydrogen with temperature at a pressure of 1 atm, and with pressure at a temperature of 800°C, respectively, for the water–gas reaction ($C + H_2O = CO + H_2$). It can be seen that the forward reaction is favored by high temperature and low pressure. Since hydrogen and carbon monoxide are produced at the same molar rate, they have the same molar (or volumetric) concentration in the gas mixture at equilibrium. Thus the concentration of steam in the mixture can be estimated by subtracting the concentrations of hydrogen and carbon monoxide (two times the value obtained from Figure 3.4a or b) from the whole.

3.3.3 METHANATION

Figure 3.5a,b shows the variation of methane in the flue gas with temperature at a given pressure (1 atm), and with pressure at a given temperature (800°C), respectively. From these figures for the methane formation reaction ($C + 2H_2 = CH_4$), it can be seen that high methane concentration is favored by low temperature and high pressure (Figure 3.6).

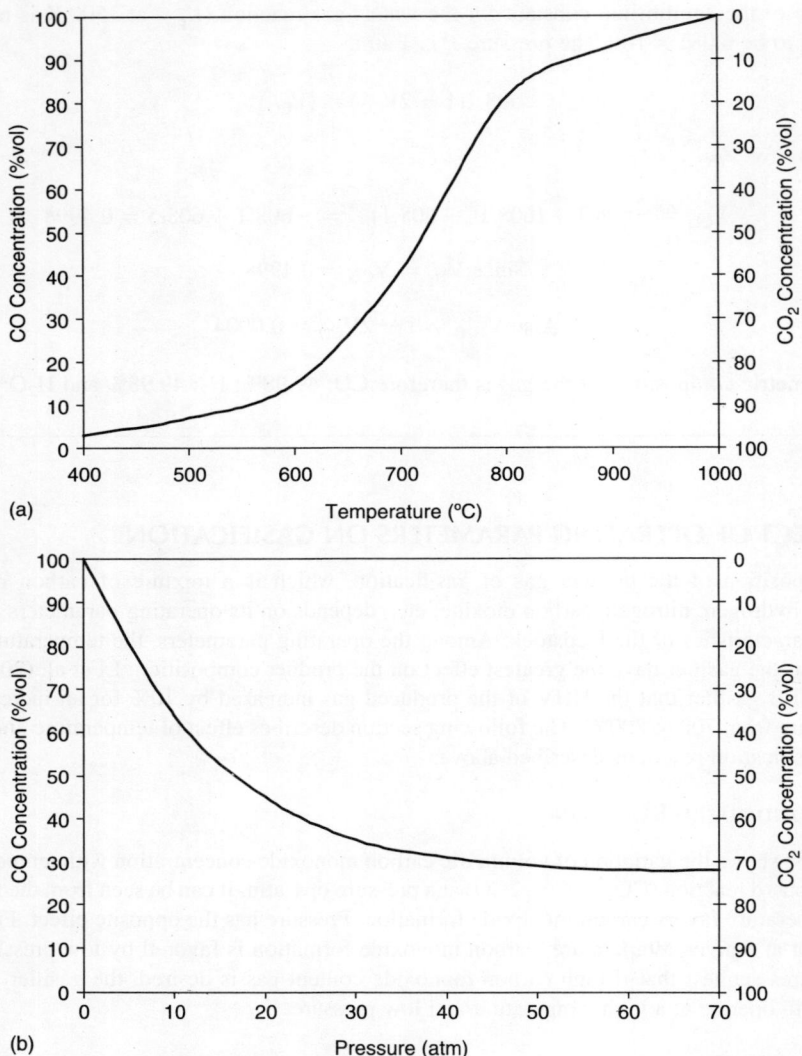

FIGURE 3.3 Boudouard reaction equilibrium: variation of carbon monoxide and carbon dioxide concentrations for gasification of carbon with oxygen (a) with temperature at a pressure of 1.0 atm, and (b) with pressure at a temperature of 800°C.

In an actual gasifier, all the reactions (Equation 3.4 to Equation 3.7) take place simultaneously. Although the conditions prevailing in a gasifier do not allow chemical equilibrium to be established, the above observations provide a useful understanding regarding the influence of operating and design parameters on gas composition.

Increasing steam supply tends to increase hydrogen production and decrease CO production through the shift reaction, (Equation 3.6). However, the effect diminishes with increasing steam supply rates. Although oxygen is normally not present in the final gas, changing the oxygen-to-fuel ratio affects the final gas composition. In general, the concentration of the products of complete oxidation, CO_2 and H_2O, the amount of heat released and the temperature levels inside the gasifier all increase with increasing oxygen supply.

Fluidized Bed Gasification

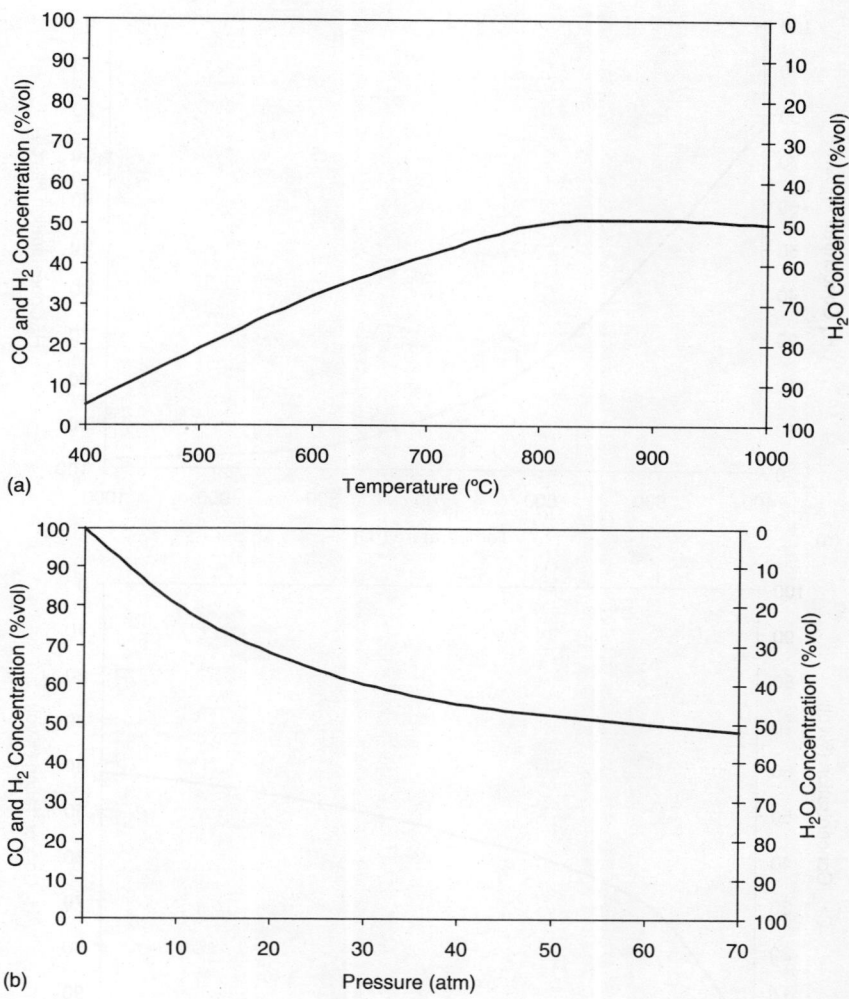

FIGURE 3.4 Water–gas reaction equilibrium: variation of carbon monoxide, hydrogen and steam (a) with temperature at a pressure of 1.0 atm, and (b) with pressure at a temperature of 800°C.

3.4 EFFECT OF FEED PROPERTIES ON GASIFICATION

The following section describes the effect of feed properties on the composition of the product gases.

3.4.1 Fuel Reactivity

In general, the reactivity of coal decreases with its rise in rank. Physical properties of coal, such as particle size and porosity also have significant effects on the kinetics of coal gasification. As the particle size becomes smaller, the specific contact area between the coal and the reaction gases increases, resulting in faster reactions (Quader, 1985).

For low- and medium-rank coals, the reactivity in gasification increases with an increase in pore volume and surface area, but for high-rank coal (C > 85%), the reactivity is not affected by changes

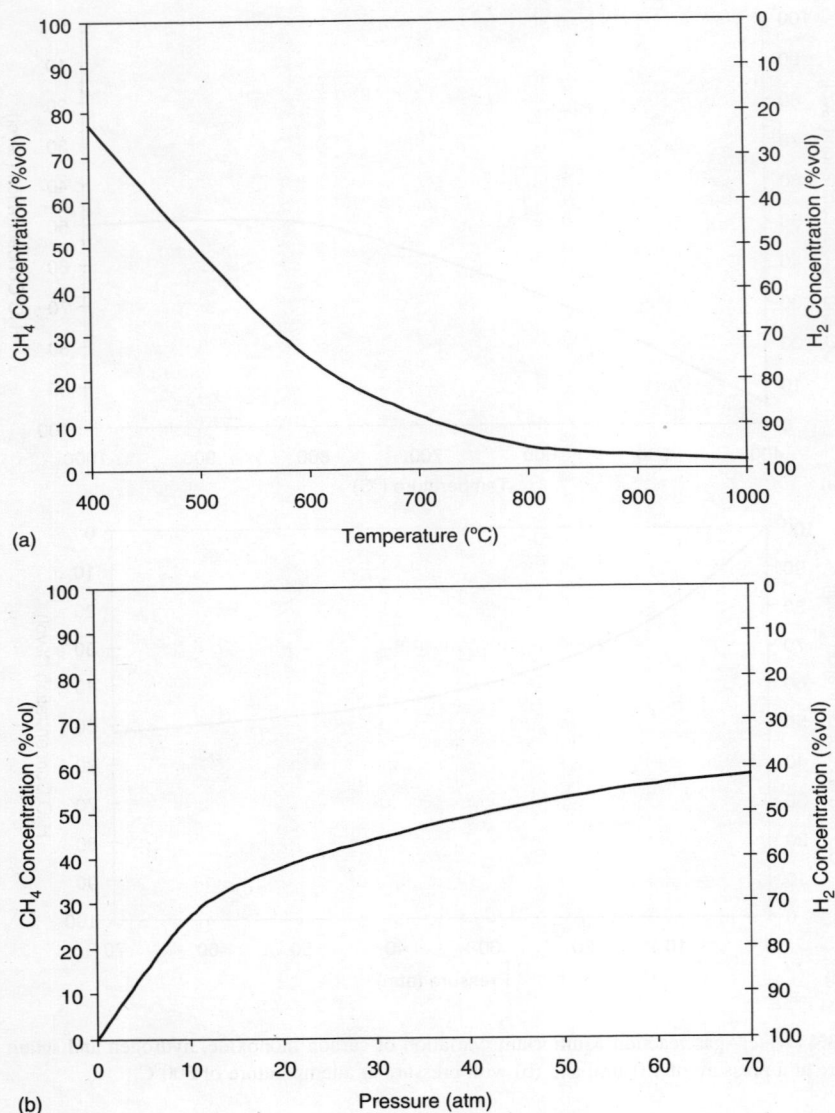

FIGURE 3.5 Variation of methane and hydrogen concentration at equilibrium (a) with temperature at a pressure of 1.0 atm, and (b) with pressure at a temperature of 800°C.

in pores. In high-rank coals, the pore sizes are so small that the reaction is diffusion-controlled (see Section 4.1.4).

3.4.2 VOLATILE MATTER

On a dry basis, the volatile matter content of a fuel varies from less than 5% in the case of anthracite to more than 75% in the case of wood. The reactivity of a solid fuel and its conversion to char inside the gasifier depend on its volatile matter content. Fuels with higher volatile matter content are more reactive and therefore can be converted more easily into gas, producing less char.

Fluidized Bed Gasification

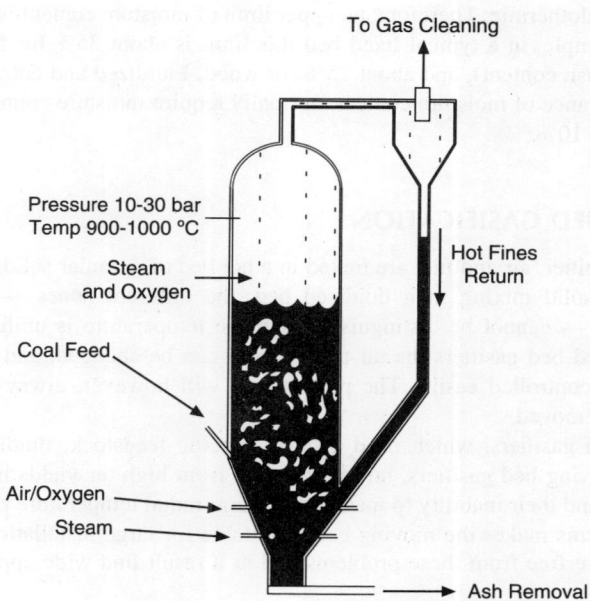

FIGURE 3.6 A bubbling fluidized bed gasifier for use in an IGCC system.

In the case of high-rank coals, which are less reactive and have low volatile matter content, char is the major product of pyrolysis. Thus, an efficient utilization of the char inside the gasifier is a major concern for the gasification of high-rank coals.

Biomass feedstock, such as wood, is highly reactive and characterized by high volatile content. Thus, these fuels produce relatively small amounts (~25% of the weight of the original dry fuel) of char. Also, the char is more porous and reacts relatively fast. For this reason, fuels with high volatile matter content are easier to gasify. However, they produce gas with a high tar content, which makes gas clean-up more difficult. More details on tar removal are given in Section 3.7.2.

3.4.3 Ash

The ash content of the fuel does not have much direct influence on the composition of the product gases, but it affects the practical operation of the gasifier. For example, in a gasifier operation, ash is an unavoidable nuisance that has to be separated from the product gases and suitably disposed of. Depending on the design, the ash is removed from a gasifier in either solid or liquid form. In the case of fixed and fluidized bed gasifiers, the ash is normally removed as a solid, requiring the peak temperature to be well below the melting point of the ash. Some units like slagging gasifiers are deliberately operated at sufficiently high temperatures to remove the ash as liquid slag. Fuels used in a gasifier designed to remove ash in dry solid form should have high ash-melting/softening temperatures to avoid ash agglomeration inside the gasifier.

3.4.4 Moisture

The moisture content of solid fuels varies widely depending on their type, ranging from below 5% for anthracite to about 40% for low-rank sub-bituminous coal and lignite. The moisture content of wood from a freshly harvested tree is typically around 50%. High moisture content of the feed lowers the temperature inside the gasifier since evaporation as well as the chemical reaction of steam with char is endothermic. Therefore, an upper limit of moisture content is set for satisfactory gasification. For example, in a typical fixed bed this limit is about 35% for a good quality coal (moderate rank and ash content), and about 25% for wood. Fluidized and entrained bed gasifiers, with their lower tolerance of moisture content, normally require moisture content of the feedstock to be reduced to 5 to 10%.

3.5 FLUIDIZED BED GASIFICATION

In a fluidized bed gasifier, air and fuel are mixed in a hot bed of granular solids such as sand. Due to the intense gas–solid mixing in a fluidized bed, the different zones — drying, pyrolysis, oxidation, reduction — cannot be distinguished, but the temperature is uniform throughout the bed. Contrary to fixed bed gasifiers the air-to-fuel ratio can be changed, and as a result the bed temperature can be controlled easily. The product gas will however, always contain some tar, which needs to be removed.

Unlike fixed bed gasifiers, which need a fairly specific feedstock, fluidized beds are more tolerant. Updraft moving bed gasifiers, however, suffer from high tar yields in the product gases (Beenackers, 1999) and their inability to maintain uniform radial temperature profiles and to avoid local slagging problems makes the moving bed unsuitable for large installations. Fluidized beds, on the other hand, are free from these problems and as a result find wide application in biomass gasification.

A bubbling fluidized bed, however, cannot achieve high solids conversion, due to the back-mixing of solids. The high degree of solid mixing helps gasification, but owing to the intimate mixing of fully gasified fuels with partially gasified any waste solid stream will contain some partially gasified solids reducing the solid conversion. Particle entrainment from a bubbling bed also contributes to the loss in a gasifier. The other important problem with FB gasifiers is the slow diffusion of the oxygen from the bubbles to the emulsion phase, which creates oxidizing conditions in the whole bed decreasing the gasification efficiency.

The circulating fluidized bed (CFB) can get around this problem by providing longer solid residence time within its solid circulation loop.

Fluidized bed gasifiers have several advantages over other types of gasifiers, which include:

1. Higher throughput than fixed bed gasifiers
2. Improved heat and mass transfer from fuel
3. High heating value
4. Reduced char

Another important advantage of fluidized bed gasifiers is that the ash does not melt, which makes its removal relatively simple. A comparison of fluidized bed gasifiers with fixed bed gasifiers is given in Table 3.3.

In a typical fluidized bed gasifier, solid fuel particles are brought into contact with a restricted supply of oxygen by feeding them into an oxygen- or air-starved fluidized bed. The fuel particles are quickly heated to the bed temperature and undergo rapid drying and pyrolysis.

Fluidized bed is especially good for biomass gasification. So far as coal gasification is concerned, fluidized beds have found only limited application because of their low carbon-conversion efficiency, which results from the relatively low bed temperatures (800 to 1000°C) needed to avoid ash agglomeration. However, fluidized beds are attractive for other reactive fuels, such as, municipal solid waste and lignite, which can be gasified satisfactorily at lower temperatures compared to high-rank coal.

Since fluidized bed gasifiers operate at relatively low temperatures, most high-ash content fuels, depending on ash chemistry, can be gasified in such beds without the problem of ash sintering and agglomeration. Also, fluidized bed gasifiers can be operated on different types of fuels or a mixture of different fuels. This feature is especially attractive for biomass fuels, such as agricultural residues and wood, that may be available for gasification at different times of the year. Because of these advantages, a great deal of current development activities on large-scale biomass gasification are focused on fluidized bed technologies.

3.5.1 Comparison of Bubbling and Circulating Fluidized Bed Gasifiers

There are two types of fluidized bed gasifiers: bubbling and circulating. Due to the back-mixing of solids as well as particle entrainment, a single bubbling fluidized bed (BFB) cannot achieve high solid conversion (Li et al., 2004). A circulating fluidized bed (CFB), which is known for its excellent heat and mass transfer and longer residence time, can achieve better conversion. Advantages of CFB over BFB for gasification are listed as follows:

1. High gas–solid slip velocity ensures good mixing and excellent heat and mass transfer.
2. A CFB can process a wider range of feed particles without the penalty of entrainment loss. In a CFB, small particles are converted in single-pass, or are entrained, separated from the gas, and returned to the gasifier via an external recycle loop. Large particles, which are converted slowly, are recycled internally inside the fast bed until they are small enough for external recycling (Higman and Burgt, 2003).
3. High recirculation rates of solids provide CFBs with a solid heating rate higher than that in bubbling fluidized beds, which in turn reduces the tar production during the heating of the fuels.
4. CFBs have less restriction on the size and shape of fuel than that for bubbling or entrained bed gasifiers. For this reason CFB gasifiers are preferred for biomass or waste-product gasification.
5. CFBs use relatively fine (<400 μm) particles. Such fine particles provide the gasification reactions with very large specific gas–solid contact-surface areas.
6. Gas by-passing through the bubbles in a bubbling bed does not occur in CFB gasifiers. Thus, in a CFB gasifier, excellent gas–solid contact, large contact surface area, and long residence time all provide favorable conditions for efficient gasification.
7. A CFB gasifier operates at high velocities (4 to 7 m/sec), several times larger than those (1 to 1.5 m/sec) used in a bubbling bed gasifier. Therefore, for a given bed area, one can expect a much larger throughput in a CFB unit. For example, for sawdust a CFB gasifier had a throughput of 28 GJ/m^2/h in compared to 4.5 GJ/m^2/h in a fixed bed. (http://europa.eu.int/comm/energy/res/sectors/doc/bioenergy/final_report_for_publication.pdf).
8. The continuous formation and breakdown of clusters in the fast bed enhances the gas–solid contact and therefore provides good carbon conversion in a CFB boiler.
9. CFB gasifiers are easy to scale-up and are reliable over a wider range of feedstock.

3.5.2 Types of Fluidized Bed Gasifiers

In some utility, the fuel is partially gasified or simply pyrolyzed to drive away the gaseous and liquid components. Such gasifiers are called *pyrolyzer, carbonizer* or *partial-gasifiers*. Besides these, all processes involve complete gasification of the fuel. Such gasifiers could be classified based on the following criteria:

3.5.2.1 Mode of Gas–Solid Contacting

Based on the gas–solid contacting process, a fluidized bed gasifier may belong to either of the following two types:

1. Bubbling fluidized bed gasifier (BFBG)
2. Circulating fluidized bed gasifier (CFBG)

3.5.2.1.1 Bubbling Fluidized Bed Gasifier

Bubbling fluidized bed gasification was introduced in 1926. The overall simplicity of the bubbling bed design and a number of other features made its application more attractive for coal gasification than spouted or circulating fluidized beds. In biomass industries it finds use in small to medium (<25 MWth) capacities.

3.5.2.1.2 Circulating Fluidized Bed Gasifier

Circulating fluidized bed gasifiers have advantages over BFBGs such as higher carbon conversion efficiency and smaller size. As a result, they are emerging as the major choice for small (a few MWth) to large (100 MWth or more) biomass gasification projects. Although, they found only limited application for coal or lignite so far, the CFB technology is likely to be the prime choice for large-scale coal gasification in the future.

3.5.2.1.3 Hybrid Fluidized Bed Gasifier

A hybrid system, using CFB and BFB is described by Kaiser et al. (2003). The circulating fluidized bed burns char, received from the gasifier in an oxidizing atmosphere. Hot solids from the CFB provide the heat for the endothermic gasification reaction in the bubbling bed. Hot flue gas containing nitrogen from the CFB does not dilute the gas produced in the BFB gasifier. The two gases have separate exits as shown in Figure 3.15. Thus, this system is capable of producing a high-grade product gas (LHV 10 to 14 MJ/nm^3)(Kaiser et al., 2003).

3.5.2.2 Gasifying Medium Used

Based on the gasifying medium used, fluidized bed gasifiers can be divided into the following types:

- Oxygen-blown
- Air-blown
- Steam-blown

Air gasification produces a low heating value gas containing about 50% N_2, while oxygen or steam gasification produces a medium heating value gas containing very little or no nitrogen. The volume of the gas obtained from gasification of a given amount the fuel is therefore significantly lower in these cases. Oxygen gasification requires an air-separation unit for producing oxygen, while steam gasification requires an indirect heat source for driving the endothermic gasification reactions.

3.5.2.3 Operating Pressure

Based on the operating pressure, fluidized bed gasifiers can be of the following types:

- Atmospheric pressure
- Pressurized

Atmospheric pressure gasification is an old and relatively established well technology. Pressurized gasification is a new development and a promising option for IGCC applications.

Fluidized Bed Gasification

3.5.2.4 Mode of Heating the Solids

Based on this criterion, FBGs can be of the following types:

- Directly heated
- Indirectly heated

In a directly heated gasifier, part of the fuel is oxidized to provide the heat needed for the endothermic gasification reactions. Most gasifiers in use today are of this type. In an indirectly heated gasifier, heat needed for gasification is supplied by a hot inert medium, which, in turn, is heated by the combustion of char produced from biomass gasification in a separate reactor.

3.6 EXAMPLES OF SOME FLUIDIZED BED GASIFIERS

In a typical application, the gas from the gasifier is cleaned of particulate matter, normally in a cyclone, and then burned for running engines or for heat and steam generation.

The gas can also be cofired with other fuels like coal, oil, or natural gas. Biomass cofiring in an existing fossil fuel-fired boiler is an attractive means of reducing the greenhouse gas emission of a power plant. Since the gas is burnt directly in the existing furnace, there is no need for gas cleaning or cooling, reducing the capital cost and improving the energy efficiency of fuel conversion. Furthermore, since the gasifier works in tandem with the main coal or oil burner of the boiler, the biomass can be fired as it is available without affecting the operation of the power plant.

Besides the above applications, gasified fuel can also be used for power generation through a combined-cycle system. This system is called *integrated gasification combined-cycle*, popularly known by its acronym *IGCC*. In an IGCC plant, the gas is compressed, cleaned and delivered to the combustion chamber of the gas turbine-cycle section (Figure 1.10, Chapter 1). A high-pressure bubbling bed gasifier like the one shown in Figure 3.7 produces the gas to drive the IGCC plant. The gasifying medium for the high-pressure gasifier is normally air obtained from the compressor of the combined-cycle section of the IGCC system. The gas from the gasifier is cooled before entering the hot gas cleaning system because of the temperature limitations of available particulate filters. The cleaned gas expands through the gas turbine, producing power. A heat-recovery steam generator (boiler) utilizes the residual sensible heat of the low-pressure waste gas from the gas turbine to produce steam, which drives a steam turbine producing additional power.

Gasification technology is currently in an important stage of development. Some major fluidized bed gasifiers used in the industry are briefly described in the section below. This list of gasifiers is by no means complete.

3.6.1 WINKLER GASIFIER

As mentioned earlier, this is the oldest fluidized bed gasifier, and has been in commercial use for many years. Depending upon its operating conditions, we can group Winkle gasifiers under low-temperature and high-temperature types.

3.6.1.1 Low-Temperature

Winkler gasifiers can operate at either atmospheric or elevated pressures. Figure 7.1 shows a schematic of an atmospheric pressure, low-temperature Winkler gasifier. Here, coal crushed to particles less than 10 mm in diameter, is fed into the gasifier while ash is removed from its bottom using screw conveyors. The temperature of the bed is normally kept below 980°C to avoid ash melting.

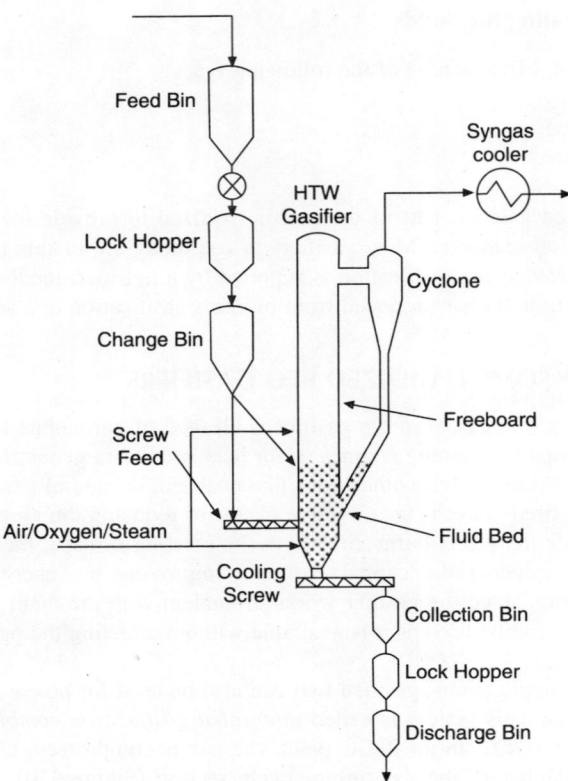

FIGURE 3.7 Schematic of the high temperature Winkler gasifier.

The gasifying medium, a mixture of steam and air or oxygen, is supplied in two stages. The first-stage supply is adequate to maintain the fluidized bed at the desired temperature, while the second-stage, supplied above the fluidized bed, serves to convert entrained unreacted char particles and hydrocarbons into useful gas.

3.6.1.2 High-Temperature

The high-temperature Winkler (HTW) gasification process (Figure 3.7) is an improvement over of the original Winkler gasification process described above. The process, developed by Rheinbraun AG of Germany employs a pressurized fluidized bed operating below the ash melting point. In order to improve carbon-conversion efficiency, small particles in the raw gas are separated using a cyclone and returned to the bottom of the main reactor.

The gasifying medium (steam and oxygen) is introduced into the fluidized bed at different levels, as well as above the fluidized bed (Bucko et al., 1999). The bed is maintained at a pressure of 10 bar and the temperature of the bed is maintained at about 800°C to avoid ash fusion. The over-bed supply of the gasifying medium raises the local temperature to about 1000°C to minimize production of methane and other hydrocarbons.

The HTW process produces better quality gas than the traditional low-temperature Winkler process and is suitable for lignite and other reactive fuels like biomass and treated municipal solid waste. It is claimed that for IGCC application, HTW plants of capacities up to 400 and 600 MW can be built for gasification with air/steam and oxygen/steam media, respectively.

3.6.2 KRW Gasifier

The Kellogg–Rust–Westinghouse (KRW) gasifier (Figure 3.8) is a pressurized ash-agglomerating bubbling fluidized bed gasifier developed for the gasification of coal with steam/air or oxygen. The bed has a conical section near the bottom. Crushed coal and limestone enter the bed with air (or oxygen) through a concentric high-velocity jet. Steam is supplied to the gasifier through a number of jets in the conical section. The inlet air jet, and the quickly evolving volatile matter from the entrained coal particles set up a pattern of circulation inside the bed similar to that occurring in a spouted bed such that the char and sorbent particles move down the sides of the reactor and reenter the central jet at the bottom. The ash particles from spent char agglomerate due to melting/softening while passing through the flame above the central jet, and are removed from the bottom of the bed.

Char particles, elutriated from the bed, are captured in a cyclone (not shown in the figure), and returned to the bottom of the bed. Part of the sulfur in the coal is removed inside the bed by limestone fed along with the coal, while most of the remaining sulfur is removed from the product gas in an external hot gas cleanup system.

3.6.3 Spouted Bed Partial Gasifier

The air-blown gasification cycle (ABGC) originally developed by British Coal, employs a pressurized spouted fluidized bed to gasify coal with a mixture of air and steam (Figure 3.10). The fluidizing air/steam mixture is supplied partly as a central jet, which carries crushed coal and

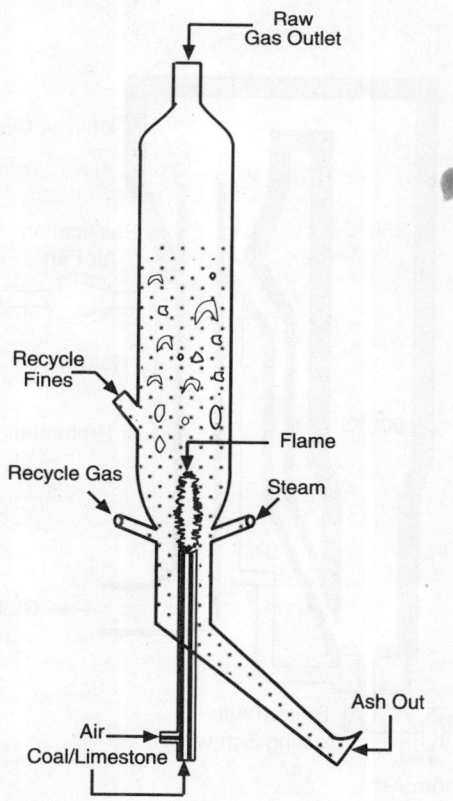

FIGURE 3.8 Schematic of the KRW fluidized bed gasifier.

limestone into the gasifier. The remaining gasifying mixture is supplied through nozzles in the wall of the conical base section of the gasifier.

The fuel is only partially gasified inside the gasifier, and the limestone serves to retain sulfur. Unreacted char and the sorbent particles are removed from the bottom of the gasifier and transferred to an atmospheric pressure circulating fluidized bed combustion boiler to generate steam by burning the char. The pressurized gas from the gasifier runs a combustion turbine–generator unit, while the steam produced in the boiler runs a steam turbine–generator unit.

The gasifier is essentially a refractory-lined cylindrical vessel with a diameter depending on its capacity, and a height, including a transport disengagement section, of about 20 m.

3.6.4 FOSTER WHEELER ATMOSPHERIC CFB GASIFIER

This is an atmospheric-pressure air-blown gasifier operating in a circulating fluidized bed (Figure 3.9). Depending on the fuel and the application, it operates at a temperature within the range of 800 to 1000°C. The hot gas from the gasifier passes through a cyclone, which separates most of the solid particles associated with the gas and returns them to the bottom of the gasifier. An air preheater is located below the cyclone to raise the temperature of the gasification air and indirectly the temperature level inside the gasifier (Figure 3.9).

At least four commercial gasifiers of this type have been installed so far in different countries. The biggest among these is a 60 MWth unit installed at a coal and natural gas fired power plant in Lahti, Finland, to provide a cheap supplementary fuel by gasifying waste wood and refuse-derived fuels.

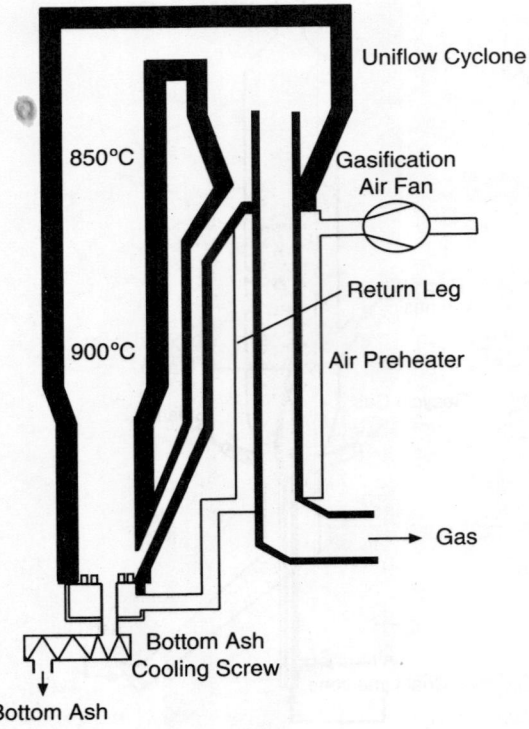

FIGURE 3.9 A circulating fluidized bed gasifier.

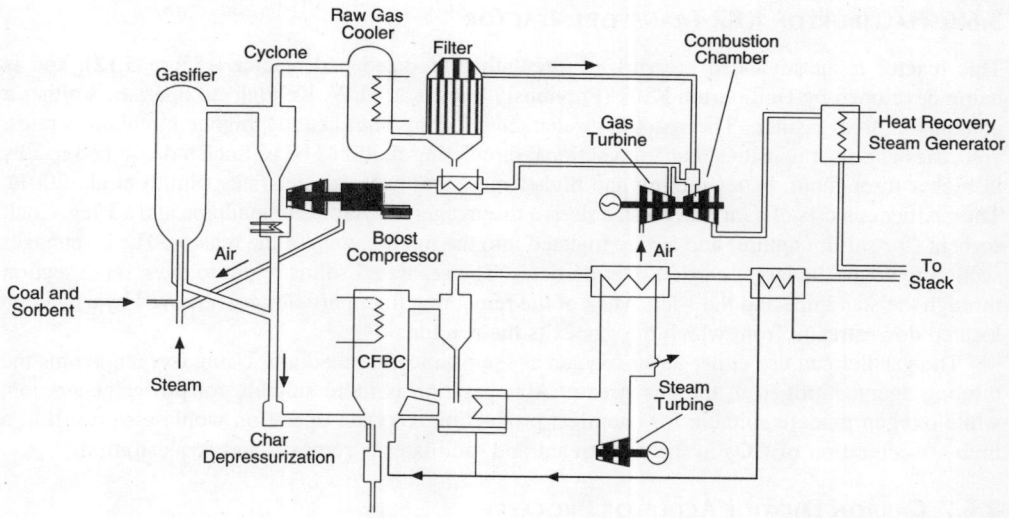

FIGURE 3.10 Application of the spouted bed gasifier in an air-blown gasification cycle.

3.6.5 BATTELLE/COLUMBUS INDIRECT GASIFIER

This atmospheric-pressure gasifier (Figure 3.11) was developed to gasify biomass indirectly to produce a medium heating value gas. The design employs two CFBs — one acting as a char combustor and the other serving to gasify the biomass. In the gasifier section, the solid biomass particles come into contact with hot sand or other bed material, and are converted to gas and char. The char particles carried over from the gasifier burn inside the char combustor and heat the bed material accompanying the char particles. The hot sand particles separated by the cyclone of the CFB combustor enter the gasifier to heat the biomass particles fed to it. The gas from the gasifier is cooled, cleaned and compressed before entering the combustion chamber of the IGCC plant or any other application point. The circulation of solids between the two CFB legs is a major challenge in this type of gasifier.

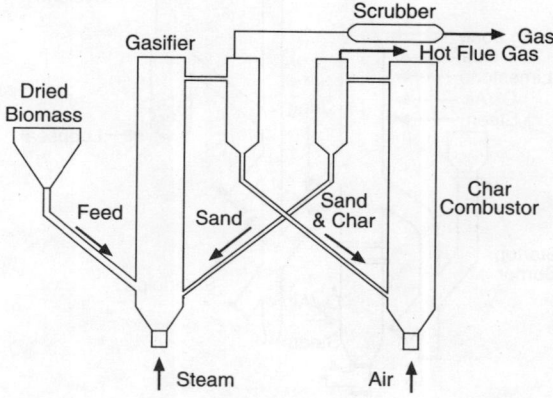

FIGURE 3.11 Schematic of atmospheric-pressure partial gasifier and char combustor.

3.6.6 HALLIBURTON KBR TRANSPORT REACTOR

This reactor is an advanced pressurized circulating fluidized bed reactor (Figure 3.12), and is being developed by Halliburton KBR (Previously known as M.W. Kellogg) to operate as either a combustor or a gasifier. The reactor is characterized by significantly higher circulation rates, velocities and riser densities than conventional circulating fluidized beds. Such a dense bed results in higher throughput, better mixing and higher mass and heat transfer rates (Smith et al., 2001). The gasifier consists of a mixing zone, a riser, a disengager, a cyclone, a standpipe and a J-leg. Coal, sorbent (for sulfur capture) and air are injected into the mixing zone of the reactor. The disengager section removes the larger carried-over particles. The separated solids return to the mixing section through the standpipe and the J-leg. Most of the remaining finer particles are removed by a cyclone located downstream, from which the gas exits the reactor.

The gasifier can use either air or oxygen as the gasification medium. Using oxygen avoids the diluting agent, nitrogen in the gas stream. Air operation is more suitable for power generation, while oxygen is more suitable for chemical production. Oxygen operation would also result in a high concentration of CO_2 in the gas stream and facilitate its removal and sequestration.

3.6.7 CARBON DIOXIDE ACCEPTOR PROCESS

Another novel approach to gasifier design is the *CO_2 acceptor process*, shown schematically in Figure 3.13. The gasifier consists of two bubbling fluidized beds, coupled together with a stream of solid particles circulating between them. Low-rank fuel is fed into the gasifier, which is an indirect gasifier and uses steam as the gasification medium. The reactor operates at 10 bar and 825°C. Heat for the endothermic gasification reaction is provided mainly by the exothermic CO_2 acceptor reaction (Hebden and Stroud, 1981) on the circulating, partially calcined dolomite from the regenerator.

$$CaO + CO_2 \rightarrow CaCO_3 + 176,784 \text{ kJ/kmol} \tag{3.9}$$

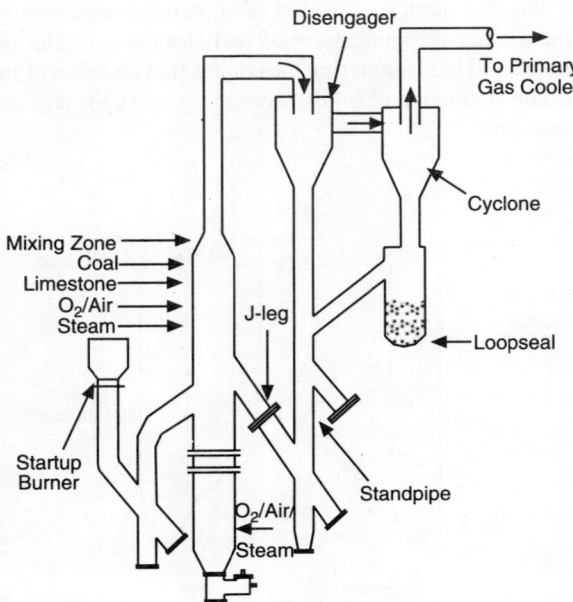

FIGURE 3.12 Halliburton advanced pressurized circulating fluidized bed reactor.

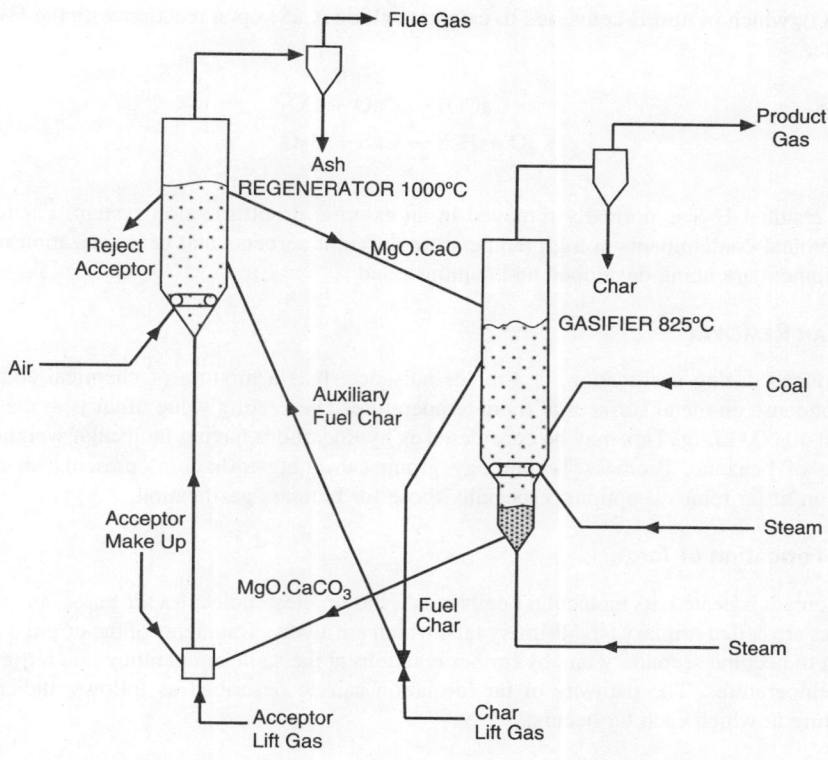

FIGURE 3.13 Pilot plant flow diagram of the CO_2 acceptor process.

Here calcined dolomite, CaO, serves as the CO_2 acceptor. Thus, besides providing heat, the reaction removes CO_2 from the gas produced so that its heating value improves. The carbonated acceptor $CaCO_3$, and residual char from the gasifier are transferred to the second fluidized bed, called the regenerator, operating at 1010°C. Here the endothermic heat requirement is provided by the combustion of residual char from the gasifier, which uses air as to burn the char. Additional details are given elsewhere (Elliott, 1981).

3.7 GAS CLEANING

The gas from a gasifier contains, besides the desired combustible components, (CO, H_2, and CH_4) entrained soot, ash, a certain amount of H_2S (depending on the sulfur content of the feed), tar, and trace quantities of NH_3, COS, HCl, and HCN. The gas needs to be cleaned to remove the contaminants. Tar makes it particularly difficult to use gasified fuels in engines. Thus, depending on the application, type of gasifier, and contaminants in the fuel, a certain level of gas conditioning (cleaning/cooling) is required.

3.7.1 SULFUR AND OTHER CHEMICAL CONTAMINANTS REMOVAL

In a gasifier, nearly the entire sulfur in the feed is converted to H_2S, which must be effectively removed to ensure that the sulfur content of the final gas is within acceptable limits. In the case of fluidized bed gasifiers, limestone can be fed into the gasifier along with coal to capture most of the H_2S produced within the bed itself. The limestone ($CaCO_3$) calcines inside the gasifier to produce lime (CaO), which in turn is converted to calcium sulfide (CaS) upon reaction with the H_2S inside the gasifier.

$$CaCO_3 \rightarrow CaO + CO_2$$
$$CaO + H_2S \rightarrow CaS + H_2O$$
(3.10)

Any residual H_2S is normally removed in an external desulfurization system. For removing other chemical contaminants at high temperature different sorbents and desulfurization/regeneration equipment are being developed and demonstrated.

3.7.2 Tar Removal

Tar, produced during gasification, is a major nuisance. It is a mixture of chemical compounds, which condense on metal surfaces at room temperature. The heating value of tar is in the range of 20,000 to 40,000 kJ/kg. Tars may be considered as hydrocarbons having molecular weights higher than that of benzene. Biomass Technology group (www.btgworld.com) presents an excellent description of tar removal options, especially those for biomass gasification.

3.7.2.1 Formation of Tar

When biomass is heated, its molecular bonds break; the smallest molecules are gases, and the larger molecules are called primary tars. Primary tars, which are always fragments of the original material, can react to become secondary tars by further reactions at the same temperature and tertiary tars at higher temperatures. The pathway of tar formation can be described as follows, indicating the temperature at which each tar occurs:

Mixed Oxygenates $\rightarrow$	Phenolic Ethers $\rightarrow$	Alkyl Phenolics $\rightarrow$	Heterocyclic Ethers $\rightarrow$	PAH $\rightarrow$	Larger PAH
400°C	500°C	600°C	700°C	800°C	900°C

3.7.2.2 Removal Options

Tar can be removed from the product gas by chemical and physical methods. Chemical methods destroy the tar, converting it to smaller molecules. Physical methods only remove the tar yielding a tar waste stream. Several devices are available for tar conversion and removal. Most removal devices are meant primarily for particle removal

(Chemical Methods)	(Physical Methods)
• Catalytic cracking	• Cyclone
• Thermal cracking	• Filters (baffle, fabric, ceramic, granular beds)
• Plasma reactors (Pyroarc, Corona, Glidarc)	• Electrostatic precipitators
• Use of catalytic bed materials	• Scrubbers

If the gas temperature is maintained above 400°C, the tar does not condense. So, in applications of direct combustion such as cofiring, one can avoid having to strip the gas of its tar.

Catalytic cracking is preferred to other forms of cleaning as it maintains the heating value of the tar by converting it to other gases. Calcined dolomite and nickel-based catalysts are used downstream of the gasifier for catalytic cracking of tar.

Use of bed materials as catalyst for tar reduction is an attractive option. Amongst a range of bed materials used in a gasifier the olivine was found most effective in tar reduction in a steam-reforming biomass gasifier (Pfeifer et al., 2005). Another group (Ross et al., 2005) studied commercial activated clay (130 m^2/g), acidified bentonite (92 m^2/g), raw bentonite (10 m^2/g) and clay housebrick (15 m^2/g). While the activated clay captured the greatest amount of tar, the ordinary housebrick captured more than twice that captured by sand or raw bentonite in a steam-reforming CFB gasifying biomass. Thus, biomass gasification with clay-derived catalysts shows good potential for use in fluidized bed gasifiers.

3.7.3 Particulate Removal

The particulate matter in the gas is removed mostly by cyclones located next to the gasifier. Further removal occurs in either hot, dry filters located downstream of the cyclone or in water scrubbers after the gas is cooled in a heat-recovery device. Water scrubbers remove the particulate matter as slurry, which is subsequently dewatered, and also remove some of the other contaminants.

The growing importance of IGCC systems as an option for improving the efficiency of electricity generation has generated a great deal of interest in hot gas cleaning techniques that remove chemical contaminants as well as particulate matter from the gas without cooling it significantly. Hot gas cleaning avoids the loss of thermal energy associated with cold gas cleaning and, In the case of IGCC, improves the efficiency of electricity generation by, approximately, 2% or more.

A number of techniques are being developed for removing particulate matter at high temperatures. These include high-temperature fabric filters, granular-bed filters, ceramic-barrier filters, etc. Of these, ceramic filters in the form of candles appear to be most promising and are ready for commercialization for cleaning at temperatures up to 500°C.

3.8 DESIGN CONSIDERATIONS

The design of the structural and mechanical components, such as the distributor, main reactor body, insulation, and cyclone of fluidized bed combustors and gasifiers are similar. Design of these components is discussed in other chapters in the context of both boilers and gasifiers. This section will therefore not consider these details, but will instead present the basic stoichiometric and heat/mass-balance calculations involved in designing an air fluidized bed gasifier.

3.8.1 Gasifier Efficiency

The performance of a gasifier is often expressed in terms of its efficiency, which can be defined in two different ways: cold gas efficiency and hot gas efficiency. The cold gas efficiency is used if the gas is used for running an internal combustion engine in which case the gas is cooled down to the ambient temperature and tar vapors are removed. The *cold gas efficiency*, η_{ceff} is defined as:

$$\eta_{ceff} = \frac{(V_g q_g)}{(M_b C_b)} \tag{3.11}$$

where

V_g = Gas generation rate (m^3/sec)
q_g = Heating value of the gas (kJ/m^3)

TABLE 3.5
Heat of Combustion of Gases

Gas	Higher Heating Value (MJ/kg mol)	Lower Heating Value (MJ/kg mol)
CO	282.99	282.99
H_2	285.84	241.83
CH_4	890.36	802.34
H_2S	562.59	518.59

M_b = Fuel consumption rate (kg/sec)
C_b = Heating value of fuel (kJ/m^3)

q_g can be estimated from the heating values of the constituents of the gas (Table 3.5) and its composition. For thermal applications, the gas is not cooled before combustion and the sensible heat of the gas is also useful (Table 3.6). The *hot gas efficiency*, η_{geff} is used for such applications and is defined as:

$$\eta_{geff} = \frac{(V_g q_g + H_{sensible})}{(M_b C_b)} \tag{3.12}$$

where

$H_{sensible} = C_p V_g (T_g - T_a)$
T_g = Gas temperature
T_a = Ambient temperature

3.8.2 Equivalence Ratio

In a fluidized bed gasifier, the fuel feed-rate and air supplies can be independently controlled, if necessary. In a combustor, the amount of air supplied is basically determined by the stoichiometry, which depends on the fuel composition, and excess air requirement. In a gasifier on the other hand, the air supply is only a fraction of the stoichiometric rate. The term *equivalence ratio* (ER) is often used in connection with gasifier air supply.

> Equivalence ratio is defined as the ratio of actual air fuel ratio to the stoichiometric air fuel ratio.

The quality of gas obtained from a gasifier depends strongly on the value of ER employed, which must be significantly below 1.0 to ensure a condition far from complete combustion.

An excessively low value of ER (<0.2) results in several problems including incomplete gasification, excessive char formation and low heating value of the product gas. On the other hand, too high a value of ER (>0.4) results in excessive formation of products of complete combustion, such as CO_2 and H_2O at the expense of desirable products like CO and H_2. This causes a decrease in the heating value of the gas. In practical gasification systems, the value of ER is normally maintained at 0.20 to 0.30. Figure 3.14 shows the variation of the carbon-conversion efficiency of a circulating fluidized bed gasifier for wood dust. The carbon-conversion efficiency increases for an equivalence ratio of up to 0.26, after which it starts declining.

The bed temperature of a fluidized bed gasifier increases with equivalence ratio because the higher the amount of air, the greater the combustion and amount of heat released. Figure 3.16 shows

TABLE 3.6
Enthalpy (kJ/kg mole) of Selected Gases at Different Temperatures above the Datum of 298 K (25°C) and 1 atm

Temperature (K)	Nitrogen	Oxygen	Carbon Dioxide	Carbon Monoxide	Methane	Hydrogen	Steam	Hydrogen Sulphide
300	54.39	54.39	66.94	54.39	66.94	54.39	62.76	62.80
400	2970.64	3029.21	4008.27	2974.82	3861.83	2958.08	3451.80	3554.59
500	5911.99	6087.7	8313.60	5928.72	8200.64	5882.70	6920.33	7192.92
600	8891.00	9246.64	12916.01	8941.20	13129.39	8811.50	10497.66	11002.91
700	11936.95	12501.79	17761.08	12020.63	18635.54	11748.67	14183.76	14988.74
800	15045.66	15840.62	22815.35	15175.36	24673.05	14702.57	17991.20	19154.61
900	18221.32	19246.40	28041.17	18397.04	31204.27	17681.58	21924.16	23496.32
1000	21459.74	22706.57	33405.06	21685.67	38179.00	20685.69	25978.46	27997.13
1100	24756.73	26216.94	38894.46	25032.87	45551.21	23723.28	30166.64	32362.85
1200	28108.11	29764.98	44484.29	28426.09	53270.69	26794.33	34476.16	37442.55
1300	31501.34	33350.66	50157.79	31865.34	61303.97	29907.23	38902.83	42349.48
1400	34936.40	36965.64	55906.61	35338.06	69609.21	33061.96	43446.66	47361.08
1500	38404.94	40609.90	61714.00	38848.44	78152.94	36266.91	48095.08	52468.98
1600	41902.76	44279.27	67579.97	42383.92	86910.05	39522.06	52843.92	57660.61
1700	45429.87	47969.56	73491.96	45940.32	95855.44	42814.87	57684.81	62927.60
1800	48982.09	51689.14	79441.61	49521.82	104959.8	46149.52	62609.38	68257.40

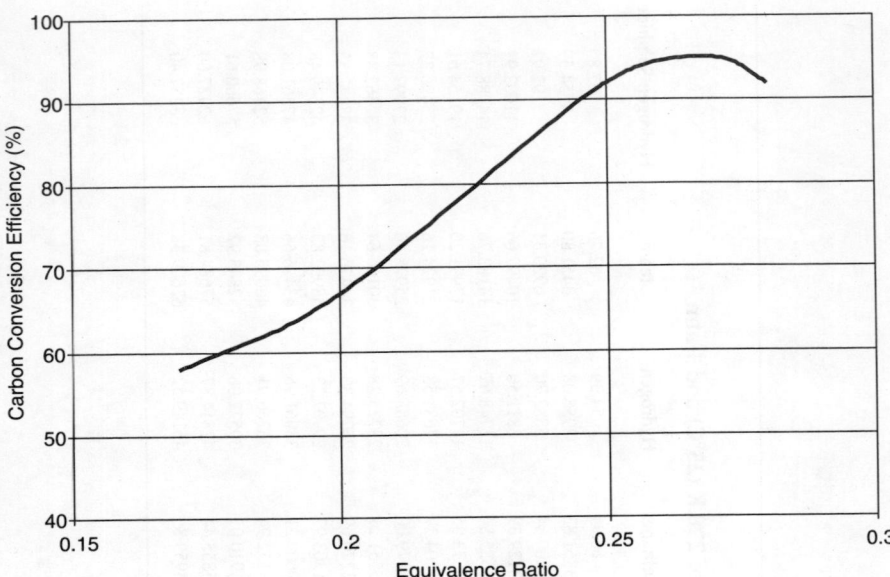

FIGURE 3.14 Variation of carbon conversion efficiency with equivalence ratio for a CFB gasifier using wood dust. (Replotted using data from Bingyan et al., 1994.)

how the resulting bed temperature of a CFB gasifier increases with increasing equivalence ratio within the range of 0.15 to 0.3.

3.8.3 Gas Composition

In commercial gasifiers, the composition of the product gas depends on a number of design and operating parameters and is difficult to predict. For design purposes, the approximate gas composition can be either assumed or roughly estimated, considering the gas to be at chemical equilibrium at the design values of temperature, T and pressure, P (see Section 3.2.4).

3.8.3.1 Equilibrium Gas Composition

The equilibrium composition of the gas, obtained from a gasifier, can be estimated using the equilibrium equations considering the ultimate analysis of the fuel and the operating temperature and pressure. It also depends on the amount of the gasifying medium supplied per unit mass of the solid fuel, and the reactions involved. Appendix 3A illustrates a simple case of estimating the equilibrium composition of gas produced from a sulfur-free solid fuel.

3.8.3.2 Typical Composition

Equilibrium conditions are rarely achieved in practical gasifiers (Li et al., 2004), so the equilibrium composition can only be regarded as indicative. Also, the determination of the equilibrium gas composition requires numerical techniques. An alternative approach in designing gasifiers may be to assume a typical composition for the type of gasifier in question, based on results reported in the literature or observed in pilot/laboratory units. (DOE, 1978), and proceed with subsequent calculations on this basis.

Fluidized Bed Gasification

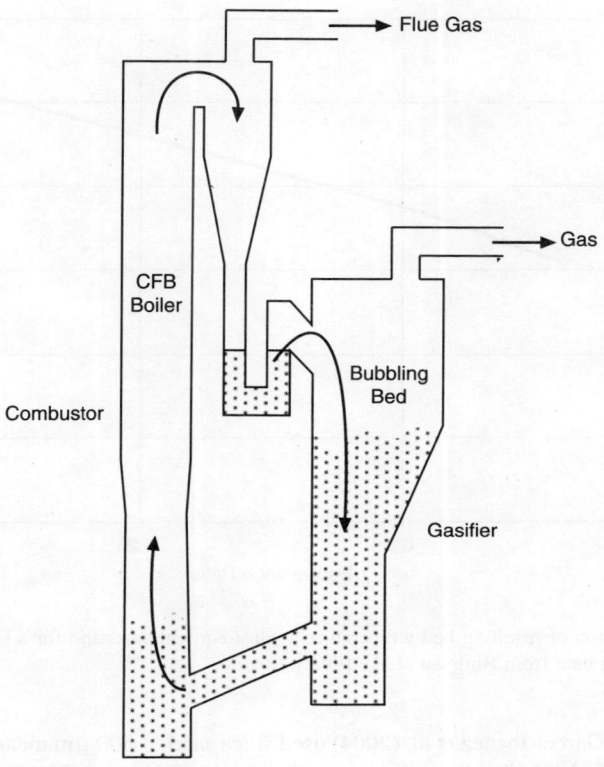

FIGURE 3.15 A fluidized bed gasifier of hybrid design.

3.8.4 Energy and Mass Balance

The mass balance of a thermodynamic system attempts to establish equality of its input and output material streams. In the case of gasifiers, a mass balance of the chemical elements involved is often useful. This can be used as a part of testing an existing gasifier. The mass balance considers the flow-rate of the gasifying medium (air or oxygen, with or without steam) and that of the fuel to be gasified. A mass-balance can also be used for design purposes if the expected performance of the gasifier can be estimated from initial pilot/laboratory-scale studies.

An example of the mass-balance calculation is given in Example 3.1.

3.8.5 Bed Materials

The bed material in a fluidized bed gasifier consists mostly of inert solid particles and some fuel particles at different stages of gasification. In the case of coal gasification, the inert bed particles are basically granular ash produced from the coal gasification process. Sometimes, limestone is added with coal particles to remove sulfur. The limestone, at different stages of calcination and sulfurization, forms a part of the bed material in this case.

In the case of biomass gasification, silica sand is normally used as the inert bed material. Magnesite (MgO) was successfully used in the first biomass based IGCC plant in Vernamo, Sweden

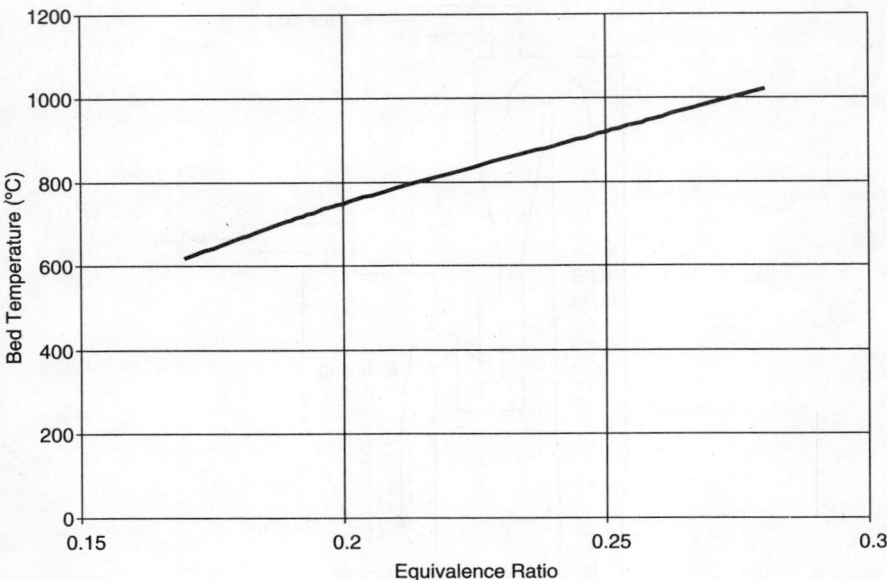

FIGURE 3.16 Variation of resulting bed temperature against equivalence ratio for a CFB gasifier for wood dust. (Replotted using data from Bingyan et al., 1994.)

(Stahl et al., 2001). Garcia-Ibanez et al. (2004) used silica sand of 500 μm mean size as the start up bed material to gasify 1.89 mm olive oil waste (bulk density 659 kg/m^3). The bed materials, besides serving as a heat carrier, can catalyze the gasification reaction by increasing the gas yield and reducing the tar. As explained in Section 3.7.2 bentonite, activated clay or even ordinary housebrick clay can reduce tar formation.

Bingyan et al. (1994) reported to have used wood powder (sawmill dust) as the bed material in a circulating fluidized bed gasifier. This reactor was, therefore, operated at a very low velocity of 1.4 m/sec, which is 3.5 times the terminal velocity of the wood particles. Chen et al. (2005) operated 1 MWe CFB gasifiers with rice husk alone. The system had difficulty with fluidization in the loop-seal with low-sphericity husk ash, but the main riser operated in fast bed without any major difficulty.

3.8.6 Gasifier Sizing

No hard and fast design method is available in published literature for the design of gasifiers. The subject, though nearly a century old, is still in evolving stages. Based on available information the following section develops a tentative method for determining size of a fluidized bed gasifier.

3.8.6.1 Cross-Sectional Area

Design of the internal cross-sectional area of the reactor column of a gasifier involves the following steps:

1. Select the throughput of fuel
2. Assume a value of equivalence ratio ($\sim$ 0.25 to 0.35)

Fluidized Bed Gasification

3. Estimate the required rate of air supply for the rate of gasification of the fuel
4. Select the inert particle size, which is similar to that used in fluidized bed combustors
5. Estimate the inside cross-sectional area so the superficial gas velocity is within acceptable limits for the selected particle size. This will ensure satisfactory fluidization and avoid excessive erosion

3.8.6.2 Fluidization Bed Volume

For bubbling fluidized beds the bed volume can be found from the equation given below using the back-flow models (DOE, 1978):

$$V = \frac{w_{out}\theta}{\rho_b} \quad (3.13)$$

where

w_{out} = Char flow-rate out of the gasifier = char feed-rate into the gasifier − rate of gasification of carbon
θ = Char residence time
ρ_b = Bed density

Design charts are available for residence time θ for test coals for different feed conversions and steam/carbon or oxygen/carbon ratios. This residence time is adjusted for the reactivity of the coal in question and that for partial gasification of the char before it enters the gasifier. Details are given in the Coal Conversion Systems Technical Data Book (DOE, 1978).

3.8.6.3 Fluidization Velocity

The range of fluidizing velocity depends on the mean particle size of the bed materials. The choice is made in the same way as for a fluidized bed combustor. If it is a circulating fluidized bed then the velocity must be within the limit of fast fluidization. In the case of bubbling beds it should reside within the minimum fluidization and terminal velocities of mean bed particles.

3.8.6.4 Gasifier Height

Since gasification involves only partial oxidation of the fuel, the heat released inside a gasifier is only a fraction of the heating value of the fuel. Also, a part of the heat released is absorbed by endothermic reactions taking place inside the gasifier. It is, therefore, uncommon to recover heat from the main gasifier column. Thus, the height of a fluidized bed gasifier is not determined by heat transfer considerations as done for fluidized bed boilers. Both gas and solid residence time is influenced by the gasifier height.

Limestone is normally fed into a fluidized bed gasifier along with the fuel for in-bed sulfur removal. The height of the gasifier should be adequate to allow a gas residence time needed for the desired extent of sulfur capture.

Tar produced in a gasifier should be thermally cracked inside the gasifier as far as possible. Therefore, the height of the gasifier should be such that the gas residence time is adequate for the desired degree of tar conversion/cracking in the gasifier.

While the above considerations are relevant to all fluidized bed configurations, In the case of bubbling or spouted fluidized beds, the total gasifier height should be sufficiently high to accommodate the dense (bubbling or spouted) bed and a transport disengagement section, which would avoid excessive loss of bed materials. Another consideration that imposes a minimum limit on gasifier height is the poor carbon conversion resulting from low residence time of the char particles.

An example of a fluidized bed biomass gasifier is given in Appendix 3B to illustrate its operation.

Example 3.2

Data for an air- and steam-blown coal gasifier are given below:
Coal analysis: C — 66.5%, O — 7%, H — 5.5%, N — 1%, S — 2%, moisture — 7.3%, ash — 10.7%.
HHV: 30.24 MJ/kg
Composition of cleaned gas: CO — 27.5%, CO_2 — 3.5%, CH_4 — 2.5%, H_2 — 15%, N_2 — 51.5%
Dry air supply rate: 2.76 kg/kg coal
Steam supply rate: 0.117 kg/kg coal
Specific humidity of supply air: 0.01 kg moisture per kg dry air
Estimate the carbon conversion efficiency and the cold gas efficiency of the gasifier.

Solution

Air contains 76.9% nitrogen by mass. So, the nitrogen supply from air

$$= 0.769 \times 2.76$$
$$= 2.121 \text{ kg } N_2/\text{kg coal}$$

Total nitrogen supplied by the air and the fuel, which carry 1% nitrogen:

$$= 2.121 + 0.01 = 2.131 \text{ kg } N_2/\text{kg coal} / 28 \text{ kg/kg.mol}$$
$$= 0.0761 \text{ kg mol } N_2/\text{kg coal}$$

Since we know that the cleaned and cooled gas produced in the gasifier contains 51.5% nitrogen, the amount of the product gas produced

$$= 0.0761/0.515$$
$$= 0.1478 \text{ kg mol gas/kg coal}$$

Oxygen flow to the gasifier
Oxygen supply from dry air, which contains 23.1% oxygen:

$$= 0.2315 \times 2.76$$
$$= 0.640 \text{ kg/kg coal}$$

0.117 kg steam is supplied per kg fuel, so the oxygen associated with steam supply

$$= 0.117 \times (8/9)$$
$$= 0.104 \text{ kg/kg coal}$$

Fluidized Bed Gasification

Oxygen associated with moisture in fuel:

$= 0.073 \times (8/9)$
$= 0.065$ kg/kg coal

Oxygen associated with moisture in air supply:
The air contains 1% moisture and (8/9) fraction of that moisture is oxygen, so, the oxygen associated is:

$= 0.01 \times 2.76 \times (8/9)$
$= 0.0245$ kg/kg coal

Total oxygen flow to the gasifier with air, steam, moisture in fuel, moisture in air and fuel:

$= 0.640 + 0.104 + 0.065 + 0.0245 + 0.07$
$= 0.9035$ kg O_2/kg coal

Hydrogen balance
Similarly total hydrogen inflow to the gasifier with fuel, steam, moisture in fuel and moisture in air:

$= 0.055 + 0.117/9 + 0.073/9 + 0.0276/9$
$= 0.0792$ kg/kg coal

Hydrogen associated with H_2 and CH_4 in dry gas

$= (0.15 + 2 \times 0.025) \times 0.1478$
$= 0.02956$ kg mol/kg coal
$= 0.02956 \times 2 = 0.0591$ kg Hydrogen/kg coal

Assuming that all sulfur is converted to hydrogen sulfide and removed by the gas cleaning system, hydrogen associated with hydrogen sulfide in the raw product gas

$= 0.02/16 = 0.00125$ kg/kg coal

Total hydrogen in the uncleaned dry product gas including that in hydrogen sulphide

$= 0.0591 + 0.00125 = 0.06035$ kg/kg coal

To find the moisture in the product gas we deduct the hydrogen in dry gas from the total hydrogen in flow obtained earlier using the hydrogen balance.

$=$ Hydrogen inflow $-$ hydrogen in dry product gas
$= 0.0792 - 0.06035 = 0.0189$ kg/kg coal

Steam associated with this hydrogen in the gas

$= 0.0189 \times 9$
$= 0.1701$ kg/kg coal

Oxygen balance
Oxygen associated with CO and CO_2 in dry gas which have half a mole and 1 mol of oxygen, respectively,

$$= (0.5 \times 0.275 + 0.035) \times 0.1478$$
$$= 0.0255 \text{ kg mol/kg coal}$$
$$= 0.0255 \times 32$$
$$= 0.8158 \text{ kg/kg coal}$$

Oxygen associated with the steam in the gas

$$= 0.1701 \times (8/9)$$
$$= 0.1512 \text{ kg/kg coal}$$

Total oxygen in gas

$$= 0.8158 + 0.1512$$
$$= 0.9670 \text{ kg/kg coal}$$

Here we note that this is slightly more than the oxygen inflow of 0.9035 kg/kg coal calculated earlier. This must be due to measurement errors in the given data on flue gas composition.

Carbon balance

The total carbon associated with CO, CO_2 and CH_4 in dry gas, whose production rate has been computed earlier as 0.1478 kg mol/kg fuel, is:

$$= (0.275 + 0.035 + 0.025) \times 0.1478$$
$$= 0.0495 \text{ kg mol/kg coal}$$
$$= 0.0495 \times 12 = 0.594 \text{ kg/kg coal}$$

Carbon input is found from the composition of the coal:

$$= 0.665 \text{ kg/kg coal}$$

Carbon conversion efficiency is found by dividing the carbon in the product gas with that in the fuel:

$$= (0.594/0.665) \times 100$$
$$= \mathbf{89.3\%}.$$

Energy Analysis

Table 3.5 gives heats of combustion for different gas constituents:
CO — 282.99 MJ/kg mol, Hydrogen — 285.84 MJ/kg mol; Methane — 890.36 MJ/kg mol.
Using this, we compute the heat of combustion of the gas.
Energy output with CO:

$$= 0.275 \times 0.1478 \text{ (kg mol CO/kg coal)} 282.99 \text{ (MJ/kg mol CO)}$$
$$= 11.50 \text{ MJ/kg coal}$$

Energy output with H_2:

$$= 0.15 \times 0.1478 \times 285.84$$
$$= 6.33 \text{ MJ/kg coal}$$

Energy output with CH_4:

Fluidized Bed Gasification

$$= 0.025 \times 0.1478 \times 890.36 = 3.29/\text{kg coal}$$

Total energy output or higher heating value of the product gas

$$= 11.5 + 6.33 + 3.29 = 21.12 \text{ MJ/kg coal}$$

Total energy input is equal to the heat in 1 kg of coal

$$= 30.24 \text{ MJ/kg coal}$$

Cold gas efficiency is found using Equation 3.11

$$= (21.12/30.24) \times 100$$
$$= 69.8\%.$$

3.8.7 Alternative Design Approach

An alternate method for the design of a fluidized bed gasifier is presented in the Coal Conversion Technical Data book (1978) of U.S. DOE. It is based on a series of design charts prepared for different pressures, temperatures and ratios of steam and carbon in the feed stream.

Example 3.3

> The gas produced by the gasifier in the previous problem is supplied to the burner at the gasifier temperature, 900°C to be burnt directly for cofiring in a boiler to generate steam. Find the hot gas efficiency of the gasifier.

In the previous problem, no information about sulfur in the gas is given as the sulfur is almost entirely removed from the gas during the cooling and cleaning process. For direct combustion of the gas, the sulfur content of the fuel appears as mostly hydrogen sulfide (H_2S) in the product gas. Since the gas is burnt directly without removing the H_2S, contribution of this component to the total energy of the gas has to be taken into account.

Sulfur content of the fuel = 0.02 kg/kg coal

Assuming all of it to be converted into hydrogen sulfide (H_2S), we find the number of moles of H_2S in the product hot gas:

$$= 0.02/32 = 0.000625 \text{ kg mol/kg coal}$$

Using Table 3.5, we get the heating value of H_2S in the gas:

$$= 0.000625 \times 562.6$$
$$= 0.35 \text{ MJ/kg coal}$$

The total heating value of the gas is the sum of the heating value of the gas itself and that of the H_2S:

$= 21.12 + 0.35$

$= 0.35 = 21.47$ MJ/kg coal

The product gas enters the burner at 900°C (1173 K). Using this temperature the enthalpy of the different components of the product gas can be estimated (by interpolation) from Table 3.6 and then added to give the enthalpy of the product gas at 900°C:

- CO: 27.52 (MJ/kg mol)0.1478 × 0.275 (kg mol/kg coal) = 1.119
- CO_2: 42.99 (MJ/kg mol)0.1478 × 0.035 (kg mol/kg coal) = 0.222
- H_2: 25.97 (MJ/kg mol)0.1478 × 0.015 (kg mol/kg coal) = 0.576
- N_2: 27.21 (MJ/kg mol) × 0.1478 × 0.0515 (kg mol/kg coal) = 2.094
- CH_4: 53.30 (MJ/kg mol)0.1478 × 0.025 (kg mol/kg coal) = 0.197
- H_2O: 33.32 (MJ/kg mol)(0.1584/18) (kg mol/kg coal) = 0.293
- H_2S: 36.07 (MJ/kg mol)(0.000625) (kg mol/kg coal) = 0.023

Total enthalpy of the gas = 4.524

To find the hot gas efficiency we use Equation 3.12.

Total thermal energy = heating value + enthalpy = 21.47 + 4.524 = 25.994 MJ/kg coal
Total gasifier efficiency = total thermal energy/heat in feedstock

$= 25.994/30.24$
$= 85.9\%$

3.9 MODELING OF FLUIDIZED BED GASIFICATION

Modeling is an important means for scale-up of a gasifier. It also helps the design of a unit based on results obtained from another gasifier operating on different feedstock. A good model will help identify the sensitivity of the performance of a gasifier to variation in different operating and design parameters. The designers can speculate the effects of many parameters even without any experimental data on them.

The models developed generally belong to one of the following three categories:

1. Kinetic
2. Equilibrium
3. Other

The kinetic model predicts the progress and product composition at different positions along a reactor. Tsui and Wu (2003) developed a kinetic model using kinetic data from previous literature. The results of their kinetic analysis show that all species approach their equilibrium values after traveling a certain distance, which was set as the height of the gasifier.

Souza-Santos (2004) has given a comprehensive account of the modeling of bubbling fluidized bed gasifiers using essentially the two-phase theory of fluidization.

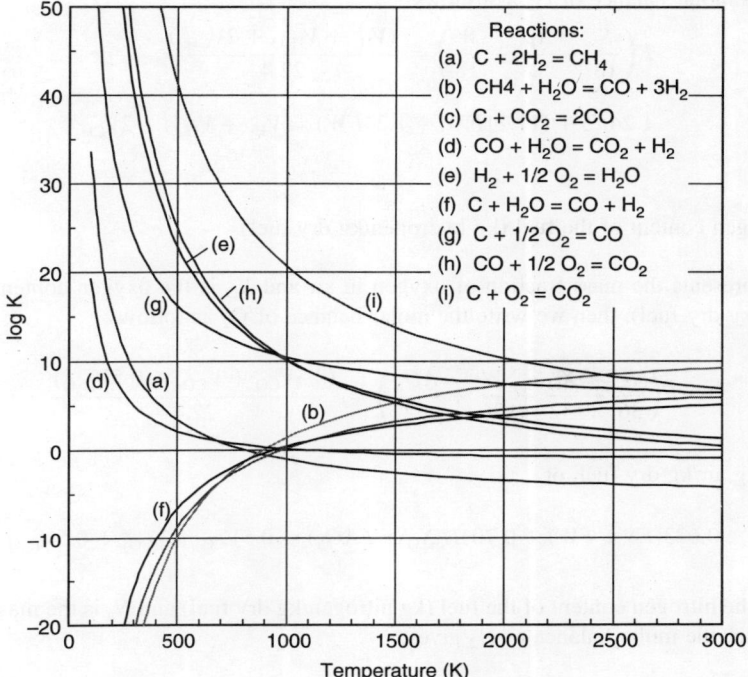

FIGURE 3.17 Calculated equilibrium constants for a number of gasification reactions. (From Li, Xuantian, personal communication, May 2005. With permission.)

A nonstoichiometric equilibrium model for CFB, developed by Li et al. (2004), predicts that the product gas composition depends primarily on the air ratio. Their experimental data suggest that real gasification processes deviate from chemical equilibrium. Li (2005) calculated thermodynamic equilibrium constants for a number of gasification reactions using standard JANAF thermodynamic data (Chase et al., 1985). His data were validated against data from Smith and Van Ness (1987). Equilibrium constants are plotted against temperature in Figure 3.17.

APPENDIX 3A

ESTIMATING EQUILIBRIUM GAS COMPOSITION

Let A denote the air supply in kg dry air/kg dry fuel, F the amount of dry fuel required to obtain one normal cubic meter of the gas, and X_c the carbon content of the fuel (kg carbon/kg dry fuel). Carbon is split between CO, CO_2, and CH_4. For 1 normal cubic meter of gas produced, one can write the carbon molar balance between inflow and outflow streams corresponding to:

$$\frac{FX_c}{12} = \frac{(V_{CO} + V_{CO_2} + V_{CH_4})}{22.4} \quad \text{or} \quad 1.866 FX_c = V_{CO} + V_{CO_2} + V_{CH_4} \quad \text{(A.1)}$$

where V represents the volumetric fraction of a constituent of the gas.

Hydrogen balance

Let S represent the total steam supplied as humidity associated with air and added steam (kg steam/kg of dry fuel), and W represent the moisture content of fuel (kg water/kg dry fuel).

We write the molar balance of H_2 as follows:

$$F\left(\frac{S}{18} + \frac{X_H}{2} + \frac{W}{18}\right) = \frac{(V_{H_2} + V_{H_2O} + 2V_{CH_4})}{22.4} \quad \text{or}$$

$$1.24FS + (11.21FX_H + 1.24FW) = V_{H_2} + V_{H_2O} + 2V_{CH_4} \quad \text{(A.2)}$$

where

X_H = Hydrogen content of the fuel (kg hydrogen/kg dry fuel)

O_2 balance

If O_a represents the mass fraction of oxygen in air and X_O is the oxygen content of the fuel (kg oxygen/kg dry fuel), then we write the molar balance of O_2 as follows:

$$F\left(\frac{S}{36} + \frac{X_O}{32} + \frac{W}{36} + \frac{AO_a}{32}\right) = \frac{(0.5V_{CO} + V_{CO_2} + 0.5V_{H_2O})}{22.4}$$

where A is kg air/kg dry fuel, or,

$$0.623(FS + FW) + 0.701(FX_O + FAO_a) = 0.5V_{CO} + V_{CO_2} + 0.5V_{H_2O} \quad \text{(A.3)}$$

N_2 balance

If X_N is the nitrogen content of the fuel (kg nitrogen/kg dry fuel) and N_a is the mass fraction of nitrogen in air, the molar balance of N_2 gives:

$$F\left(\frac{X_N}{28} + \frac{AN_a}{28}\right) = \frac{V_{N_2}}{22.4} \quad \text{or,} \quad 0.8FX_N + 0.8FAN_a = V_{N_2} \quad \text{(A.4)}$$

The volume fractions of all constituents of the product gas add up to 1.0. We, therefore, also have:

$$V_{CO} + V_{CO_2} + V_{H_2} + V_{CH_4} + V_{H_2O} + V_{N_2} = 1 \quad \text{(A.5)}$$

To estimate the values of the seven unknowns: $V_{CO}, V_{CO_2}, V_{H_2}, V_{H_2O}, V_{N_2}, V_{CH_4}$, and F we need a total of seven equations. For this purpose, besides the above five equations (Equation A.1 to Equation A.5), we can assume any two of the equations listed in Section 3.2.3 to be in equilibrium. We choose to work with Equation 3.4 and Equation 3.5.

For the Boudouard reaction Equation 3.5, the equilibrium constant is:

$$K_{pb} = \frac{P_{CO}^2}{P_{CO_2}}$$

where P_{CO} is the partial pressure of CO, which is equal to volume fraction of CO, ($V_{CO} \times$ the pressure of the reactor, P)

$$K_{pb} = \frac{(V_{CO}P)^2}{V_{CO_2}P} = \frac{(V_{CO})^2 P}{V_{CO_2}} \quad \text{(A.6)}$$

Similarly, for the water–gas reaction (Equation 3.4):

$$K_{pw} = \frac{P_{H_2}P_{CO}}{P_{H_2O}} = V_{H_2}V_{CO}\frac{P}{V_{H_2O}} \quad \text{(A.7)}$$

Solving equations (A.1) to (A.7) equilibrium concentrations of gases are found.

APPENDIX 3B

AN EXAMPLE OF A FLUIDIZED BED BIOMASS GASIFIER

The following data are taken from a circulating fluidized bed wood dust gasifier operating in a compress wood factory in China (Bingyan et al., 1994)

Fuel: Sawmill dust
Average size of fuel: 329 μm
Density: 430 kg/m^3

Analysis of fuel:

Carbon (%)	Hydrogen (%)	Oxygen (%)	Ash (%)	Moisture (%)	HHV (kJ/kg)
47.25	6.04	46.71	2.46	5	18,320

Operating Conditions:

Terminal velocity of particle — 0.4 m/sec
Fluidization velocity — 1.4 m/sec
Bed diameter — 0.41 m
Riser height — 4.0 m
Solid circulation — 1.73 kg/sec
Average residence time of solids — 8 sec
Equivalence ratio — 0.2
Fuel feed rate — 250 kg/h

Product:

ER	Gas Production (m^3/kg)	Temperature (°C)	CO$_2$ (%)	CO (%)	CH$_4$ (%)	C$_n$H$_m$ (%)	H$_2$ (%)	O$_2$ (%)	N$_2$ (%)	LHV (kJ/m^3)	Thermal Conversion (%)
0.224	1.72	919	15.8	16.1	7.32	1.2	13.24	1.1	44.8	7190	70

NOMENCLATURE

C_b: heating value of fuel (kJ/m^3)
C_p: specific heat of gas (kJ/kg K)
ER: equivalence ratio
$H_{sensible}$: sensible heat of flue gas (kJ/kg)
K_{ps}: equilibrium constant for water–gas shift reaction
K_{pw}: equilibrium constant for water–gas reaction
K_{pb}: equilibrium constant for Boudouard reaction
K_{pm}: equilibrium constant methanation reaction
M_b: fuel consumption rate (kg/sec)
P: pressure (Pa)
q_g: heating value of gas (kJ/m^3)
T_a: ambient temperature (°C)
T_g: gas temperature (°C)
T: temperature (°C)
V: bed volume (m^3)

V_{CO}: volume fraction of CO in the product mixture
V_g: gas generation rate (m³/sec)
V_{H_2}: volume fraction of H_2 in the product mixture
V_{H_2O}: volume fraction of H_2O in the product mixture
w_{out}: char flow-rate out of the gasifier (kg/sec)

GREEK SYMBOLS

η_{ceff}: cold gas efficiency
η_{geff}: hot gas efficiency
θ: char residence time (sec)
ρ_b: bed density (kg/m³)

REFERENCES

Beenackers, A. A. C. M., Biomass gasification in moving beds, a review of European technologies, *Renew. Energy*, 16, 1180–1186, 1999.

Bingyan, Xu, Zengfan, L., Chungzhi, W., Haitao, H., and Xiguang, Z., Circulating fluidized bed gasifier for biomass, In *Integrated Energy Systems in China — The Cold Northeastern Region Experience*, Nan, L., Best, G., Coelho, C., and De Carvalho, O., Eds., FAO, The United Nations, Rome, 1994.

Biomass Technology Group, http://www.btgworld.com/technologies/gasification.html#fluidized-bed.

Bucko, Z., Engelhard, J., and Wolff, A. G. J., 400 MWe IGCC Power Plant with HTW Gasification in the Czech Republic Z, *Gasification Technologies Conference*, San Francisco, CA, October 17–20, 1999 (also available at http://www.gasification.org/98GTC/gtc99240.pdf).

Chase, M. W. Jr., Davies, C. A., Downey, D. J. Jr., Frurip, D. J., McDonad, R. A., and Syverud, A. D., JANAF thermodynamic tables, 3rd ed. *J. Phys. Chem. Ref. Data*, 14(suppl. 1), 1–1856, 1985.

Chen, P., Zhao, Z., Wu, C., Zhu, J., and Chen, Y., Biomass gasification in circulating fluidized bed, In *Circulating Fluidized Bed Technology VIII*, Cen, K., Ed., International Academic Publishers, Beijing, pp. 507–514, 2005.

DOE, *Coal Conversion Systems Technical Data Book*, U.S. Department of Energy, Contract no. EX-76-C-0102286, HCP/T2286-01. p. IIIA.52.3, 1978.

DOE (Department of Energy, U.S.A.), *Gasification Technologies*, available at http://www.fe.gov/coal_power/gasification/index.html, 2001.

Elliott, M. A., Ed., *Chemistry of Coal Utilization*, J. Wiley and Sons, Toronto, pp. 1799, 1981, 2nd supplementary volume.

Garcia-Ibanez, P., Cabanillas, A., and Sanchez, J. M., Gasification of leached Ourjillo in a pilot plant circulating fluidized bed reactor — preliminary results, *Biomass Bioenergy*, 27, 183–194, 2004.

Hebden, D. and Stroud, H. J. F., Coal gasification processes, In *Chemistry of Coal Utilization*, Elliott, M. A., Ed., Wiley, Toronto, pp. 1642–1648, 1981.

Higman, C. and Burgt, M., *Gasification*, Elsevier, U.S.A., p. 105, 2003.

Kaiser, S., Loffler, G., Bosch, K., and Hofbauer, H., Hydrodynamics of a dual fluidized bed gasifier, Part II: simulation of solid circulation rate, pressure loop and stability, *Chem. Eng. Sci.*, 58, 4215–4223, 2003.

Li, Xuantian, Personal Communication, May, 2005.

Li, Xuantian, Grace, J. R., Lim, C. J., Watkinson, A. P., Chen, H. P., and Kim, J. R., Biomass gasification in a circulating fluidized bed, *Biomass Bioenergy*, 26, 171–193, 2004.

Maniatis, K, "Progress in Biomass gasification — an overview", EU Thermie Project. http://europa.eu.int/comm/energy/res/sectors/doc/bioenergy/km_tyrol_tony.pdf

Pfeifer, C. M., Pourmodjib, M., Rauch, R., and Hofbauer, H., Bed material and additive variations for a dual fluidized bed steam gasifier in laboratory, pilot and demonstration scale, In *Circulating Fluidized Bed Technology VIII*, Cen, K., Ed., International Academic Publishers, Beijing, pp. 482–489, 2005.

Quader, S. A., Natural gas substitutes from coal and oil, *Coal Sci. Technol.*, vol. 8 1985.

Ross, D., Noda, R., Adachi, M., Nishi, K., Tanaka, N., and Horio, M., Biomass gasification with clay minerals using a multi-bed reactor system, In *Circulating Fluidized Bed Technology VIII*, Cen, K., Ed., International Academic Publishers, Beijing, pp. 468–473, 2005.

Schimmoller, B. K., Coal gasification: striking while the iron is hot, *Power Eng.*, 30–40, March, 2005.

Smith, J. M. and Van Ness, H. C., *Introduction to Chemical Engineering Thermodynamics*, 4th ed., McGraw-Hill, Toronto, in Figure 15.3, p. 509, 1987.

Smith, P. V., Vimalchand, P., Gbordzoe, E., and Longanbach, J., Initial Operation of the PSDF Transport Gasifier, *16th International Conference on Fluidized Bed Combustion*, May 13–16, 2001, Reno, NV, U.S.A., 2001.

de Souza-Santos, M. L., *Solid Fuels Combustion and Gasification*, Marcell Dekker, New York, 2004, pp. 264–365.

Squires, A. M., Three bold exploiters of coal gasification: Winkler, Godel and Porta, In *Fluidised Beds: Combustion and Applications*, Howard, J. R., Ed., Applied Science Publishers, London, 1983.

Stahl, K., Neergaard, M., and Nieminen, J., "Progress Report: Varnamo Biomass Gasification Plant", *Proceedings Gasification Technologies Conference*, October 30 http://www.gasification.org/98GTC/gtc99290.pdf, 2001.

Stultz, S. C. and Kitto, J. B., *Steam — Its Generation and Uses*, 40th ed., Babcock & Wilcox, Barberton, chap. 17, 1992.

Tsui, H. and Wu, C., Operating concept of circulating fluidized bed gasifier from the kinetic point of view, *Powder Technol.*, 132, 167–183, 2003.

Williams, A., Pourkashanian, M., Jones, J. M., and Skorupska, N., *Combustion and Gasification of Coal*, Taylor & Francis, New York, pp. 222–225, 2000.

4 Combustion

The cost of fuel constitutes a major part of the operating cost of a fluidized bed (FB) boiler. A gain of even 0.5–1.0% in the combustion efficiency can save a large amount of money in terms of the operating cost of the boiler over its lifetime. Since the expenditure on fuel is much greater than that on sorbents, the impact of combustion efficiency on the operating cost is greater than that of the sorbent-utilization performance of the boiler. A qualitative as well as quantitative understanding of the process of combustion is indeed very important for a rational boiler design.

The combustion process may be defined as follows:

> Combustion is an exothermic oxidation process occurring at a relatively high temperature.

According to the old rule, Time, Temperature and Turbulence are three major requirements of a good combustion process. These three requirements are adequately met in an FB combustor, whose excellent internal and external recirculation of hot solids at the combustion temperature provides a long residence time and adequate temperature to the fuel particles. The high degree of gas–solid mixing in the FB furnace also provides the turbulence necessary for good combustion.

Since coal is the principal fuel fired in most FB boilers, the present discussion will focus primarily on coal. The combustion process of coal particles will be discussed with specific reference to the FB furnace. This chapter also presents a brief note on design considerations for bubbling and circulating FB boilers.

4.1 STAGES OF COMBUSTION

A particle of solid fuel injected into an FB undergoes the following sequence of events:

1. Heating and drying
2. Devolatilization and volatile combustion
3. Swelling and primary fragmentation (for some types of coal)
4. Combustion of char with secondary fragmentation and attrition

These processes are shown qualitatively in Figure 4.1, which shows the order of magnitude of time taken by each step.

4.1.1 HEATING AND DRYING

The combustible material generally constitutes around 0.5–5.0% by weight of the total solids in an FB combustor. The remaining solids, known as bed materials, are fuel ash, sorbents or some other noncombustible solids; and they constitute 95–99.5% of the bed. Thus, a large body of noncombustible hot solids engulfs a fresh fuel particle as soon as it drops into an FB combustor. These particles heat the cold fuel particles to a temperature close to that of the bed. The rate of heating may vary from 100°C/sec to more than 1000°C/sec, depending upon several factors, including the fuel particle size. The heat transfer to a fuel particle of size d_V from a fast bed of fine

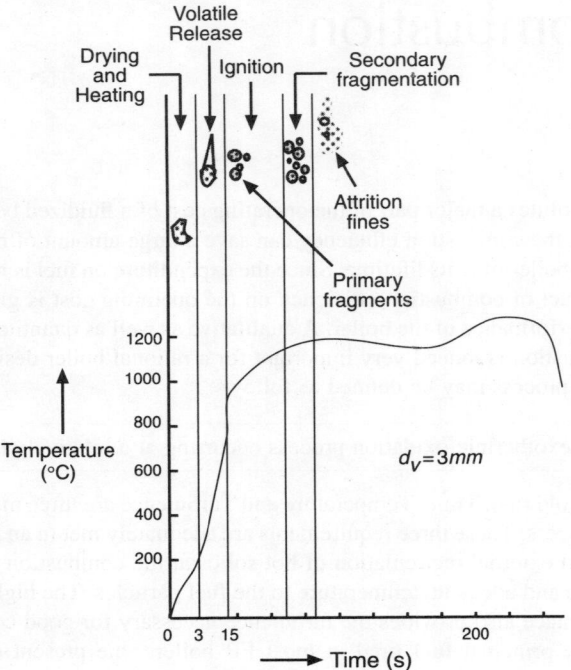

FIGURE 4.1 Sequence of events in the combustion of a coal particle.

solids with size d_P may be estimated using the following expression from Halder (1989):

$$\mathrm{Nu} = 0.33\, \mathrm{Re}^{0.62} \left(\frac{d_v}{d_p}\right)^{0.1} + K_g e_p \frac{\sigma(T_b^4 - T_s^4)}{d_v(T_b - T_s)}; \quad 5 < d_v < 12 \text{ mm}, \ 900 < \mathrm{Re} < 2500 \quad (4.1)$$

where: e_p is the emissivity of the fuel; σ, the Stefan–Boltzman constant; T_s and T_b, the temperatures of the coarse fuel particle and bed, respectively.

4.1.2 Devolatilization

Devolatilization or pyrolysis is the process where a wide range of gaseous products are released through the decomposition of fuel. The volatile matter (VM) comprises a number of hydrocarbons, which are released in steps. Figure 4.2 shows the constituents of the VM of a coal particle as they are released in stages. The first steady release usually occurs at around 500–600°C and the second release occurs at around 800–1000°C. Although the proximate analysis provides an estimate of the VM, the actual yield of VM and its composition may be affected by a number of factors like:

- Rate of heating
- Initial and final temperature
- Exposure time at the final temperatures
- Particle size
- Type of fuel
- Pressure

The weight fraction (based on the original weight of the fuel) of VM "V" evolved up to a time t is obtained by integrating the amount of individual components having different activation

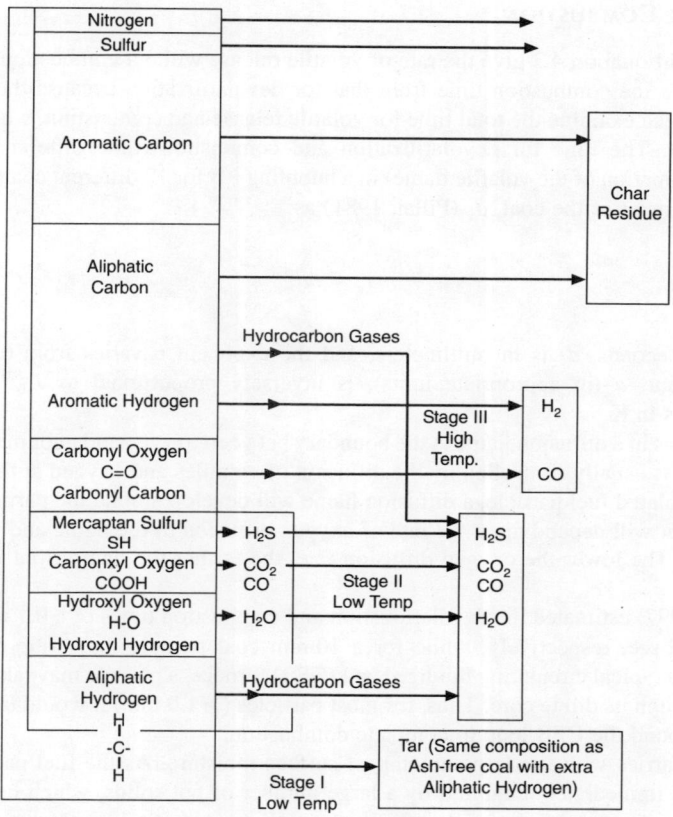

FIGURE 4.2 Sequence of volatile release showing the release of different constituents of volatiles during different stages of devolatization.

energies. Assuming the activation energies to have a Gaussian distribution with a mean value E_0 and a standard deviation δ the VM released at a time t is given by (Anthony et al., 1975);

$$\frac{V^* - V}{V^*} = \frac{1}{\delta\sqrt{2\pi}} \int_{-\infty}^{\infty} \exp\left[K_0 \int_0^t \exp\left(\frac{-E}{RT_s}\right) dt - \frac{(E - E_0)^2}{2\delta^2}\right] dE \quad (4.2)$$

This equation correlates fuel devolatilization data using four parameters which include the asymptotic value of the volatile yield V^*, the mean activation energy E_0, the standard deviation of Gaussian distribution of activation energy δ and the preexponential factor of the volatile release rate K_0. The VM content of the coal, as determined by proximate analysis, may give the asymptotic value V^*. LaNauze (1985) used the shrinking core model to present a simpler expression to find the fraction of VM released x at time t:

$$\frac{t}{\Gamma} = 1 - 3(1 - x)^{2/3} + 2(1 - x) \quad (4.3)$$

Here, Γ is the time for complete devolatilization, which is proportional to the square of the diameter of the coal particle. The total yield of volatiles increases with temperature.

4.1.3 VOLATILE COMBUSTION

Equation 4.2 and Equation 4.3 give the rate of volatile release without combustion. However, it is difficult to isolate the combustion time from that for devolatilization because the two processes overlap. We instead examine the total time for volatile release and combustion, which is of greater practical interest. The time for devolatilization and combustion (time difference between the initiation and extinction of the volatile flame) in a bubbling FB for 12 different coals t_v is correlated to the initial diameter of the coal, d_v (Pillai, 1981) as;

$$t_v = a d_v^p \qquad (4.4)$$

where: t_v is in seconds, d_v is in millimeters, and the exponent p varies from 0.32 to 1.8. The empirical constant, a (in appropriate units), is inversely proportional to $T_b^{3.8}$ when the bed temperature T_b is in K.

The VM burns in a diffusion flame at the boundary between oxygen and unburned volatiles. The combustion rate is usually controlled by the diffusion of volatiles and oxygen at their interface. In the case of an isolated fuel particle, a diffusion flame will develop around the particle, the position of the flame-front will depend upon the rate of oxygen diffusion to the flame and upon the rate of volatile release. The lower the oxygen diffusion rate, the farther the flame-front is from the coal surface.

Brereton (1997) estimated the devolatilization and combustion times of a 0.1 mm coal particle to be 22 and 32 sec, respectively, while for a 10 mm coal particle they were 50 and 655 sec, respectively. In a typical circulating fluidized bed (CFB) furnace, a particle may take between 3 and 6 sec to rise through its dilute core. Thus, for most particles (>1.0 mm) it would take at least more than one trip around the CFB loop to complete combustion.

Fuel often carries with it a large amount of surface moisture. As the fuel particles drop into the bed they are immediately engulfed by a large number of hot solids, which heat up and drive the moisture away from the fuel particles. Since, the particle-to-particle heat transfer is very high in an FB, the drying and heating processes are very rapid.

The VM in coal contains many chemicals whose release rates are different. The slowest releasing species in the volatiles might be the CO (Keairns et al., 1984), yet this rate is at least an order smaller than that of the volatiles combustion. A small coal particle, say 3 mm in diameter, could take about 14 sec to devolatilize at 850°C while the released VMs burn in just a few seconds (Basu and Fraser, 1991). Thus, the combustion rate of volatiles is usually controlled or defined by its release rate.

Example 4.1

Estimate the time taken by a 3 mm coal particle to complete its devolatilization and volatile combustion in an FB operating at 850°C. Given $p = 1.5$ and $a = 3.5$ in Equation 4.4 at 775°C.

Solution

We assume that the particle is heated to the bed temperature before devolatilization starts.

$$775°C = 775 + 273 = 1048 \text{ K}$$

$$850°C = 580 + 273 = 1123 \text{ K}$$

We interpolate the value of a at 850°C to find;

$$a_{850°C} = 3.5 \times (1048/1123) = 2.69$$

From Equation 4.4

$$t_v = 2.69 \times 3^{1.5} = 14 \text{ sec}$$

Volatiles from the coal complete combustion in 14 sec.

4.1.4 CHAR COMBUSTION

The devolatilized fuel, known as char, burns rather slowly. For example, it would take 50–150 sec for a char of size less than 0.2 mm to burn out (Keairns et al., 1984). Since it takes this long to burn completely, there is a chance that some of the particles may not burn out in the bubbling bed before leaving. The elutriation of these unburnt, fine char particles results in combustion losses. As the combustion of char takes much longer than volatilization, the char particle has adequate time to spread around and thus burn throughout the body of the bubbling bed. All these concerns make the study of char combustion very important.

The combustion of a char particle generally starts after the evolution of volatiles from the parent fuel particle, but sometimes the two processes overlap. During the combustion of a char particle, oxygen from the bulk stream is transported to the surface of the particle. The oxygen then undergoes an oxidation reaction with the carbon on the char surface to produce CO_2 and CO. The char, being a highly porous substance, has a large number of internal pores of varying size and tortuosity. Surface areas of the pore walls are several orders of magnitude greater than the external surface area of the char. Oxygen, under favorable conditions, diffuses into the pores and oxidizes the carbon on the inner walls of the pores.

The mechanism of combustion of char is fairly complex. Some of the factors influencing it are discussed below.

4.1.4.1 Reaction Product

As mentioned earlier, the primary products of combustion on the carbon surface could be both carbon monoxide and carbon dioxide, according to the following equations:

$$C(12 \text{ kg}) + O_2(32 \text{ kg}) = CO_2(44 \text{ kg})$$

$$C(12 \text{ kg}) + \tfrac{1}{2}O_2(16 \text{ kg}) = CO(28 \text{ kg})$$

(4.5)

Ratio of the production rates of carbon monoxide and carbon dioxide, $[CO]/[CO_2]$, depends on the surface temperature of the fuel and is given by the following relation (Arthur, 1951):

$$\frac{[CO]}{[CO_2]} = 2400 \exp\left[\frac{-51830}{8.31 T_s}\right]$$

(4.6)

where: T_s is the temperature of the carbon surface.

At temperatures above 1000°C, the reaction product is generally CO, which in turn burns to CO_2 at some distance away from the char surface. However, at atmospheric pressures, CO burns so close to the surface of char particles for ($d_c > 1$ mm) that the primary combustion product may be considered to be CO_2.

4.1.4.2 Regimes of Combustion

The combustion of char is a two-step process:

1. Transportation of oxygen to the carbon surface
2. Reaction of carbon with oxygen on the carbon surface

The combustion of char may occur on its external surfaces as well as inside its pores depending on the regime under which it is burning. There are three regimes of combustion, which are governed by the relative rates of oxygen transport (mass-transfer rate) and the chemical reaction (kinetic rate) on the carbon surface. These rates are influenced by the combustor's operating condition, as well as on the characteristics of the char particle. These regimes are compared in Figure 4.3 and Table 4.1. Figure 4.3 plots the reaction rate qualitatively against combustion temperature while showing the probable spatial distribution of oxygen concentration in and around a burning char particle.

4.1.4.2.1 Regime I

In this regime, the chemical kinetic rate is much slower than the mass-transfer (diffusion) rate. Therefore, the combustion is kinetic-controlled and the oxygen gets adequate time to enter all the pores of a porous char particle, where it burns uniformly. As a result, the density of the particle, rather than its diameter, decreases as the combustion progresses, making it a diminishing-density burning. It can be seen in Figure 4.3 that in Regime I, the oxygen concentration is uniform throughout the char.

On nonporous particles, this regime would occur at temperatures around 900°C, while on porous coarse particles, it would occur at temperatures above 600°C. For fine porous particles, where the mass-transfer rate is high, combustion could be in this regime. Typical situations when this regime is prevalent are:

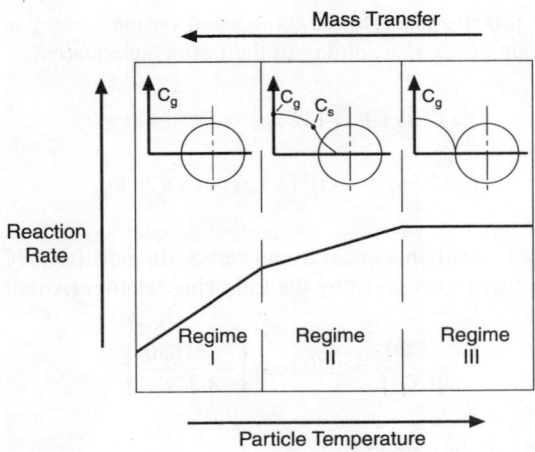

FIGURE 4.3 Rate-controlling regimes for heterogeneous char oxidation.

TABLE 4.1
Regimes of Combustion of Char

Combustion Regime	I	II	III
Reaction order	$n = m^a$ (Smith, 1982)	$n = (m + 1)/2$	
Controlling rate	Kinetic	Pore diffusion	External diffusion
Apparent activation energy	True value	(True value)/2	
Probable situation of occurrence in FB furnace	Start-up	Fast and bubbling	Fast and bubbling

[a] n and m are the apparent and true orders of reactions, respectively.

1. During light-up in an FB boiler when the temperature is low and consequently the kinetic rate is very slow
2. In fine particles where the diffusion resistance is very small

4.1.4.2.2 Regime II

Here, both reaction and pore diffusion rates are comparable, allowing a limited penetration of oxygen into the pores near the external surface.

This condition of combustion occurs for medium-sized char particles in bubbling FBs and also in some parts of the CFB where mass transfer to the pores is comparable to the reaction rate.

4.1.4.2.3 Regime III

This condition occurs when the mass-transfer rate is very slow compared to the kinetic rate, or the kinetic rate is so fast that oxygen reaching the external surface is immediately consumed by the char on the surface. This type of combustion is sometimes called diffusion-controlled combustion. It occurs in large particles and where the mass transfer is small compared to the reaction rate.

4.1.5 BURNING RATE OF CHAR IN FLUIDIZED BEDS

The burning rate of a char particle is the instantaneous rate of oxidation of the char per unit time and may be expressed in kg/sec. The specific burning rate (kg/m^2 sec), gives the rate of oxidation of the char particle per unit external surface area of the particle.

The burning rate depends on a number of factors including the particle size, reactivity of the char, and oxygen concentration. The bed hydrodynamics affect the mass transfer of oxygen and therefore the burning rate. A circulating FB is a great deal more agitated and better mixed than a bubbling FB. For that reason, the burning rate of char particles in a fast bed is higher than that in a bubbling bed (Basu and Halder, 1989). Figure 4.4 compares the burning rates of a given sized char particle in a CFB (fast bed) with that in a bubbling bed.

From basic mass transfer we know that the mass-transfer rate to a particle size increases with decreasing size of the particle (see Equation 4.8), but the reaction rate is independent of the particle size. As combustion progresses, the size of the char reduces due to combustion. Therefore, the mass-transfer rate increases, but the kinetic rate remains steady. Thus, according to Figure 4.3 the combustion mechanism moves from Regime III to Regime II and then to Regime I conditions.

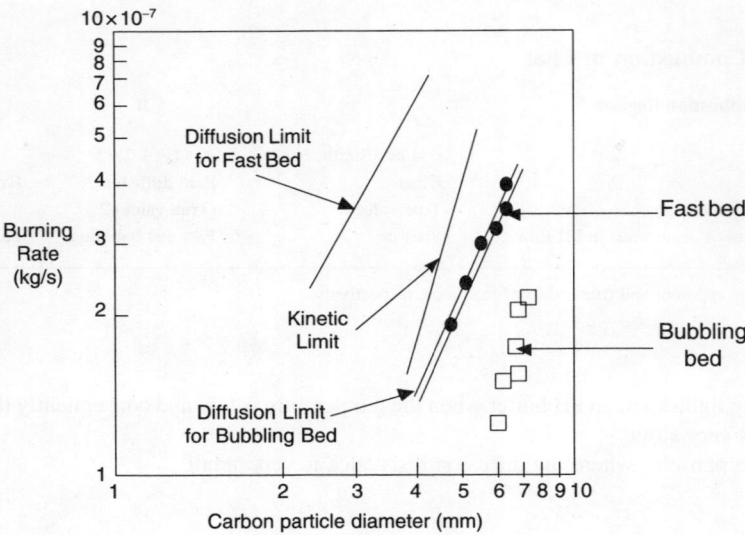

FIGURE 4.4 Comparison of burning rates of carbon in a bubbling bed with those in a fast fluidized bed.

Due to this, a single expression for the burnout time of char particles may be misleading. However, the following expressions derived for the instantaneous burning rate of char of a given size d_c is valid for all regimes.

4.1.5.1 Mass-Transfer Rate

The mass of carbon oxidized (through transfer of oxygen to the carbon surface) per unit external area of the particle, per unit time (specific burning rate) q is given as;

$$q = h_m(C_g - C_s) \qquad (4.7)$$

where:

$$h_m = \frac{12\phi Sh D_g}{d_c R T_m} \qquad (4.8)$$

and C_s and C_g are partial pressures of oxygen (kPa) on the external surface of the char and in the bulk gas, respectively; h_m is the mass-transfer coefficient, kg(C)/m² kPa² s; D_g, the diffusivity of oxygen in the flue gas, m²/sec; R, the universal gas constant (8.314 kPa m³/kmol K); Sh, the Sherwood number; 12 is the molecular weight of carbon [kg(C)/kmol]; ϕ is the mechanism factor, which is equal to 1 when the primary product of combustion is CO_2 and 2 when the primary combustion product is CO.

If the combustion is controlled by diffusion, as in Regime III, the maximum oxygen transfer rate m_d can be obtained by setting the oxygen concentration on the carbon surface C_s to zero. This rate is called the diffusion limit;

$$m_d = h_m C_g \qquad (4.9)$$

No comprehensive information on the Sherwood number Sh in fast FBs is available at the moment. The experimental correlation of Halder and Basu (1988) for coarse particles may be used for a first

Combustion

approximation in the absence of more comprehensive one:

$$\text{Sh} = 2\varepsilon + 0.69\left(\frac{\text{Re}}{\varepsilon}\right)^{0.5} \text{Sc}^{0.33} \qquad (4.10)$$

where: $\text{Re} = [(U_g - U_c)d_c\rho_g/\mu]$; Schmidt number, $\text{Sc} = (D_g\rho_g/\mu)$; ε, the local bed voidage; μ, the viscosity of gas; U_g and U_c, the velocities of the gas and the char particle, respectively.

The above relation is valid for char particle size d_c between 4 and 12 mm and average bed particle size d_p between 100 and 350 μm. Until a more reliable expression for U_c is available, one could use Equation 2.40 to calculate the slip velocity between the char particle and gas, $(U_g - U_c)$. For $d_c \approx d_p$, the slip velocity may have a value between U_t and U_{slip} defined by Equation 2.40.

4.1.5.2 Reaction Rate

The specific rate of chemical reaction of carbon with oxygen per unit external area, or the specific burning rate of the char particle may be written as:

$$q = R_c C_s^n \qquad (4.11)$$

where: n is the apparent order of the reaction and R_c is the reaction rate of carbon, based on the external surface area of the char [kg(C)/m^2 sec (kPa)^n].

The specific burning rate depends on several parameters including the intrinsic rate R_i of chemical reaction on internal pore surfaces of the char, the quantity of pore surfaces, and the extent to which oxygen diffuses into the pores. This relationship is given later by Equation 4.15.

By eliminating C_s from Equation 4.7 and Equation 4.11, one can obtain an expression for burning rate q in terms of C_g, reaction rate R_c, and the maximum diffusion rate m_d:

$$R_c = \frac{q}{\left[C_g\left(1 - \dfrac{q}{m_d}\right)\right]^n} \qquad (4.12)$$

For a first-order reaction ($n = 1$) the burning rate, q can be simplified in terms of oxygen concentration in the bulk stream C_g as:

$$q = \frac{C_g}{\dfrac{1}{h_m} + \dfrac{1}{R_c}} \qquad (4.13)$$

The chemical reaction rate coefficient R_c is generally given in the Arrhenius form;

$$R_c = A' \exp\left(-\frac{E}{RT_s}\right) \text{ kg/m}^2\text{sec (kPa)}^n \qquad (4.14)$$

where: A' is the frequency factor; E, the apparent activation energy; R, the universal gas constant (8.314 kJ/kmol K); and T_s, the absolute temperature (K) of the particle.

For average coals, the mean values of A' and E for a first-order reaction ($n = 1$) may be roughly taken as 8 kg/m^2 sec (kPa)^n and 90,000 kJ/kmol, respectively. However, because of the influence of numerous parameter on these values for reliable combustion predictions one should carry out experiments to measure the reactivity of the coal used. Table 4.2 lists experimental values of rate parameters for several fuels indicating the wide variation in the reactivities.

Under Regime I conditions (typically $d_c < 100$ μm and $T_s < 700$–800°C) the combustion takes place on the interior pore surfaces of the char. To express this burning rate in terms of the

TABLE 4.2
Reactivities, R_c of Various Fuel Particles Reactivity as Defined in Equation 4.14

Reference	Density of Carbon (kg/m³)	A' kg/m² sec (kPa)^n	E/R (K)	Order of Reaction n	Fuel
Hamor et al. (1973)	440	0.918	8200	0.5	Brown coal char
Smith (1971a)	1360	2.013	9600	1.00	Semi-anthracite
Smith and Tyler (1972)	1320	5.428	20,100	1.00	Semi-anthracite char
Sergeant and Smith (1973)	760	2.902	10,300	1.00	Bituminous char
Howard and Essenhigh (1967)	N/A	—	3000–6000	1.0	Bituminous char
Daw and Krishnan (1983)	N/A	0.0404	5787	0.6	Illinois coal (Bit.)
Basu and Wu (1993)	N/A	20.57	9600	0.5	Prince coal (Bit.) ($T_s < 1000$ K)
Basu and Wu (1993)	N/A	0.0263	3106	0.5	Prince coal (Bit.) ($T_s > 1000$ K)
Young and Smith (1981)	1640 and 1850	7.0	9911	0.5	Petroleum coke
Field et al. (1967)	N/A	859.0	17,976	1.0	Various char
Smith (1971b)	1640–1850	1.97	9058	1^a	Petroleum coke
Smith (1971b)	1500	0.99	8555	1^a	Anthracite char
Smith (1971b)	450–7900	0.79	8052	1^a	Bituminous char
Halder and Basu (1987)	1640	1.10	8125	0.4	Electrode carbon
Essenhigh et al. (1965)	N/A	—	20,000	0 ($T_s < 1000$ K) 1.0 ($T_s > 1000$ K)	Carbon

N/A — not available.

TABLE 4.3
Porosity of Some Typical Char and Carbon

Type of Char	Char density (ρ_{ch}) (kg/m³)	Carbon density (ρ_c) (kg/m³)	Porosity	Pore surface area per unit char volume (m² × 10⁶/m³)
Activated carbon	1478	1393	0.128	535
Lignite char	1883	1130	0.132	147
Illinois 5 char (1)	1800	1368	0.120	17.8
Merthyr Vale char	1765	1483	0.121	5.9
Illinois 5 char (20)	1689	1385	0.125	1.5
Phurnacite breeze	1741	1393	0.128	0.95
Tymawr char	1917	1476	0.137	0.50
Coke (Ross[2])	1608	1447	0.210	0.42

$\rho_c = \rho_{ch} \times$ (% C as received)/100. Pore surface area is determined by absorption of N_2 at 77 K (NCB).
Source: Adapted from Turnbull et al., *Chem. Eng. Res. & Dev.*, 62, p. 225, 1984.

external surface area, we use the following expression of specific burning rate, q as:

$$q = R_i \beta A_g \tau \rho_c C_g^m \text{ kg/m}^2 \text{s} \qquad (4.15)$$

where: R_i is the intrinsic reaction rate based on the pore surface area (kg/m² sec (kPa)^n); β, the effectiveness factor denoting the degree of penetration into the pores (= 1.0 for complete penetration); τ, the ratio of char volume and its external surface; ρ_c, the char density, (kg/m³); and A_g, the pore surface area per unit mass of the char (m²/kg). The pore surface area increases as the pores enlarge with combustion. Thus, the specific burning rate will initially increase, but it will soon decrease due to coalescence of the pores. Typical pore surface areas of some char particles are given in Table 4.3.

A wide range of apparent reactivities are observed in different fuels due to variation in their pore structures and impurities. Smith (1982) correlated them by expressing the reaction rate per unit surface area (internal), q_i and by reducing all data to oxygen partial pressure of 1 atm:

$$q_i = 3050 \exp(-43,000/RT_s) \text{ kg/m}^2 \text{ sec} = R_i C_s^m \qquad (4.16)$$

Here, R is 1.986 kcal/kmol K and m is the true order of the reaction.

Although Equation 4.16 is the best fit for a wide range of data on reactivity, the scatter of the data should be kept in mind while using this equation.

Example 4.2

> One 8.4 mm char particle, produced from Prince coal of Nova Scotia, burns in a fast FB at 1073 K and at 7.8% oxygen concentration. The measured surface temperature of the particle is 1143 K and the fluidization velocity is 5 m/s. The average velocity of the char particle is 0.5 m/s. The bed voidage is 0.989. The diffusivity of oxygen in nitrogen is 1.88×10^{-4} m²/sec at 1100 K.
> Estimate the burning rate of the particle using kinetic rate for Prince Coal from Table 4.2.

Solution

First, we find the apparent kinetic rate based on the external surface area using Equation 4.14 with data for Prince coal from Table 4.2:

$$R_c = 0.0263 \exp(-3106/1143) = 1.73 \times 10^{-3}$$

The concentration of oxygen in freestream is:

$$C_g = 7.8\% = 0.078 \text{ atm} = 0.078 \times 101 \text{ kPa} = 7.87 \text{ kPa}$$

For the kinetic reaction limit,

$$q_c = R_c C_g^n = 1.73 \times 10^{-3} \times 7.87^{0.5} = 4.87 \times 10^{-3} \text{kg/m}^2 \text{ sec}$$

For the diffusion limit (from Equation 4.9)

$$m_d = h_m C_g$$

To calculate, h_m, we need to use Equation 4.8 and Equation 4.10.
The Reynolds number is calculated below by taking property values of the gas at 1073 K.

$$\mu = 4.5 \times 10^{-5} \text{ kg/m sec}$$

$$\rho_g = 0.316 \text{ kg/m}^3$$

$$D_g = 1.88 \times 10^{-4} \text{ m}^2/\text{sec}$$

$$\text{Re} = \frac{(U_g - U_c)d_c \rho_g}{\mu} = (5 - 0.5) \times 0.0084 \times 0.316/(4.5 \times 10^{-5}) = 265$$

The average gas film temperature, T_m is $T_m = (1073 + 1143)/2 = 1108$. For using Equation 4.10 we calculate the Schmidt number:

$$\text{Sc} = \frac{D_g \rho_g}{\mu} = \frac{0.000188 \times 0.316}{0.000045} = 1.32$$

From Equation 4.10:

$$\text{Sh} = 2 \times 0.989 + 0.69 \times (265/0.989)^{0.5} \times 1.32^{0.33} = 14.36$$

Assume CO_2 to be the surface product for coarse particles, so, $\phi = 1$. Using Equation 4.8

$$h_m = \frac{12 \phi \text{Sh} D_g}{d_c R T_m} = \frac{12 \times 1 \times 14.36 \times 0.000188}{0.0084 \times 1108 \times 8.314} = 4.18 \times 10^{-4}$$

The diffusion limit is found from Equation 4.9 to be:

$$m_d = 4.18 \times 10^{-4} \times 7.87 = 3.29 \times 10^{-3} \text{kg/m}^2 \text{ sec}$$

We note that the diffusion limit $m_d = 3.29 \times 10^{-3}$ and kinetic limit $q_c = (4.87 \times 10^{-3})$ are of the same order of magnitude. This suggests that the reaction occurs in Regime II. For Prince coal the overall reaction is of half order. (Table 4.2). The overall burning rate was found earlier from

Equation 4.12:

$$R_c = \frac{q}{[C_g(1 - q/m_d)]^n} = 1.73 \times 10^{-3} = \frac{q}{[7.87(1 - q/3.29 \times 10^{-3})]^{0.5}}$$

Solving for q, we get the specific burning rate:

$$q = 2.45 \times 10^{-3} \text{ per m}^2 \text{ sec}$$

4.1.6 Communition Phenomena during Combustion

The size of a large coal particle could be reduced through combustion and other comminution processes, including fragmentation and attrition. These processes generate fines that generally escape through the cyclone or the freeboard, constituting a major combustibles loss. These concurrent processes (Figure 4.5) exert an important influence on the combustion process in an FB. A brief discussion of these phenomena is presented below.

4.1.6.1 Swelling

Caking coals of intermediate rank pass through a plastic phase at 420–500°C during volatilization, when their pores break down. The particle surface area is, therefore, lowest at the onset of the devolatilization, but gases released from the interior of the coal particle cause it to swell. In some cases, a balloon-like censophere is formed due to uniform swelling.

After this, the particle could undergo the process of fragmentation, which could be of three possible types: primary, secondary or percolative.

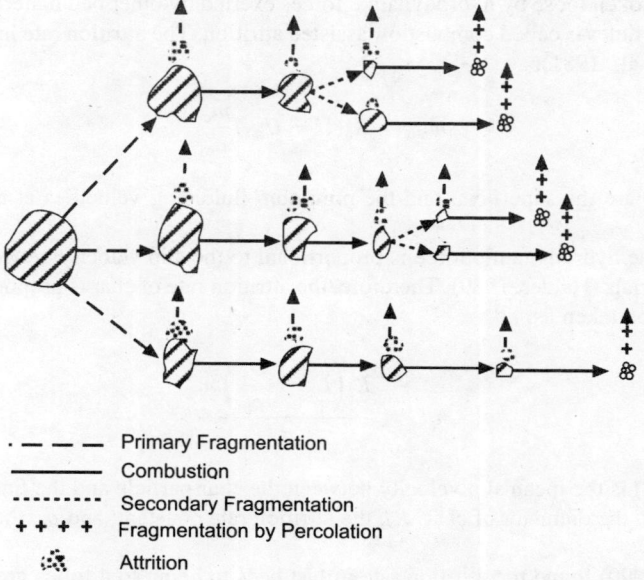

- — — — Primary Fragmentation
- ———— Combustion
- ·········· Secondary Fragmentation
- + + + + Fragmentation by Percolation
- ∴ Attrition

FIGURE 4.5 Communition process during the combustion of coal particles.

4.1.6.2 Primary Fragmentation

The volatile gases released inside nonporous coal particles exert a high internal pressure that sometimes breaks the coal into fragments. This phenomenon is called primary fragmentation. Here, a coal particle is broken into several pieces that are smaller but comparable to the size of the parent coal particle.

4.1.6.3 Secondary Fragmentation

When char burns under Regime I or II conditions, the internal pores of the char increases in size, thus weakening the bridge connecting carbon islands inside the char. When a bridge is too weak to withstand the hydrodynamic force on the char, a fragment breaks loose. This process is called secondary fragmentation. The fragments produced by secondary fragmentation are an order of magnitude larger in size than the attrited carbon fines.

4.1.6.4 Percolative Fragmentation

If the coal burns under the Regime I condition, i.e., combustion occurring uniformly throughout the char body, a sudden collapse of all internal bridges may occur, giving rise to the special type of secondary fragmentation known as percolative fragmentation.

The fragmented char particles may also experience further division into finer particles in an FB due to the abrasive actions of bed materials.

4.1.6.5 Attrition

This is the process of production of fines (typically <100 μm) from relatively coarse particles through mechanical contact like abrasion with other particles.

Attrition is greatly enhanced by combustion. Macerals of varying reactivities are present in the char. This causes an uneven oxidation or combustion on the char surface (D'Amore et al., 1989). Thus, some parts of the char surface burn faster than others, leaving fine ridges on the surface. These ridges are broken loose by hydrodynamic forces exerted by other bed materials. This type of attrition of char particles is called combustion-assisted attrition. The attrition rate in bubbling FBs is given as (Donsi et al., 1981):

$$m_{attr} = K_a(U - U_{mf})\frac{m_c}{d_c} \qquad (4.17)$$

where: U and U_{mf} are the superficial and the minimum fluidizing velocities at bed temperature, respectively.

In a fast FB, the hydrodynamic force is proportional to the slip velocity between char and the abrading bed materials (Halder, 1989). Therefore, the attrition rate of char is proportional to the slip velocity and may be taken as:

$$m_{attr} = \frac{K_a\left(U_c - \dfrac{G_s}{\rho_b}\right)m_c}{d_c} \qquad (4.18)$$

Here, $(U_c - G_s/\rho_b)$ is the mean slip velocity between the char particle and the fine bed solids; m_c, the mass of char; d, the diameter of char; K_a, the attrition rate constant; and ρ_b; the bulk density of the bed.

Arena et al. (1990) found the attrition rate in fast beds to be up to 4 times greater than that in bubbling beds. If the attrition rate constant is calculated by Equation 4.17, its value is of the same order of magnitude as that in a bubbling bed where the slip velocity between the char and bed particles is taken as $(U_g - U_{mf})$ (Halder, 1989). This suggests that the attrition rate constant is independent of the types of fluidization.

4.2 FACTORS AFFECTING COMBUSTION EFFICIENCY

The combustion efficiency of a bubbling fluidized bed (BFB) boiler is typically up to 90% without fly-ash recirculation and could increase to 98–99% with recirculation (Oka, 2004). The efficiency of a circulating fluidized bed (CFB) boiler is generally higher due to its tall furnace and large internal solid recirculation. The efficiency depends to a great extent on the physical and chemical characteristics of the fuel as well as on the operating condition of the furnace. The following section discusses different factors that could influence the combustion efficiency. Factors affecting the combustion efficiency can be classified into three categories:

1. Fuel characteristics
2. Operational parameters
3. Design parameters

4.2.1 FEED STOCK

In conventional boilers, the fuel is characterized by a number of parameters including heating value and grindability. These characteristics are not adequate or even entirely relevant for fluidized bed (FB) boilers, because the combustion mechanism in these systems is different.

4.2.1.1 Fuel Ratio

The fuel ratio of a fuel is the ratio of fixed carbon (FC) and VM contents of the fuel. This ratio has an important effect on the combustion efficiency of coal in a CFB boiler (Tang and Engstrom, 1987) with higher ratios possibly leading to lower combustion efficiencies (Makansi, 1990). Experiments with seven different coals in a 50-MWe BFB boiler (Yoshioka and Ikeda, 1990) and 20 different coals in a 0.3 m × 0.3 m BFB pilot plant (Oka, 2004) provide clear confirmation of the above dependence. A high rank fuel like anthracite has a higher fuel ratio than a low rank fuel like lignite. For this reason one can see in Figure 4.6 that low-rank fuels (or low fuel ratio) like lignite and bituminous have higher efficiencies than anthracite.

The fuel ratio is easily computed from the proximate analysis of a fuel.

4.2.1.2 Attrition

It was explained earlier that the attrition of burning char particles makes a major contribution to the combustible loss. Sorbents, used for sulfur retention, also undergo attrition in the furnace. This affects the utilization efficiency of the sorbents for sulfur capture. Attrition also occurs during transportation of the fuel and sorbent from storage to the furnace. The combustion-assisted attrition of char and the mechanical attrition of sorbents could both be measured in a specially designed high-velocity FB as shown in Figure 5.7. For measuring the mechanical attrition during transportation through fuel-handling systems a separate instrument or facility is required.

4.2.1.3 Intrinsic Reactivity

Carbon fines, whose combustion rate controls the combustion efficiency, burn under the kinetic-controlled Regime I condition. Here, the intrinsic reactivity controls the burning rate and thus has a major influence on the combustion efficiency, sometimes irrespective of the volatile content of the coal.

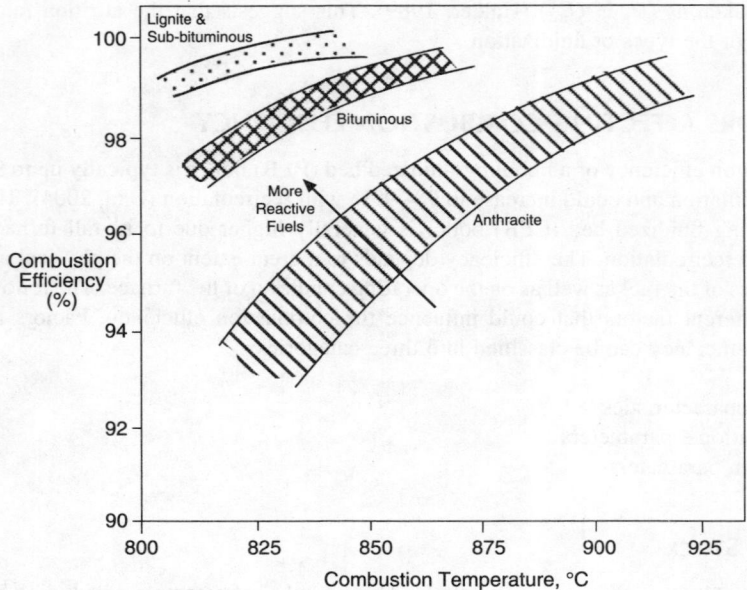

FIGURE 4.6 Effect of bed temperature and reactivity of fuel on the combustion efficiency of a bubbling fluidized bed. (Adapted from Tang and Engstrong, 1987.)

More reactive fuels have higher combustion efficiencies. The combustion temperature also has a positive influence on combustion efficiency. Figure 4.6 shows that at a given temperature, less reactive anthracite results in a lower combustion efficiency than that of the more reactive bituminous and lignite fuels burning in a CFB. It also shows that for all fuels, the combustion efficiency increases with combustion temperature, supporting the hypothesis that particles smaller than 100 μm, which are mainly responsible for unburnt combustible losses, burn under the kinetic-control regime.

The reactivity can be measured by burning a small sample of coal in a TGA (thermo-gravimetric analyzer), as shown in Figure 5.8. D'Amore et al. (1989) give details of the measurement technique.

4.2.1.4 Volatile Matter and Devolatilization

The volatile matter (VM), though it burns faster than char, could result in lower combustion efficiency due to improper mixing, prompting escape of CO from the furnace. High-volatile fuels would need very efficient mixing in the bed to ensure good combustion. Bituminous fuel would, therefore, give higher combustion efficiency than anthracite (Figure 4.6).

The yield, composition and rate of combustion of volatiles can be measured in an apparatus similar to that in Figure 5.7, which can simulate the rate of heating of the BFB or CFB furnace. Among other parameters, the particle size and temperature conditions affect the volatilization yield.

4.2.1.5 Fuel Particle Size

Many manufacturers of CFB boilers specify an upper limit for the fraction of fuel particles below 50 μm, because these particles, especially in anthracite or other less reactive fuels, escape the cyclone unburnt.

In bubbling bed boilers the feed size could be less than 10 mm, but the effect of feed size is not clear because finer particles have a higher probability of entrainment and a shorter residence time, but also have a shorter burnout time. For over-bed feeding, particles smaller than 1 mm exert the greatest negative influence on the combustion; as these particles are entrained instead of reaching the bed surface. Under-bed feedings may give higher efficiency, but some fines could still by-pass the bed through the bubbles.

4.2.1.6 Swelling

The standard ASTM swelling test, as used for coals to be burned in conventional coal-fired boilers, may not be adequate for FB boilers due to their different hydrodynamic and heat-transfer conditions. Specialized tests for large pieces of coal in a high-velocity FB must be used.

4.2.1.7 Fragmentation

The fragmentation of coal and sorbents greatly affects the combustion and sulfur-capture behavior of the furnace. Mechanical and thermal shock may influence theses properties, so this characteristic must be measured in a turbulent FB, which to some extent simulates the physical condition of the fast bed (Figure 5.7).

4.2.1.8 Ignition and Agglomeration

The ignition temperature influences the bed start-up condition. The lower the ignition temperature, the shorter the start-up time and the lower the auxiliary fuel consumption.

Bed agglomeration, leading to clinkering, could be a problem in bubbling bed boilers. Clinker formation leads to particle growth and hence alters the size distribution of particles in the beds. Thus, ignition as well as agglomeration temperatures of the coal (and its ash) need to be measured with an appropriate apparatus. The details of the measurement of the agglomeration temperature are given in Basu and Sarkar (1983). The initial sintering temperature of the fuel ash determines the agglomeration potential of the bed, thus, a proper design calls for data on this temperature.

4.2.2 OPERATING CONDITIONS

4.2.2.1 Fluidizing Velocity

The combustion efficiency generally decreases with increasing fluidizing velocity due to higher entrainment of the unburnt fines and oxygen by-passing. However, lower velocities run the risk of defluidization and therefore clinkering. The fluidization velocity is specified taking these things into account.

4.2.2.2 Excess Air

The mixing between fuel and air is never perfect. Some areas will be oxygen-deficient and some areas even oxygen-starved. Ultimately, all fuel particles must have the necessary oxygen to complete their burning; thus, extra oxygen is always provided in FB boilers in the form of excess air. The combustion efficiency improves with excess air, but this improvement is less significant above an excess air of 20%. Bubbling bed boilers may need a slightly higher amount of excess air than CFB boilers.

4.2.2.3 Combustion Temperature

The combustion efficiency generally increases with bed temperature because the carbon fines burn faster at high temperatures. The effect of temperature is especially important for less reactive particles, which burn under kinetic-controlled regimes. This effect is more prominent in the 700–850°C range than above it. The less reactive the coal, the higher the combustion temperature should be, from a combustion efficiency stand-point. Figure 4.6 shows the increase in combustion efficiency with temperature in a bubbling bed combustor burning different fuels.

4.3 COMBUSTION IN BUBBLING FLUIDIZED BED BOILERS

The two most important sources of combustible losses in an FB boiler are:

1. Unburnt carbon
2. Unburnt CO

The unburnt carbon leaves the boiler along with the fly ash and the bed-ash and thus constitutes an important combustible loss.

The bed-ash is coarse and is drained directly from the bed or furnace. The carbon content of the bed-ash is an order of magnitude less than that of the fly ash because the coarse particles reside in the combustion zone much longer to complete their combustion than the fine ones could.

The fly ash may contain large amounts of carbon exceeding 3-10% as a result of poor operation. If the fuel is less reactive, like anthracite or petroleum coke, the unburned carbon content in the ash could rise correspondingly.

A typical solid fuel contains both fine (<100 μm) and coarse particles. The fines are produced mainly from:

1. Parent coal
2. Attrition of char

In the case of bubbling FB boilers, the mode of feeding of coal also affects the combustion efficiency. For over-bed feeding, coal fines are entrained before hitting the bed surface. Since the freeboard above the bed does not provide a very conducive combustion environment, a large fraction of the entrained fines leave the furnace unburned. In the case of under-bed feeding, the fines are injected directly into the bed. As such, they get well mixed in the bed and often enjoy adequate residence time in the dense bubbling bed to complete combustion. For this reason, under-bed fed bubbling bed boilers have higher combustion efficiencies than over-bed fed boilers. However, under-bed feed requires the use of more finely ground coal, which means more fines in the feed coal. Furthermore, the under-bed feed option is expensive and more prone to plugging.

4.3.1 RECIRCULATION OF FLY ASH

In the case of BFB boilers, the recirculation of fly ash, captured at the back of the boiler, results in reduced carbon burnout as the unburnt carbon in the fly ash gets another chance to complete its combustion. Tang and Engstrom (1987) showed (Figure 4.7) that by increasing the recirculation ratio (solid recirculation/feed) in a bubbling FB from zero to four, the combustion efficiency of anthracite increased from 86 to 95%. For bituminous coal, raising it from 0 to 2 the combustion efficiency increased from 88 to 98%. The recirculation is more effective in the case of less reactive fuels.

4.3.2 EFFECT OF DESIGN PARAMETERS ON COMBUSTION EFFICIENCY

The combustion efficiency of bubbling FBs is affected by several design parameters as described below:

- Bed height
- Freeboard height
- Recirculation of unburnt solids
- Fuel feeding
- Secondary air injection

Combustion

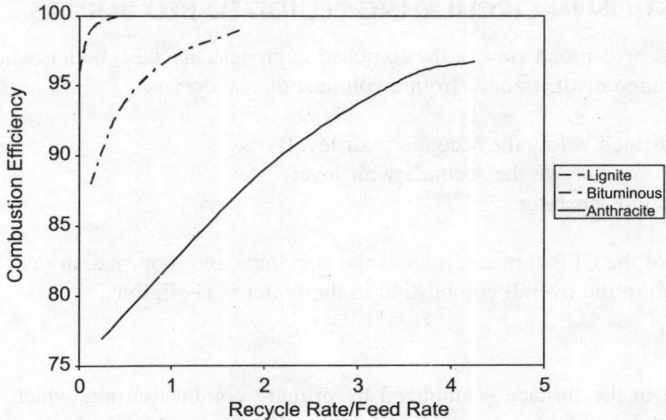

FIGURE 4.7 Effect of fly ash recirculation on the combustion efficiency. (Adapted from Tang and Engstrom, 1987.)

4.3.2.1 Bed Height

A deeper bubbling bed would give higher combustion efficiency as it provides longer residence time for combustion, but it increases the fan power requirement and entrainment rate of solids.

4.3.2.2 Recirculation of Unburnt Solids

The recirculation of fly ash has a positive influence on the combustion efficiency as one can see from Figure 4.7.

4.3.2.3 Freeboard Height

The freeboard height increases the combustion efficiency as it allows longer time for combustion. A heavily cooled freeboard, however, may not be as effective.

4.3.2.4 Fuel-Feeding

The fuel ratio, defined earlier, is an important parameter affecting the combustion efficiency. A low fuel-ratio is often responsible for low combustion efficiency especially in a BFB furnace, but under-bed feeding gives higher efficiency for low fuel-ratio. Over-bed feeding is more effective for higher fuel ratio feed stock.

4.3.2.5 Secondary Air Injection

Secondary air is more commonly used in bubbling bed boilers, especially for biomass fuels. It helps burn the volatiles and also destroys NO_x more effectively. For low VM fuels, secondary air may not improve the combustion efficiency (Oka, 2004), but it could reduce the requirement of in-bed tubes, which are prone to erosion.

4.4 COMBUSTION IN CIRCULATING FLUIDIZED BED BOILERS

This section presents a broad view of the combustion process in a CFB boiler, whose furnace could be divided into three distinct zones from a combustion standpoint:

1. Lower (located below the secondary air level)
2. Upper (located above the secondary air level)
3. Hot gas/solid separator

Other parts of the CFB furnace, such as the standpipe and loop seal, are not included here, as their contribution to the overall combustion in the boiler is negligible.

(i) Lower Zone

The lower zone in the furnace is fluidized by primary combustion air, which constitutes about 40–80% of the stoichiometric quantity of air required for the coal feed. This section receives fresh coal from the coal feeder and unburned char from the hot cyclone. Char particles that are captured by the hot cyclone are returned to this section by means of a loop-seal or L-valve. Devolatilization and partial combustion occur in this zone, which is usually oxygen deficient. Therefore, to protect the boiler tubes from possible corrosion attack, this zone is refractory lined.

The lower zone of a CFB furnace is much denser than the upper zone. Thus it serves also for insulated storage of hot solids, providing the CFB boiler with a thermal "fly-wheel."

(ii) Upper Zone

The secondary air is added at the interface between the lower and upper zones of the furnace. Sometimes, when staged combustion is not essential, as in the case of low-volatile coal, secondary air may be added close to the grid.

In any case, the entire combustion air passes through the upper furnace. Char particles, transported to the upper zone, are exposed to an oxygen-rich environment, where most of the combustion occurs. The upper zone is usually much taller than the lower zone. Char particles, transported upwards through the core, slide down the wall or through the core, often entrapped in falling solid clusters. The char particles could make many trips around the height of the furnace before they are finally entrained out through the top of the furnace.

(iii) Cyclone

Unburned char particles that are entrained out of the furnace go around the refractory-lined cyclone. The residence time of char particles within the cyclone is short and the oxygen concentration in it is low. Thus the total extent of combustion inside the cyclone is expected to be small compared to that in the rest of the combustion loop. However, there could be exceptions. For example, data from an operating unit visited by the author showed a temperature rise as high as 20–200°C between the furnace exit and the loop-seal suggesting strong combustion in the cyclone. Table 4.4 shows that

TABLE 4.4
Temperature Rise across the Cyclone of Two Commercial CFB Boilers

Capacity of boiler	Temperature at bottom furnace (°C)	Cyclone entrance (°C)	Cyclone gas-exit (°C)	Loop-seal (°C)
75 t/h, 4.8 MPa, 480°C	983	867	930	977
82.4 t/h, 4.8 MPa, 472°C	978	911	966	1022

the gas temperature increased by 50°C between the cyclone inlet and exit, while the solid temperature increased by about 100°C between the cyclone entry and the loop-seal. This suggests the presence of significant combustion in the cyclone as well as the standpipe. This may have been due to the excess amount of oxygen in the cyclone and that dragged by the solids in the standpipe.

4.4.1 COMBUSTION OF CHAR PARTICLES

Char particles burning in a CFB may be broadly classified into three groups based on their mode of combustion and size. This classification is essentially based on the Archimedes number of the char with respect to that of bed solids. All other parameters being the same, the classification reduces to the following:

Fines	$[d_c < d_p]$
Fragments	$[d_c \approx d_p]$
Coarse char	$[d_c > d_p]$

where: d_c and d_p are the diameters of char and average bed particles, respectively.

(i) Fines

Attrition and combustion of coarser char particles produces carbon fines. Typical fines are less than 50–100 μm in size and are generally below the cut-size for 50% cyclone efficiency. The fines burn under Regime I conditions. Some of them, being trapped inside larger clusters of bed solids, are captured by the cyclone, but others escape through the cyclone. Thus, fines constitute the bulk of the carbon in the fly ash that leaves the boiler.

(ii) Burnout Time

The burnout time is the time a fuel particle takes to complete its combustion. It is a critical parameter in determining the carbon loss from an FB. For complete combustion, the total residence time of a fuel particle must be longer than its burnout time. The burnout time depends on the burning rate of a particle as well as on a number of other factors, including the combustion temperature, particle size and the reactivity of the fuel.

Using data from Basu and Fraser (1991) and Pillai (1981), Brereton (1997) calculated the devolatilization and combustion time for an anthracite particle. Luo et al. (1995) calculated the burnout time of particles of a high-ash coal. Both data are given in Table 4.5.

In CFB boilers, the cyclones have the most important influence on fly-ash loss. The residence time of char particles in a CFB furnace is difficult to calculate as the bed is neither fully back-mixed nor in plug flow. Particles' trajectories are complex in the bed. The fine particles are most likely to follow the gas moving in plug-flow in the furnace core, resulting in the shortest residence time, τ_{fine}. The residence time of a particle for a single pass through the furnace, known as single-pass

TABLE 4.5
Typical Combustion Times Calculated for Anthracite Particles of Different Sizes in a CFB Combustor

	Brereton (1997) for anthracite		Luo et al. (1995) for high ash coal
Particle diameter (mm)	Devolatilization time (sec)	Char burning time (sec)	Burnout time (sec)
0.01	—	0.2	—
0.1	10	2.5	0.68
1.0	22	32	23.1
2.0	—	—	50.1
10	50	655	—

residence time can be approximated as:

$$\tau_{fine} = \frac{H}{U_{core} - (U_t/\varepsilon_{core})} \tag{4.19}$$

where: H is the furnace height and U_{core} is the gas velocity in the core. Although the core velocity is higher than the superficial velocity due to the presence of a boundary layer on the wall, it may be taken as the superficial velocity for the first approximation. The voidage in the core, ε_{core} is higher than the choking voidage of 0.9976 calculated in one of the examples in Chapter 2.

The single-pass residence time of coarse particles, τ_{coarse} can be calculated, assuming the CFB furnace or bubbling bed to be back-mixed;

$$\tau_{coarse} = \frac{W}{G} \tag{4.20}$$

where: W is the total mass of bed solids and G is the sum of the feed and external solid recycle rates.

Experimental data published (Oka, 2004) to date, suggests that fuel particles less than 1 mm in diameter burn in FBs in the kinetic regime, while particles larger than 1–3 mm burn in the diffusion-controlled regime. As the carbon in fly ash is typically less than 1 mm in size, its burnout is controlled by the kinetic rate.

In addition to their movement around the furnace, cyclone and solid circulation loop, the solids also travel around within the furnace. Fines are carried upward in the core and then descend along the wall. Thus, the residence time of the fines is governed by internal circulation, height of the furnace and the cyclone performance. For complete combustion, the burnout time of the fines should be shorter than their residence time.

The burnout time of the fines is an important parameter affecting the combustion efficiency. It depends on the particle surface areas, its intrinsic reactivity and the order of reaction. Figure 4.8 presents computed values of 99% burnout time of char fines having different internal surface areas assuming two different orders of reactions, n. It is apparent from this that char fines with fewer internal pores need longer time to complete their combustion. The difference is significant between internal surface areas of 100 and 200 m²/g, where the burnout time could double. This explains why a simple change in fuel can result in a significant drop in combustion efficiency.

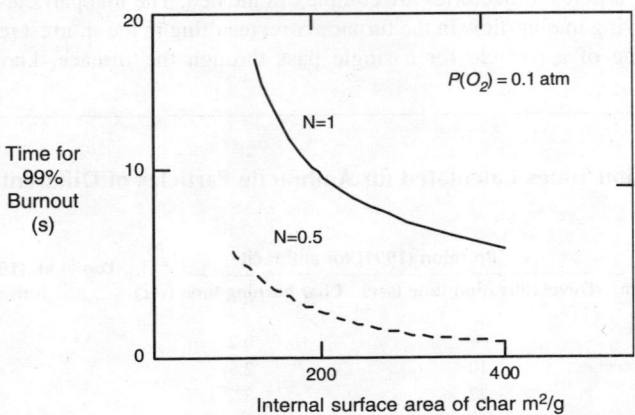

FIGURE 4.8 Estimated burnout time of fine char particles.

4.4.2 EFFECT OF CYCLONE DESIGN ON COMBUSTION EFFICIENCY

An increase in gas temperature across the cyclones in some CFB boilers suggests the occurrence of combustion inside them. This increase could be as high as 100°C (Boemer et al., 1993). Carbon monoxide is produced by char particles as they travel in the cooler wall region of a CFB boiler furnace. As gas mixing in a CFB furnace is rather poor, the CO does not have the opportunity to mix with oxygen until it reaches the cyclone. Here, unused oxygen from the bed mixes with CO, generating heat. Also, fine char particles utilize some of this unused oxygen to raise the temperature of the gas and solid exiting the cyclone. This often creates a clinkering problem in the solid return system. High temperature could also cause corrosion problems in the back-pass of the boiler.

Unburnt carbon escapes primarily through the cyclone. If one could design an ideal cyclone with 100% efficiency for all sizes of particles, the only source of loss a CFB boiler would have is the loss through the bed drain, which is generally very low anyway. Thus, the design of the hot cyclone is critical for combustion efficiency.

Except for very fine particles, all char particles would need multiple trips around the CFB loop to complete their combustion. If the cyclone is not sufficiently efficient to capture the particles with burnout times longer than the single-pass residence time (as given by Equation 4.19), these particles would most likely leave the boiler unburnt. Thus, the cut-off size of the cyclone must be smaller than d_c the size of char particles with a burnout time longer than single pass residence time:

$$D_{\text{cut-off}} < d_c \tag{4.21}$$

4.4.3 SOME COMBUSTION DESIGN CONSIDERATIONS

In previous sections, a general understanding of the processes of combustion was developed. Here, we examine how some of the information can be utilized in the design of an FB combustion furnace.

(i) Grate Heat Release Rate

The amount of combustion heat that can be released per unit cross-section of the furnace is another important design criterion of an FB boiler and determines the furnace cross-section area. Grate heat release rate is a function of the mass flow rate of combustion air passing through the furnace. For typical fossil fuels, there is a crude relationship between combustion heat and superficial gas velocity (Waters, 1975):

$$\text{Grate heat release rate} = \frac{3.3 \, U_0}{[\text{EAC}]} \, \text{MW/m}^2 \tag{4.22}$$

where: U_0 is the superficial gas velocity through the furnaces, referred to 300 K and [EAC] is the excess air coefficient.

The superficial velocity, however, is limited by the hydrodynamic conditions necessary for fast fluidization (see Chapter 2). For typical CFB furnaces using 20% excess air (EAC = 1.2) and gas velocity of 3.5–5.0 m/sec at 850°C, the grate heat release rate is in the range of 3.0–4.5 MW/m² (see Table 8.6). In the case of bubbling beds, the velocity is generally around 1 m/sec.

The volumetric heat release rate is not commonly used for the design of CFB boilers, because the height of the furnace is dictated by the requirements of heat absorption by the furnace walls. However, there are other design considerations. Table 8.6 analyzes a large number of CFB boilers, showing that the volumetric heat release rate is generally in the range of 0.08–0.14 MW/m³. In commercial boilers, the fuel and secondary air enter the furnace only from the sides. The depth of the furnace is often too great to allow good mixing of volatiles. In some furnaces, volatile flames up to 20 m high may be seen. The lateral mixing of fuel and secondary air is a limiting factor on the depth, as well as on the height. A wider furnace would also necessitate a taller furnace to facilitate complete mixing of the volatiles and the secondary air. In the absence of any rule of thumb, the designer may have to go for performance modeling to determine this limit.

(ii) Effect of Fuel on the Design

The fuel has a significant effect on the design and operation of FB boilers. For preliminary designs, the heating value and ultimate and proximate analyses of the fuel are required. The heating value governs the coal feed rate, while the proximate analysis specifies components such as cyclone and downstream equipment. Their use in the design was discussed earlier in Section 4.2.1. Higher reactivity of the fuel will result in higher combustion efficiency, as will coarser and less friable fuel particles.

The rank of the fuel or its volatile content has an important effect on the unburnt carbon in fly ash. Lijun et al. (2005) studied results from as many as seventeen 50-MWe CFB boilers and concluded that the unburnt carbon content in the fly ash decreases with increasing ratio of the volatile content and the heating value of the fuel (Table 4.6).

If the boiler does not use sorbents, the average size of bed materials, which governs the bed hydrodynamics and heat transfer, is greatly influenced by the size distribution of ash in the fuel and their attrition properties.

In the case of bubbling FB boilers, the fuel governs the heat balance of the dense bed, and thus decides the amount of heat absorbing tubes necessary in it.

(iii) Combustion Temperature

The choice of combustion temperature is an important issue in the design of FB boilers. Typical FB furnaces are operated in the temperature range 800–900°C. The combustion temperature is maintained at around 850°C for the following four reasons:

1. Ash does not fuse at this temperature
2. Sulfur capture is optimal at around 850°C
3. Alkali metals from the coal, which cause fouling, are not vaporized at such low temperatures
4. Nitrogen in the combustion air is not readily converted into NO_x at this low temperature

A high combustion temperature also has some advantages. The combustible loss is primarily due to the unburned carbon fines that escape through the cyclone before they can complete their combustion. Fine carbon particles burn under kinetic-controlled regime. Thus, a higher temperature may help reduce the burnout time and hence the combustible loss. A temperature in the neighborhood of 900–950°C may be acceptable for combustion and attractive from a greenhouse gas emission standpoint, because higher temperatures reduce the generation of N_2O (see Chapter 5). Lijun et al. (2005) operated the furnace of a 100-MWe CFB boiler in the range of 900–1050°C to reduce the unburnt carbon in fly ash. This, combined with fly-ash recirculation, greatly reduced the carbon in fly ash. High temperature combustion is beneficial especially for low-volatile, less

TABLE 4.6
Effect of Volatile Matter Content and Heating Value of Fuel on the Carbon Content of Fly Ash in a Number of CFB Boilers.

Steam capacity (t/h)	Type of fuel	Volatile (%)	LHV (MJ/kg)	Carbon in fly ash (%)
75	Lignite	44.4	13.8	6.4
75	Bituminous	28.2	16.8	8.9
75	Anthracite	5.2	12.7	27.2
120	Bituminous	27.1	21.7	18

Source: Adapted from Lijun et al., 2005.

reactive fuels, but it should never be so high as to soften the ash or to exceed the limit of NO_x emission or compromise sulfur capture capability if the fuel contains sulfur.

Some biomass fuels like rice husks could attract value-added use for their ash if the combustion temperature is low. It could be burnt at a temperature as low as 600–700°C. However, a low combustion temperature may increase the carbon in fly ash compromising its use in fly ash.

4.4.4 Performance Modeling

A performance model of a combustor helps define the relationship between design and operating parameters and performance. It also tells the designer whether the specified furnace height and other dimensions will allow the boiler to meet its combustion and emission requirements. The performance of the combustor may be generally judged by:

- Unburnt carbon loss
- Distribution of volatiles, oxygen and carbon along the height and across the cross-section of the furnace
- Flue gas composition at the exit of the cyclone, especially the emission of SO_2 and NO_x
- Heat release and absorption pattern in the furnace
- Solid waste generation

During the design stage, a prediction of the boiler performance would help determine the most economic size of the boiler for the best performance. After the boiler is built, performance modeling can do the following:

- Predict the performance of the boiler operating under off-design conditions or with a nonspecified feed stock
- Help improve the performance of the boiler through adjustments of process parameters

A number of models are available in the literature, and are being continually upgraded. This aspect of the FB boiler is still developing, thus, no specific model is described here. Readers are instead directed to recent references.

4.5 BIOMASS COMBUSTION

Biomass is a renewable fuel source and is considered greenhouse gas-neutral. Compositions of some popular biomass fuels are given in Table 4.7. These fuels are generally more reactive than fossil fuels such as anthracite, lignite etc. As a result, combustion is not such a major issue as are problems like bed sintering, slagging, fouling and corrosion. The first three problems are direct results of the relatively high alkali content of biomass fuels, while the corrosion is due to their chlorine content.

Another characteristic of biomass fuels is its relatively high volatile content (60–80%). This makes staged combustion particularly suitable because combustion in the lower bed is relatively low. Staged combustion avoids the need for heat extraction by tubes immersed in the bed, which is better from the boiler-availability standpoint because erosion of in-bed tubes is a major problem of bubbling bed boilers.

Relatively high moisture content (10–65%) increases the flue gas volume per unit heat release. This calls for larger cyclone size and back-pass width in CFB boilers. Low ash content creates the problem of bed inventory. Periodic topping up of bed materials may be needed.

It is evident from Table 4.7 that, unlike fossil fuels, the oxygen content of biomass fuel could be very high: in the 30–40% range. Thus, the combustion-air requirement of a biomass-fired boiler is

TABLE 4.7
Composition of Some Biomass Fuels

Fuel	Unit	Rice Husk	Waste Wood	Saw Dust
HHV (dry basis)	MJ/kg	16.2	20.14	20
Moisture	Wt%	9.7	14.2	53.9
Volatiles	—	66	79.1	39.19
C	Wt% (ash-free basis)	40.2	49	50.6
H	—	5	5.9	6
O	—	36.3	40.7	42.78
N	—	0.3	2.5	0.2
S	—	0.05	0.05	0.01
Cl	—	0.1	0.03	0.01
Ash	—	18.2	1.9	0.41
Ash composition				
SiO_2	% of ash	89	32	4.9
Al_2O_3	—	0.41	6.2	0.93
TiO_2	—	—	14	0.2
Fe_2O_3	—	0.25	2.8	1.3
CaO	—	0.75	17	32
MgO	—	0.43	2.7	4.8
Na_2O	—	0.06	2.8	2.4
K_2O	—	2.2	4.7	8.7
P_2O_5	—	—	6.3	—
SO_3	—	0.49	—	19

relatively low, which means there would be less nitrogen in the flue gas, which will help reduce the dry flue gas loss from the boiler.

Table 4.8 lists some of the problems an FB boiler might face due to different constituents of a biomass fuel. Agglomeration of bed materials, fouling of boiler tubes, and hot corrosion are some of the significant problems.

TABLE 4.8
Effect of Biomass Fuel Properties on Boiler Performance.

Elements in fuel	Effect
Alkalis like sodium, potassium	Bed agglomeration
	Fouling of back-pas tubes and hot corrosion
Chlorine	Fouling of back-pas tubes and hot corrosion
	HCl emission
	Dioxin formation
Heavy metals	Emissions
	Furnace corrosion
	Ash handling
Sulfur	SO_2 emission
	Cold end corrosion
	Hot corrosion (S/Cl ratio)
Nitrogen	NO_x emission

Source: Adapted from Hulkkonen et al., 2003.

4.5.1 AGGLOMERATION

A major problem of biomass combustion in FBs is agglomeration of bed materials. Many biomass fuels contain alkali salts which react preferentially with silica in the common bed material (sand) to form a low-melting, eutectic mixture

$$< \text{Silica} > + < \text{alkali} > = < \text{Eutectic mixture of silicate} >$$

$$2SiO_2 + Na_2CO_3 = 2Na_2O \cdot 2SiO_2 + CO_2 \tag{4.23}$$

Although the bed material, silicon dioxide, melts at 1450°C, the eutectic mixture formed melts at 874°C. For potassium salt, the eutectic mixture will melt at 754°C. This makes the particle surface sticky disturbing the fluidization and generating local hot spots. Hot spots lead to agglomeration and sintering.

The sintering tendency of a fuel is indicated by the following two parameters (Hulkkonen et al., 2003). The sintering tendency is severe when the fuel constituent parameters exceed the following limits:

$$\frac{(K_2O + Na_2O)}{SiO_2} > 1; \quad \frac{(K_2O + Na_2O)}{HHV} > 0.34 \tag{4.24}$$

In some biomass like straw, potassium, the more damaging alkali, is much higher than sodium.

4.5.1.1 Options for Avoiding the Agglomeration Problem

Formation of agglomerates could be detected by a sharp temperature gradient in the bed and by pressure drop fluctuations in the bed (Korbee et al., 2003). Early detection of the onset of agglomeration can help corrective measures to avoid agglomeration. There are, however, situations where early detection does not help. In those cases, one of the following approaches can be adopted to prevent bed agglomeration:

Use of Additives

Some additives, like china clay, dolomite or limestone, could work well with bed materials (Silvennoines, 2003). Soils, especially those containing kaoline, could help reduce the agglomeration potential.

Preprocessing of Fuels

Preprocessing could involve reduction of harmful constituents like alkali. In Denmark, straw is left in fields over the winter and cut in late spring. This allows rain water of leach alkali (Silvennoines, 2003).

Use of Alternative Bed Materials

If the bed material contains a sufficient amount of iron oxide (Fe_2O_3), the iron will react preferentially with alkali salt to produce $X_2Fe_2O_4$ which melts at a higher temperature of 1135°C (Werther et al., 2000). Other materials like feldspar, dolomite, magnesite and alumina, if used as bed material, can also reduce the agglomeration problem because they too form eutectic mixtures, but with high melting temperatures. Diabase materials could also serve as good bed material for biomass combustion.

Since the key elements causing agglomeration are sodium and potassium, co-firing of biomass with high ash coal with sulfur helps reduce the concentration of the potassium and thereby the risk of agglomeration (Glazer et al., 2005). Alkali metal of biomass interacts with clay mineral present in the coal to form alkali–alumina silicate. With high-sulfur coal the equilibrium shifts towards alkali sulfates. These have a higher melting point. Thus clay containing kaolin could help reduce the agglomeration potential.

Co-Firing with Coal

The co-firing of coal with biomass has another benefit of reduced N_2O. Biomass contains more volatile nitrogen which is more likely to end up being NH_3 than HCN from where N_2O is formed. (Gulyurtlu et al., 1995). The sulfur in coal helps reduce the agglomerate formation (Yrjas et al., 2005).

Reduction in Bed Temperature

Reducing the combustion temperature could reduce the vaporization of alkali salts and hence the resulting agglomeration or hot corrosion.

4.5.2 FOULING

Fouling is the sticky deposition of ash constituents on boiler tubes. It reduces heat transfer to the tubes, reducing generation or steam temperature and increases the metal temperature, which could accelerate corrosion. The fouling of downstream tubes is due to evaporation of alkali salts. Therefore, the boilers can either be designed for low gas velocity in the back-pass or superheater placement in a cooler region.

4.5.3 CORROSION POTENTIAL IN BIOMASS FIRING

Chlorine contributes to the hot corrosion potential, which is especially a concern for boilers operating at high steam temperature and pressure. Hot corrosion becomes a problem with high chlorine fuels especially on superheaters with high steam temperature ($>460-480°C$, $P > 55$ bar). Chlorine reacts with alkali metals to form low temperature melting alkali chlorides, which form a sticky deposit on superheater tubes reducing their heat absorption and causing high temperature corrosion.

If the chlorine in fuel exceeds 0.1% (Hulkkonen et al., 2003), provisions to guard against corrosion are required. The temperature of the flue gas entering the superheater section should be less than the ash melting temperature. Presence of sulfur could deprive the flue-gas chlorine of the alkali metals and instead form alkali sulfates, leaving the chlorine to escape as HCl (Yrjas et al., 2005). Thus, the gas path can be designed to allow sufficient time for completion of the gas-phase reactions producing alkali sulfates and gaseous chlorine-containing compounds.

NOMENCLATURE

A': frequency factor in Equation 4.14, kg/m^2 sec (kPa)^n
A: external surface area of char, m^2
A_g: pore surface area/mass of char, m^2/kg
A: constant in Equation 4.4
a: constant in Equation 4.4
C_p: specific heat of char, kJ/kg K
C_g, C_s: oxygen concentration away from the char and on its surface, respectively, in partial pressure, kPa
D_g: molecular diffusivity of oxygen, m^2/sec
D_p: pore diffusion coefficient, m^2/sec
D: diameter of char particle, m
d_v: initial diameter of a coal particle, mm
d_c, d_p: diameters of char and average bed particles, respectively, m
$D_{\text{cut-off}}$: cut-off size of the cyclone, m
e_p: emissivity of coal particle
E: activation energy, kJ/kmol
E_0: mean activation energy for the release of volatiles, Equation 4.2, kJ/kmol
F_{dg}: drag force due to gas per unit volume of particle, kN/m^3
F_{ds}: drag force due to solid–solid interaction per unit volume of particle, kN/m^3
G: some of feed-stock and recirculating solids, kg/sec
G_s: circulation rate in fast bed, kg/m^2 sec
g: acceleration due to gravity, 9.81 m/sec^2

H:	height of CFB furnace or dense section of bubbling bed, m
HHV:	heating value of char, kJ/kg
h_m:	mass-transfer coefficient (carbon flux), kg/m² kPa
h_m:	mass-transfer coefficient expressed in terms of the amount of carbon removed, kg(C)/m² kPa sec
h_{gp}:	gas-particle heat transfer coefficient, kW/m²K
K_a:	attrition rate constant for combustion-assisted attrition
K_g:	thermal conductivity of gas, kW/m K
K_0:	preexponential factor in Equation 4.2, sec^{-1}
m_c:	mass of a char particle, kg
m_{attr}:	rate of combustion-assisted attrition, kg/sec
m:	true order of reaction
m_d:	diffusion limit, kg/m² sec
n:	apparent order of reaction
p:	exponent in Equation 4.4
q:	specific burning rate of char based on external surface, kg(C)/m² sec
q_c:	kinetic reaction rate limit, kg/m² sec
q_i:	reaction rate based on internal surface area, g(C)/cm² sec
R:	universal gas constant, 8.314 kPa m³/kmol K
R_c, R_i:	apparent and intrinsic reaction rate based on external and internal surface area, respectively, kg(C)/m²sec (kPa)^n
T:	temperature of the coal, K
T_b, T_s:	temperatures of the bed and of the surface of the char, respectively, K
T_m:	mean temperature of the diffusion layer around the char particle, K
t:	time, sec
t_v:	time for devolatilization and combustion, sec
U:	superficial or fluidizing velocity at bed temperature, m/sec
U_{mf}:	minimum fluidization velocity at bed temperature, m/sec
U_c, U_g:	velocities of char and gas, respectively, m/sec
U_p:	velocity of average bed particles, respectively, m/sec
U_{core}:	gas velocity in the core of a fast bed, m/sec
U_0:	superficial gas velocity through bed or furnace, referred to 300 K, m/sec
U_{slip}:	slip velocity, m/sec
U_t:	terminal velocity of char particle, m/sec
V:	amount of volatile yet to be released, %
V^*:	asymptotic value of the extent of volatilization, %
W:	total mass of bed solids, kg
x:	fraction of volatile matter released

GREEK SYMBOLS

ρ_{ch}, ρ_p:	density of char, and bed particle, respectively, kg/m³
ρ_g:	density of gas, kg/m³
ρ_b:	bulk density of the fast bed, kg/m³
ρ_c:	density of carbon in char, kg/m³
η:	ratio of actual combustion rate and rate attainable if no pore diffusion resistance existed
β:	effectiveness factor (=1, for Regime I)
ε:	local voidage
ε_{core}:	voidage in the furnace core
σ:	Stefan–Boltzman constant, W/m⁴ K

ϕ: mechanism factor ($= 1$ for CO_2; 2 for CO)
τ: ratio of volume to surface area of char, m
τ_{fine}: residence time of fine particles in the bed, sec
τ_{coarse}: residence time of coarse particles in the bed, sec
Γ: time for complete devolatilization, sec
θ: porosity of char
EAC: excess air coefficient
δ: standard deviation of Gaussian distribution of activation energy, kJ/mol
μ: viscosity of gas, kg/m sec

DIMENSIONLESS NUMBERS

Nu: Nusselt number, $h_{gp}d_v/K_g$
Re: Reynolds number $(U_g - U_c)d_v p_g/\mu$
Sc: Schmidt number, $D_g p_g/\mu$
Sh: Sherwood number, as in Equation 4.8

REFERENCES

Anthony, D. B., Howard, J. B., Hottel, H. C., and Meissner, H. P., Rapid devolatilization of pulverized coal, *15th Symposium (International) on Combustion*, Combustion Institute, Pittsburgh, pp. 1303–1315, 1975.

Arastoopour, H., Wang, C. H., and Weil, S. A., Particle–particle interaction force in a dilute gas–solid system, *Chem. Eng. Sci.*, 37, 1379, 1982.

Arena, U., Massimilla, L., Cammarota Sciliano, L., and Basu, P., Carbon attrition during the combustion of char in a circulating fluidized bed, *Combust. Sci. Technol.*, 73, 383–394, 1990.

Arthur, J. R., Model reaction between carbon and oxygen, *Trans. Faraday Soc.*, 47, 164–178, 1951.

Basu, P. and Fraser, S., *Circulating Fluidized Bed Boilers*, Butterworth Heinemann, Stoneham, 1991.

Basu, P. and Halder, P. K., Combustion of single carbon particles in a fast fluidized bed of fine solids, *Fuel*, 68, 1056–1063, 1989.

Basu, P., Halder, P. K., and Greenblatt, J., Fragmentation of high swelling coal in a high velocity fluidized bed, Presented at the Fall Meeting of Society of Chemical Engineering of Japan, Tokyo, October, 1989.

Basu, P. and Sarkar, A., Agglomeration of coal-ash in fluidized beds, *Fuel*, 62, 924–926, 1983.

Basu, P. and Wu, S., Surface reaction rate of coarse sub-bituminous char particles in the temperature range 600–1340 K, *Fuel*, October 1993.

Boemer, A., Braun, A., and Renz, U., Emission of N_2O from four different large scale circulating fluidized bed combustors, *Proceedings of the 12th International Conference on Fluidized Bed Combustion*, Vol. I, Rubow, L., Ed., ASME, New York, pp. 585–598, 1993.

Brereton, C. M. H., Combustion performance, In *Circulating Fluidized Beds*, Grace, J. R., Avidan, A. A., and Knowlton, T. M., Eds., Chapman and Hall, London, pp. 369–411, 1997.

D'Amore, M., Masi, S., and Salatino, P., The relevance of macerals in the combustion of a coal char in the chemical kinetic regime, *Combust. Sci. Technol.*, 63, 63–73, 1989.

Daw, C. S. and Krishnan, R. P., ORNL Report no ORNL/TM-8604, 1983.

Donsi, G., Massimilla, L., and Miccio, M., Carbon fines production and elutriation from the bed of a fluidized coal combustor, *Combust. Flame*, 41, 57–64, 1981.

Essenhigh, R. H., *Chemistry of Coal Utilization*, 2nd Supplement Volume, Elliott, M. A., Ed., John Wiley, New York, pp. 1198, 1981.

Essenhigh, R. H., Froberg, R., and Howard, J. B., Combustion behavior of small particles, *Ind. Eng. Chem.*, 57, 32–43, 1965.

Field, M. A., Gill, D. W., Morgan, B. B., and Hawksley, P. G. W., *BCURA Mon. Bull.*, 31, 285–345, 1967.

Furusawa, T. and Schimizu, T., Analysis of circulating fluidized bed combustion technology and scope for future development, In *Circulating Fluidized Bed Technology II*, Basu, P. and Large, J. F., Eds., Pergamon Press, Oxford, p. 51–62, 1988.

Glazer, M. P., Schürmann, H., Monkhouse, P., Jong, de W., and Spliethoff, H., Co-combustion of coal with high alkali straw. Measuring of gaseous alkali metals and sulfur emissions monitoring, In *Circulating Fluidized Bed Technology—VIII*, Cen, K., Ed., International Academic Publishers, Beijing, pp. 630–636, 2005.

Gulyurtlu, T., Bordaloc, C., Penha, E., and Cabrita, I., Co-combustion of coal with various types of biomass in a CFB, In *Combined Combustion of Biomass/Sludge and Coal, Final Report EC-Research Project APAS Contract COAL-CT92-0002*, Bemtgen, J. M., Heink, R. G., and Minchener, A. J., Eds., 1995.

Halder, P. K., Combustion of single coal particles in circulating fluidized bed, Ph.D. Thesis, Technical University of Nova Scotia, 1989.

Halder, P. K. and Basu, P., Kinetic rate of electrode carbon in the temperature range 700–900°C, *Can. J. Chem. Eng.*, 65, 696–699, 1987.

Halder, P. K. and Basu, P., Mass transfer from a coarse particle to a fast bed of fine solids, *AIChE Symp. Ser.*, 84(262), 58–64, 1988.

Hamor, R. J., Smith, I. W., and Tyler, R. J., *Combust. Flame*, 21, 153, 1973.

Howard, J. B. and Essenhigh, R. H., Mechanism of solid-particle combustion with simultaneous gas-phase volatile combustion, In *11th Symposium (International) on Combustion*, Combustion Institute, Pittsburgh, pp. 399–408, 1967.

Hulkkonen, S., Fabritius, M., and Enestam, S., Application of BFB technology for biomass fuels: technical discussions and experience from recent project, In *Proceedings of the 17th International Conference on Fluidized Bed Combustion*, Pisupati, S., Ed., ASME, New York, 2003, Paper no. FBC2003-132, 2003.

Hyppanen, T., Lee, Y., and Rainio, A., A three dimensional model for CFBC, In *Circulating Fluidized Bed Technology III*, Basu, P., Hasatani, M., and Horio, M., Eds., Pergamon Press, Oxford, 1991.

Keairns, D. L., Newby, R. A., and Ulrich, N. H., Fluidized bed combustion design, In *Fluidized Bed Boilers: Design & Applications*, Basu, P., Ed., Pergamon Press, Toronto, p. 107, 1984.

Korbee, R., Van Omen, J. R., Lensselink, J., Nijenhuis, J., Kiel, J. H. A., and Van dan Bleek, C. M., Early agglomeration recognition system, In *Proceedings of the 17th International Conference on Fluidized Bed Combustion*, Pisupati, S., Ed., ASME, New York, 2003, Paper no. FBC2003-151.

LaNauze, R. D., *Fluidization*, Davidson, J. F., Harrison, D. H., and Clift, R., Eds., Academic Press, New York, p. 642, 1985.

Lijun, D., Jing, M., Liou, Z., Sun, X., Li, G., and Li, Z., Economic ways to improve CFB boiler efficiency, In *Circulating Fluidized Bed Technology VIII*, Cen, K., Ed., International Academic Publishers, Beijing, pp. 602–608, 2005.

Luo, Z., Li, X., Mengxiang, F. et al., Design and performance of a CFBB firing stone coal, In *Proceedings of the 13th International Conference on Fluidized Bed Combustion*, ASME, New York, p. 619, 1995.

Makansi, J., Fuel type and preparation emerge as critical to FBC design, *Power*, January, 642, 1990.

Mehta, B. N. and Aries, R., Communication on the theory of diffusion and reaction via the isothermal path order reaction, *Chem. Eng. Sci.*, 26, 1699–1712, 1971.

Oka, S. N., *Fluidized Bed Combustion*, Marcel Dekker, New York, pp. 331–334, see also pp. 426–428, 435, 2004.

Pillai, K. K., *J. Inst. Energy*, 54, 142–150, 1981.

Ross, I. B. and Davidson, J. F., The combustion of carbon particles in a fluidized bed, *Trans. Inst. Chem. Eng.*, 60, 108–114, 1982.

Sergeant, G. D. and Smith, I. W., *Fuel*, 51, 52–57, 1973.

Silvennoines, J., A new method of inhibiting bed agglomeration problems in fluidized bed boilers, In *Proceedings of the 17th International Conference on Fluidized Bed Combustion*, Pisupati, S., Ed., ASME, New York, 2003, Paper no. FBC2003-081.

Smith, I. W., *Combust. Flame*, 17, 421–428, 1971a.

Smith, I. W., *Combust. Flame*, 17, 303–314, 1971b.

Smith, I. W., The combustion of rates of coal char: a review, *19th Symposium (International) on Combustion*, Combustion Institute, Pittsburgh, pp. 1045–1065, 1982.

Smith, I. W. and Tyler, R. J., *Fuel*, 51, 312–321, 1972.

Solomon, P. R. and Colket, M. B., Coal devolatilization, *17th Symposium (International) on Combustion*, Combustion Institute, Pittsburgh, 131–140, 1978.

Tang, J. and Engstrom, F., Technical assessment of ahlstrom pyroflow and conventional bubbling fluidized bed combustion systems, In *Proceedings of the 9th International Conference on Fluidized Bed Combustion*, Mustonen, J. P., Ed., ASME, New York, pp. 38–54, 1987.

Turnbull, E., Kossakowski, E. R., Davidson, J. F., Hopes, R. B., Blackshaw, H. W., and Goodyer, P. T. Y., Effect of pressure on combustion of char in fluidized beds, *Chem. Eng. Res. Dev.*, 62, 225, 1984.

Young, B. C. and Smith, I. W., *Eighteenth Symposium (International) on Combustion*, Combustion Institute, Pittsburgh, 1981, 1249–1255.

Waters, P. L., Factors influencing fluidized combustion of low-grade solid and liquid fuels, *Symposium Series. 1.* Institute of Fuel, London, p. C6-3, 1975.

Weiss, V., Scholer, J., and Fett, F. N., Mathematical modeling of coal combustion in a circulating fluidized bed reactor, In *Circulating Fluidized Bed Technology II*, Basu, P. and Large, J. F., Eds., Pergamon Press, Oxford, pp. 289–298, 1988.

Werther, J., Saenger, M., Hartge, E.-U., Ogada, T., and Siagi, Z., Combustion of agricultural residues, *Prog. Energy Combust. Sci.*, 26, 1–27, 2000.

Wu, S., Sengupta, S. P., and Basu, P., Generalized mathematical model for CFB boiler furnace, In *Proceedings of the 10th International Conference on Fluidized Bed Combustion*, Manaker, A., Ed., ASME, New York, pp. 1295–1302, 1991.

Yoshioka, T. and Ikeda, S., Wakamatasu 50 MWe atmospheric fluidized bed combustion test results and future test plan, Presented at 20th IEA-AFBC Technical Meeting, Lisbon, 1990.

Yrjas, P., Skrifvas, B. J., Hupa, M., Reppo, J., Nylund, M. P., and Vaimikku, P., Chlorine in deposits during co-firing of biomass peat and coal in full scale CFBC boiler, In *Proceedings of the 18th International Conference on Fluidized Bed Combustion*, Jia, L., Ed., ASME, New York, 2005, Paper no. FBC2005-78097.

5 Emissions

The combustion of fossil fuels in stationary and transportation systems is the main source of air pollution. Various boilers, furnaces, and engines burning fossil fuels emit gaseous pollutants, such as SO_2, NO_x, CO, N_2O, Hg, and volatile organic compounds (HC). Winds carry these pollutants over great distances, sometimes creating transboundary pollution problems. Some gaseous pollutants, such as SO_x, enter complex chemical reactions with moisture, catalyzed by sunlight, to form acids. These acids are then precipitated on the earth through rain as acid rain (Figure 5.1). Besides these, fossil fuel-fired plants also emit greenhouse gases like CO_2 and N_2O, adding to global climate change.

In an environmentally conscious society, it is desirable, if not essential, to minimize the emission of these pollution-causing gases from fossil fuel-fired boilers. Concerns about climate change have heightened the need for the reduction of greenhouse gases. The present chapter discusses the impact of some of these pollutants and the means for minimizing their emissions using fluidized bed gasifiers or boilers.

5.1 AIR POLLUTION

Air pollution is a major threat to human society. It affects our health as well as the ecosystem that surrounds us. There are two types of air pollution: global and regional. Some of the global pollutant gases, whose effects are not limited to a region but span the entire globe, are also called greenhouse gases. The following gases belong in this category:

- Carbon dioxide (CO_2)
- Methane (CH_4)
- Nitrous oxide (N_2O)

Regional pollutant gases affect the region around their source either directly or indirectly through the formation of secondary pollutants. Gases responsible for regional pollution include:

- Sulfur dioxide (SO_2)
- Nitric oxide (NO)
- Mercury (Hg)
- Volatile organic compounds (VOC)

Air pollution is partly due to natural causes like volcanic action, dust from the desert and forest fires, and partly due to human activities (anthropogenic). Fossil fuel combustion and transportation are major sources of anthropogenic pollution. The following section describes the formation of pollutants and their effects.

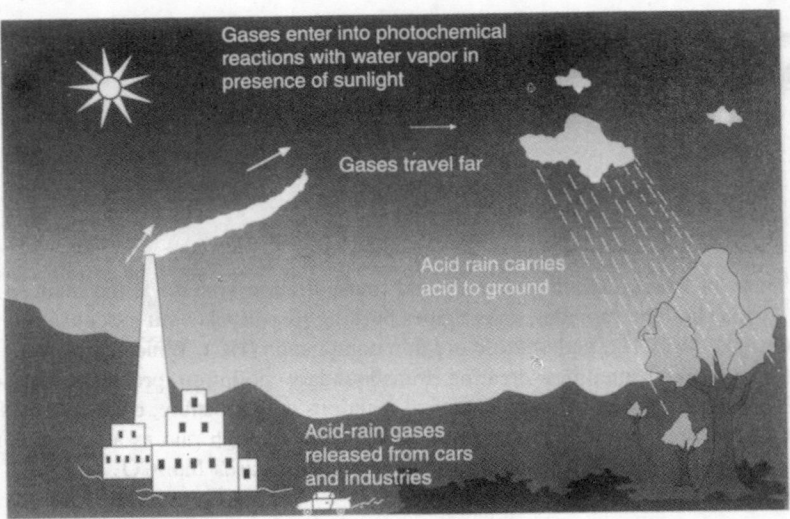

FIGURE 5.1 Formation of acid rain.

5.1.1 POLLUTANTS

Table 5.1 presents an estimate of three major pollutants from different sources. It suggests that combustion in stationery sources like boilers is the primary source of sulfur and nitrogen oxide emissions. Table 5.2 compares the emission levels of several major air pollutants from pulverized coal (PC) and circulating fluidized bed (CFB) fired boilers. It shows that the firing system has an important influence on the level of pollution produced by the boiler. The effect of air pollution is divided into three categories:

1. Urban pollution
2. Acidification
3. Climate change

Urban pollution is generated in urban areas and affects the people living there. Pollution due to acid rain travels much further, endangering plants, vegetation and living species in a geographical region. Climate change is a global phenomenon, unrestricted by national or regional boundaries. The following section discusses these pollution types in more detail.

TABLE 5.1
Air Pollution Emissions in the United States

	Emission (million tons/year)		
Source	HC	NO_x	SO_x
Combustion in stationary sources (boilers, etc.)	1.1	9.10	18.2
Transportation	8.0	8.5	0.8
Industrial processes	11.1	0.7	4.3
Others	3.0	1.0	0.8

Source: From US/Canada Memorandum of Intent on Transboundary Air Pollution, Feb., 1983, 1–7.

TABLE 5.2
Emissions from Typical Circulating Fluidized Bed and Pulverized Coal Fired Boilers Firing Bituminous Coal

	Peak Temperature	NO_x	SO_2	Fly Ash	N_2O
Circulating fluidized bed	862	133	397 at Ca/S 2.5	5–65 opacity[a]	74
	901	197	366 at Ca/S 3.5		26
Pulverized coal-fired	1600	480–950			<10

[a] Makansi, 1994.

5.1.1.1 Urban Pollution

Smog is a major result of air pollution that affects city life and is a rapidly increasing problem. It irritates the eyes, causes mild to severe respiratory problems and affects vegetation. In a particularly serious case in 1952, severe smog stayed over the city of London for 5 days from December 5th to 9th causing the deaths of nearly 4000 people. There are two types of smog:

1. Smoke
2. Photochemical

5.1.1.1.1 Smoke Type

This is caused by the emission of sulfur dioxide, dust, and moisture. It produces droplets of sulfuric acid. Such smog is more common in countries with uncontrolled smoke-belching industries, polluting automobiles, and dusty roads. Moisture from air or fog creates the smog.

5.1.1.1.2 Photochemical Type

This occurs in relatively clean places with little or no visible pollution, especially on sunny days due to intense traffic. Nitrogen oxides in exhaust gases and hydrocarbons from various human activities and biogenic sources react in the presence of sunlight to produce a noxious mixture of aerosols and gases. Photochemical smog contains ozone, formaldehyde, ketone, and peroxyacetyl nitrates.

Unlike carbon monoxide, volatile organic compounds (HC) are not poisonous themselves, at least in the concentration normally found in the atmosphere, but they are considered major pollutants due to their indirect effects. In the presence of sunlight, HC reacts with nitrogen oxide to form a complex variety of secondary pollutants, called photochemical oxidants, the chief constituents of smog.

Colorless nitric oxide (NO) is sometimes oxidized by ground level ozone to a brownish nitrogen dioxide (NO_2), which absorbs ultraviolet radiation from sunlight, splitting itself into NO and atomic oxygen, which in turn forms ozone. The presence of ozone at ground level is potentially dangerous, especially to plants and people with respiratory problems. Should its concentration exceed 80 to 120 ppb, it may be a significant human health hazard.

5.1.1.2 Acid Rain

Acid rain is caused by the presence of sulfur dioxide and nitrogen oxides in rain or snow. Over the past two decades, rain has increased in acidity in North–Eastern USA from a normal pH of 5.6 to one of 4.5 and sometimes as low as 3.6 (Cogbill and Likens, 1974). Acid rain has adverse effects on:

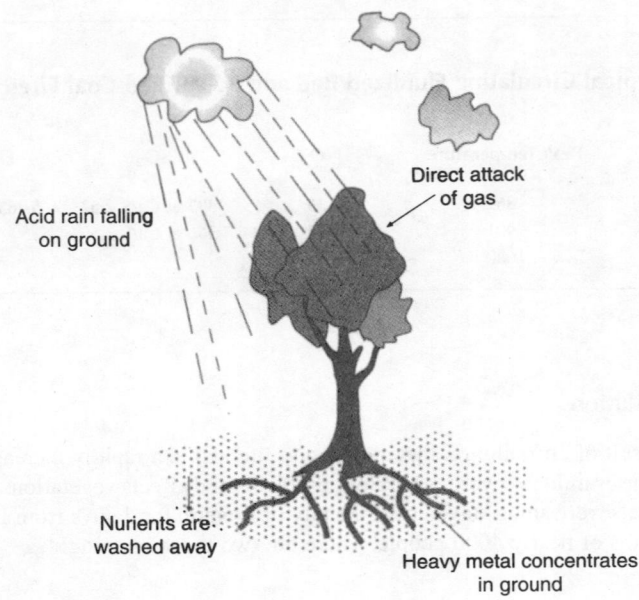

FIGURE 5.2 Gaseous pollutants attack plants in three ways.

- Trees and crops
- Water and aquatic species
- Buildings
- Human health

Acid rain affects plants and animals at a great distance from its sources. Figure 5.2 shows how it attacks plants. A direct attack on the needles and leaves of trees comes from pollutants in the air. Acid rain, falling on a tree, causes damage to it. Rain falling on to the ground acidifies the topsoil and enriches it with heavy metals while nutrients are washed away. Subsequent disturbances in water and nourishment absorption cause damage to the fine roots of plants.

Controlled laboratory and field studies have demonstrated decreased productivity of plants, necrosis of leaves, nutrient loss from foliage and soils, weathering of leaf surfaces, and a change in the pathogenicity of plant parasites (Likens and Borman, 1974). Certain species of fish (salmon and trout) are particularly sensitive to acid rain. Acid rain adversely affects microorganisms and changes the yields of various plants and aquatic species.

A visible effect of acid rain is the damage to paints and metals. The plasters and stonework of many monuments are damaged through the reaction of acids on their surfaces. The pollution from a 1000 MWe power plant can cause damage to property as high as $8 to $16 million per year. An exceptionally high concentration of sulfur may also affect human health. The United States Environmental Protection Agency has set the upper limit of sulfur dioxide for safety to human health as 80 $\mu g/m^3$ on an annual average and 365 $\mu g/m^3$ on a 24-h average exposure (Elliot, 1981).

5.1.1.3 Climate Change

Many experts believe that the emission of certain gases may bring about irreversible changes in the world's climate. Global warming, for example, is an important example of climate change. The sun's radiation, as shown in Figure 5.3, is mostly of short wavelength allowing it to penetrate

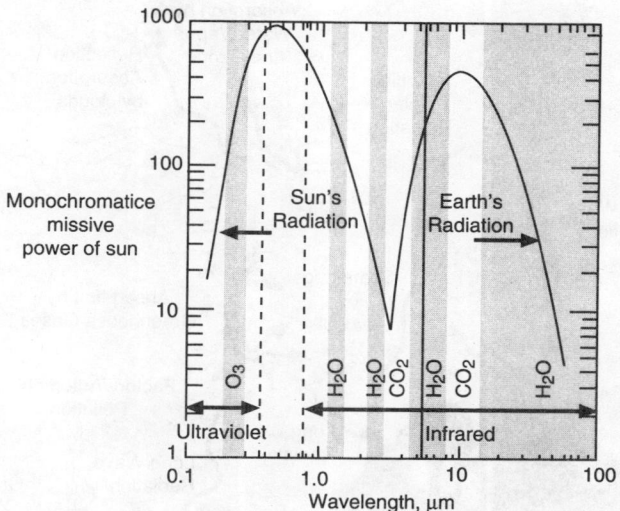

FIGURE 5.3 Absorption bands of both CO_2 and H_2O covers the infrared radiations and as a result these two gases do not allow the thermal radiation from the earth to pass through.

through the earth's atmosphere to reach the ground, but the radiation reflected from the earth is in the longer, infrared range of wavelength, which is absorbed by gases like water vapor, carbon dioxide, methane, and nitrous oxide. These gases trap the heat as a greenhouse does.

Figure 5.4 shows how the greenhouse gases contribute to the warming of earth, a phenomenon known as global warming. General thermal equilibrium keeps the earth's surface warm at an average (over the entire surface of the earth) temperature of 17°C. Excessive emission of the above heat trapping gases, known as greenhouse gases, could trap more heat, raising the earth's average temperature, which in turn could cause permanent change in climates. Some of the implications of global warming are demonstrated through:

- Impact on plant metabolism.
- Higher annual temperatures causing melting of polar ice with a resulting rise in sea level.
- Shift of agro-climatic regions due to changes in rainfall pattern.
- Possible changes in weather patterns.
- Impact of possible increased cloudiness on the growth of crops.

The combustion of fossil fuels is a major source of carbon dioxide released to the atmosphere. Nitrous oxide (N_2O) is another major greenhouse gas emitted from fossil fuel-fired plants. Thermal power plants do not produce as much methane (CH_4), which is at least 21 times more damaging than CO_2 as far as global warming is concerned, as by natural sources like forests and paddy fields.

Carbon dioxide and other gases such as methane, nitrous oxide and chlorofluorocarbons (CFCs) contribute to the greenhouse effect, which traps heat inside the Earth's atmosphere as shown in Figure 5.4. Water vapor traps a considerable amount of heat like CO_2, but it is not classified as a greenhouse gas because its lifespan as vapor is very small.

Different gases absorb heat to different extents depending upon their absorption spectra and other factors. To simplify everything and bring all of them to a common denominator,

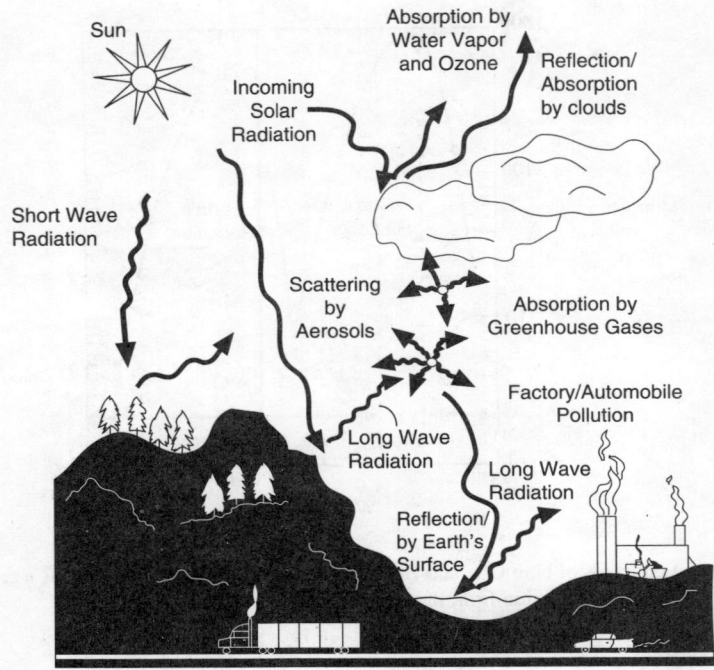

FIGURE 5.4 A part of the energy received from the sun is trapped by greenhouse gases to cause global warming.

an equivalent amount, known as global warming potential (GWP), is defined. Table 5.3 lists the GWP values of some gases.

5.1.1.4 Other Effects

Besides smog, acid rain, and global warming, gases emitted from power plants could cause other types of damage as well. Emission of mercury (Hg) is a good example. Mercury pollution contaminates fish in lakes, reservoirs and rivers. Eating mercury-contaminated fish can result in damage to the human nervous system, causing developmental delays and learning disabilities, and posing special risks for children and women of child-bearing age.

TABLE 5.3
Global Warming Potential (GWP) of Some Greenhouse Gases

Gas	20-Year GWP	100-Year GWP	500-Year GWP
Carbon dioxide (CO_2)	1	1	1
Nitrous oxide (N_2O)	280	310	170
Methane (CH_4)	56	21	6.5
CFC-11		4600	

Source: U.S. Greenhouse Gas Inventory Program, 2002.

5.1.2 EMISSION STANDARDS

Many countries have imposed ceilings on the emission of harmful gases and particles from pollution sources. These ceilings are in different forms. Some stipulate maximum emissions from individual units while others put an overall ceiling across a region or an entire country. Some countries impose a tax on the amount of pollutant generated. New or larger capacity boilers may have more stringent emission standards than smaller or older ones. Individual states or provinces may also have a ceiling lower than the national ceiling. Emission limits for most developed countries are generally as follows (Oka, 2004):

- Particulates < 50 mg/m^3
- Sulfur dioxide < 400 mg/m^3
- Nitric oxide < 200 mg/m^3
- Carbon monoxide < 40 mg/m^3

The specific emission limits of some countries are given in Table 5.4 for comparison. These values depend on the capacity of the plant as well as whether it is a new or old plant. Some countries allow higher emission limits for older plants for economic reasons. The USA stipulates a cap and allows trading of emission allowances within the cap (EPA, 2005).

While a significant reduction in sulfur dioxide emissions has been achieved in many industrialized countries, significant growth in industrial activities in some large countries is adding to the global sulfur dioxide emission at an increasing rate. Since gaseous pollutants do not follow any national boundary, the rising emission in one country is a concern for all, and especially for neighboring countries.

5.2 SULFUR DIOXIDE EMISSION

This section describes the mechanism of sulfur dioxide production and its absorption in CFB boilers.

TABLE 5.4
National Standards of Emission from New Coal-Fired Boilers for a Typical 150 MWe Capacity Boiler in mg/m^3

Country	SO$_2$	NO$_2$	Particulate
Australia		500–800	80–250[1]
Austria	200–400	200	50
Belgium	1200	800	50
Canada[#]	700	614	116
France	local	local	95
Germany	400 + 85%	800/1000	50
Italy	400	650/1200	50
Japan	133–267	400	100
Netherlands	700	500/400	50
Spain	5000	none	150
Switzerland	2000	200	50
European Community	2000	100/50	none
USA	740–1480	750	35

[1] – For 12% CO$_2$; # – guideline.
Source: Adapted from Oka, 2004.

The sulfur content of coal varies widely in the range of 0.1 to 10.0% and may occur in three forms:

1. Pyritic
2. Organic
3. Sulfate

5.2.1 Chemical Reactions

The following chemical reactions play an important role in the formation and capture of sulfur dioxide in fluidized beds typically operating in the temperature range of 800 to 900°C.

5.2.1.1 Formation of Sulfur Dioxide

When coal burns, the sulfur is oxidized primarily to sulfur dioxide.

$$S + O_2 = SO_2 + 296 \text{ kJ/g mol.} \tag{5.1}$$

The emission of sulfur dioxide is often expressed in parts per million (ppm). Since the heating value of the coal varies, the sulfur released from a plant with a given thermal input will also vary. Therefore, the SO_2 emission is sometimes also expressed in terms of pollutants released per unit of energy released (g/MJ or lb/million BTU).

The mineral matter in coal may contain some calcium oxide, CaO, which may absorb a part of the sulfur dioxide as calcium sulfate immediately after the former is released.

$$CaO + \tfrac{1}{2}O_2 + SO_2 = CaSO_4 + 486 \text{ kJ/g mol.} \tag{5.2}$$

The rest of the sulfur dioxide escapes into the atmosphere. Part of the sulfur dioxide may be converted into sulfur trioxide.

$$SO_2 + \tfrac{1}{2}O_2 = SO_3 \tag{5.3}$$

The formation of sulfur trioxide depends on gas residence time, temperature, excess air, and the presence of catalytic surfaces in the furnace. The reaction shown in Equation 5.3 is favored by high temperature and pressure. Since the reaction is slow, only a small part of the sulfur dioxide has time to convert into sulfur trioxide (Burdett et al., 1983, 1984). The sulfur trioxides formed comes in contact with moisture in the flue gas and readily forms sulfuric acid, which is condensed on cold surfaces.

$$SO_3 + H_2O = H_2SO_4 \tag{5.4}$$

The condensation of the moisture in the flue gas below the dew point leads to its collection on carbonaceous materials, which forms agglomerates that flake off and are emitted through the chimney as acid smog.

5.2.1.2 Retention of SO_2

Limestone ($CaCO_3$) and dolomite ($CaCO_3 \cdot MgCO_3$) are two principal sorbents used for the absorption of sulfur dioxide in fluidized bed combustors. In addition to these, some synthetic sorbents are also being developed, but their use is still restricted.

The overall chemical reaction of sulfur dioxide with limestone that leads to its retention is:

$$CaCO_3 + SO_2 + \tfrac{1}{2}O_2 = CaSO_4 + CO_2 \tag{5.5}$$

5.2.1.3 Calcination

Equation 5.5, which absorbs SO_2, does not take place in one step. The first step is calcination, where the limestone decomposes into CaO and CO_2 through an endothermic reaction:

$$CaCO_3 \rightleftharpoons CaO + CO_2 - 183 \text{ kJ/g mol}. \tag{5.6}$$

The direct reaction of calcium carbonate with sulfur dioxide (Equation 5.5) at fluidized bed temperatures (800 to 900°C) is so slow that the sulfur dioxide absorbed through direct reaction with calcium carbonate is insignificant. Reaction 5.5 has to take place through the decomposition (calcination) of $CaCO_3$ to calcium oxide and carbon dioxide (Reaction 5.6). The carbon dioxide released during the calcination creates and enlarges many pores in the limestone (Figure 5.5), which exposes greater surface area for the subsequent sulfation reactions, Equation 5.10 and Equation 5.11.

The calcination Reaction 5.6 is relatively fast. For example, the calcination of a 0.5-mm particle would take around 50 sec to complete. The sulfation Reaction 5.2 on the other hand is relatively slow. For a 0.5-mm calcined particle, it would take about 1200 sec to complete (Arsic et al., 1988).

The calcination Reaction 5.6 occurs only if the partial pressure of carbon dioxide is less than the equilibrium partial pressure of carbon dioxide, P_e, at the calcination temperature (Stantan, 1983). The equilibrium partial pressure, P_e is given as:

$$P_e = 1.2 \times 10^7 \exp(-E/RT) \text{ bar} \tag{5.7}$$

where E is the activation energy (159,000 kJ/kmol), and R is the universal gas constant (8.314 kJ/kmol K).

From Equation 5.7 we see that for each partial pressure of CO_2 there is a corresponding temperature known as the equilibrium calcinations temperature. No reaction will take place if the combustor temperature is higher than the equilibrium calcination temperature for that partial pressure of CO_2.

The partial pressure of CO_2 in a gas depends on the amount of excess air and the total pressure. Figure 5.6 shows the equilibrium calcination temperature for a limestone at different partial pressures of carbon dioxide. It also shows how this limit changes with excess air and combustor pressure. For example, if the carbon dioxide in the flue gas is 18.5% by volume, its partial pressure at combustor pressures of 1 and 10 bar will be 0.185 and 1.85 bar, respectively. From Figure 5.6, one can find that the partial pressures of CO_2 at 0.185 and 1.85 bar will correspond to equilibrium temperatures of 787 and 943°C, respectively. If the excess air increases at any of the pressures, the minimum temperature for calcination decreases.

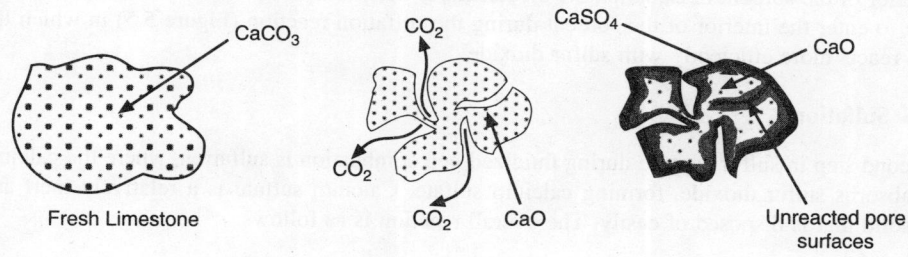

FIGURE 5.5 Absorption of sulfur dioxide by sorbents.

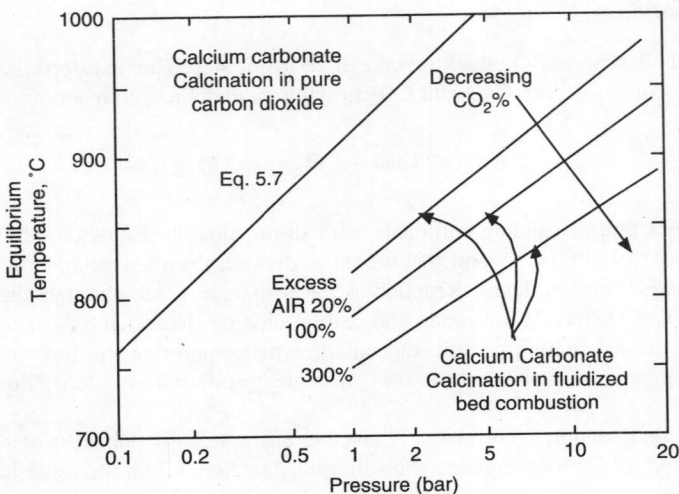

FIGURE 5.6 Effect of combustor pressure and excess air (or carbon dioxide partial pressure) on the equilibrium calcinations temperature of $CaCO_3$.

From this discussion one can see that the limestone could calcine in an atmospheric pressure fluidized bed but not in a pressurized fluidized bed. For that reason, one must use dolomite in a pressurized bubbling fluidized bed involving half-calcination.

5.2.1.4 Half-Calcination

Dolomite, a compound of calcium carbonate and magnesium carbonate, is another type of sorbent used especially in pressurized fluidized bed combustion because limestone does not calcine at typical combustion temperatures at pressures above 3 atmospheres. When heated, it forms half-calcined dolomite as below:

$$CaMg(CO_3)_2 = CaCO_3 \cdot MgO + CO_2 - 128 \text{ kJ/g mol} \tag{5.8}$$

$$CaCO_3 + MgO + SO_2 + \tfrac{1}{2}O_2 = CaSO_4 \cdot MgO + CO_2 \tag{5.9}$$

The magnesium oxide reacts so slowly with SO_2 at temperatures between 540 and 980°C that it hardly makes any contribution to the sulfur capture. However, the carbon dioxide released from the interior of the sorbent in Equation 5.9 creates pores in the sorbent particles. This allows sulfur dioxide to enter the interior of the sorbent during the sulfation reaction (Figure 5.5) in which the $CaCO_3$ reacts more efficiently with sulfur dioxide.

5.2.1.5 Sulfation

The second step in sulfur capture during fluidized bed combustion is sulfation, where the calcium oxide absorbs sulfur dioxide, forming calcium sulfate. Calcium sulfate is a relatively inert and stable solid that is disposed of easily. The overall reaction is as follows:

$$CaO + SO_2 + \tfrac{1}{2}O_2 = CaSO_4 + 486 \text{ kJ/g.mol} \tag{5.10}$$

Emissions

This reaction involves an increase in the molar volume as:

$$\text{CaO molar volume} = 16.9 \text{ cm}^3/\text{mol}$$

$$\text{CaSO}_4 \text{ molar volume} = 52.2 \text{ cm}^3/\text{mol}$$

$$SO_2 + \tfrac{1}{2}O_2 = SO_3$$

$$CaO + SO_3 = CaSO_4 \tag{5.11}$$

The second route of reaction through SO_3 is followed only with assistance from some heavy metal salt catalysts (Yates, 1983).

5.2.1.6 Pore Plugging

The calcium oxide produced by calcination (Equation 5.6) cannot be fully converted into calcium sulfate through the above reactions. The sulfation reactions involve an increase in the volume of the solid phase as shown in Equation 5.10. One mole of $CaCO_3$ produces 1 mol of $CaSO_4$. Under normal conditions, however, one mole of $CaCO_3$ occupies 36.9 cm^3, while 1 mol of $CaSO_4$ has a volume of 52.2 cm^3. Thus, the pores and pore entrances of this sorbent are plugged due to the larger volume of reaction products built up before the interior of the sorbent has a chance to react completely (Figure 5.5).

Therefore, only a fraction of the sorbent can be utilized, creating a major problem with sulfur capture in fluidized bed boilers. For this reason, a fluidized bed boiler plant must use two or three times the amount of limestone it actually needs for sulfur capture, and dispose of equal amounts of extra solid waste.

5.2.1.7 Reverse Sulfation

It has been observed that under certain conditions the calcium sulfate may be reduced to release SO_2 (Talukdar et al., 1996):

$$CaSO_4 + CO = CaO + SO_2 + CO_2 \tag{5.12}$$

$$CaSO_4 + 4CO = CaS + 4CO_2 \tag{5.13}$$

$$CaO + SO_2 = CaS + \tfrac{3}{2}O_2 \tag{5.14}$$

The lower section of a CFB combustor, which is rich in CO, operates under substoichiometric conditions due to the staged addition of combustion air. The reaction in Equation 5.12, though favored at very high temperatures, may still occur at a relatively low temperature of about 850°C if sufficient CO is available (Ghardashkhani et al., 1989). The staged addition of combustion air, a characteristic of CFB boilers, is thus not necessarily favorable to efficient sulfur capture.

5.2.1.8 Improved Utilization of Sorbents

Owing to the pore-plugging phenomenon, only a fraction of the calcined limestone is utilized in capturing sulfur. The interior of the sorbent particle (40 to 70%) remains as unreacted CaO. This greatly increases the cost of limestone and the disposal of spent sorbent. Several options have been suggested (Ahlstrom Pyropower, 1992) to improve the utilization of limestones including the following:

- Physical grinding of particles
- Dehydration of spent sorbents with steam or water
- Slurrying and reinjection of ash
- ADVACAT process

The first process involves grinding of the ash from the bed drain as well as from the first pass of particulate collection equipment. This exposes the unreacted core of the spent sorbent before it is reinjected.

The next two processes involve hydration. Water enters the interior of the sorbent and reacts with CaO to produce $Ca(OH)_2$. The high molar volume of hydroxide as well as the release of steam upon decomposition of this hydroxide after reinjection breaks open the particles and exposes the fresh sorbent surface for reaction in the furnace.

The ADVACAT process involves production of reactive hydrated calcium silicate for cold-end absorption of sulfur dioxide.

5.2.2 REACTIONS ON SINGLE SORBENT PARTICLES

The sulfation reaction is a relatively slow process. For example, a 0.5-mm particle would take about 1200 sec to sulfate while taking only about 50 sec to calcine. Thus, the sulfation reaction controls the rate of sulfur capture. The sulfation rate of calcium oxide particles $R(t)$, through the reaction ($CaO + SO_2 + \frac{1}{2}O_2 = CaSO_4$) is found to decrease exponentially with time (Lee et al., 1980).

$$R(t) = R_0 \exp\left(-\frac{t}{t_p}\right) \tag{5.15}$$

where R_0 is the initial reaction rate, i.e., the moles of SO_2 absorbed by one particle in unit time (kmol/sec particle). The pore-plugging time constant, t_p is inversely proportional to the sulfur dioxide concentration, C_{SO_2} (Chrostowski and Georgakis, 1978):

$$t_p = \frac{\alpha}{C_{SO_2}} \tag{5.16}$$

where α is the proportionally constant.

The initial reaction rate, R_0 is proportional to the sulfur dioxide concentration, C_{SO_2} and the volume of the particle, V_p and inversely proportional to the sulfation time constant, t_{sf}. The sulfation reaction is first-order in sulfur dioxide and zero-order in oxygen (Borgwardt, 1970):

$$R_0 = \frac{V_p C_{SO_2}}{t_{sf}} \tag{5.17}$$

The sulfation time constant depends on reactivity, size, pore characteristics, and composition of the sorbent.

The extent of sulfation or the fractional weight of a sulfated sorbent particle is found by integration of the rate of sulfation of a particle (Equation 5.15):

$$\delta(t) = \frac{1}{M_{Ca}} \int_0^t R(t) dt = \delta(\infty)\left[1 - \exp\left(-\frac{tC_{SO_2}}{\alpha}\right)\right] \tag{5.18}$$

where M_{Ca} is moles of calcium in the sorbent particle and $\delta(\infty)$ is the asymptotic or maximum extent of conversion. The latter is found to be related to other variables as (Lee et al., 1980):

$$\delta(\infty) = \frac{\alpha M_{CaCO_3}}{t_{sf} \rho_p X_{CaCO_3}} = \frac{PV_p}{M_{Ca} t_{sf}} \qquad (5.19)$$

where M_{CaCO_3} is the molecular weight of calcium carbonate, X_{CaCO_3} is the weight fraction of calcium carbonate in the sorbent particle, and ρ_p is the density of uncalcined sorbent.

However, experimental data on the extent of sulfation, $\delta(t)$ is found to show that the dependence on time is slightly different from that in Equation 5.18. It is of the form:

$$\delta(t) = \delta(\infty)[1 - \exp(-k't^n)] \qquad (5.20)$$

Here, k' and n are fitting constants. From similarity, $k' = C_{SO_2}/\alpha$. The value of n is found to lie between 0.6 to 1.2 for sorbents from the United States (Fee et al., 1983) and Canada (Hamer, 1986).

Fee et al. (1983) suggested that over the range of normal operation of fluidized bed combustors, n may be taken as equal to 1.0. This assumption does not compromise the prediction accuracy to any great extent.

Now, if we define a reaction rate constant, K such that when multiplied by the SO_2 molar concentration it will give the rate of change of the sulfation extent of sorbent particles. One can write this from Equation 5.20 as:

$$K = \frac{1}{C_{SO_2}} \frac{d\delta(t)}{dt} = \frac{[\delta(\infty) - \delta(t)]}{\alpha} \qquad (5.21)$$

The reactivity of limestone, K may be expressed through the above equation. The significance of this is that when multiplied by the total moles of calcium in the bed ($K\,C_{SO_2}$), it directly gives the molar rate of sulfation of calcium or absorption of SO_2 (since 1 mol of SO_2 is absorbed by 1 mol of calcium oxide to form 1 mol of calcium sulfate).

Some typical values of α and $\delta(\infty)$ for some Canadian limestones are presented in Table 5.5. These were obtained by linearizing the data of Hamer (1986). Fee et al. (1982) generated similar data for several American sorbents.

TABLE 5.5
Reactivity of Some Limestone $K = (1/\alpha)[\delta(\infty) - \delta]$ m³ SO_2 per kmol of Calcium in Sorbent per sec

Sorbent Source	CaCO₃ (%)	MgCO₃ (%)	$\delta(\infty)$	1/α (m³/kmol)
Irish Cove	94.40	1.00	0.284	14.48
Calpo	95.60	1.99	0.340	13.46
Calpo	96.10	0.90	0.535	10.23
Brookfield	83.90	1.97	0.164	27.25
Glencoe	92.10	3.84	0.221	21.88
Glendale	92.10	5.77	0.236	22.97
Havelock	94.60	1.83	0.330	15.14
Carlisle	98.40	1.00	0.267	16.09
Elmtree	91.40	1.00	0.198	21.68
Clausen	90.10	2.89	0.272	19.12
Syncrude	91.30	2.49	0.229	18.89
Exshaw	97.00	—	0.227	13.52

5.2.3 REACTIVITY OF SORBENTS

A limestone particle reacts much more slowly than a coal particle. For example, a 3.0-mm lignite particle can take between 10 and 340 sec to complete its devolatilization and combustion at 850°C, depending on the oxygen concentration, but a sorbent particle can take up to 40 min for calcination alone and up to several hours to complete the sulfation process, depending on the size of the sorbent particle and its properties.

The rate $R(t)$ in Equation 5.15 is the instantaneous sulfation rate (kmol/sec) for a particle of a given size. It can be seen from the above equations that with a knowledge of α and $\delta(\infty)$, one can determine the reactivity, K and initial reaction rate, R_0. The reaction rate, $R(t)$, will depend on:

- Particle size
- Level of sulfation
- Temperature
- Combustor pressure
- CO_2 concentration (affects only the calcination)
- Pore size distribution after calcination

5.2.3.1 Measurement of Reactivity

The reactivity of a sorbent is an important parameter used in sorbent selection, as well as in the prediction of the sulfur capture performance of a combustor. Two methods can be used to measure the reactivity:

- Fluidized bed reactor (FBR)
- Thermo-gravimetric apparatus (TGA)

5.2.3.1.1 Fluidized Bed Reactor

A bed of inert ash (or sand) in a reactor (Figure 5.7) is fluidized by a synthetic mixture of sulfur dioxide, nitrogen, oxygen, and carbon dioxide gases at a typical temperature for combustion (800 to 900°C). A small, but measured amount of the sorbent of the required size is dropped into the bed.

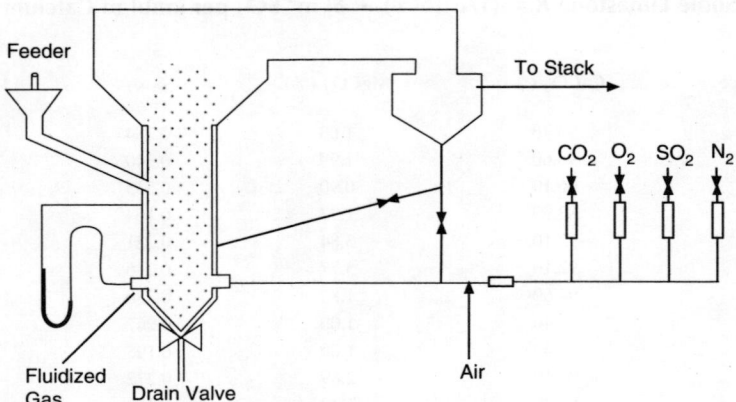

FIGURE 5.7 A turbulent fluidized bed apparatus for measurement of reactivity in fluidized beds.

Emissions

The concentration of sulfur dioxide at the exit of the reactor is monitored continuously. When no further change in the concentration is noted, the bed material is removed for analysis. From the time trace of SO_2 concentration in the flue gas, the sorbent reactivity is determined in the same way as is done for standard TGA.

This gives reliable data for bubbling fluidized bed combustors when the reactor is operated in bubbling regime. For the simulation of sulfation reactions in CFB combustors, the reactor should be operated in the turbulent bed regime. Though the expanded section (Figure 5.7) returns substantial fraction of the high level of entrainment from the bed, long-duration experimentation is difficult. For such simulations it may be necessary to use the quartz wool matrix (QWM) reactor described below.

5.2.3.1.2 Quartz Wool Matrix Apparatus

The QWM apparatus is an electronic balance with a weighing pan maintained in a controlled environment (Figure 5.8). It is an alternative to the thermo-gravimetric apparatus popularly used for many fixed bed reactions.

In CFB combustors, the sorbent is highly dispersed. Reaction products from one particle do not affect the other. The reacting gases also have free, uniform access to all sides of the reacting particles. This does not happen in a TGA apparatus, where particles are ground and stacked in a pan with restricted passage to particles inside the heap.

In a QWM apparatus, a small sample (a few milligrams to a few grams) of sorbent particles of the required size is sprinkled over highly expanded quartz wool (Figure 5.8). This allows the particles to be supported and yet dispersed as they are in a fast bed. The wool is suspended from an electronic balance and exposed to a temperature similar to that anticipated in a CFB combustor. The sample is first calcined in a stream of carbon dioxide and nitrogen. When further weight loss of the sample ceases, a synthetic mixture of sulfur dioxide, oxygen, and carbon dioxide is passed through the apparatus.

The composition of the synthetic mixture should be approximately the same as expected in the combustor. The weight of the sorbent changes as it is sulfated. A continuous trace of the sorbent's weight gives the change in sulfation level with time (Figure 5.9). From this graph, the rate of

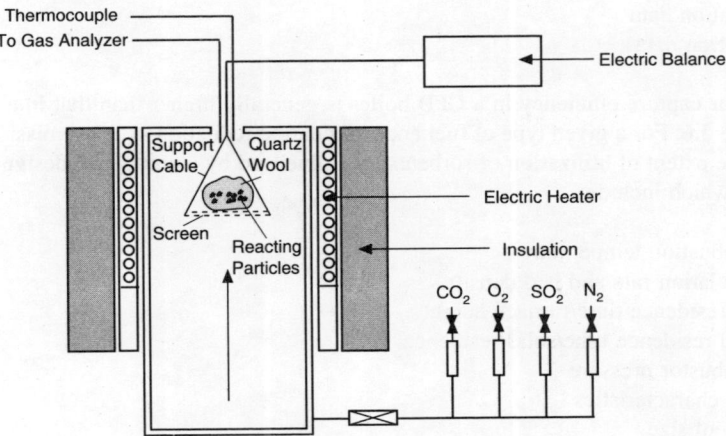

FIGURE 5.8 A schematic of the quartz wool matrix for measurement of reactivity apparatus.

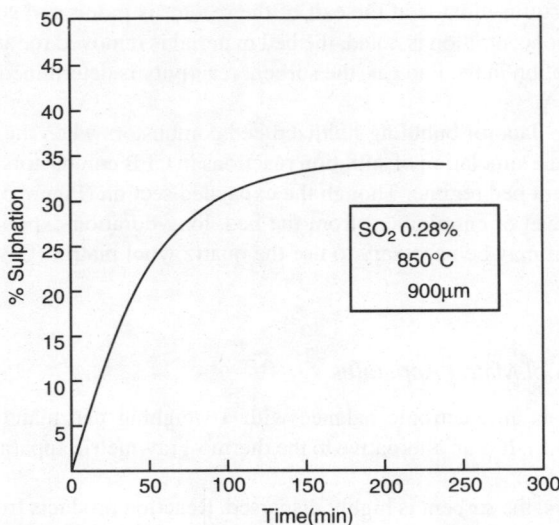

FIGURE 5.9 A typical TGA (thermo-gravimetric apparatus) curve for sorbents.

sulfation, $d\delta(t)/dt$ is taken at the required level of sulfation, $\delta(t)$. The reactivity can be found using Equation 5.20 and Equation 5.21.

5.2.4 SULFUR CAPTURE IN FLUIDIZED BEDS

The mechanism of sulfur capture in the foregoing discussion is common to all combustion processes. The major characteristics of limestone that affect sulfur capture in a fluidized bed system include:

- Chemical, petrographic, and mineral composition
- Physical properties, such as bulk and particle densities, crushing strength, pore volume and size, surface area, particle size, and shape and hardness
- Sulfation data
- Attrition data

The sulfur capture efficiency in a CFB boiler is generally higher than that in a BFB boiler as seen in Table 5.6. For a given type of fuel and sorbent, the reduction in SO_2 emission from a CFB boiler and the extent of utilization of sorbents are influenced by a number of design and operating parameters, which include:

- Combustion temperature
- Circulation rate and bed density
- Gas residence time/furnace height
- Solid residence time/solid residence
- Combustor pressure
- Pore characteristics
- Sorbent size

The effects of these physical parameters on sorbent utilization are briefly described below.

TABLE 5.6
Comparison of Sulfur Dioxide and Nitric Oxide Emission from Bubbling and Circulating Fluidized Bed Boilers

	Ca/S	SO₂ Capture Efficiency (%)	
		CFB	BFB
Sulfur capture efficiency. High reactivity limestone	1	70	—
	1.2	75	—
	1.3	80	—
	1.5	85	—
	1.7	90	—
	2	93	66
	2.2	—	70
	2.5	—	75
	3.0	—	78
	3.5	—	84
		—	90
NO_x emission (Nitrogen in coal 0.29 mg/MJ)	CO emission 40–125 ppm	Emission of NO_x @6% O_2	
		250–320 ppm	100–160 ppm

Source: Data collected from Tang and Engstrom, 1987.

5.2.4.1 Combustion Temperature

Reactivity increases with temperature, reaching an optimum value at around 800 to 850°C (Ehrlich, 1975) after which it drops. It drops because with rising temperature the sulfation rate increases quickly, plugging the pores of the calcium oxides, further inhibiting the utilization of the sorbent. The drop in sulfur capture at higher temperatures is also attributed to the decomposition of $CaSO_4$ (Fields et al., 1979). Figure 5.10 shows the effect of temperature on sorbent demand for a given level of sulfur capture. The demand of sorbents for sulfation decreases between 700 and 850°C and

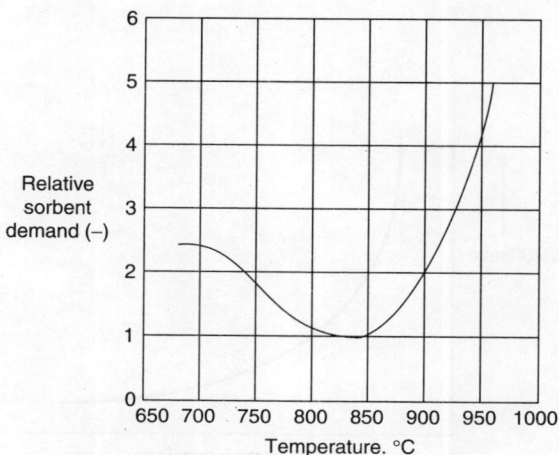

FIGURE 5.10 Effect of combustion temperature on sorbent demand for a desired sulfur capture efficiency.

then again increases above 850°C with rather sharp rise above 900°C. This is noticed in most commercial CFB boilers, where the sorbent requirement rises rapidly if the furnace temperature rises above 900°C due to any reduction in the heat absorption of the furnace.

5.2.4.2 Bed Density and Circulation Rate

Unlike in a bubbling fluidized bed, the density of the CFB is not constant. It depends on several operating parameters. The effects of bed density and circulation rate on sulfur capture are speculative at best. Some data show improvements in sulfur capture at higher bed densities (Basu et al., 1989), while there is a possibility that at higher bed densities the average extent of sulfation increases, which tends to reduce the sulfur capture rate of individual sorbent particles. This may result in lower sulfur capture at higher bed densities.

5.2.4.3 Gas Residence Time and Furnace Height

The sulfation reaction is slow, as such longer gas residence is favorable for sulfur capture. The gas in a fluidized bed is generally in plug flow. The gas residence time may therefore be taken as H/U, where H is the furnace height and U is the superficial velocity. Thus, H and U affect sulfur emission from the furnace. In a bubbling fluidized bed, a higher fluidizing velocity reduces the residence time and increases the sorbent entrainment. It may, however, improve the mixing. As a result, the effect of fluidizing velocity is not straightforward for BFB boilers.

Sulfur capture in a CFB boiler takes place mainly above the secondary air level of the furnace. A longer residence time of SO_2 allows a greater fraction of CaO to convert into $CaSO_4$, but the rate of sulfation is proportional to the concentration of sulfur dioxide. Hence, the improvement in sulfur capture with furnace height decays exponentially, as predicted by Equation 5.25. This feature is common to both circulating and bubbling fluidized bed combustors.

Figure 5.11 shows that the Ca/S ratio needed to achieve 90% sulfur capture decreases quickly with increasing gas residence time, but for a longer residence time this improvement is much less. In commercial CFB furnaces, where the gas residence time is already quite high (in the range of 2 to 5 sec), a further increase in furnace height would only marginally improve the sulfur capture efficiency.

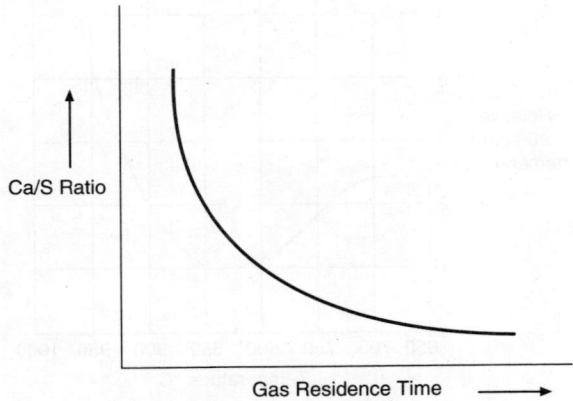

FIGURE 5.11 Limestone demand as a function of the residence time for a specific level of sulfur capture.

5.2.4.4 Solid Residence Time and Cyclone Performance

The sulfation of sorbents, being a relatively slow process, is also influenced by the residence time of solids in the CFB loop. The longer a sorbent particle stays in the CFB loop, the greater the extent of its conversion to $CaSO_4$, subject to its maximum sulfation level. The residence time of solids and the performance of the cyclone are interrelated.

The more efficient the cyclone, the smaller is the cut-off size of particles. This allows finer particles to stay longer in the system and therefore participate in the sorption process. However, particles finer than the cut-off size of the cyclone do not usually return to the bed, but escape partially unused. The average residence time of these particles, t_{fs} is found from:

$$t_{sf} = \int_0^H \frac{dx}{U_s(x)} \qquad (5.22)$$

where the average solid velocity at a height x, $U_s(x)$, is given by $G_s/\rho_b(x)$.

Coarser particles, which are captured by the cyclone, make many trips around the combustor. Their residence time, t_{cs} is found by assuming ideal mixing of solids in the combustor:

$$t_{cs}(d) = \frac{Wf(d)\Delta d}{\Sigma m_{d,out}} \qquad (5.23)$$

where $\Sigma m_{d,out}$ is the sum of all out-flows of sorbents of size d and $d + \Delta d$, and $Wf(d)\Delta d$ is the mass of solids within the size range of d and $d + \Delta d$.

An improvement in cyclone efficiency could greatly enhance the sulfur capture performance of a CFB boiler. For example, Lalak et al. (2003) noted an improvement in sulfur capture efficiency by 5.4% with the use of 40% less limestone when the cyclone of a 450 t/h CFB boilers was modified.

In a bubbling fluidized bed, the recirculation of fly ash from the cyclone, bag-house or electrostatic precipitators improves the sulfur capture efficiency. For example, at a Ca/S ratio of 3.2, the sulfur capture efficiency improved from 90 to 94% when the recirculation ratio was increased from 0 to 2.0 (Saroff et al., 1989).

5.2.4.5 Effect of Pressure on Calcination

Figure 5.6 and Equation 5.7 show that the equilibrium temperature for calcination of limestone decreases with decreasing pressure and a decreasing concentration of CO_2. For example, in atmospheric pressure, bubbling FBC limestone can easily calcine because the partial pressure of carbon dioxide is low and the corresponding equilibrium calcination temperature (Figure 5.6) is below 770°C for the normal condition of excess air. In pressurized beds, the partial pressure of CO_2 is much higher at the same excess air, and consequently the equilibrium calcination temperature is much higher. For the calcination reaction to occur, either the combustion temperature must be very high or the partial pressure of CO_2 must be very low. Limestones are therefore not easily calcined in pressurized bubbling fluidized beds, which are typically operated at around 850°C. Pressurized bubbling beds use dolomite, which easily calcines, as explained earlier in Equation 5.9.

In a CFB combustor, however, the partial pressure of CO_2 is very low in the lower part of the combustor because it normally operates under a substoichiometric condition. Limestone, fed at this level, is easily calcined at normal operating temperatures of 800 to 850°C, though the pressure of the combustor may be very high. The calcined limestone then absorbs sulfur in oxidizing conditions further up in the bed.

5.2.4.6 Sorbent Size

The advantage of using smaller sorbents is obvious because, as explained earlier (Figure 5.5), finer sorbents leave a relatively small unreacted core (on a weight-fraction basis) after sulfation. The hydrodynamics of a fast bed allow CFB boilers to use sorbent particles as fine as 100 to 300 μm, while bubbling bed boilers cannot use particles smaller than 500 to 1500 μm due to excessive entrainment. This is why CFB boilers require a Ca/S ratio of only 1.5 to 2.5 for 90% sulfur capture, while bubbling bed boilers require a ratio of 2.0 to 3.5 for the same level of sulfur retention. However, the sorbent size in a CFB boiler cannot be reduced much without a penalty. Finer sorbents easily escape through the cyclone before they are completely utilized. For a specific CFB combustor and a given limestone, there may be an optimum size of limestone particles for best utilization. However, this will depend on the pore characteristics of the limestone and the cyclone characteristics.

5.2.4.7 Pore-Size Distribution and Reactivity of Limestone

Both reactivity and pore-size distribution have an important influence on the sulfation rate of limestone. A more reactive sorbent has greater utilization than a less reactive one. Fine pores provide a large area of pore surfaces per unit weight of the sorbent, but their entrances are easily blocked. Large pores give free access to the interior of the sorbent but do not provide as much reaction surface area as fine pores. A good reactive sorbent will have a proper balance between large pores and fine pores, as well as a high pore volume. Once the pores are completely penetrated, there is no further gain in sulfur capture due to the reduction in the sorbent size. On the contrary, sulfur capture may be reduced due to the increased entrainment of finer particles.

5.2.4.8 Coal Characteristics

The ash in coal sometimes contains CaO, which is very effective in capturing sulfur from the coal. As it is well mixed within the coal matrix, it readily reacts with the sulfur in the coal when exposed to combustion conditions, forming $CaSO_4$. The amount of this calcium is generally small, but could be high in high-ash low-sulfur coal. However, utilization of this calcium is very high and sometimes even complete.

Highly volatile fuels generally have lower sulfur capture. The release and formation of SO_2 and H_2S during devolatilization may be the cause. Another contributing factor could be the over-bed feeding systems that are generally used for high volatile-content fuels.

5.2.4.9 Excess Air

Higher excess air reduces the oxygen deficient pockets in the bed where $CaSO_4$ is reduced, releasing further SO_2. Thus, one finds better sulfur capture with higher excess air.

5.2.4.10 Staging of Combustion Air

The combustion air is often split into primary and secondary air for better NO_x reduction. Such staging adversely affects the sulfur capture as it allows reduction of calcium sulfate with regeneration of SO_2.

5.2.5 SIMPLE MODEL FOR SULFUR CAPTURE IN CFB BOILERS

For the sake of overall assessment of sulfur capture in a CFB boiler, a simplified model is discussed below. It can demonstrate, at least qualitatively, the effects of feedstock characteristics, and some design and operating parameters on sulfur capture. Such a model may also be effective in extrapolating performance data from a known datum.

Model assumptions:

1. Gas is in plug flow.
2. Solids in the CFB furnace are well mixed.
3. Cross-sectional average bed density decreases exponentially from the secondary air level to the top of the furnace and is expressed as:

$$\rho_b(x) = \rho_b(\alpha) + [\rho_b(0) - \rho_b(\alpha)]\exp(-ax) \tag{5.24}$$

where $\rho_b(\alpha)$ is the asymptotic value of bed density. The exponent a can be found from a fitting of the given (or calculated) axial voidage distribution to this form.

4. The amount of sulfur dioxide capture in the substoichiometric region below the secondary air is negligible. All sulfur in the coal is converted into the sulfur dioxide within this region.
5. The kinetic rate of sulfation is proportional to the first-order sulfur dioxide concentration and volume of sorbents.

A mass balance of sulfur dioxide over a horizontal slice of the combustor gives:

$$U\frac{dC_{SO_2}}{dx} = m - KC_{SO_2}C_c \tag{5.25}$$

where m is the molar rate of formation of sulfur dioxide per unit volume of the bed, C_{SO_2} is the concentration of sulfur dioxide, U is the superficial velocity of gas, and C_c is the molar concentration of calcium in the bed given by

$$C_c = \frac{f_c X_{CaCO_3} \rho_b(x)}{M_{CaCO_3}}$$

The reaction rate (KC_{SO_2}), as explained with Equation 5.21, is the molar rate of SO_2 absorption per mole of calcium in the sorbent. It is a function of the maximum $\delta(\infty)$ and current $\delta(t)$ extent of conversion of CaO into CaSO$_4$ (sulfation), and is given by Equation 5.21. The maximum extent of sulfation, $\delta(\infty)$ is found through sorbent characterization experiments.

Since sulfur dioxide is assumed to be released only below the secondary air entry level, the concentration of SO_2 at $x = 0$ is:

$$C_{SO_2} = C_{SO}$$

and

$$m = 0; \quad 0 < x < H$$

The concentration of SO_2 in the exit gas may be found by integrating Equation 5.25 with Equation 5.24 using the above-mentioned boundary condition:

$$C_{SO_2}(H) = C_{SO}\exp\left[\frac{-KX_{CaCO_3}f_c}{100U}\left(\frac{\{\rho_b(0) - \rho_b(\infty)\}\{1 - \exp(-aH)\}}{a} + \rho_b(\alpha)H\right)\right]$$

The sulfur capture efficiency, E_{sor} can be derived as:

$$E_{sor} = 1 - \frac{C_{SO_2}(H)}{C_{SO}} = 1 - \exp\left[\frac{-KX_{CaCO_3}f_c}{100U}\rho_{bav}H\right] \tag{5.26}$$

where ρ_{bav} is the average bed density and 100 is the molecular weight of CaCO$_3$.

The exit section of the furnace may be imagined to serve as a gas–solid separator with efficiency, E_e, approximated by the ratio of cross-sectional areas of the exit and the main furnace. Thus, a part of the upward flux, however small, is returned as the downward flux at the exit. For a first approximation (Bolton and Davidson, 1988):

$$E_e = 1 - \left(\frac{A_e}{A}\right) \qquad (5.27)$$

where A_e/A is the ratio of the exit and main furnace cross sections.

$$G_d = E_e G_u \qquad (5.28)$$

The net solid flux leaving the furnace, G_s is equal to $G_u - G_d$:

$$G_u = \frac{G_s}{(1 - E_e)} \qquad (5.29)$$

The suspension density at the exit of the furnace, $\rho_b(H)$ should ideally be calculated by Equation 5.24, but, since $U \gg U_t$, $G_d \ll G_u$, $U_s \approx U$, and the voidage is close to 1.0, the bed density, $\rho_b(H)$ is approximated as:

$$\rho_b(H) = \frac{G_s}{U} \qquad (5.30)$$

The molar concentration of sulfur dioxide at $x = 0$ can be found from a mass balance of sulfur using assumption 5.4 as:

$$C_{SO} = \frac{F_c S}{32 U A} \qquad (5.31)$$

where F_c is the coal feed rate, S is the sulfur fraction in coal, U is the superficial gas velocity and A is the cross section of the furnace.

If the solids in the bed are a perfect mixture of coal-ash and sorbents, and there is negligible entrainment or drainage of solids, the fraction of sorbent in the bed solids (f_c), is equal to the ratio of the sorbent feed-rate to the total inert solid fed into the bed:

$$f_c = \frac{F_{sor}}{F_c \text{ASH} + F_{sor}} \qquad (5.32)$$

where F_{sor} is the sorbent feed rate and (ASH) is the ash fraction in the coal feed.

The mean extent of sulfation, δ of a sorbent particle, which makes many trips around the CFB combustor loop before leaving it, is expressed by Basu et al. (1989) as Appendix 3:

$$\delta = \frac{E_{sor}}{E_c [\text{Ca/S}]} \qquad (5.33)$$

where E_c is the overall cyclone efficiency, E_{sor} is the sulfur capture efficiency, and Ca/S is the calcium to sulfur molar ratio in the coal and limestone fed to the combustor. The calcium to sulfur molar ratio may also be expressed as

$$\frac{F_{sor} X_{CaCO_3}}{100} \cdot \frac{32}{F_0 S}$$

where $X_{CaCO_3}/100$ is the number of moles of calcium carbonate (and hence of calcium) per unit weight of the sorbent, and $S/32$ is the number of moles of sulfur in unit weight of the coal.

The above equations can be rearranged and simplified (Appendix 3) to give the ratio of the feed rate of coal to sorbent for a given level of sulfur capture, cyclone efficiency, and sorbent characteristics as:

$$\frac{F_{sor}}{F_c} = \frac{3.12 E_{sor} \rho_{bav} HS - 100\alpha U[ASH]E_c \ln(1 - E_{sor})}{E_c[\delta_{max} X_{CaCO_3} \rho_{bav} H + 100\alpha U \ln(1 - E_{sor})]} \quad (5.34)$$

5.2.6 SELECTION OF SORBENT

The selection of sorbent is a critical choice the operators of a fluidized bed boiler must make. The choice depends on the rank of the sorbent and the operating conditions.

5.2.6.1 Sorbent Ranking

The ranking is important, especially for the selection of sorbents for a new plant, since it identifies the degree of effectiveness of the sorbents. The following method of ranking has been suggested by Westinghouse (Keairn and Newby, 1981) for ranking sorbents. The criterion chosen for ranking was the extent of sulfation under a standard test condition, as given below, when the reaction rate reaches 0.001 per min.

- Sorbent size (surface mean diameter): 1.0 to 1.19 mm
- Calcination condition: 840°C at 15% CO_2
- Sulfation condition: 840°C at 4% O_2 and 0.5% SO_2

5.2.6.2 Ranks

For a tentatively defined effectiveness of the sorbents, the ranking criteria as given in Table 5.7 may be used. An alternative ranking is available where the Ca/S molar ratio required for 85% sulfur retention, and the attrition loss, both under standard operating conditions, are provided. Data for 34 representative American limestones and dolomites were presented by Fee et al. (1982). Another, more comprehensive, but expensive, method of ranking is available where all aspects of the limestone's use starting from its handling right up to its sulfation are considered (Arsic et al., 1991). Once the sorbent is selected, its requirement to meet a certain emission standard can be estimated using the simplified model given above in Equation 5.34. The following example illustrates this method.

TABLE 5.7
Ranking of Sorbents

Rank	Sulfur absorbed (mg SO_3/mg raw sorbent)
High	>0.3
Medium–High	0.26–0.3
Average	0.16–0.25
Medium–Low	0.11–0.15
Low	<0.10

Example 5.1

Given:

A CFB boiler consumes 5 t/h coal having the following composition: sulfur, 4.34%; ash, 19%; HHV, 24.4 MJ/kg.
Other details are:
Height, $H = 30$ m
Fluidizing velocity, $U = 5$ m/sec
Suspension density, $\rho_b(\chi) = 2 + 480 \exp(-0.328x)$
From TGA data, reactivity of limestone, $K = 50.5(0.41 - \delta)$ per second, where δ is the average rate of sulfation
SO_x limit, $L = 0.52$ g/MJ or 90% (whichever gives lower pollution)
Weight fraction of calcium in limestone, $C_{Ca} = 0.35$
Cyclone efficiency on average size sorbents = 99%
$X_{CaCO_3} = 0.875$,
$M_{CaCO_3} = 100$

Find the limestone consumption rate.

Solution

1. Equivalent sulfur in coal to meet the SO_x limit (S')

$$S' = L \times HHV \times 100 = (0.52/1000) \times 24.4 \times 100 = 1.26\%$$

2. Sulfur retention required for this coal to meet the SO_x limit

$$E_{sor} = \frac{(S - S')}{S} = \frac{(4.34 - 1.26)}{4.34} = 0.71\%$$

So use $E_{sor} = 90\% = 0.9$. The average bed density is found by integration:

$$\rho_{bav} = \frac{1}{30} \int_0^{30} [2 + 480 \exp(-0.328x)]dx = 50.7 \text{ kg/m}^3$$

3. The ratio of sorbent to the coal feed rate is found by substituting values of U and $(1/\alpha)$ in Equation 5.34

$$= \frac{3.12 \times 0.9 \times 50.7 \times 30 \times 0.0434 \times 50.5 - 100 \times 5 \times 0.19 \times 0.99 \times \ln(1 - 0.9)}{0.99[0.41 \times 0.875 \times 50.7 \times 30 \times 50.5 + 100 \times 5 \times \ln(1 - 0.9)]}$$

$$= 0.366$$

4. From above sorbent feed rate = 0.366 × coal feed rate = 0.366 × 5 = 1.83 t/h.

5.3 NITROGEN OXIDE EMISSION

Nitrogen oxides are some of the major air pollutants emitted by coal-fired boilers. Uncontrolled NO_x emissions from conventional PC-fired boiler are in the range of 0.8 to 1.6 mg/MJ. Strict legislation (Table 5.4) is being considered or has been implemented by many countries that restrict the emission of nitrogen oxides. The designer of a fluidized bed boiler is required to ensure that the emission level is below the regulatory guidelines. If the operator uses innovations like a low NO_x burner, over-fire air or reburn, the NO_x might be reduced by 20 to 79%. If ammonia is injected as in selective noncatalytic reactors (SNCR), the NO_x emissions may reduce by 20 to 50%. The greatest reduction is obtained by the use of a selective catalytic reducer (SCR) downstream of the boiler, where ammonia is injected just prior to passing the flue gas over a stack of catalyst. The SCR can reduce the emission by 80 to 95%, but it is a relatively expensive retrofit and needs replacement of expensive catalysts. The following section describes the mechanism of formation (or destruction) and the effect of operating parameters on nitrogen oxide emission from BFB or CFB boilers. Section 5.4 discusses emission of the other oxide of nitrogen, N_2O.

5.3.1 SOURCES OF NO_x

The symbol NO_x represents nitric oxide (NO) and nitrogen dioxide (NO_2). Amongst these, nitric oxide is the major product of coal combustion, and therefore most of the discussion will center on this compound.

Nitric oxide is formed through oxidation of the following:

- Atmospheric nitrogen (giving thermal NO_x)
- Fuel-bound nitrogen (giving fuel NO_x)

During combustion, the nitrogen of the combustion air is oxidized to thermal NO_x, but it is significant only above 1540°C (Morrison, 1980). Thus, it is a minor contributor (< 10%) to the NO_x generated in fluidized bed boilers, where the combustion temperature rarely exceeds 900°C.

The nitrogen content of coal is typically 1 to 2% on a dry, mineral-free basis and it comes from the volatiles and the char of the coal. The possible reaction path towards the formation of NO from the coal is shown in Figure 5.12. It assumes that the fuel nitrogen is equally (50%) distributed between the char and volatiles.

The char nitrogen is oxidized to NO through a series of reactions. The volatile nitrogen appears as NH_3 or HCN. Ammonia (NH_3) may decompose into NO being catalyzed by CaO or char, while the HCN is primarily converted into N_2O (Moritomi and Suzuki, 1992). Approximately 77% of the fuel nitrogen is oxidized to NO by the above reactions (Johnsson, 1989), and the rest appears as NH_3, which in turn is partly converted to nitrogen (Figure 5.12). Thus with a series of parallel-consecutive reactions, NO is formed by the oxidation of volatile nitrogen (Sarofim and Beer, 1979). Part of the NO formed above is also reduced back to nitrogen.

A large number of complex chemical reactions are involved in the formation and destruction of nitric oxide from either char or volatiles. Some of these reactions are catalyzed by calcined limestone (CaO), spent limestone ($CaSO_4$), and char. A list of the probable reactions and their rate constants is given by Johnsson (1989). The contributions of each reaction to the formation of NO or its subsequent destruction are not equal.

5.3.2 METHODS OF REDUCTION OF NO_x EMISSIONS

The generation of NO_x in a combustion system can be reduced somewhat through suitable modifications of the system. These modifications are discussed below.

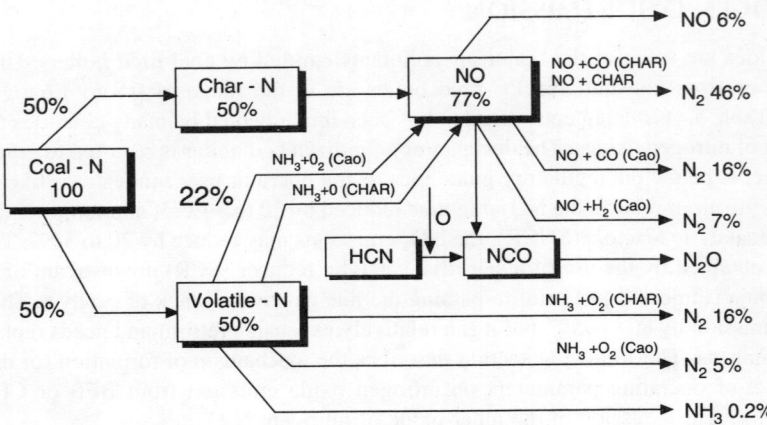

FIGURE 5.12 Relative importance (percentages of total fuel nitrogen involved) of different reaction paths in the formation and reduction of nitric oxide are indicated through numerical figures. Char and calcium oxide catalyzes some reactions as indicated within brackets.

5.3.2.1 Lowering of Combustion Temperature

Low combustion temperature inhibits the oxidation of the nitrogen in the combustion air to thermal NO_x. Thus, the generation of thermal NO_x is negligible in the temperature range of 750 to 900°C. The amount of NO_x generated from fuel nitrogen is also found to decrease with temperature (Figure 5.13 and Figure 5.14).

5.3.2.2 Staging Air

Instead of feeding all the combustion air through the bottom of the furnace, part of it can be added at a section further downstream in the furnace. Such staging of combustion air has a significant beneficial influence on the reduction of NO_x emissions, especially for highly volatile coals.

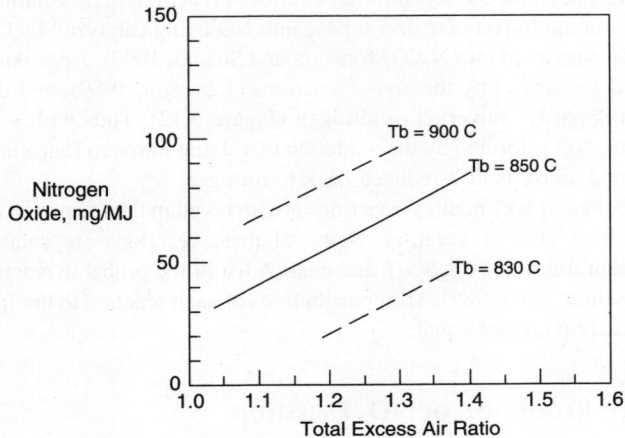

FIGURE 5.13 NO_x emission increases with excess air ratio as well as with combustion temperature.

Emissions

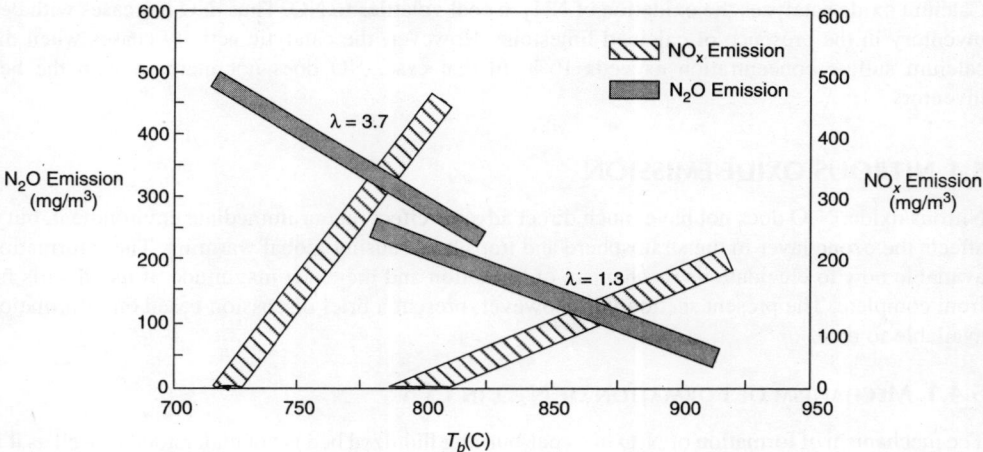

FIGURE 5.14 Dependence on the emissions of N_2O and NO_x of the combustion temperature of a fluidized bed burning low-volatile bituminous coal. Results are shown for two excess air $\lambda = 1.3$ and $\lambda = 3.7$. Thickness of the line shows the scatter of the data points. (Adapted from Braun, 1987.)

Insufficient combustion air passing through the bottom of the furnace helps the nitric oxide to be reduced by char and CO in the lower zone. A low NO_x burner in a PC boiler, working on this principle, can reduce NO_x by 40 to 50%.

5.3.2.3 Injecting Ammonia

Injections of ammonia (NH_3) into the upper section of the furnace or the cyclone of a CFB boiler have proven successful in further reduction of NO_x emissions. However, there is some danger of NH_3 escaping into the solid waste or flue gas creating additional hazards. Combustion of chlorine-bearing coal may emit ammonium chloride through the stack. Therefore, the NH_3 injection should be carefully monitored.

5.3.2.4 Lowering Excess Air

The effect of excess air on the NO_x emission from a commercial CFB boiler is shown in Figure 5.13. The NO emissions are reduced significantly when the excess air is reduced. For example, Hirama et al. (1987) noted that the NO emission from a CFB boiler reduced from 150 to 80 ppm when the excess air was reduced from 30 to 10%.

5.3.3 NO_x Emission from CFB

The mechanism of NO_x emission from CFB boilers is different from that of BFB boilers, and this difference is responsible for a lower level of emission (Hirama et al., 1987) from the former. Even without the staging of air, the NO is found to be progressively reduced to N_2 by the unburnt carbon in the furnace of a CFB boiler. Table 5.6 compares the capture efficiency of NO_x and SO_2 from CFB and BFB boilers. The nitric oxide emission from the BFB boiler is about twice that from the CFB boiler.

Very little ammonia or CO is available in the furnace to assist the reduction of NO. It is interesting to note that if the bed material contains calcined limestone, the trend is reversed.

Calcium oxide catalyzes the oxidation of NH_3 in coal volatiles to NO. Thus, NO increases with bed inventory in the presence of calcined limestone. However, the catalytic activity ceases when the calcium sulfate concentration exceeds 10%. In that case, NO does not increase with the bed inventory.

5.4 NITROUS OXIDE EMISSION

Nitrous oxide, N_2O does not have much direct adverse effect on our immediate environment, but it affects the ozone layer in the stratosphere and traps heat causing global warming. The information available now to elucidate its mechanism of formation and the exact magnitude of its effect is far from complete. The present section will, however, present a brief discussion based on information available to date.

5.4.1 MECHANISM OF FORMATION OF N_2O IN CFB

The mechanism of formation of N_2O in a coal-burning fluidized bed is not understood as well as it is for a gaseous flame. In gaseous flames, the intermediate combustion compound HCN is an important source of N_2O,

$$HCN + O \rightarrow NCO + H$$

$$NCO + NO \rightarrow N_2O + CO \tag{5.35}$$

However, the nitrous oxide formed is immediately destroyed by its reaction with hydrogen radicals (Amand and Andersson, 1989).

$$N_2O + H \rightarrow N_2 + OH \tag{5.36}$$

The extent of destruction of N_2O increases with the reaction temperature.
Nitrous oxide may also be produced through:

1. Reduction of nitric oxide by carbon in char:

$$Char + 2NO \rightarrow N_2O + CO$$

2. Direct oxidation of char nitrogen during combustion:

$$Char\ N + O_2 \rightarrow N_2O \tag{5.37}$$

In the range of 800 to 900°C, the rate of formation of nitrous oxide from char nitrogen is proportional to the combustion rate of char (deSoete, 1989). However, experiments at higher temperatures (>900°C) in pulverized coal flames found negligible contribution of char nitrogen oxidation or NO reduction by char carbon to the total nitrous oxide emission (Kramlich et al., 1988).

5.4.1.1 Level of Emission

The nitrous oxide emission from CFB boilers is in the range of 50 to 200 ppm (Moritomi et al., 1990). In a typical pulverized coal-fired boiler, N_2O emission is in the range of 1 to 20 ppm because of the higher combustion temperature of the PC boiler.

Harada (1992) observed that the emission of N_2O in a bubbling fluidized bed firing low-volatile coal decreased from 250 to 75 mg/m^3 when the combustion temperature increased from 800 to

900°C, but the NO_x emission increased from near 0 to 200 mg/m^3. Interestingly, the sum of NO_x and N_2O decreases only slightly with increasing temperature.

5.4.2 Effects of Operating Parameters on N_2O

5.4.2.1 Combustion Temperature

Combustion temperature has the dominant effect on the emission of NO_x and N_2O. High temperature is favorable to the thermal decomposition of N_2O, so at high temperature (800 to 900°C) one finds a rise in NO_x and a fall in N_2O emission. At higher temperatures, the char concentration is lower due to its higher burning rates. Thus, there is less reduction of NO_x to molecular nitrogen. Figure 5.14 shows that while the NO_x emission from a bubbling fluidized bed increases with temperature, the N_2O decreases continuously.

An important observation made in a CFB furnace (Amand and Andersson, 1989) was that the concentration of N_2O increased along the height of the CFB combustor, while the concentration of NO decreased continuously along the height of the combustor. Unlike the emission of NO, that of N_2O is greater at lower combustion temperatures.

5.4.2.2 Effect of Volatiles

Higher volatile content of the fuel decreases the N_2O formation, but increases the NO_x formation. Harada (1992) observed that in a bubbling fluidized bed, the sum of N_2O and NO_x remains nearly constant in the range of 800 to 900°C. Low-volatile coal gives higher carbon content in beds, creating a more favorable condition for the reduction of NO_x into molecular nitrogen, which is a precursor for N_2O.

Biomass fuel was found to have exceptionally low emissions of nitrous oxide in both CFB and BFB boilers (Leckner et al., 1992). Co-firing of coal, however, produced high N_2O emission.

5.4.2.3 Effect of Excess Air

For a given excess air, splitting the air into primary and secondary parts generally results in a decrease in emissions of both NO_x and N_2O because the oxygen deficient, substoichiometric condition helps reduce nitrogen oxides into molecular nitrogen. The NO_x emission also shows significant dependence on the total amount of excess air as shown in Figure 5.14. Additionally, this figure shows that although N_2O emission also increases with excess air, the increase is not significant between excess air coefficients of 1.3 and 3.7 at the same operating temperature.

5.4.2.4 Effect of Limestone

The presence of CaO helps reduce the NO_x in a CFB furnace, but it is not very effective in the case of a BFB furnace (Leckner and Amand, 1987). It is apparently due to the presence of a substoichiometric zone in CFB boilers and its absence in BFB boilers.

The higher catalytic effect of CaO on the destruction of N_2O at higher temperatures is another reason for lower N_2O at higher temperatures.

5.4.2.5 Effect of NH_3 Injection

Ammonia injection helps reduce NO_x emission when used in a narrow temperature range around 870°C (McInnes and Van Wormer, 1990) and is therefore used when very low NO_x emission is required. However, it increases N_2O formation. The injection of ammonia on catalysts brings about significant reduction in NO_x emissions. It is in fact the basis of operation for selective catalytic reducers (SCR)s.

5.4.3 REDUCTION OF N_2O

Afterburning in the cyclone of a CFB boiler is a good technique for reduction in N_2O emission. Gustavsson and Leckner (1995) burned gas in the cyclone to raise the cyclone flue gas temperature and it reduced the N_2O emission from 150 to 30 ppm.

Only a small part (5%) of the nitrogen in char is converted to nitrous oxide, but its reduction on the char surface is faster than that of NO reduction (deSoete, 1989). Unlike NO reduction, N_2O reduction is independent of CO concentration. The conversion of the coal nitrogen into N_2O depends on the devolatilization process. It has been observed (Freihaut and Seery, 1981) that when coal is heated at a moderate rate up to 900°C, only a small part of the coal nitrogen appears as HCN, which is the major source of N_2O. Thus, control of the coal devolatilization rate may provide some clue to the reduction of nitrous oxide in CFB boilers. Feeding the coal in areas of the CFB loop having a lower heat-transfer rate may delay the devolatilization process.

5.5 MERCURY EMISSION

The emission of mercury has received much attention. Some countries are contemplating enforcing 90% capture from coal-fired power plants within a specified period of time.

Several options for mercury capture are used. The options can be classified under three groups (ICAC, 2005): sorbent injection, electro-catalytic oxidation, and precombustion technique.

5.5.1 SORBENT INJECTION

This is the most popular method. Sorbents are injected into the combustion gas to absorb gaseous mercury. These sorbents capture the gaseous mercury to retain it in solid form, and are then captured downstream in a particulate capture device like a bag-house or ESP. The sorbents could be:

1. activated carbon
2. bromine
3. polysulfide

In some cases, SCRs (used for NO_x control) or wet scrubbers (used for SO_x control) are also found to capture significant amounts of mercury.

Another approach involves the addition of some chemicals that will oxidize elemental gaseous mercury into solid compounds and make it easier to capture. Such mineral-based reagents would be cheaper than the activated carbon traditionally used for capturing mercury.

5.5.2 ELECTRO-CATALYTIC OXIDATION

This technique is a multicontrol technology. It can reduce the emissions of NO_x, SO_2, particulate matter, and mercury simultaneously. It consists of a discharge reactor, ammonia-based wet scrubber, and a wet electrostatic precipitator. The whole unit is located downstream of the existing ESP or fabric filter of the plant. The barrier discharge reactor oxidizes NO_x, SO_2, and H, while the ammonia-based wet scrubber removes NO_x, SO_2, and oxidized mercury creating an ammonium sulfate/nitrate solution. The wet ESP captures acid aerosols, fine particulates, and the oxidized mercury. The solution is sent to a filtration system to separate spent mercury and spent activated carbon for disposal as hazardous waste. The other part of the solution can be used in fertilizer.

5.5.3 PRECOMBUSTION TECHNOLOGIES

Some proprietary processes are available (K-fuel) which can wash the coal of its pollutants, but the mercury removal is in the range of 70% instead of 90% as in other methods.

There is another consideration for the mercury in solid residues. Its concentration is highest in fly-ash with about 0.33 ppm, followed by FGD waste with about 0.22 ppm. Thus the disposal of these solid residues from a power plant using mercury control technology needs careful planning to prevent reemission of the mercury into the atmosphere.

5.6 CARBON MONOXIDE EMISSION

When carbon monoxide is inhaled, it displaces oxygen in the blood. Oxygen deprivation can be especially dangerous for the heart and brain tissues. CO, particularly in urban areas, comes primarily from automobiles. The emission of carbon monoxide from fluidized bed boiler plants is not generally perceived to be a major problem, and is normally below the statutory limit. The emission depends on the fuel composition and the combustion temperature. The CO level in the flue gas increases with decreasing combustion temperature, particularly below 800°C. Typical levels of emission are in the range of 40 to 300 ppm (Brereton, 1997), as compared to 200 to 400 ppm in stoker-fired boilers.

5.7 CARBON DIOXIDE EMISSION

Carbon dioxide, being the most important greenhouse gas, has received the greatest attention in terms of emission control. Release of carbon dioxide per unit power produced is an important index of a plant's performance. Table 1.3 compares the CO_2 emissions of plants using different technological options.

Carbon intensity, which is the amount of carbon released during combustion, is highest for coal, followed by oil and gas. For efficient combustion, all the carbon is oxidized into carbon dioxide. Sulfur capture requires the use of limestone in a fossil fuel-fired boiler, which produces additional amounts of carbon dioxide from the calcination reaction of limestone (Equation 5.6). The operation of a BFB boiler involves calcination of 3 to 4 times the limestone actually consumed in capturing the sulfur, while emitting an equivalent amount of extra carbon dioxide into the atmosphere. A CFB, on the other hand, needs about twice the stoichiometric amount. This aspect, as well as higher nitrous oxide emissions, makes fluidized bed boilers less attractive from a climate change point of view especially if they use limestone for sulfur capture. The problem is not as severe for fluidized bed boilers firing low-sulfur fuels which do not use limestone. If one takes a total picture of the powerplant, the comparison with a pulverized coal-fired boiler with FGD, SCR, etc., may not be as pessimistic. Higher auxiliary power consumption by mills and FGD also require additional CO_2 emission for the same net power output (Miller, 2003).

Less use of limestone means less production of CO_2 as every mole of limestone produces one mole of CO_2. Considerable effort has been made to reduce the calcium-to-sulfur ratio by using techniques like rehydration of sorbents, injection of water or steam into flue gas leaving the boiler with fly ash, and the use of synthetic sorbents. There is also the potential of using a carbonate cycle where the reverse of Equation 5.6 could be forced at a lower temperature ($<750°C$) to absorb CO_2 at $CaCO_3$.

$$CaO + CO_2 = CaCO_3 + heat \tag{5.38}$$

The chemical looping cycle, though still in the stage of laboratory-scale research, holds high promise of reduction of CO_2 emission from conventional fluidized bed plants. The best and the least expensive approach for reduction would be the use of higher efficiency combined- or hybrid-cycle discussed in Chapter 1.

5.8 EMISSION OF TRACE ORGANICS

Fluidized beds are also used for incinerating waste products, which are sometimes hazardous. Dichlorinated benzene and PCBs are some of the items that can be destroyed in fluidized bed incinerators. In order to provide the necessary high-temperature residence time, the incinerators are operated at high temperatures.

Desai et al. (1995) found the dioxin concentration in the flue gas from a PCB incinerator to be in the range of 3 to 42 ng/m^3. This level, being above the statutory limit of 0.1 to 1 ng/m^3 a dry scrubber is needed, which allows removal of the dioxin and furan by 99.9% (Brereton, 1997). Scrubbers are also required for removal of HCl, which can be captured by the CaO in the cooler (<650°C) section of the incinerator.

Co-combustion of coal and waste products could help reduce the emission of dioxin below the statutory limit especially in presence of sulfur in the fuel.

5.9 PARTICULATE EMISSION

In any boiler, ash of a fuel is split into two streams. The coarser fraction is drained from the bed as bottom ash, and the finer fraction leaves the furnace as fly ash to be collected by downstream particulate collection equipment like a fabric filter or electrostatic precipitator. The amount of fuel ash that appears as fly ash in a fluidized bed boiler is generally lower than that in a pulverized coal-fired boiler because a sizeable part of the ash in the former appears as bed drain. In a typical PC boiler, more than 80% of the fuel ash appears as fly ash, while only a small to negligible fraction appears as bottom ash. The fly ash from fluidized bed boilers without limestone addition is in the range of 40 to 80%. The split of ash between bed ash and fly ash depends on:

- feed-size distribution
- dispersion of ash in the coal (intrinsic or extraneous)

The ash in a fluidized bed is generated at 800 to 900°C. Therefore, it does not melt but remains rough edged, maintaining the original shape of the mineral matter in the ash. The ash in a PC-fired boiler, on the other hand, is produced at 1000 to 1500°C. PC ash therefore melts and, owing to its surface tension, assumes spherical shape. Figure 5.15 shows scanning electron microscope photographs of fly-ash samples from both CFB and PC boilers. The spherical shapes of the PC ash and irregular shapes of the CFB ash are clearly visible from this picture.

5.9.1 ASH CHARACTERISTICS

Oka (2004) reasoned that since the ash generated in a BFB boiler is entrained at a lower velocity than that for PC boiler only the finest particles appear as fly ash in a BFB boiler. The mean diameter of fly ash particles in a BFB boiler is in the range of 3.0 to 5.0 μm, while that in a PC boiler is in the range of 5.0 to 8.5 μm. Table 5.8 presents a comparison of physical characteristics of fly ash from PC and BFB boilers as outlined below:

Higher specific surface area and higher porosity favor the use of fluidized bed fly ash for concrete and cement applications. Finer size also helps its application in cement.

Oka (2004) suggests that the rough edges of the fluidized bed boiler ash increase the hydrodynamic resistance of the cakes formed on bag filters. Thus a fluidized bed boiler may need lower filtration velocity in the bag than would a PC boiler with smooth ash particles. Table 5.8 shows the filtration velocities recommended by Oka (2004) for BFB and PC boilers. The filtration velocity for BFB fly ash is less than half of that for PC fly ash, so a BFB boiler might need a larger bag-house than a PC boiler.

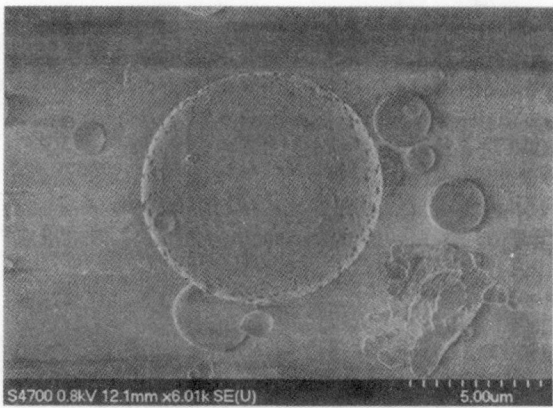
Fly ash from Pulverized coal fired boiler

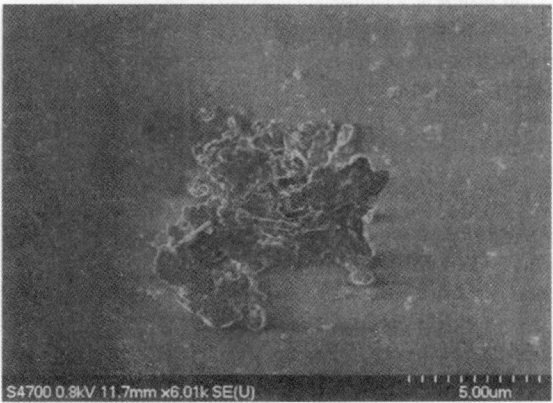
Fly ash from Circulating fluidized bed boiler

FIGURE 5.15 Scanning electron microscope photographs of samples of fly ash from both CFB boiler and a PC boiler.

Fluidized bed boilers with limestone injection are likely to have less sulfur in the ash. Thus, in the design of electrostatic precipitators, one needs to take note of the higher resistivity of fluidized bed boilers with sulfur-capture systems. This would require large plate area.

TABLE 5.8
Comparison of Design Parameters for Particulate Control Devices for Fluidized Bed with Limestone Injection and Pulverized Coal Fired Boiler

Properties	BBF boiler	PF boiler
Specific surface area, m^2/gm	5–20	1.0–4.0
Bulk porosity ε	82–84%	58–75%
Filtration velocity, $m^3/min/m^2$	0.15–0.4	0.5–0.8
Specific dust loading, Kg/m^2	1.5–2	2.5–5
Ash resistivity, Ohm. cm	1011–1013	1010

Source: Data collected from Oka, 2004.

NOMENCLATURE

A: cross section of the furnace (m^2)
A_e: area of the furnace exit (m^2)
a: exponent in bed density profile, Equation 5.21 (m^{-1})
ASH: weight fraction of ash in coal
(Ca/S): calcium to sulfur molar ratio
C_c: molar concentration of sorbents in the bed (kmol/m^3)
C_{Ca}: weight fraction of calcium in limestone
C_{SO_2}: concentration of sulfur dioxide (kmol/m^3)
C_{SO}: concentration of sulfur dioxide at $x = 0$ (kmol/m^3)
$C_{SO_2}(x)$: concentration of sulfur dioxide at x (kmol/m^3)
d: diameter (m)
E: activation energy (kJ/kmol)
E_c: average cyclone efficiency
E_e: solid collection efficiency of the furnace exit
E_{sor}: sulfur capture efficiency
F_c: coal feed rate (kg/sec)
F_{sor}: sorbent feed rate (kg/sec)
f_c: weight fraction of unreacted sorbent in bed materials
G_d: downward solid flux (kg/m^2 sec)
G_u: upward solid flux (kg/m^2 sec)
G_s: solid recycle rate (kg/m^2 sec)
H: height of the furnace above the secondary air level (m)
HHV: higher heating value (kJ/kg)
k': fitting constant for Equation 5.20
$K(t)$: reactivity of sorbent particles at time t (sec^{-1})
K: reaction rate of sulfation (m^3/kmol sec)
L: SO$_x$ limit (kg/kJ)
M_{CaO}: molecular weight of calcium oxide (56 kg/kmol)
m: local rate of SO$_2$ formation (kmol/m^3 sec)
$\dot{m}_{d,out}$: sum of all outflows of sorbents of size d
M_{CaCO_3}: molecular weight of limestone (100 kg/kmol)
n: index of reaction rate, Equation 5.30
p: moles of calcium per unit volume of sorbent particle (kmol/m^3)
P: Constant in Equation 5.19
P_e: equilibrium partial pressure of CO$_2$
R: universal gas constant (8.314 kJ/kmol K)
R_0: initial reaction rate (kmol/sec particle)
$R(t)$: reaction rate of sulfation at time t (kmol/sec particle)
r: sorbent radius (m)
S: sulfur fraction in coal
S': equivalent sulfur in coal to meet the SO$_x$ limit
T: temperature (K)
t: time (sec)
$t_{cs}(d)$: total residence time of a particle of diameter d (sec)
t_{fs}: average particle residence time during single trip through the bed (sec)
t_p: pore plugging time constant (sec)
t_{sf}: sulfation time (sec)
U: superficial gas velocity (m/sec)
$U_s(x)$: net upward velocity of solids at height x (m/sec)

U_s, U_d: upward and downward velocity of solid particles near the exit of the furnace (m/sec)
U_t: terminal velocity of single particle (m/sec)
V_p: volume of a sorbent particle (m³)
$Wf(d), \Delta d$: mass of bed materials of size between d and $d + \Delta d$ (kg)
x: height in the furnace above the secondary air level (m)
X_{CaCO_3}: weight fraction of calcium carbonate in the sorbent particle

GREEK SYMBOLS

α: proportionality constant in Equation 5.16 (sec kmol/m³)
ε: average bed voidage near the furnace exit
δ: current extent of sulfation
δ_{max}: maximum extent of sulfation
$\delta(t)$: extent of sulfation at time t
$\delta(\infty)$: final extent of sulfation of the sorbent
ρ_p: density uncalcined sorbent (kg/m³)
$\rho_b(0)$, $\rho_b(x)$: density of bed at secondary air level, distance, x above it (kg/m³)
$\rho_b(\alpha)$: asymptotic value of bed density (kg/m³)
ρ_{bav}: average bed density (kg/m³)

REFERENCES

Ahlstrom Pyropower, Activation/reuse of fluidized bed waste: phase I — literature review, *Report to the Canadian Electrical Association*, CEA Project 9131 G 891, 1992.

Amand, L. E. and Andersson, S., *Emissions of Nitrous Oxide from Fluidized Bed Boilers, Proceedings of the 10th International Conference on Fluidized Bed Combustion*, Manaker, A., Ed., ASME, New York, pp. 49–56, 1989.

Arsic, B., Oka, S., and Radovanovic, M., Characterization of limestones for SO_2 absorption in fluidized bed combustion, *Presented at Fifth International Conference on Fluidised Bed Combustion*, London, pp. 171–177, 1991.

Arsic, B., Radovanovic, M., and Oka, S., Comparative analysis of efficiency of sulphur retention of Yugoslav limestones (in Serbian), *Report of the Institute of Nuclear Sciences Bosris Kidric*, Vinca, Belgrade, IBK-ITE-717, 1988.

Basu, P., Greenblatt, J., Wu, S., and Briggs, D., Effect of Solid Recycle Rate, Bed Density and Sorbent Size on the Sulfur Capture in a Circulating Fluidized Bed Rivers, *Proceedings of the 10th International Conference on Fluidized Bed Combustion*, Manaker, A., Ed., ASME, New York, pp. 701–708, 1989.

Bolton, L. W. and Davidson, J. F., Recirculation of Particles in a Fast Fluidized Bed Risers, In *Circulating Fluidized Bed Technology II*, Basu, P. and Large, J. F., Eds., Pergamon Press, Oxford, pp. 139–146, 1988.

Borgwardt, R. H., *Environ. Sci. Technol.*, 4, p. 855, 1970.

Brereton, C., Combustion Performance in Circulating Fluidized Bed, Grace, J. R., Avidan, A. A., and Knowlton, T. M., Eds., Chapman & Hall, London, pp. 392, 1997.

Burdett, N. A., Gliddon, B. J., Hotchkiss, R. S., and Squires, R. T., SO_3 in coal-fired fluidized bed combustors, *J. Inst. Energy*, 56, 119–124, 1983.

Burdett, N. A., Longdon, W. E., and Squires, R. T., Rate coefficients for the reaction $SO_2 + O_2 = SO_3 + O$ in the temperature range 900–1350 K, *J. Inst. Energy*, 57, 373–376, 1984.

Chrostowski, J. W. and Georgakis, C., *ACS Symp. Sr.*, 65, 2251978.

Cogbill, C. V. and Likens, G. E., Acid precipitation in northeastern US, *Water Resour. Res.*, 10, 11331974.

deSoete, G. G., *Heterogeneous NO and N_2O formation from Bound Nitrogen during Char Combustion paper presented at the Joint Meeting of the British & French Section of Combustion Institute*, Rouen, April pp. 18–21, 1989.

Ehrlich, S., A coal-fired fluidized bed boiler, *Inst. Fuel Symp. Sr. no. 1*, 1, C41975.
Elliot, M. A., *Chemistry of Coal Utilization — Second Supplementary Volume*, Wiley, New York, p. 1462, 1981.
EPA, Clean Air Markets–Cap and Trade, www.epa.gov/airmarkets/capandtrade/index.html, 2005.
Fee, D. C., Wilson, W. I., Myles, K. M., Johnson, I., and Fan, L. S., Fluidized bed coal combustion in bed sorbent sulfation model, *Chem. Eng. Sci.*, 38(11), 1917–1925, 1983.
Fee, D. C., Wilson, W. I., Myles, K. M., Johnson, I., Fan, L. S., Smith, G. W., Wong, S. H., Shearer, J. A., and Lenc, J. F., *Sulfur Control in Fluidized Bed Combustors: Methodology for Predicting the Performance of Limestone and Dolomite Sorbents*, Argonne National Laboratory Report no ANL/FE-8-10, 1982, NTIS USA.
Fields, R. B., Burdett, N. A., and Davidson, J. F., Reaction of sulfur dioxide with limestone particles: the influence of sulfur-trioxide, *Trans. Inst. Chem. Eng.*, 57, 276–280, 1979.
Freihaut, J.D. and Seery, D.J., Evolution of Fuel Nitrogen during the Vacuum Thermal Devolatilization of Coal, *Presented at American Chemical Society Division of Fuel Chemistry*, New York, August 23–28, 1981.
Ghardashkhani, S., Ljungstrom, E., and Lindquist, O., *Release of Sulfur Dioxide from Calcium Sulfate under Reducing Atmosphere*, Proceedings of the 10th International Conference on Fluidized Bed Combustion, Manaker, A., Ed., ASME, New York, pp. 611–615, 1989.
Gustavsson, L. and Leckner, B., Abatement of N_2O emissions from circulating fluidized bed combustion through afterburning, *Ind. Eng. Chem. Res.*, 34, 1419–1427, 1995.
Hamer, C. A., Evaluation of SO_2 Sorbent Utilization in Fluidized Beds, *Energy Mines and Resources*, Canada, CANMET Report 86-9E, 1986.
Harada, M., N_2O emission from FBC, *Presented at 24th IEA–FBC Technical Meeting*, Turku (Finland), 1992.
Hirama, T., Takeuchi, H., and Horio, M., *Nitric Oxide Emission form Circulating Fluidized Bed Coal Combustion*, Proceedings of 9th International Conference on Fluidized Bed Combustion, Vol. 1, Mustonen, J., Ed., ASME, New York, pp. 898–903, 1987.
Institute of Clean Air Companies, www.icac.com/hgcontrol010305.pdf, January, 2005.
Johnsson, J. E., A kinetic model for NO_x formation in fluidized bed combustion, Proceedings of 10th International Conference on Fluidized Bed Combustion, Manaker, A., Ed., ASME, New York, pp. 1112, 1989.
Keairn, D. L. and Newby R. A., FBC Sulfur Control Perspective on Understanding, Modeling and Application, *Presented at MIT Summer Institute on Fluidized Bed Combustion*, July, 1981.
Kramlich, J. C., Lyon, R. R., and Lanier, W. S., eds. EPA/NOAA/NASA/USDA/N_2O Workshop, v1, EPA-600/8-88-079, 1988.
Lalak, I., Seeber, J., Kluger, F., and Krupka, S., *Operational Experience with High efficiency Cyclones: Comparison between Boilers A and B in the Zora Power Plant — Warsaw, Poland*, Proceedings of 17th International Conference on Fluidized Bed Combustion, Pisupati, S., Ed., ASME, New York, 2003, paper: FBC2003-146.
Leckner, B. and Amand, L. E., *Emission from a Circulating & a Stationery Fluidized Bed Boiler a Comparison*, Proceedings of 9th International Conference on Fluidized Bed Combustion, Mustonen, J., Ed., ASME, New York, pp. 891–897, 1987.
Leckner, B., Karlsson, M., Mjornell, M., and Hagman, U., Emissions from a 165 MWth Circulating fluidized bed boiler, *J. Inst. Energy*, 65(464), 122–130, 1992.
Lee, D. L., Hodges, J. L., and Georgakis, C., Modeling of SO_2 emission from fluidized bed coal combustors, *Chem. Eng. Sci.*, 35, 302–306, 1980.
Likens, G. E. and Borman, F. H., Acid rain: a serious regional environmental problem, *Science*, 184, 1176–1179, 1974.
Makansi, J., Few new options emerge for complying with CAA Phase II Power, v. 138, n7, July 1994, p. 21–27.
McInnes, R. and Van Wormer, M. B., Cleaning up NO_x emissions, *Chem. Eng. Mag.*, September, 131–135, 1990.
Miller, L., Analysis of thermal power generation projects considering current trends in emission policy, M.Sc. Thesis, Dalhousie University Mechanical Engineering Department, Halifax, Canada, 2003.
Moritomi, H. and Suzuki, Y., *Nitrous oxide Formation Fluidized Bed Combustion Conditions*, Proceedings of Seventh International Fluidization Conference, Brisbane, Potter, O. E. and Nicklin, D. J., Eds., United Engineering Foundation, Park Avenue, NY, pp. 495–507, 1992.

Moritomi, H., Suzuki, Y., Kido, N., and Ogisu, Y., NO_x Emission and Reduction from Circulating Fluidized Bed Combustor, In *Circulating Fluidized Bed Technology III*, Basu, P., Hasatani, M., and Horio, M., Eds., Pergamon Press, Oxford, pp. 339–404, 1990.

Morrison, G.F., Nitrogen Oxides from Combustion — Abatement and Control, *Report no ICTIS/TRII; IEA Coal Research*, London, p. 13, 1980.

Oka, S. N., *Fluidized Bed Combustion*, Marcell Dekker, New York, see pages 510, 569, 2004.

Sarofim, A. F. and Beer, J. M., Modeling of fluidized bed combustion, in *17th Symposium (International) on Combustion, Combustion Institute*, pp. 189–204, 1979.

Saroff, L., Zitterbart, M. T., Ahart, A. K., and Hooper, H. M., *A relationship between solids recycling and sulfur retention in fluidized bed combustors, Proceedings of 10th International Fluidized Bed Combustion Conference, San Francisco*, Vol. 2, pp. 1003–1008, 1989.

Stantan, J. E., *Fluidized Beds-Combustion and Applications*, Howard, J. R., Ed., Applied Science, London, p. 209, 1983.

Tang, J. T. and Engstrom, F., Technical assessment on the Ahlstrom Pyroflow Circulating and conventional Bubbling fluidized bed combustion systems Proc, *9tj Int. Conf. on FBC, Boston*, 1, 38–53, 1987.

Talukdar, J., Basu, P., and Greenblatt, J. H., Reduction of Calcium Sulfate in Circulating Fluidized Bed Furnace, *Fuel*, 75(9), 1115–1123, 1996.

Yates, J. G., *Fundamentals of Fluidized bed Chemical Processes*, Butterworths, see pages 183, 185, 188, 1983.

6 Heat Transfer

The magnitude of heat transfer and location of heating surfaces of a boiler greatly influence its thermal efficiency and output. Therefore, the designer of a fluidized bed boiler must have a good understanding of the heat transfer mechanism within the boiler. Figure 6.1 shows the location and arrangement of different types of heat transfer surfaces in a circulating fluidized bed (CFB) boiler. It shows that this can be divided into two sections: primary and secondary. The primary section is comprised of the CFB loop while the secondary section includes the convective or back-pass section of the boiler. The secondary section is same as in any conventional boiler. Table 6.1 presents the typical range of heat transfer coefficients on different sections of a bubbling fluidized bed (BFB) and circulating fluidized bed (CFB) boiler. In a fluidized bed (CFB or BFB) boiler one can identify several modes of heat transfer including:

- Between the gas and particles
- Between the bed and walls of a fast fluidized bed furnace
- From the fast fluidized bed to tube surfaces immersed in it
- To the boiler tubes immersed in the BFB furnace or heat exchanger
- To the walls of cyclone or other primary separator
- Between the freeboard and tubes

The present chapter discusses different modes of heat transfer in both circulating and BFB boilers, and presents some equations for the estimation of heat transfer coefficients. The application of heat transfer knowledge to the design of a boiler surface and load control are also illustrated. Heat transfer to the cyclone enclosure is complex, and only limited information is available. For this reason, a discussion on this type of heat transfer is beyond the scope of this chapter.

6.1 GAS-TO-PARTICLE HEAT TRANSFER

Because of the very large particle surface area per unit mass of solids, gas to particle heat transfer has rarely been an issue, even though the heat-transfer coefficient based on actual particle surface area is rather small (6 to 23 W/m^2 C) (Botterill, 1986). Gases and solids in regions close to the distributor, the solid feed-points and the secondary air-injection ports are at temperatures different from those in the bulk of the bed. This thermal lag can be very important. In the case of coal, for example, the heating rate of coal particles will likely influence its volatile release. Rates of combustion, attrition and fragmentation are also affected by the rate of heat transfer to coal particles (Salatino and Massimilla, 1989). The gas-to-particle heat transfer also governs the transient response of the boiler furnace.

6.1.1 GAS-TO-PARTICLE HEAT TRANSFER EQUATIONS

Fine particles show very high rates of gas-to-particle heat transfer near the grid of a bubbling or CFB. This occurs because the temperature difference between the gas and the solids, as well as the

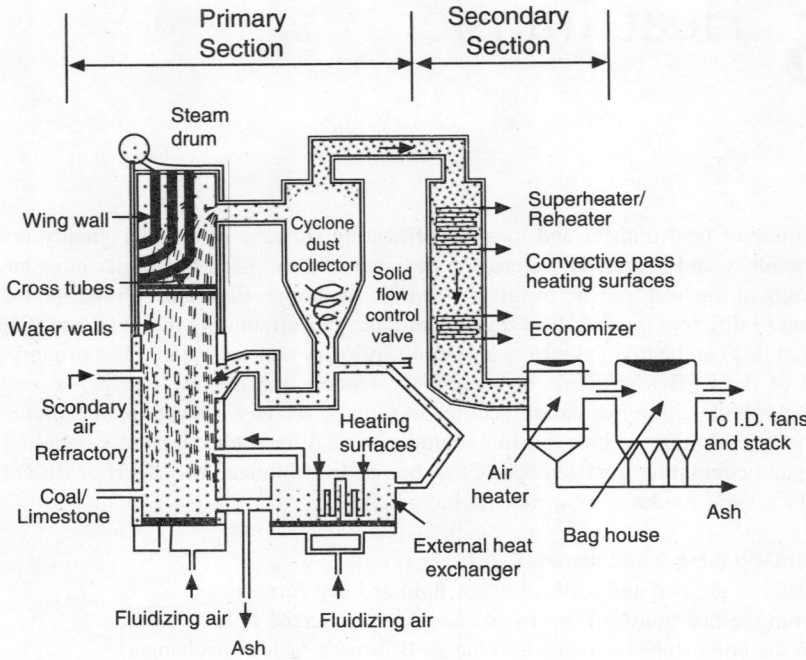

FIGURE 6.1 Heat-absorbing sections of a circulating fluidized bed boiler.

TABLE 6.1
Heat-Transfer Processes in Circulating Fluidized Bed Boiler

Locations	Type of Sections	Typical Heat-Transfer Coefficients (W/m² K)
Water wall tubes above refractory walls in furnace (890–950°C)	Evaporators	150–190
Wing inside furnace (850–950°C)	Evaporator, reheater and superheater	~70–75% of water wall
Tubes placed across furnace (800–900°C)	Superheater	180–220
Cyclone enclosure (800–900°C)	Superheater	–
Horizontal tubes in external heat exchanger (600–850°C)[a]	Evaporator, reheater and superheater	340–510
Horizontal cross-tube heat exchanger in back-pass (800–200°C)	Economizer, superheater and reheater	70–80
Gas-to-bed material (420–177 μm, 50°C)[b]	Furnace	30–200

[a] Gottung and Darling (1989).
[b] Watanabe et al. (1991).

slip velocity between them, is very high. Further up the bed, the gas–solid heat transfer stabilizes to a lower value due to a smaller temperature difference between them.

The heat transfer coefficient or the particle Nusselt number $(h_{gp}d_p)/K_g$ increases with average particle Reynolds number $[(U - U_p)d_p\rho_g]/\mu$ based on the gas–solid slip velocity, and decreases with increasing solid concentration (Watanabe et al., 1991).

6.1.1.1 Heat Transfer Coefficient in Fixed and Bubbling Beds

Ranz and Marshall (1952) presented the following correlation for a small, isolated sphere passing through a gas with a relative velocity $(U - U_p)$:

$$Nu_p = 2 + 0.6 Re_p^{0.5} Pr^{0.33} \qquad (6.1)$$

where Re_p is $[(U - U_p)d_p\rho_g]/\mu$.

For a gas passing at a superficial velocity U through a fixed bed with a large number of particles, Ranz and Marshall (1952) modified the above equation to get:

$$Nu_p = 2 + 1.8 Re_p^{0.5} Pr^{0.33} \qquad (6.2)$$

The above two expressions give heat transfer coefficients on a single particle dropped into bed of particles at average bed temperature. This situation is different from one in which the whole bed of particles is involved in exchanging heat with the fluidizing gas. At a low Reynolds number, the gas–particle heat transfer coefficient on bulk solids in the bed is lower than that on single (isolated) particles. It can be as much as three orders of magnitude below the single-particle heat transfer coefficient (Kunii and Levenspiel, 1991). However, in coarse-particle beds the gas–particle heat transfer coefficients for both bulk solids and isolated particles are closer.

Experimental data on gas–particle heat-transfer coefficients in bubbling beds (Figure 6.2) shows that they are much smaller than those predicted from Equation 6.1 and Equation 6.2.

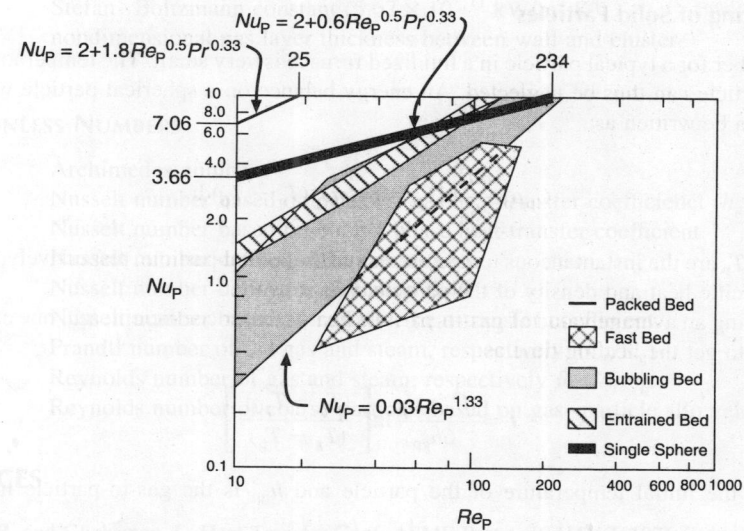

FIGURE 6.2 Effect of particle Reynolds number on gas to particle Nusselt number in different types of beds. (Adapted from Watanabe et al., 1991.)

Kothari (1967) correlated the data for the gas–particle whole-bed coefficient with the following equation:

$$Nu_{bed} = \frac{h_{bp}d_p}{K_g} = 0.033 Re_p^{1.3} \quad \text{for } Re_p = 0.1 - 100 \tag{6.3}$$

6.1.1.2 Gas–Particle Heat Transfer in Fast Beds

Figure 6.2 shows the range of gas–particle Nusselt number for bubbling bed, fast fluidized bed and entrained bed systems. It is apparent from the equation that for a given particle Reynolds number, the entrained bed would have the highest heat-transfer coefficient followed by that in bubbling and fast fluidized beds, respectively. Data presented on fast fluidized beds for the whole bed coefficient, h_{bp} are taken from Watanabe et al. (1991), who based the Reynolds number on slip velocity $(U - U_p)$ as opposed to superficial velocity U. The particle velocity is computed as $G_s/(1-\varepsilon)\rho_p$. They did not provide any empirical relation for the whole bed coefficient.

An empirical correlation for heat transfer to coarse particles (>5 mm) in a fast bed of fine solids (<300 μm) developed by Halder (1989) may be used for a first approximation of the convective component of the heat transfer to coarse particles. The gas–particle heat-transfer coefficient, excluding radiation, is

$$h_{gp} = \frac{K_g}{d_{cp}} 0.33 Re_{cp}^{0.62} \left[\frac{d_{cp}}{d_p} \right]^{0.1} \tag{6.4}$$

where Re_{cp} is $[(U - U_{cp})d_{cp}\rho_g]/\mu$, d_{cp} is the diameter of the coarse particle suspended in a bed of fine particles of average diameter d_p, and U_{cp} is the average velocity of coarse particle. This relation is valid for 5 mm < d_{cp} < 12 mm; 50 μm < d_p < 350 μm; 900 < Re < 2500.

6.1.2 HEATING OF GAS AND SOLIDS IN THE FAST BED

6.1.2.1 Heating of Solid Particles

The Biot number for a typical particle in a fluidized furnace is very small. The temperature gradient within the particle can thus be neglected. An energy balance on a spherical particle without heat generation can be written as:

$$C_p \rho_p \frac{\pi d_p^3}{6} dT_p = h_{gp} \pi d_p^2 (T_g - T_p) dt \tag{6.5}$$

where T_g and T_p are the instantaneous temperatures of the gas and particle, respectively, and C_p and ρ_p are the specific heat and density of the particle, respectively.

By assuming an average value of gas-to-particle heat-transfer coefficient, h_{gp}, one can integrate Equation 6.5 to get the heating time as:

$$t = \frac{C_p \rho_p d_p}{6 h_{gp}} \ln \left[\frac{(T_g - T_{p0})}{(T_g - T_p)} \right] \tag{6.6}$$

where T_{p0} is the initial temperature of the particle and h_{gp} is the gas-to-particle heat transfer coefficient.

One can see from above that as the gas enters a bed of particles, it approaches the particle temperature asymptotically. Thus, to get a design value we define $t_{99\%}$ as the time required for the gas to reach 99% of the particle temperature, i.e., when $[(T_g - T_p)/(T_g - T_{p0})] = 0.01$. This can be

Heat Transfer

computed from the expression:

$$t_{99\%} = \frac{0.768 C_p \rho_p d_p}{h_{gp}} \tag{6.7}$$

6.1.2.2 Heating of Gas

When cold gas or air enters the bottom of a fluidized bed, it is rapidly heated by a large mass of hot solids. In a fluidized bed, this heating of cold gas brings about a negligible cooling of the hot particles. Thus, one can calculate the length of furnace (X_{99}) the gas must traverse to reach within 99% of the furnace temperature as:

$$X_{99\%} = \frac{\rho_g U}{\rho_b S} \int_{T_{g0}}^{T_{99\%}} \frac{C_g dT_g}{h_{gp}(T_p - T_g)} \tag{6.8}$$

where $T_{99\%} = T_{g0} + 0.99(T_p - T_{g0})$. Also, S is the surface area per unit mass of the particles, U is the superficial gas velocity, ρ_b is the bed density, and C_g is the specific heat of the gas.

For precision calculation, the variation of C_g and ρ_g with temperature should be considered in the above integration.

Example 6.1

> Cold primary air at 30°C enters a bed of 200 μm sand (2500 kg/m^3) fluidized at 6 m/sec. The specific heats of gas and solid particles are 1.15 and 1.26 kJ/kg K, respectively. Other parameters are as follows: The local bed density is 200 kg/m^3 and the average solid velocity is approximately 0.15 m/sec. The viscosity, density and thermal conductivity of the gas at 850°C are 43.2 × 10^{-6} N sec/m^2, 0.3177 kg/m^3, and 0.072 W/m K, respectively.
>
> Find:
>
> 1. The height within which the primary air will reach within 99% of a bed temperature of 850°C.
> 2. The time required for a 6000-μm fresh coal particle (ρ_p = 1350 kg/m^3) to be heated to within 99% of the above bed temperature.

Solution

The voidage,

$$\varepsilon = 1 - \frac{200}{2500} = 0.92$$

From Equation 6.8

$$X_{99\%} = \frac{\rho_g U}{\rho_b S} \int_{T_{g0}}^{T_{99\%}} \frac{C_g dT_g}{h_{gp}(T_p - T_g)}$$

Surface area per unit mass of particles,

$$S = \frac{6}{\rho_p d_p} = \frac{6}{2500 \times 0.0002} = 12$$

1. $T_{99\%} = 30 + 0.99(850 - 30) + 273 = 1114$ K.

$$U_{cp} = 0.15 \text{ m/sec}$$

$$\text{Re}_{cp} = \frac{(U - U_{cp})\rho_g d_{cp}}{\mu} = \frac{(6 - 0.15) \times 0.317 \times 0.0002}{43.2 \times 10^{-6}} = 8.6$$

Although the validity of Equation 6.4 for fine particles is uncertain, we use this in absence of a better correlation. Substituting $d_{cp} = d_p = 0.0002$ m we get:

$$h_{gp} = 0.33 \frac{0.072}{0.0002} 8.6^{0.62} \left[\frac{0.0002}{0.0002} \right]^{0.1} = 450 \text{ W/m}^2 \text{ K}$$

From Equation 6.8:

$$X_{99\%} = \frac{0.3177 \times 6}{200 \times 12} \int_{303}^{1114} \frac{1.15 \times 10^3 dT_g}{450(1123 - T_g)} = 0.0093 \text{ m} = 9.3 \text{ mm}$$

2. Here, $d_{cp} = 0.006$ m, $d_p = 0.0002$ m.

By assuming the carbon particle to have the same velocity as the bed materials and by neglecting the radiation effect, we use Equation 6.4 to calculate the gas–particle convective heat-transfer coefficient:

$$\text{Re}_{cp} = (6 - 0.15) \times 0.317 \times 0.006/(43.2 \times 10^{-6}) = 257$$

$$h_{gp} = \frac{0.072}{0.006} \times 0.33 \times 257^{0.62} \left[\frac{0.006}{0.0002} \right]^{0.1} = 173 \text{ W/m}^2 \text{ K}^*$$

The time required to reach 99% of the bed temperature is calculated from

$$t_{99\%} = \frac{0.765 C_p \rho_p d_{cp}}{h_{gp}} = \frac{0.765 \times 1.26 \times 1350 \times 0.006}{0.173} = 45 \text{ sec}$$

*The above calculation is only tentative because the radiative heat transfer has not been considered.

6.2 HEAT TRANSFER IN CIRCULATING FLUIDIZED BEDS

This section presents a brief discussion to help understand the influence of design and operating parameters on heat transfer to the walls and other surfaces of a CFB boiler.

6.2.1 MECHANISM OF HEAT TRANSFER

In a CFB, fine solid particles (Geldart Group A and B) agglomerate and form clusters or strands in a continuum of generally up-flowing gas containing sparsely dispersed solids. The continuum is called the *dispersed phase*, while the agglomerates are called the *cluster phase*. The majority of the bed particles move upwards through the core of the bed, but flow downwards along the wall in the form of clusters of particles or strands as shown in Figure 6.3 and are discussed in detail in Sections 2.2.1 and 2.4.1, 2.4.2. The heat transfer to the furnace wall occurs through conduction from particle clusters, convection from the dispersed phase, and radiation from both phases (Figure 6.3).

Solid clusters sliding down the wall experience unsteady-state heat conduction to the wall. The clusters cool down while losing heat to the wall through conduction and radiation. In commercial

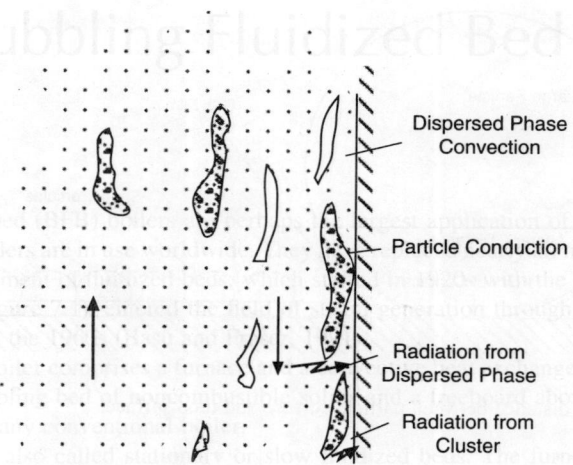

FIGURE 6.3 Schematic of the mechanism of heat transfer to the walls of a circulating fluidized bed.

boilers, heat-absorbing surfaces are very long, so in spite of a lateral exchange of solids, a cooled layer of clusters may be formed on the furnace wall creating a thermal boundary layer.

A comprehensive theory of heat transfer in CFB boilers is still awaited. The following sections present a simple mechanistic model followed by experimentally observed effects of different variables on the heat transfer.

6.2.2 Theory

A simplified treatment based on the cluster renewal model (Subbarao and Basu, 1986; Dutta and Basu, 2004) is presented here. This model can explain most experimental observations of heat transfer in fast-fluidized beds.

6.2.2.1 Heat Transfer Mechanism

The heat transfer in a fluidized bed is made up of three components: gas convection, particle convection and radiation. These components are not strictly additive, but it is a common practice to treat them separately.

If f is the average fraction of the wall area covered by clusters, the time-averaged overall heat-transfer coefficient, h may be written as the sum of convective, h_{conv} and radiative, h_r heat transfer coefficients:

$$h = h_{conv} + h_r = f(h_c + h_{cr}) + (1-f)(h_d + h_{dr}) \qquad (6.9)$$

where h_c and h_d are the convective heat-transfer coefficients due to the cluster and dispersed phase, respectively. The heat-transfer coefficients h_{cr} and h_{dr} are the radiative contributions of the cluster and dispersed phase, respectively.

Since the contact area between the particles and the wall is small, the direct heat transfer from the particles to the wall through the point of contact is negligible. The majority of the heat is transferred through conduction across the thin layer of gas between the first layer of particles and the wall as shown in Figure 6.4. Thus, the thermal conductivity of the gas and the thickness of the gas layer determine the heat transfer between the gas–particle suspension and the wall.

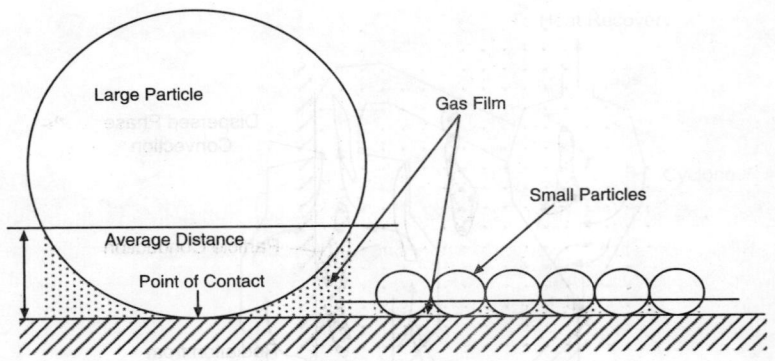

FIGURE 6.4 Average distance between particle surface and tube surface.

This thickness being very small, the heat transfer from clusters is significantly higher than the convective heat transfer from the gas or dispersed phase.

The particle volume fraction on a flat wall is about three times the riser's cross-sectional average value. Since the particle volume fraction at the wall is proportional to the cross-sectional average suspension density, the latter represents the condition at the wall.

Agglomeration of solid particles into clusters is a characteristic feature of a CFB furnace. Figure 6.3 and Figure 6.5 show the mechanism of cluster renewal in the core annulus structure of a CFB riser column.

The clusters travel down the wall a certain distance, disintegrate and reform periodically in the wall region of the furnace (Figure 6.3). When the clusters slide over the wall, unsteady-state heat conduction between the clusters and the wall takes place. The time-averaged heat-transfer coefficient due to cluster conduction is given by the following equation (Mickley and

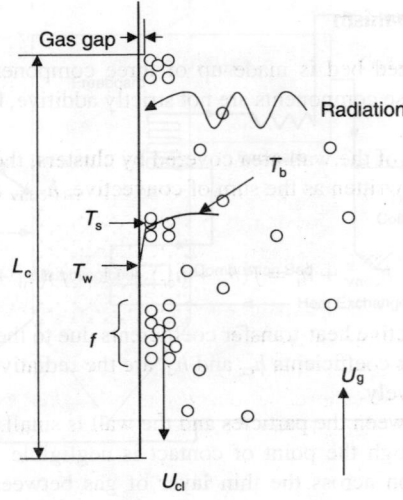

FIGURE 6.5 Heat-transfer mechanism in a fast fluidized bed.

Heat Transfer

Fairbanks, 1955):

$$h_{\text{cluster}} = \left[\frac{4K_c(\rho c)_c}{\pi t_c}\right]^{0.5} \quad (6.10)$$

where t_c is the residence time of clusters on the wall.

6.2.2.2 Cluster Properties

The thermal conductivities of a cluster, K_c and the gas, K_g are calculated from the equation developed by Gelperin and Einstein (1971) for packet heat transfer:

$$\frac{K_c}{K_g} = 1 + \frac{M}{N} \quad (6.11)$$

When the ratio of thermal conductivities of particle and gas, K_p/K_g is less than 5000 and the particle size is below 0.5 mm the parameters M and N are given by:

$$M = (1 - \varepsilon_c)\left(1 - \frac{K_g}{K_p}\right); \qquad N = \left(\frac{K_p}{K_g}\right) + 0.28\varepsilon_c^{0.63(K_g/K_p)^{0.18}}$$

The specific heat of a cluster, $(\rho C)_c$ is calculated as a lumped property (Basu and Nag, 1996) as below:

$$(\rho C)_c = (1 - \varepsilon_c)\rho_p C_p + \varepsilon_c \rho_g C_g \quad (6.12)$$

where ε_c is voidage in the cluster and is calculated as $\varepsilon_c = 1 - C_{sf}$. The solid fraction in the cluster, C_{sf} depends on the cross-sectional average bed voidage ε_{avg}. For an atmospheric pressure CFB, it is given by (Lints and Glicksman, 1994):

$$C_{sf} = 1.23(1 - \varepsilon_{avg})^{0.54} \quad (6.13)$$

After coming in contact with the wall surface, a cluster sweeps down the wall. Initially it accelerates to a steady velocity and decelerates in the vertical direction before moving away from the wall. For simplification, we assume that the cluster travels on the wall with an average velocity U_{cl} over a characteristic length L_c, and then dissolves or returns to the core (Figure 6.5). The residence time for each cluster (t_c) at the wall surface is given by:

$$t_c = \frac{L_c}{U_{cl}} \quad (6.14)$$

The cluster velocity can be estimated by the following correlation (Noymer and Glicksmann, 2000):

$$U_{cl} = 0.75\sqrt{\frac{\rho_p}{\rho_g}gd_p} \quad (6.15)$$

The characteristic length L_c increases with suspension density (Wu et al., 1990):

$$L_c = 0.0178(\rho_{sus})^{0.596} \quad (6.16)$$

The value of the characteristic length varies from 0.5 m (Lints and Glicksman, 1994) to 2.0 m. A value between 1.2 and 1.5 m would be a reasonable estimate.

6.2.2.3 Heat Transfer from Cluster Phase

Besides the transient heat conduction resistance within the cluster (Equation 6.10), there is another resistance for heat conduction across the thin gas layer between the cluster and the wall. The heat transfer coefficient due to this conduction through the gas layer is given as:

$$h_w = \frac{k_g}{\delta d_p} \tag{6.17}$$

where δ is the nondimensional gas layer thickness between the wall and cluster, and is calculated using the expression given by Lints and Glicksman (1994):

$$\delta = 0.0282(1 - \varepsilon_{avg})^{-0.59} \tag{6.18}$$

where ε_{avg} is the cross-sectional average voidage.

By assuming the contact resistance and the transient conduction through a cluster of particles, to act independently in series, we find the combined cluster convective heat transfer coefficient, h_c as:

$$h_c = \left[\frac{1}{h_{cluster}} + \frac{1}{h_w}\right]^{-1} = \frac{1}{\left[\frac{\pi t_c}{4k_c(\rho c)_c}\right]^{0.5} + \frac{\delta d_p}{k_g}} \tag{6.19}$$

The radiation heat transfer from the cluster to the wall may be considered as that between two parallel plates. Thus, the cluster radiation component of the heat transfer coefficient is estimated from the equation given below:

$$h_{cr} = \frac{\sigma(T_s^4 - T_w^4)}{\left(\frac{1}{e_c} + \frac{1}{e_w} - 1\right)(T_s - T_w)} \tag{6.20}$$

where T_s is the cluster layer temperature, and e_c and e_w are the emissivities of the cluster and wall, respectively.

The cluster emissivity e_c, is higher than the emissivity of the particles e_p. It is estimated from the following relation (Grace, 1982):

$$e_c = 0.5(1 + e_p) \tag{6.21}$$

6.2.2.4 Heat Transfer from Dispersed Phase

The dispersed phase contains a small concentration of particles, but that affects the heat transfer coefficient. Therefore, the expression for flow through air heater tubes (Basu et al., 1999) is modified with a correction factor for particle presence (C_{pp}):

$$h_d = 0.023 C_{pp} C_l C_t \frac{k_g}{D_b} Re^{0.8} Pr^{0.4} \tag{6.22}$$

where D_b is the equivalent furnace diameter and k_g is the conductivity of gas. The coefficients, C_{pp}, C_l, and C_t are described below.

The correction factor C_{pp} for the presence of particles in the dilute phase can be assumed to be 1.1. The correction factor for the temperature difference between wall and medium, C_t can be calculated as (Basu et al., 1999):

$$C_t = \left(\frac{T_b}{T_w}\right)^{0.5} \quad (6.23)$$

The correction factor of tube length, $C_l = 1$, when $L/D_b > 50$, but CFB furnaces are generally not as tall as this. So, to account for the entrance effect one can correlate C_l using the following equations (Perry and Green, 1997):

$$C_l = 1 + F\left(\frac{D_b}{L}\right) \quad (6.24)$$

and F is 1.4 for fully-developed flow.

Radiation between the suspension (dilute phase) and the bare wall is estimated from the usual expression for parallel surfaces, (Basu, 1990) i.e.:

$$h_{dr} = \frac{\sigma(T_b^4 - T_w^4)}{\left(\frac{1}{e_d} + \frac{1}{e_w} - 1\right)(T_b - T_w)} \quad (6.25)$$

where e_w and e_d are the emissivities of the wall and dilute phase at T_w and T_b, respectively, for a large boiler estimated from the correlation of Brewster (1986) as shown below.

In a CFB boiler the contribution of combustion gas as well as particle cloud is considered to find the emissivity of the dilute phase e_d, as:

$$e_d = \lfloor e_g + (1 - e_g)e_{pc} \rfloor \quad (6.26)$$

The emissivity of the particle cloud is

$$e_{pc} = \left[1 - \exp\left(-\frac{1.5 e_p F' L_b}{d_p}\right)\right] \quad (6.27)$$

where F' is the volume fraction of solids in dilute phase, and e_p is the emissivity of particle surface.

The mean beam length, L_b is given by:

$$L_b = \frac{3.5V}{A}$$

If the L_b and F' are such that e_{pc} is in excess of 0.5 to 0.8, the scattering must be considered (Andersson et al., 1987). Such is the case of large CFB boilers where the dilute phase emissivity e_d, may be calculated according to Brewster (1986) by the following equation:

$$e_d = \left[\frac{e_p}{(1-e_p)B}\left\{\frac{e_p}{(1-e_p)B} + 2\right\}\right]^{0.5} - \frac{e_p}{(1-e_p)B} \quad (6.28)$$

For isotropic scattering $B = 0.5$ and for diffusely reflecting particles $B = 0.667$.

6.2.2.5 Wall Fraction Covered by Clusters (f)

Experimental data on the wall coverage plotted by Glicksman (1997) shows a strong influence of column diameter, which was taken into account by Golriz and Grace (2002) in the following expression:

$$f = 1 - \exp\left\{-25,000\left[1 - \frac{2}{e^{0.5D} + e^{-0.5D}}\right](1 - \varepsilon)\right\} \quad (6.29)$$

where D is the equivalent diameter of the furnace cross section.

The above equation, based on data from laboratory units (0.09 to 0.30 m), predicts the wall coverage for small units well, but fails to do the same for large units. For example, in the case of large (equivalent bed diameter >3 m) CFB boilers, which often operate with a suspension density of 2 kg/m^3 (solid volume fraction is 0.0004 or below, Equation 6.29 gives a value of wall coverage f, approaching unity.

Dutta and Basu (2004) modified the above equation, noting that the average solid concentration, the most dominant contributor for the wall coverage, varies exponentially with the height of the furnace.

$$f = 1 - \exp\left(-a\{1 - \varepsilon\}^b \left\{\frac{D}{H}\right\}^c\right) \quad (6.30)$$

Data from several commercial boilers were analyzed to derive values for the coefficients a, b, and c as follows: $a = 4300$, $b = 1.39$, $c = 0.22$.

6.2.3 Experimental Observations in a CFB Boiler

The bed-to-wall heat transfer in a fast-fluidized bed furnace is influenced by a number of design and operating parameters. The nature of those effects is discussed below.

6.2.3.1 Effect of Suspension Density

The time-averaged suspension density on the wall is the most important factor influencing the bed-to-wall heat transfer in a fast or circulating fluidized bed (CFB). The suspension density on the wall is proportional to its cross-sectional average value as shown in Equation 2.40. Figure 6.6 shows the

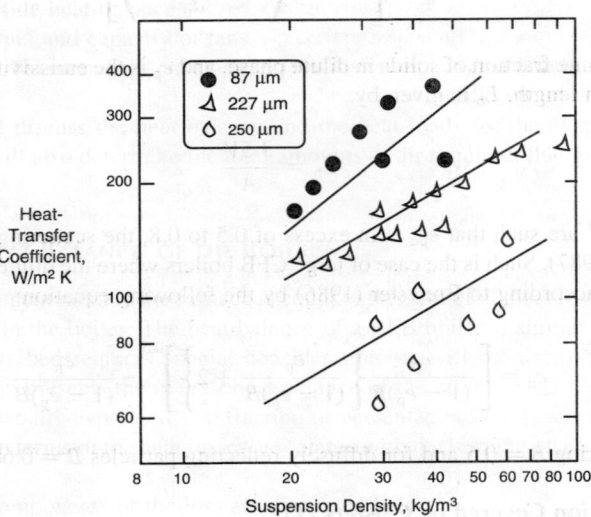

FIGURE 6.6 Effect of suspension density on heat-transfer coefficients measured on a small heat-transfer probe at room temperature.

Heat Transfer

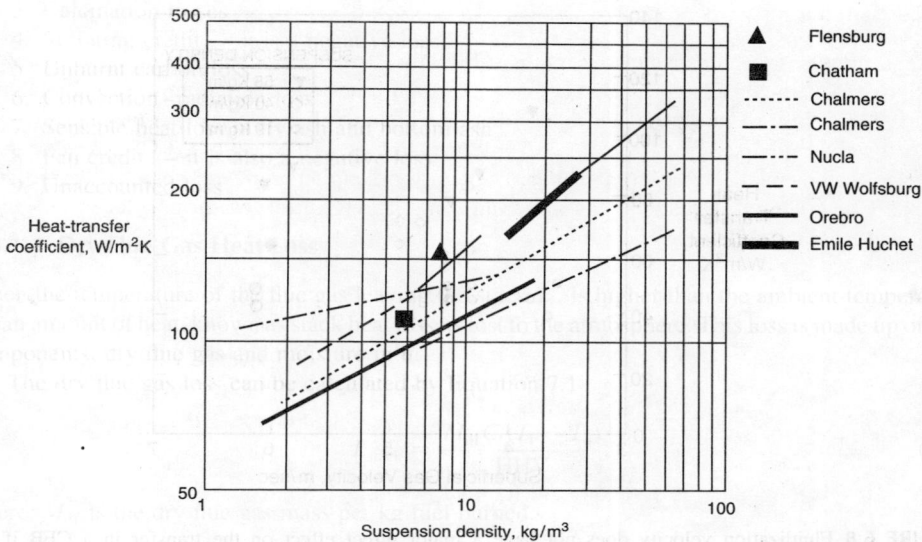

FIGURE 6.7 Heat-transfer coefficient vs. suspension density for large CFB.

effect of cross-sectional average suspension density on the heat-transfer coefficient for different particles in a laboratory unit operating at room temperature. The effect of suspension density in larger industrial-sized units operating at elevated temperature is shown in Figure 6.7. In both cases, the heat-transfer coefficient increases with suspension density. The slope of the curves suggests that the heat-transfer coefficient is proportional to the square root of the suspension density (Glicksman, 1988).

The heat-transfer coefficient increases from the bottom to the top of a boiler furnace and is influenced by a number of factors, including air flow (primary and secondary), solid circulation, solid inventory and particle size distribution. Much of the effect of these parameters on the heat transfer is due to their influence on the suspension density.

In commercial CFB boilers, where the solid circulation rate cannot be varied at will, the ratio of primary to secondary air is adjusted to control the suspension density in the upper section of the furnace.

6.2.3.2 Effect of Fluidization Velocity

Unlike in BFBs, the fluidizing velocity does not have any direct influence on the heat transfer in fast beds, except through the suspension density. This is why heat-transfer coefficients at varying fluidizing velocities but at a fixed suspension density show only a minimal effect of the fluidizing velocity (Figure 6.8). In most situations, when the velocity is increased leaving the solid circulation rate constant, the heat-transfer coefficient drops. This is an effect of reduced suspension density, rather than of increased gas velocity. However, in exceptionally dilute beds, a positive effect of fluidizing velocity may be present.

6.2.3.3 Effect of Vertical Length of Heat-Transfer Surface

The heat-transfer coefficient on a vertical surface decreases down its length, but with a diminishing rate as shown in Figure 6.9. Though the heat-transfer coefficient decreased continuously with length, it changed very little beyond 0.7 m in this case. As a cluster descends along the heat-transfer surface, its temperature approaches that of the wall and thereby reduces the temperature difference between the wall and the first layer of particles in the cluster. Consequently, the heat-transfer

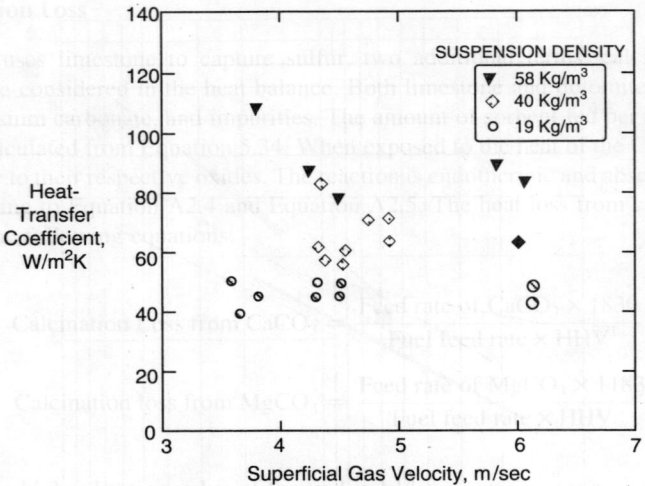

FIGURE 6.8 Fluidization velocity does not have a major direct effect on the transfer in a CFB if the suspension density does not change with the fluidization velocity (Basu and Fraser, 1991).

coefficient calculated on the basis of temperature difference between the wall and the bulk of the bed decreases along the length of the heat-transfer surface.

After falling through a certain height, the cluster either returns to the core or dissolves into dispersed solids, while a fresh cluster takes its place on the wall. This renewal of clusters is similar to the renewal of individual particles on the wall of a BFB as shown later in Figure 6.14. Thus, beyond a certain vertical length of the heat-transfer surface, the heat-transfer coefficient approaches an asymptotic value, as shown in Figure 6.9.

6.2.3.4 Effect of Bed Temperature

The overall heat-transfer coefficient increases with bed temperature as is apparent in Figure 6.10 for a large commercial CFB boiler furnace.

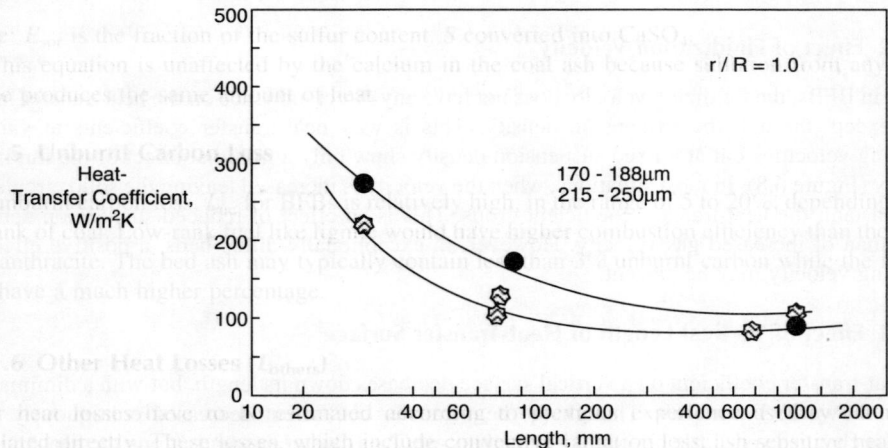

FIGURE 6.9 Effect of vertical length of heat-transfer section on heat-transfer coefficient.

Heat Transfer

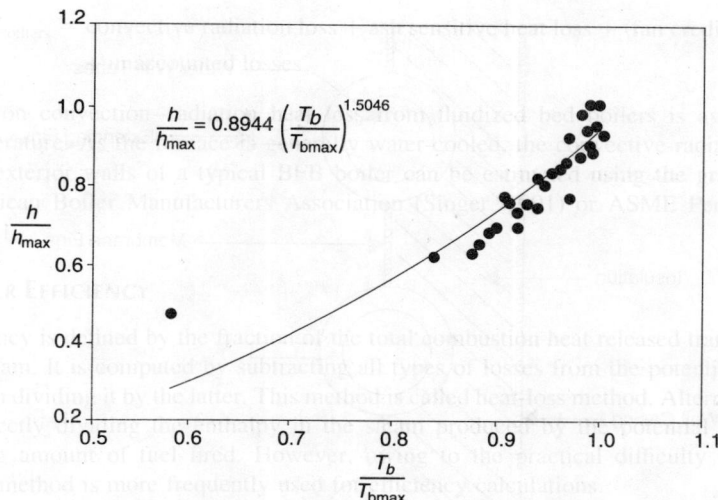

FIGURE 6.10 Water wall heat-transfer coefficient with non-dimensional bed temperature (170 MW$_e$). (From Dutta and Basu, 2002.)

High temperature reduces the thermal resistance of the first layer of particles, as well as that of the rest of the cluster, due to the increase in thermal conductivity of the fluidizing gas with temperature. The rise in the heat-transfer coefficient with the bed temperature is also attributed to the increase in radiation from the bed. Dutta and Basu (2002), who plotted the data on a nondimensional scale, found the heat transfer to be proportional to $T_b^{1.5}$.

At a given suspension density, the cooling of clusters sliding down the wall tends to reduce the effect of high temperature on the overall heat transfer. Consequently, the effect of bed temperature on the total heat transfer is more pronounced in shorter heat-transferring surfaces. In the case of heavily-cooled surfaces, such as the water-cooled wall of commercial boilers, the solids sliding down the wall may be sufficiently cooled to form a thermal boundary layer. This may again reduce the effect of temperature on the bed-to-wall heat transfer. The effect of bed temperature is more significant in the case of a more dilute bed, where radiation is dominant.

6.2.3.5 Effect of Particle Size

For short surfaces, like those in bench scale units, finer particles result in higher heat-transfer coefficients (Figure 6.6) primarily due to an increase in the convective component of heat transfer when the particle size decreases. This feature of fast beds is similar to that of BFBs.

The effect of particle size is, however, less pronounced for longer heat-transferring surfaces, like those in industrial units. When the heat-transfer surface is long, the first layer of particles, in contact with the wall, has adequate time to cool down. Thus, the thermal resistance of these particles plays a less important role. For the longer surfaces of commercial boilers, the effect of particle size on the total heat-transfer coefficient is therefore less significant.

6.2.4 EFFECT OF VERTICAL FINS ON THE WALLS

If the area of the furnace wall can be increased through the use of fins projecting into the furnace, it may be possible to provide a much greater heat absorption in the furnace wall (Basu et al., 1991). This will allow the boiler furnace to be more compact or even less tall. In some cases it may even be possible to avoid the use of an external heat exchanger (EHE) or heating surfaces suspended inside

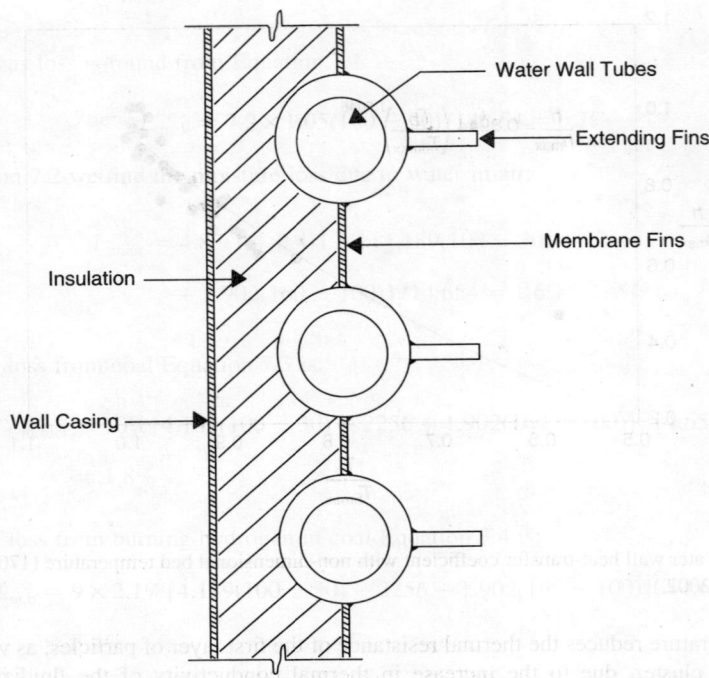

FIGURE 6.11 Two types of fins on the walls of a CFB boiler.

the furnace. It would also provide some flexibility to the boiler designer in apportioning different heating surfaces for the boiler. A typical wall fin is shown in Figure 6.11.

The heat transfer from a bed at T_b to finned wall at T_w may be written as:

$$Q_f = A_f \eta_f h(T_b - T_w) \qquad (6.31)$$

where A_f is the surface area of the fins, η_f is the fin efficiency, and h is the heat-transfer coefficient on the base area (area in the absence of fins). The fin efficiency, measured at room temperature, is found to be generally in the range of 40 to 100% (Basu and Cheng, 2000). Lower efficiencies were found in laboratory beds while higher efficiency was found in a hot demonstration boiler. A value of the efficiency close to 100% suggests some improvement in the heat transfer coefficient on the fin tip due to enhanced local solids motion due to the presence of the fins.

Example 6.2

A boiler panel wall of membrane type (50-mm diameter tube arranged at 75-mm pitch) contains one 26-mm high and 3-mm thick vertical fin on each tube. Find the enhancement of heat absorption by the wall.

Solution

Panel area per unit length without the fin = $3.14 \times 0.05 + 0.025 = 0.182$ m²/m
Heat absorbed by the panel without the fin = $0.182h (T_b - T_w)$.

After installation of the fins, the additional surface added to the panel per unit length would be:

$$A_f = 2 \times 0.026 + 0.003 = 0.053 \text{ m}^2/\text{m}$$

Assuming a fin efficiency, $\eta_f = 70\%$, the effective area after the installation of fins is:

$$A_{\text{effective}} = 0.182 + 0.053 \times 0.7 - 0.003 = 0.2161 \text{ m}^2/\text{m}$$

$$\text{Enhancement in heat absorption} = \frac{0.2161 h (T_b - T_w)}{0.182 h (T_b - T_w)} = 1.187$$

This suggests an enhancement in heat absorption of $(1.187 - 1)$ or 18.7%. A deeper fin may increase the heat absorption further, but it may also affect the local hydrodynamics or have an unacceptably high metal temperature.

6.3 HEAT TRANSFER IN BUBBLING FLUIDIZED BED

A typical BFB boiler furnace or the EHE of a CFB boiler operates in the bubbling fluidized bed (BFB) regime. A BFB is comprised of a dense bubbling bed and a freeboard (Figure 6.12) above it. The heat transfer mechanisms for these two sections are different, so they are described separately, starting with the discussion on bubbling bed heat transfer. Heat transfer in the freeboard is discussed later in Section 6.4.

The overall gas-side heat transfer coefficient, h_o can be written as a sum of the contributions of three components: particle convection or conduction h_{cond}, gas convection h_{conv}, and radiation h_{rad}.

$$h_o = h_{\text{cond}} + h_{\text{conv}} + h_{\text{rad}} \tag{6.32}$$

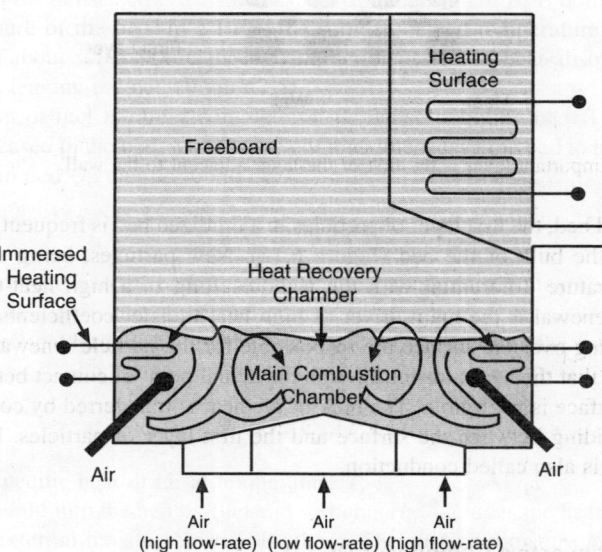

FIGURE 6.12 A bubbling fluidized bed furnace is comprised of a dense bubbling bed and a lean freeboard above it. Heat can be absorbed from both sections.

As the name suggests, the particle conduction or convection is the contribution of the particles, while gas convection is due to the gas flowing over the surface. Radiative heat transfer is relevant only at elevated temperatures.

For BFBs, it is difficult to compute the heat transfer purely from the theory, so design equations are developed based on many experiments. Mechanistic models are developed primarily for data extrapolation and qualitative understanding of the heat transfer process. One of the following two approaches can be used to calculate the heat transfer coefficient for a BFB.

1. Based on mechanistic model or empirical equation
2. Based on design graphs

6.3.1 MECHANISTIC MODEL

Heat transfer in a BFB is best explained by the packet-renewal model developed by Mickley and Fairbanks (1955). Recent experimental data lend support to this packet-renewal model (Chandran and Chen, 1982). According to this model, packets of particles (or parcels of emulsion phase) at the bulk bed temperature are swept to the heat-transfer surface due to the action of bubbles. The packets stay in contact with the surface for a short time τ_p, and are then swept back to the bed.

During this time of contact, unsteady state heat transfer takes place between the packet and the surface. The first layer of particles, closest to the wall temperature (Figure 6.13) is the most important layer for heat transfer, especially for particles whose thermal time constant $(d_p \rho_p C_p / 6 h_{gp})$ is larger than the residence time τ_p (Equation 6.42), which typically happens for Group B and D particles. While in contact, the particles cool down, so the temperature difference between the heat-absorbing surface and heat giving particles reduces. This results in a lower heat-transfer coefficient.

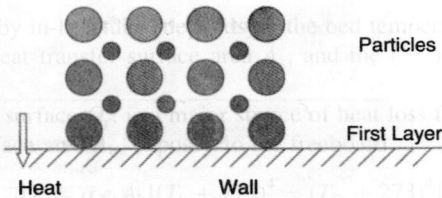

FIGURE 6.13 Most important layer is the first or the layer adjacent to the wall.

Unlike in a fixed bed, the first layer of particles in a fluidized bed is frequently replaced by fresh hot particles from the bulk of the bed (Figure 6.14). New particles, being at bed temperature, increase the temperature differential with the wall resulting in a high heat-transfer coefficient. Thus, the particle renewal is the main driver of high heat-transfer coefficients in a fluidized bed. The bubbles sweeping over the surface are responsible for the particle renewal.

It may be noted that the heat transfer through the actual point of contact between the first layer particles and the surface is negligible. The bulk of the heat is transferred by conduction across the thin film of gas residing between the surface and the first layer of particles. For this reason, the particle convection is also called conduction.

6.3.1.1 Particle Convective Component (h_{cond})

The heat transfer between the tubes and the fluidized bed is made up of contributions of the bubble and the emulsion phase. Thus, h_{cond} the convective component of the heat transfer coefficient, can be

Heat Transfer

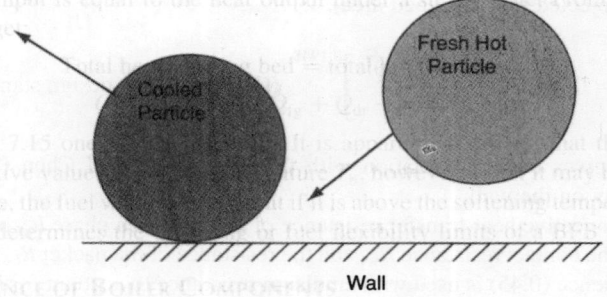

FIGURE 6.14 Particle replacement adjacent to surface in fluidized bed.

written as:

$$h_{cond} = f_b h_b + (1 - f_b) h_e \quad (6.33)$$

where f_b is the fraction of tube surface covered by bubbles, and h_e is the emulsion-phase heat-transfer coefficient. These include both radiation and convection.

6.3.1.1.1 Heat Transfer from Bubble Phase (h_b)

Heat transfer in the bubble phase is less than that in the emulsion phase. Therefore, in some calculations, the heat transfer from the bubble phase is neglected, although in some cases it is still considered. The following equation has been developed based on the laminar boundary layer on a flat plate (Glicksman, 1984):

$$(Nu)_b = 0.664 \sqrt{\frac{(U_b + 3U_{mf})\rho_g d_p}{\mu}} Pr^{0.33} \sqrt{\frac{d_p}{L}} \quad (6.34)$$

where, U_b is average bubble rise velocity (m/sec), L is one fourth the circumference of tube (m), and μ is the dynamic viscosity of gas.

The rise velocity of a bubble, U_b in a bed fluidized at velocity U is given as:

$$U_b = 0.7\sqrt{gd_b} + U - U_{mf} \quad (6.35)$$

where d_b is the diameter of the bubble.

6.3.1.1.2 Emulsion Phase Heat Transfer (h_e)

It may be noted that in most BFBs, the bubble phase heat transfer h_b is small compared to the contribution from the emulsion phase h_e, and as such it can be neglected (Chen et al., 2005). So, Equation 6.33 can be simplified as:

$$h_{cond} = (1 - f_b) h_e \quad (6.36)$$

The fractional wall surface coverage f_b, can be determined by the empirical equation:

$$f_b = 0.19 \left[\frac{U_{mf}^2 \left(\frac{U}{U_{mf}} - \alpha \right)^2}{d_p g} \right]^{0.23} \quad \text{(Kim et al., 2003)} \quad (6.37)$$

where $\alpha = 1 -$ (projected tube area/bed area).

$$f_b = 0.08553 \left[\frac{(U - U_{mf})^2}{9.81 d_p} \right]^{0.1948} \quad \text{(Chinese Boiler design standard)} \quad (6.38)$$

Equation 6.37 gives better correlation with the measurement of Chen (2003) for the time fraction clusters stay on the wall.

The particle-convective heat-transfer resistance of the emulsion phase h_e, has two components in series. They are the contact resistance R_w, and the resistance of the packet R_c. An empirical factor derived from experience (0.45) is used with emulsion resistance to account for some overlap of the components:

$$\frac{1}{h_e} = R_w + 0.45 R_c \quad (6.39)$$

The contact resistance, R_w is expressed empirically by:

$$R_w = \frac{d_p}{3.75 K_c} \quad (6.40)$$

where d_p is the average particle diameter in fluidized bed, and K_c is the effective thermal conductivity of a packet of particles.

The instantaneous packet resistance is developed here using the unsteady-state heat-conduction model of Mickley and Fairbanks (1955). This expression is similar to Equation 6.10, which was used to compute the time-averaged heat-transfer from transient clusters in a fast fluidized bed.

$$R_c = \sqrt{\frac{\pi \tau_p}{K_p \rho_b C_b}} \quad (6.41)$$

$\tau_p =$ contact time between tube surface and clusters(s)
$\rho_b =$ bulk density of bed (kg/m^3)
$C_b =$ specific heat of bed (kJ/kg K)

The contact time, τ_p can be obtained using the empirical equation developed by Kim et al. (2003):

$$\tau_p = 1.2 \left[\frac{d_p g}{U_{mf}^2 (U/U_{mf} - \alpha)^2} \right]^{0.3} \left(\frac{d_p}{d_t} \right)^{0.225}$$

An alternative equation is offered by the Chinese boiler standard:

$$\tau_p = 8.932 \left[\frac{(9.81 d_p)}{(U - U_{mf})^2} \right]^{0.0756} \sqrt{\frac{d_p}{0.025}} \quad (6.42)$$

$d_t =$ tube's outside diameter
$\alpha = 1 -$ (projected tube area/bed area)
$U =$ superficial gas velocity through fluidized bed (m/sec)
$U_{mf} =$ minimum fluidization velocity of fluidized bed (m/sec)

TABLE 6.2
Specific Heat of Coal-Ash (kJ/kg K)

T (C)	100	300	400	500	600	700	800	900	1000
C_p (kJ/kg K)	0.810	0.880	0.900	0.916	0.935	0.947	0.960	0.971	0.984

Equation 6.42 shows a better correlation with the contact time measured in a bubbling bed, according to Chen (2003). The effective thermal conductivity K_p, can be obtained by:

$$K_p = K_c + 132 d_p U_{mf} \rho_g C_g \qquad (6.43)$$

K_c = thermal conductivity of cluster (W/m K)
ρ_g = gas density (kg/m^3)
C_g = gas specific heat (kJ/kg K)

The thermal conductivity of the cluster on particle packet K_c, can be determined by:

$$K_c = K_p \left\{ 1 + \frac{(1 - \varepsilon_{mf})\left(1 - \dfrac{K_g}{K_p}\right)}{\left[\dfrac{K_g}{K_p} + 0.025 \varepsilon_{mf}^{0.63(K_g/K_p)^{0.18}}\right]} \right\} \qquad (6.44)$$

K_p = thermal conductivity of solid particles in bed (kW/m K)
K_g = thermal conductivity of gas (kW/m K)
ε_{mf} = voidage of bed at minimum fluidization velocity (m/sec)

The following thermo-physical data of bed materials and typical flue gas in BFB boilers may be used for practical designs.

1. Thermal conductivity of bed particle, K_p = 0.30 to 0.33 (kW/m K).
2. Bed voidage at minimum fluidization, ε_{mf} = 0.5.
3. Specific heat of bed is dependent on the bed materials. Coal ash, whose specific heat is given in Table 6.2, is commonly used as the bed material.

Both the specific heat and thermal conductivity of the flue gas depend on the gas temperature. Some typical values at several temperatures are given in Table 6.3.

TABLE 6.3
Specific Heat and Thermal Conductivity of Typical Flue Gas of a Coal-Fired Fluidized Bed Boiler

T (C)	100	200	300	400	500	600	700	800	900
C_g (kJ/kg K)	1.05	1.08	1.11	1.14	1.17	1.20	1.33	1.25	1.27
K_g (kW/m K)	0.0312	0.0401	0.0483	0.0569	0.0655	0.0741	0.0826	0.0914	0.1000

6.3.1.1.3 Empirical Relation

An alternative to the above comprehensive mechanistic model is the following empirical relation of Glicksman (1984):

$$(Nu)_{cond} = \frac{h_{cond}d_t}{K_g} = 6.0 \qquad (6.45)$$

where d_t is the outside diameter of heat transfer tube.

6.3.1.2 Gas Convective Component (h_{conv})

Heat transfer in both fixed and fluidized bed share the same mechanism. The only difference between them is the temperature of the first layers particles; which is different because of the replacement of particles in the fluidized bed and the absence of the same in the fixed bed. So the equations obtained from fixed beds to calculate the convective component of heat-transfer coefficient can be used in the fluidized bed's emulsion phase without a great loss of accuracy.

The following equations, developed originally for packed beds, can be used for the convective and conductive components of the emulsion phase in fluidized beds. Gas convection may be calculated using the following equations:

Yagi and Kunii (1960):

$$(Nu)_{conv} = \frac{h_{conv}d_t}{K_g} = 0.05 Re\, Pr \qquad Re < 2000 \qquad (6.46)$$

Plautz and Johnstone (1955):

$$(Nu)_{conv} = \frac{h_{conv}d_t}{K_g} = 0.18 Re^{0.8} Pr^{0.33} \qquad Re > 2000 \qquad (6.47)$$

where Reynolds number, $Re = Ud_t\rho_g/\mu$ and Prandtl number, $Pr = C_p\mu/K_g$.

6.3.1.3 Radiation Component (h_{rad})

The radiative equation is written as below:

$$(Nu)_{rad} = \frac{K_g}{d_t}\left[\frac{\sigma(T_{eb}^4 - T_w^4)}{\left(\frac{1}{e_b} + \frac{1}{e_w} - 1\right)(T_{eb} - T_w)}\right] \qquad (6.48)$$

e_b = effective emissivity of particles in bed. It is approximately 0.7 to 1.0 (Ozkaynak and Chen, 1983)
e_w = effective emissivity of tube surface
σ = Stefan–Boltzmann constant, equal to 5.7×10^{-8}
T_{eb} = effective bed temperature, $\cong 0.85 T_b$ (°K)
T_w = tube surface temperature, $\cong T_{sat} + 30$ (°K)

The temperature of the clusters starts dropping after they contact the tube surface. Since the radiation exchange takes place between the tube and the contacting particles directly facing the tube, the effective bed temperature seen by the tubes is slightly lower than the bulk temperature T_b.

Heat Transfer

This has been taken into account by using a slightly lower temperature as the effective temperature ($0.85T_b$), which is used to calculate the radiation component.

6.3.2 Experimental Observations in Bubbling Fluidized Beds

The following section describes some experimentally observed effects of different operating parameters on the bed-to-tube heat transfer in a BFB.

6.3.2.1 Effect of Particle Size on Heat Transfer

Particle size is the most important factor influencing the heat transfer in a BFB, as one can see in Figure 6.15. The particle-conductive heat transfer is essentially thermal conduction across a thin gas film whose average thickness is around 1/6 to 1/10th the diameter of the particles. Naturally, a bed of smaller particles will have a smaller gas film thickness and hence greater conduction. Furthermore, one can note that smaller particles have a shorter time of residence on the tube surface (see Equation 6.42), so smaller particles have higher heat-transfer coefficients. A large particle, on the other hand, would have a lower heat-transfer coefficient due to a thicker gas film and longer residence time. Figure 6.15 illustrates the effect of particle size on the bed-to-tube heat-transfer coefficients using data from a number of bubbling beds operating at room temperature.

Figure 6.15 shows that in very fine particles (Group C), the heat-transfer coefficient is very low and increases with size. This happens because of reduced mobility and circulation in Group C particles where inter-particle forces are dominant over the gravitational force.

For Group A and B particles, the heat-transfer coefficient decreases with increasing particle size, which is primarily due to the decrease in particle convection with increasing size. The majority of fluidized bed boilers operate with Group A and B particles, making this part of the characteristic graph most important.

For very large particles like those belonging to Group D, the gas convection starts to dominate the particle convection. The gas convection, being a function of the U_{mf} of the particle, increases

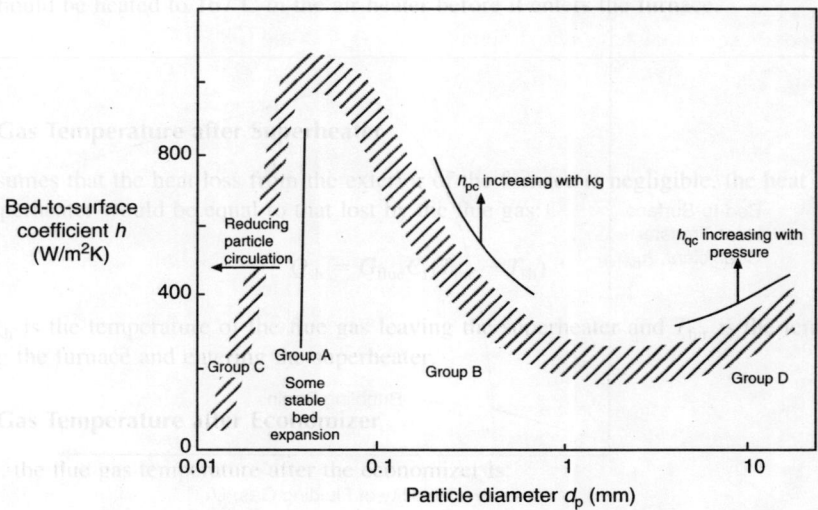

FIGURE 6.15 The bed-to-surface heat-transfer coefficient is greatly influenced by the particle size and the powder group the particles belong to.

with particle size. Thus, with increasing particle size, the heat-transfer coefficient begins to rise moderately.

In a bed of finer particles, particle conduction is more important than gas convection, but in a bed of larger particles, gas convection (Glicksman, 1984) plays a major role.

If the thermal time constant is much longer than the residence time of the particles, the temperature of the first layer may be assumed to be equal to the bed temperature. For example, the thermal time constant of a 1-mm limestone particle is around 2 sec, which is much longer than its residence time on the wall of a typical fluidized bed. Therefore, it is reasonable to assume that the temperature of the first layer of limestone particles is equal to that of the bed.

6.3.2.2 Effect of Fluidization Velocity

The effect of fluidization velocity is clearly visible in a bed of uniform particle size. Figure 6.16 shows this for Group B particles. The heat-transfer coefficient increases moderately with velocity until it reaches the minimum bubbling velocity, beyond which there is a rapid rise in the coefficient. This increase is essentially due to greater mobility of particles on the heat-transfer surface with reduced contact time. As the velocity is increased further, the heat-transfer coefficient reaches a maximum value and then starts decreasing (Figure 6.16) at a moderate pace. Higher velocity reduces the particle residence time on the wall (Chen, 2003), but also decreases the concentration of particles. These two opposing effects result in the reduction of heat transfer beyond an optimum value of velocity. Most BFB boilers run at a superficial velocity of 3 to 4 times the minimum fluidization velocity. If the fluidizing velocity is too high, both the number and size of bubbles increases, resulting in a larger fraction of the surface f_b, being exposed to the bubble phase. Since heat transfer from the bubble phase is much lower than that from the emulsion phase, the total heat-transfer coefficient drops.

6.3.2.3 Effect of Flow Pulsation

Vibration and pulsed air supply are different techniques that can be employed to enhance the heat-transfer coefficient. Flow pulsation enhances the particle mobility on the tube surface, and therefore

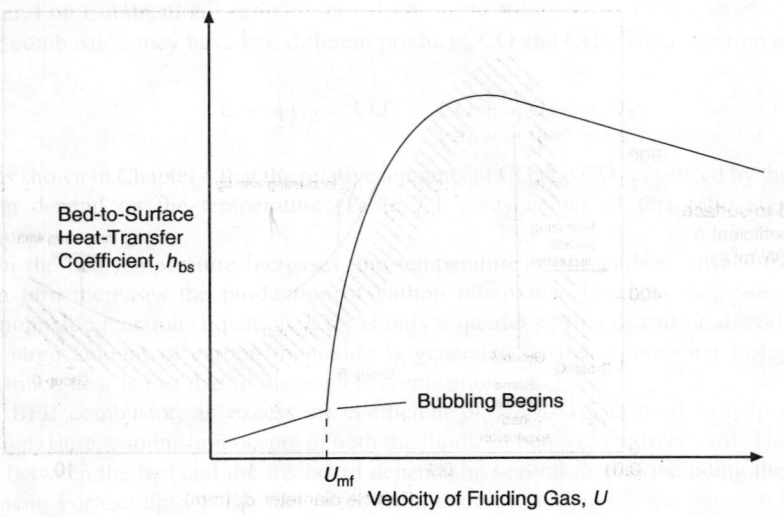

FIGURE 6.16 The bed-to-surface heat-transfer coefficient as a function of fluidizing velocity — Group B particles.

Heat Transfer

increases the heat transfer. This effect is greatest at close to or below the U_{mf}. For example, in a bed of 160-micron silica sand, Zhang (2005) found the heat transfer at $U/U_{mf} = 0.8$ to increase from 50 to 350 W/m² C when pulsation frequency increased from 0 to 7 Hz, but at $U/U_{mf} = 3.0$ the heat-transfer coefficient increased only from 550 to 600 W/m² C. The reduction in heat transfer may be due to decreased contact between the emulsion phase and the tube at high pulse frequency.

6.3.3 Heat-Transfer Correlation in Bubbling Beds

The bed-side heat-transfer coefficient h_o depends on a number of parameters, such as particle size, bed temperature and fluidizing velocity. Figure 6.15 shows the effect of particle size on heat-transfer coefficient. Bed materials in a CFB boiler (100 to 300 μm) are much finer than those in bubbling bed boilers (400 to 1500 μm). For such fine particles, the empirical correlation of Andeen and Glicksman (1976) for fine solids may be used to calculate the bed-to-tube heat-transfer coefficient. The heat-transfer coefficient is given as the sum of the convective and radiative terms:

$$h_o = 900(1-\varepsilon)\frac{K_g}{d_t}\left[\frac{Ud_t\rho_p}{\mu}\frac{\mu^2}{d_p^3\rho_g^2 g}\right]^{0.326} Pr^{0.3} + \frac{\sigma(T_b^4 - T_w^4)}{\left[\frac{1}{e_b}+\frac{1}{e_w}-1\right](T_b - T_w)} \quad (6.49)$$

α = Stefan–Boltzmann constant, 5.67×10^{-11} (kW/K⁴ m²)
T_b = temperature of the bed of EHE (K)
T_w = temperature of the tube wall (K)
e_w = emissivity of the tube wall
U = superficial gas velocity through the bubbling bed (m/sec)
ε = void fraction in the bed

Also, $(Ud_p\rho_g/\mu_g) < 10$, and the emissivity of the bed e_b is found from the relation (Grace, 1982):

$$e_b = 0.5(1 + e_p)$$

where e_p is the emissivity of bed particles.

6.3.4 Use of Design Graphs

This section presents a calculation method for heat transfer designs using tables and graphs. This approach is used by the boiler industry in China for design of BFB boilers.

Heat transfer to the tubes, Q_{fbhe} can be calculated using the following equation:

$$Q_{fbhe} = h_{tube} A_0 \text{LMTD} \quad (6.50)$$

where LMTD is the log mean temperature difference between the fluid inside the tube and the fluidized bed outside, and A_0 is the external surface area of the tube.

The overall heat transfer coefficient, h_{tube} is obtained from the relation:

$$h_{tube} = \frac{1}{\dfrac{d_t}{d_i h_i} + \dfrac{d_t \ln(d_t/d_i)}{2K_m} + \dfrac{1}{h_o}} \quad (6.51)$$

where h_o and h_i are the outside and inside heat-transfer coefficients, and K_m is the thermal conductivity of the tube materials.

6.3.4.1 Tube-to-Steam Heat-Transfer Coefficient (h_i)

The heat-transfer coefficient on the inside surface of tubes, h_i is comparable to that on the outside, h_o if the tubes carry superheated steam or air. Therefore, one can use the Sieder–Tate equation (Turek et al., 1985) to calculate the internal heat transfer coefficient:

$$\frac{h_i}{V_i \rho_f C_f} Pr^{0.66} \left[\frac{\mu_{fw}}{\mu_f} \right]^{0.14} = \frac{0.023}{Re^{0.2}} \quad (6.52)$$

where

V_i = fluid velocity inside the tube
ρ_f = fluid density
C_f = specific heat of the fluid
Pr, Re = Prandtl and Reynolds number, respectively for fluid inside the tube
μ_f, μ_{fw} = viscosity of the steam at the bulk and wall temperatures, respectively

If the tube carries water, as in the evaporator or economizer, h_i is an order of magnitude larger than h_o. In this case, the first term in the denominator of Equation 6.51 may be neglected.

6.3.4.2 Outer Heat-Transfer Coefficient (h_o)

The total heat-transfer coefficient, h_o is made up of two components: radiative (h_r) and convective (h_c).

$$h_o = \xi (h_r + h_c) \quad (6.53)$$

where ξ is a correction factor discussed in the next section.

6.3.4.2.1 Correction Coefficient for Vertical and Horizontal Tubes (ξ)

The coefficient ξ varies with the pitch of horizontal and vertical tubes as shown in Figure 6.17. The coefficient ξ for horizontal tubes can be chosen from Table 6.4 against lateral and longitudinal

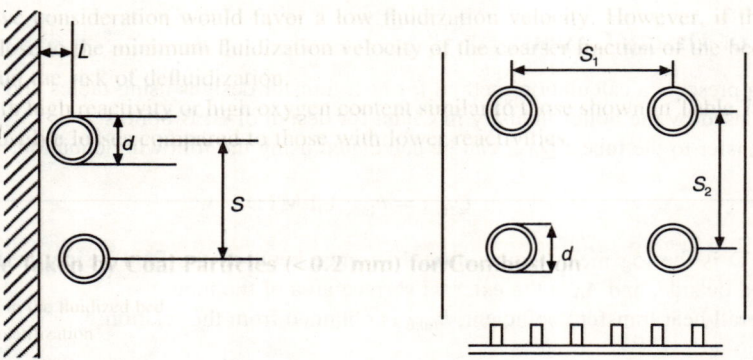

(a) Plane view of vertical furnace wall tubes (b) Cross-section elevation of horizontal tubes

FIGURE 6.17 Two arrangements of tubes in a bubbling fluidized bed.

TABLE 6.4
Correction Factor ξ for Horizontal Tubes in Bed Heat Transfer

S_1/d	In-line ($S_2/d > 2$)	Stagger ($S_2/d = 2$)	Stagger ($S_2/d = 4$)	Stagger ($S_2/d = 6$)	Stagger ($S_2/d > 8$)
1.5	0.82	0.70	0.74	0.77	0.80
2.0	0.87	0.78	0.81	0.83	0.85
2.5	0.91	0.83	0.85	0.87	0.89
3.0	0.93	0.86	0.88	0.90	0.92
3.5	0.95	0.89	0.91	0.93	0.94
4.0	0.97	0.92	0.93	0.95	0.96
4.5	0.98	0.94	0.95	0.96	0.97
5.0	0.98	0.95	0.96	0.97	0.98
5.5	0.99	0.96	0.97	0.98	0.99
6.0	0.99	0.97	0.98	0.99	0.99
7.0	0.99	0.98	0.99	0.99	0.99

S_1: Horizontal pitch of the tubes across the gas flow rate in bed (mm) (Figure 7.14). S_2: Vertical pitch of the tubes along the gas flow rate in bed (mm). d: The outside diameter of tubes (mm).

pitches. The coefficient ξ for vertical tubes is a product of two factors, ξ_1 and ξ_2. They are coefficients for the pitch and distance between the furnace wall and tubes, respectively.

$$\xi = \xi_1 \xi_2 \quad (6.54)$$

where ξ_1 and ξ_2 can be chosen from Table 6.5 and Table 6.6, respectively.

For an inclined tube, when the angle between the tube and the horizontal exceeds 60°, it can be treated as a vertical tube. In other cases, the calculation should be based on horizontal tubes.

The distance between the bottom of the in-bed tubes and the top of the fluidizing grid (distributor) should be over 400 mm. This will avoid potential erosion due to fluidizing air jets.

TABLE 6.5
Correction Factor, ξ_1 for Vertical Tubes with the Distance between Wall and the Center Line of Tube

L/d	0.5	1.0	1.5	2.0	2.5	3.0	3.5	4.0
ξ_1	0.70	0.81	0.87	0.92	0.95	0.97	0.98	0.99

TABLE 6.6
Correction Factor ξ_2 for Vertical Tubes with Tube Pitches

S/d	2.0	3.0	4.0	5.0	6.0
ξ_2	0.9	0.93	0.96	0.98	1.00

S: Pitch of the tubes (mm) (Figure 6.15); S should be minimum of (S_1, S_2). L: Distance between water tubes centerline and furnace wall (mm) (Figure 6.15).

The radiative and convective components of the heat-transfer coefficient (Equation 6.53) can be obtained using either of the following two approaches.

6.3.5 Calculation Based on Tables

The sum of convective and radiative components in Equation 6.53 is calculated as a product of four factors: particle-diameter coefficient C_d, bed-temperature coefficient C_T, superficial-velocity coefficient C_u, and particle-density coefficient C_{pd}, values of which are given in tables below.

$$h_r + h_c = C_d C_T C_u C_{pd} \qquad (6.55)$$

C_d = particle-diameter coefficient in Table 6.7
C_T = bed-temperature coefficient in Table 6.8
C_u = superficial-velocity coefficient in Table 6.9
C_{pd} = particle-density coefficient in Table 6.10

TABLE 6.7
Particle-Diameter Coefficients (C_d)

Mean Particle Diameter d_p (mm)	1.2	1.3	1.4	1.5	1.6
C_d	1.035	1.020	1.000	0.980	0.965

TABLE 6.8
Bed-Temperature Coefficients (C_T)

T_{bed} (C)	800	850	900	950	1000	1050	1100
C_T (W/m² sec K)	256	265	276	285	295	306	316

TABLE 6.9
Superficial-Velocity Coefficients (C_u)

U/U_{mf}	3.0	3.5	4.0	4.5	5.0
C_u	1.03	1.02	1.00	0.98	0.96

U: Superficial velocity through the bubbling fluidized bed (m/sec). U_{mf}: Minimum fluidization velocity (m/sec).

TABLE 6.10
Particle-Density Coefficients (C_{pd})

ρ_p (kg/m³)	2000	2100	2200	2300	2400
C_{pd}	0.98	0.99	1.00	1.01	1.02

Heat Transfer

Example 6.3

Given:

- Bed temperature $T_{bed} = 850°C$
- Mean particle diameter $d_p = 1.2$ mm
- Flue gas density $\rho_g = 0.338$ kg/m^3
- Particle density $\rho_p = 2400$ kg/m^3
- Voidage of the bed $\varepsilon_{mf} = 0.5$
- Superficial velocity through bed $U_f = 1.3$ m/sec
- Flue gas dynamic viscosity: $\mu = 5.68 \times 10^{-5}$ N sec/m^2
- Outside diameter of the in-bed tubes $d = 51$ mm
- Pitch of in-line vertical tubes $S = 120$ mm
- Distance between furnace wall and centerline of water wall tubes $L = 120$ mm

Find:

- Heat-transfer coefficient of imbedded tubes (h)

Solution

This problem will be solved using tables as below:

- Minimum fluidization velocity, calculated using equations given in Section 2.1.2: $U_{mf} = 0.44$ m/sec
- From Table 6.5 with $L/d = 2.4$, we have $\xi_1 = 0.94$
- From Table 6.6 with $S/d = 2.4$, we have $\xi_2 = 0.92$
- Factor $\xi = \xi_1 \xi_2 = 0.94 \times 0.92 = 0.87$
- From Table 6.7 with $d_p = 1.2$, we have $C_d = 1.035$
- From Table 6.8 with $T_{bed} = 850$, we have $C_T = 265$ W/m^2 K
- From Table 6.9 with $U_f/U_{mf} = 1.3/0.44 = 3.0$, we have $C_u = 1.03$
- From Table 6.10 with particle density in bed $\rho_p = 2400$ kg/m^3, we have $C_{pd} = 1.02$
- $h_r + h_c = C_d C_T C_u C_{pd} = 1.035 \times 265 \times 1.03 \times 1.02 = 288$ W/m^2 K
- From Equation 6.53, heat-transfer coefficient for evaporative tubes, $h_o = \xi(h_r + h_c) = 0.87 \times 288 = 250$ W/m^2 K

6.4 FREEBOARD HEAT TRANSFER IN BUBBLING FLUIDIZED BEDS

The freeboard of a BFB resides between the bed surface and the furnace exit. This region is split between the space above and below the transport disengaging height (TDH). Above the TDH, gas moves in a plug flow, so the heat transfer can be modeled as a single-phase heat transfer of the gas and tube. Below the TDH, the heat transfer can be calculated using the equations for in-bed tubes. Combining these two the following equation has been developed by George and Grace (1979):

$$h_{conv} = h_{TDH} + \frac{(h_{bed} - h_{TDH})}{\left[1 + 34\left(\frac{3.5z}{TDH}\right)^n\right]} \quad (6.56)$$

where $n = 3$ to 5, and is valid for particles 100 to 900 μm in diameter.

The radiative heat transfer from Equation 6.48 is to be added to h_{conv} using appropriate values of emissivity.

6.5 HEAT TRANSFER IN COMMERCIAL-SIZE CFB BOILERS

The basic concept of heat transfer in CFB boilers was discussed in Section 6.2, which also presented a mechanistic model to analyze heat transfer. Such models are ideal for parametric evaluation and understanding of the heat-absorption behavior of a fluidized bed boiler, but it is rather difficult to use these models for every-day design or to calculate heat transfer rates from first principles. The following section presents some approaches, though somewhat crude, used by design engineers to determine the size of heating surfaces in CFB boilers.

6.5.1 HEAT TRANSFER TO THE WALLS OF COMMERCIAL CFB BOILERS

By analyzing data from commercially-operating CFB boilers, practicing engineers often develop and use rough but simple correlations for overall heat transfer to the water wall h. One such empirical relation developed by Dutta and Basu (2002) is:

$$h = 5\rho_{avg}^{0.391} T_b^{0.408} \text{ W/m}^2 \text{ K} \tag{6.57}$$

where ρ_{avg} is the average suspension density of the furnace, and T_b is the average temperature of the suspension.

Figure 6.18 shows a comparison of the above correlation with data from large commercial CFB boilers.

6.5.2 WING WALLS

Most of the discussion in the foregoing sections is concerned with heat transfer to the riser walls of a CFB boiler. The information on riser wall tubes may not be directly relevant to the tubes located inside the furnace because these surfaces are not exposed to the same hydrodynamics, and therefore heat-transfer conditions, as the walls of the furnace. High-capacity boilers often need to have a part of their evaporative and or superheater tubes inside the furnace to absorb the required fraction of combustion heat in the furnace. These tubes are in the form of either wing wall (platen) or cross

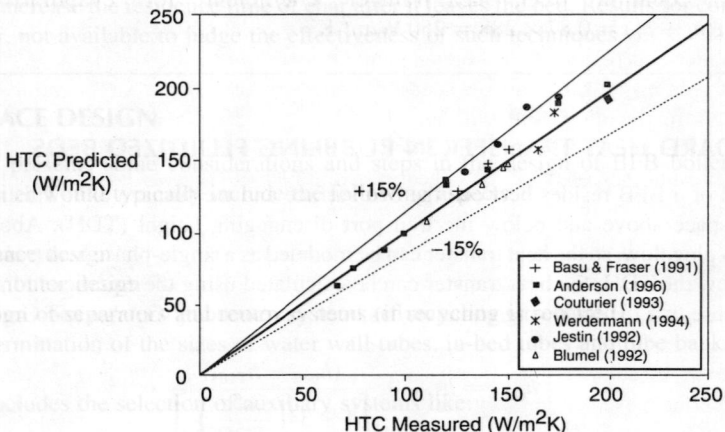

FIGURE 6.18 Plot of bed-to-wall heat-transfer coefficients (HTC) with those predicted by the correlation.

tubes (known as *Omega tubes* in some cases) as shown in Figure 6.1. The wing walls enter the furnace evaporator through the sidewall and exit from the top. Thus, they could carry either the evaporator or superheater duty of the boiler. More details of such tubes are discussed in Chapter 8.

The vertical wing wall tubes located near the exit of the furnace are exposed to a somewhat different hydrodynamic condition. The gas–solid mixture near the top of the furnace turns towards the exit. If the furnace exit is smaller than the bed cross section, there is a sudden change in both the direction and velocity of the mixture. This would result in an increase in the suspension density upstream of the exit. The higher density and the crossflow of the gas–solid suspension over the platen tubes may result in a heat-transfer coefficient higher than that on the vertical furnace walls tubes.

6.5.3 HEAT-TRANSFER VARIATION ALONG THE FURNACE HEIGHT

The heat-transfer coefficient on the furnace wall decreases along the height of the furnace (Figure 6.19), but in a large CFB boiler this decrease does not necessarily follow the exponential decay of suspension density. As we see from the data from a 135-MWe unit, the variation is gentler. This is due to the dominant influence of radiative heat transfer on the overall heat flux on the wall.

6.5.4 CIRCUMFERENTIAL DISTRIBUTION OF HEAT-TRANSFER COEFFICIENT

Unlike in pulverized coal-fired or gas/oil flame-fired boilers, the circumferential distribution of wall heat-transfer coefficient in CFB boilers is relatively uniform. This is one of the reasons why a supercritical CFB boiler does not need the boiler tubes to be wrapped around the furnace as required for supercritical pulverized coal-fired boilers. However, there is some difference in the heat-transfer coefficient as can be seen from measurements taken on the water walls of a 135-MWe CFB boiler (Zhang et al., 2005). Data measured on the water wall of this boiler (Figure 6.20) shows that heat-transfer coefficient decreases from the corner to the middle of the wall. Figure 6.20 recalculated the data presented by Zhang et al. (2005) and by taking 120 W/m² K to be a more credible value of the heat-transfer coefficient at the furnace center.

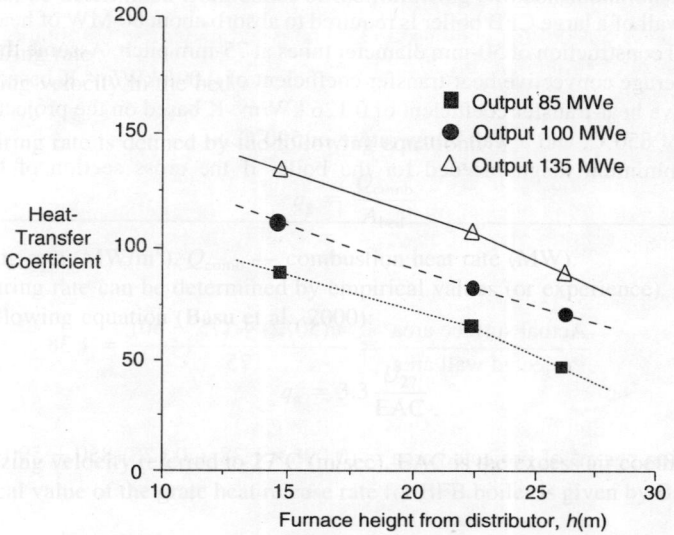

FIGURE 6.19 Axial distribution of wall heat-transfer coefficients in a typical CFB boiler.

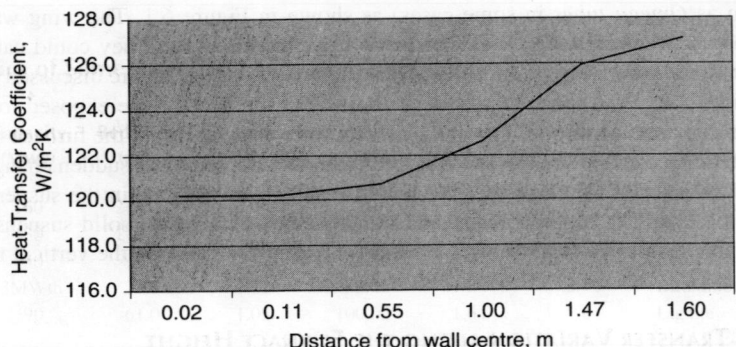

FIGURE 6.20 Circumferential distributions of heat-transfer coefficients on side wall of a 135-MWe CFB boiler's water wall. (Plotted with data from Zhang et al., 2005.)

Here we note that the heat-transfer coefficient is lowest at the wall center and it rises toward the furnace corners. The variation, however, is limited to 6% from the middle of the wall to its edge. Higher heat transfer in the corners of the furnace is due to higher solid flux or density, which one would expect in the low gas-velocity region in the corners.

Midsections of the horizontal (omega) tubes located in the core of the furnace usually contact the relatively dilute up-flowing gas and dispersed solids. Since the heat-transfer rate in a fast bed is proportional to the suspension density, the rate on the mid-section of these tubes could be lower than that on the section near the wall. Experimental data at room temperature (Bi et al., 1991) would suggest that the heat-transfer coefficient increases continuously from the core to the wall, following the variation in suspension density. Though this data is for room temperature, it supports the data gathered on the walls of a hot-operating CFB boiler (Zhang et al., 2005).

Example 6.5

> The furnace wall of a large CFB boiler is required to absorb about 80 MW of heat. The furnace wall is a panel construction of 50-mm diameter tubes at 75-mm pitch. Assume the furnace wall to have an average convective heat-transfer coefficient of 0.096 kW/m² K based on the actual area, a radiative heat-transfer coefficient of 0.126 kW/m² K based on the projected area, a bed temperature of 850°C, and a wall temperature of 590°C.
>
> Find the minimum height needed for the boiler if the cross section of the furnace is 20 m × 7 m.

Solution

$$\frac{\text{Actual surface area}}{\text{Projected wall area}} = \frac{\pi(50/2) + (75 - 50)}{75} = 1.38$$

Heat absorption per m² of projected area = $(0.096 \times 1.38 + 0.126)(850 - 590) = 67.2 \text{ kW/m}^2$

$$\text{Wall area} = \frac{80,000}{67.2} = 1190 \text{ m}^2$$

Assuming the furnace openings to have 30% of its cross section:

$$\text{Minimum furnace height} = \frac{(1190 - 0.7 \times 20 \times 7)}{2(20+7)} = 20.2 \text{ m}$$

NOMENCLATURE

A:	surface area of wall receiving radiation, m^2
A_f:	area of fin, m^2
A_0:	outer surface area of tubes in EHE, m^2
a, b, c:	coefficients in Equation 6.30
B:	parameter in Equation 6.28
C_b:	specific heat of bed, kJ/kg K
C_d:	particle-diameter coefficient
C_l:	correction factor for tube length
C_c, C_p, C_f, C_g:	specific heat of cluster, solid, steam, and gas, respectively, kJ/kg K
C_{pd}:	particle-density coefficient
C_{pp}:	correction factor in Equation 6.22 for presence of particles
C_t:	correction factor for temperature difference between wall and medium
C_u:	superficial-velocity coefficient
C_{sf}:	solid fraction in cluster
D:	equivalent diameter of furnace cross section
D_b:	equivalent diameter of bed, m
D_c:	equivalent diameter of cluster, m
d_{cp}:	diameter of coarser particles, m
d_i:	inside diameter of heat transfer tube, m
d_p:	diameter of average bed particles, m
d_t:	outer diameter of heat transfer tube, m
e_b, e_c, e_g, e_p:	emissivity of bubbling bed, cluster, gas, and particle, respectively
e_{pc}, e_d:	effective emissivity of particle cloud and dispersed phase, respectively
e_w:	emissivity of wall
F':	volume fraction of solids in dilute phase
f:	time-averaged fraction of wall area covered by cluster
f_b:	fraction of tube surface covered by bubbles
g:	acceleration due to gravity, 9.81 m/sec^2
H:	height of water wall, m
h:	overall heat-transfer coefficient, kW/m^2 K
h_b:	heat-transfer coefficient from bubble phase, kW/m^2 K
h_{bp}:	gas–particle heat-transfer coefficient for whole bed, kW/m^2 K
h_c:	convective heat-transfer coefficient due to clusters, kW/m^2 K
$h_{cluster}$:	heat-transfer coefficient due to conduction into cluster, kW/m^2 K
h_{cond}:	gas-side particle conduction heat transfer coefficient, kW/m^2 K
h_{conv}:	gas-side convection heat-transfer coefficient, kW/m^2 K
h_{cr}:	radiative heat-transfer coefficient due to clusters, kW/m^2 K
h_d:	convective heat-transfer coefficient due to dilute phase, kW/m^2 K
h_{dr}:	radiative heat-transfer coefficient due to dilute phase, kW/m^2 K
h_e:	emulsion phase heat-transfer coefficient, kW/m^2 K
h_o:	overall external heat-transfer coefficient kW/m^2 K
h_{gp}:	gas–particle heat-transfer coefficient for single particle, kW/m^2 K

h_i:	heat-transfer coefficients for the internal surfaces of the tube, kW/m² K
h_r:	total radiative heat-transfer coefficient in CFB, kW/m² K
h_{rad}:	total radiative heat-transfer coefficient in BFB, kW/m² K
h_t:	heat-transfer coefficient on a cluster after residing time, t on the wall, kW/m² K
h_{TDH}:	heat-transfer coefficient above TDH in a bubbling bed, kW/m² K
h_{tube}:	overall heat-transfer coefficient on the tube, kW/m² K
h_w:	conduction heat-transfer coefficient through gas layer between cluster and wall
J:	thermal time constant, kW/m K
K_c:	thermal conductivity of particle packet in bubbling bed, kW/m K
K_g, K_{gf}, K_c:	thermal conductivity of gas, gas–film, and cluster, respectively, kW/m K
K_m, K_p:	thermal conductivity of tube metal, kW/m K
L:	vertical length of the heat-transferring surface, m
L_b:	mean beam length, m
L_c:	characteristic length traveled by cluster on wall, m
LMTD:	log mean temperature difference between fluid on inside of tube and fluidized bed outside, K
M, N:	parameters in Equation 6.11
Q_f:	heat absorbed by the furnace wall, kW
Q_a:	heat absorbed by the furnace wall, kW
Q_{fbhe}:	heat absorbed in EHE, kW
R:	radius of the bed, m
R_c:	packet resistance, m² K/kW
R_w:	contact resistance (between packet and tube surface, m² K/kW)
r:	radial distance from the center of the bed, m
r_o, r_i:	outer and inner radius of heat-transferring tube, m
S:	surface area of particles per unit weight of particles, m²/kg
T_{bed}:	temperature of bed, K
T_b, T_g, T_w, T_p:	temperatures of bed, gas, heat-transferring wall and bed–particle, respectively, K
T_{eb}:	effective bed temperature, K
T_{ehe}:	temperature of the EHE, K
T_{p0}:	initial temperature of the particles, K
T_{g0}:	temperature of the gas entering the bed, K
T_{sat}:	saturation temperature, K
T_w:	temperature of wall, K
$T_{99\%}$:	limit in Equation 6.8 expressed as $T_{g0}+0.99(T_p-T_{g0})$, K
t:	time, sec
$t_{99\%}$:	time required for gas–particle temperature difference to reduce to 1% of its original value, sec
t_c:	mean residence time of cluster on wall, sec
U:	superficial gas velocity through bed, m/sec
U_b:	average bubble rise velocity, m/sec
U_c:	velocity of cluster on wall, m/sec
U_{cl}:	average velocity of cluster on wall, m/sec
U_{cp}:	average velocity of coarser particle, m/sec
U_m:	maximum fall velocity of clusters, m/sec
U_{mf}:	minimum fluidization velocity, m/sec
U_p:	solid velocity, m/sec
U_t:	terminal velocity of a single particle, m/sec
V:	volume of the bed, m³/sec
V_i:	velocity of steam in tube, m/sec
W:	solid circulation rate, kg/m² sec

$X_{99\%}$: distance required for gas to reach 99% of the overall bed temperature, m
x: distance along the height of the fast bed, m
z: height above the surface of bubbling bed, m

GREEK SYMBOLS

α: $1 -$ (projected tube area/bed area)
ξ: correction factor in Equation 6.53
ε: cross-sectional average voidage
ε_{avg}: cross-sectional average bed voidage
ε_c: voidage in cluster
ε_{mf}: voidage of bed at minimum fluidization velocity
ε_w: voidage near the wall
$\varepsilon(r)$: voidage at radius r from the center
ρ_{avg}: suspension density averaged over the upper CFB furnace, kg/m^3
ρ_c: density of cluster, kg/m^3
ρ_f: density of steam, kg/m^3
ρ_b: bulk density of the bed, kg/m^3
ρ_{dis}: density of dispersed phase, kg/m^3
ρ_g: density of gas, kg/m^3
ρ_p: density of bed material, kg/m^3
ρ_{sus}: cross-sectional average suspension density a height, kg/m^3
$(\rho C)_c$: thermal capacity of a cluster
η_f: fin efficiency
τ_c: time that a cluster of particles stays in contact with surface in a fast bed, sec
τ_p: time that packet of particles stays in contact with surface in a bubbling bed, sec
μ: dynamic viscosity of gas, N sec/m^2
μ_f, μ_{fw}: viscosity of steam at bulk temperature and wall temperature, respectively, N sec/m^2
σ: Stefan–Boltzmann constant (5.67×10^{-11} kW/m^2 K^4)
δ: nondimensional gas layer thickness between wall and cluster

DIMENSIONLESS NUMBERS

Ar: Archimedes number
Nu_p: Nusselt number based on particle–gas heat-transfer coefficienct ($h_{gp}d_p/K_g$)
Nu_b: Nusselt number based on bubble phase heat-transfer coefficient
Nu_{cond}: Nusselt number defined in Equation 6.45
Nu_{conv}: Nusselt number defined in Equation 6.46
Nu_{rad}: Nusselt number based on radiative heat-transfer coefficient
Pr, Pr_f: Prandtl number of the gas and steam, respectively, (C_p/K_g)
Re, Re$_f$: Reynolds number of gas and steam, respectively $ReUd_tg/\mu$
Re$_{cp}$: Reynolds number of coarse particles based on gas–particle slip velocity

REFERENCES

Andeen, B. R. and Glicksman, L. Heat Transfer Conf. ASME Paper. 76-HT-67, 1976.
Andersson, B. A., Johnsson, F., and Leckner, B., Heat flow measurement in fluidized bed boilers, In *Proceedings of the 9th International Conference on Fluidized Bed Combustion*, Mustonen, J. P., Ed., ASME, New York, pp. 592–598, 1987.

Basu, P., Heat transfer in fast fluidized bed combustors, *Chem. Eng. Sci.*, 45(10), 3123–3136, 1990.

Basu, P. and Fraser, S. *Circulating fluidized bed boiler–design and operation*, Butterworth-Heinemann, Stoneham, 1991.

Basu, P. and Nag, P. K. Heat transfer to walls of a circulating fluidized bed furnace, *Chemical Engineering Science*, 51(1), 1–26, 1996.

Basu, P. and Cheng, L. An experimental & theoretical investigation into the heat transfer of a finned water wall tube in a CFB boiler, *Int. J. Energy Research*, 24, 291–308, 2000.

Basu, P., Keta, C., and Jestin, L., Boilers and Burners-Design & Theory, Springer & Verlag, New York, p. 197, 1999.

Basu, P., Ali, N., Nag, P. K., and Lawrence, D. Heat transfer to finned surfaces in a fast fluidized bed, *Int. J. Heat Mass Transfer*, 34, 2317–2326, 1991.

Bi, H., Jin, Z., Yu, Z., and Bai, D. R., An investigation of heat transfer in circulating fluidized bed, In *Circulating Fluidized Bed Technology III*, Basu, P., Hasatani, M., and Horio, M., Eds., Pergamon Press, Oxford, pp. 233–238, 1991.

Botterill, J. S. M., Fluid bed heat transfer, In *Gas Fluidization Technology*, Geldart, D., Ed., Wiley, Chichester, pp. 219–222, 1986.

Brewster, M. Q., Effective absorptivity and emissivity of particulate medium with application to a fluidized bed, *Trans. of ASME*, New York, 108, pp. 710–713, August 1986.

Chandran, R. and Chen, J. C., *AIChE J.*, 29(6), 907–913, 1982.

Chen, J. C., Max Jacob Award Lecture, *J. Heat Transfer*, 125, 549–566, 2003.

Chen, J. C., Grace, J. R., and Golriz, M. R., Heat transfer in fluidized beds: design methods, *Powder Technol.*, 150, 1123–1132, 2005.

Dutta, A. and Basu, P., Overall heat transfer to water walls and wing walls of commercial circulating fluidized bed boilers, *J. Inst. Energy*, 75(504), 85–90, 2002.

Dutta, A. and Basu, P., An improved cluster-renewal model for estimation of heat transfer coefficient on the water-walls of commercial circulating fluidized bed boilers, *J. Heat Transfer*, 126, 1040–1043, 2004.

Gelperin, N. I. and Einstein, V. G., Heat transfer in fluidized beds, In *Fluidization*, Ch-10, Davidson, J. F., and Harrison, D., Ed., Acadamic Press, New York, 1971.

George, S. E. and Grace, J. R., Heat transfer to horizontal tubes in the freeboard region of a gas fluidized bed, *AIChE Meeting*, San Francisco, 1979.

Glicksman, L. R., Heat transfer in fluidized bed combustors, In *Fluidized Bed Boilers — Design and Applications*, Basu, P., Ed., Pergamon Press, Toronto, pp. 63–100, 1984.

Glicksman, L. R., Heat transfer in circulating fluidized beds, In *Circulating Fluidized Beds*, Grace, J. R., Avidan, A. A., and Knowlton, T. M., Eds., Chapman & Hall, London, 261–310, 1997.

Glicksman, L. R., Circulating fluidized bed heat transfer, In *Circulating Fluidized Bed Technology II*, Basu, P. and Large, J. F., Eds., Pergamon Press, Oxford, pp. 13–30, 1988.

Golriz, M. R. and Grace, J. R., Predicting heat transfer in large CFB boilers, In *Circulating Fluidized Bed Technology-VII*, Grace, J. R., Zhu, J., and Lasa, H. D., Eds., 121–128, 2002.

Gottung, E. J. and Darling, S. L., Design considerations for CFB steam generators, In *Proceedings of the 10th International Conference on Fluidized Bed Combustion*, Manaker, A., Ed., ASME, New York, pp. 617–623, 1989.

Grace, J. R., Fluidized bed heat transfer, In *Handbook of Multiphase Flow*, Hestroni, G., Ed., McGraw-Hill Hemisphere, Washington, DC, pp. 9–70, 1982.

Halder, P. K. Combustion of single carbon particles in CFB combustors, Ph.D. Dissertation, Technical University of Nova Scotia, 1989.

Kim, S. W., Ahn, J. Y., Jun, S. D., and Lee, D. H., Heat transfer and bubble characteristics in a fluidized bed with immersed horizontal tube bundle, *Int. J. Heat Mass Transfer*, 46(3), 399–409, 2003.

Kothari, A. K., M.S. Thesis, Illinois Institute of Technology, Chicago, 1967.

Kunii, D. and Levenspiel, O., *Fluidization Engineering*, Butterworths-Heinemann, Stoneham, p. 268, 1991.

Lints, M. C. and Glicksman, L. R., Parameters governing particle to wall heat transfer in a circulating fluidized bed, In *Circulating Fluidized Bed Technology IV*, Avidan, A. A., Ed., pp. 297–304, 1994.

Mickley, H. S. and Fairbanks, D. F., Mechanism of heat transfer to fluidized beds, *AIChE J.*, 1(3), 374–384, 1955.

Noymer, P. D. and Glicksman, L. R., Descent velocities of particle clusters at the wall of a circulating fluidized bed, *Chem. Eng. Sci.*, 55, 5283–5289, 2000.

Ozakaynak, T. F. and Chen, J. C. Fluidization. Kunii, D. and Toei, R. Eds., Engineering Foundation, pp. 371–378, 1983.

Perry, R. H. and Green, D. W., *Perry's Chemical Engineers' Handbook*, McGraw-Hill, New York, pp. 5–17, 1997.

Plautz, D. A. and Johnstone, H. F., Heat and mass transfer in packed beds, *AIChE J.*, 193–199, 1955.

Ranz, W. E. and Marshall, W. R., *Chem. Eng. Prog.*, 48, 247, 1952.

Salatino, P. and Massimilla, L., A predictive model of carbon attrition in fluidized bed combustion and gasification of a graphite, *Chem. Eng. Sci.*, 44(5), 1091–1099, 1989.

Subbarao, D. and Basu, P., A model for heat transfer in circulating fluidized beds, *Int. J. Heat Mass Transfer*, 39(3), 487–489, 1986.

Tang, J. T. and Engstrom, F., Technical assessment on the Ahlstrom pyroflow circulating and conventional bubbling fluidized bed combustion systems, In *Proceedings of the 9th International Conference on Fluidized Bed Combustion*, Mustonen, J. P., Ed., ASME, New York, pp. 38–54, 1987.

Turek, D. G., Sopko, S. J., and Jansesen, K. A generic circulating fluidized bed for cogenerating steam, electricity and hot air, In *Proceedings of the 8th International Conference on Fluidized Bed Combustion*, DOE/METC-856021, Vol. 1, pp. 395–405, July 1985.

Watanabe, T., Yong, C., Hasatani, M., Yushen, X., and Naruse, I., Gas to particle heat transfer in fast fluidized bed, In *Circulating Fluidized Bed Technology III*, Basu, P., Hasatani, M., and Horio, M., Eds., Pergamon Press, Oxford, pp. 283–287, 1991.

Wu, R., Lim, J., Chouki, J., and Grace, J. R., Heat transfer from a circulating fluidized bed to membrane water-cooling surfaces, *AIChE J.*, 33, 1888–1893, 1987.

Wu, R. L., Grace, J. R., and Brereton, C. M. H., Instantaneous local heat transfer and hydrodynamics in a circulating fluidized bed., *Int. J. Heat Mass Transfer*, 34, 2019–2027, 1990.

Yagi, S. and Kunii, D., Studies on heat transfer near wall surfaces in packed beds, *AIChE J.*, 5, 79–85, 1960.

Zhang, D., Hydrodynamics & heat transfer in a pulsed bubbling fluidized bed, M.Sc. Thesis, Mechanical Engineering, Dalhousie University, Canada, 2005.

Zhang, H., Lu, J., Yang, H., Yng, J., Wang, Y., Xiao, X., Zhao, X., and Yue, G., Heat transfer measurements and predictions inside the furnace of 135 MWe CFB boiler, In *Circulating Fluidized Bed Technology VIII*, Cen, K., Ed., International Academic Publishers, Beijing, pp. 254–266, 2005.

7 Bubbling Fluidized Bed Boiler

Bubbling fluidized bed (BFB) boilers are perhaps the largest application of fluidized beds. More than 10,000 BFB boilers are in use worldwide. They have replaced nearly all traditional stoker-fired boilers. The development of fluidized beds, which started in 1920s with the fluidized bed gasifier of Fritz Winkler (Figure 7.1), entered the field of steam generation through the pioneering work of Douglas Elliott in the 1960s (Basu and Fraser, 1991).

A typical BFB boiler comprises a furnace and a convective heat exchange section. The furnace is made up of a bubbling bed of noncombustible solids and a freeboard above it. The convective section is similar to any conventional boiler.

BFB boilers are also called stationary or slow fluidized beds. The furnace of a BFB boiler, operated under atmospheric pressure, is known as atmospheric fluidized bed combustion (AFBC) while that at high pressure is called pressurized fluidized bed combustion (PFBC).

The present chapter will discuss the working principles, design and operation of BFB boilers.

7.1 DESCRIPTION OF A BFB BOILER

A BFB boiler system may be divided into several subsystems, which are elucidated in the following sections with reference to Figure 7.2:

1. Feedstock preparation, transport, and flow-rate control
2. Combustion
3. Air and flue gas handling
4. Ash handling system and emission control
5. Steam generation

7.1.1 FEEDSTOCK PREPARATION AND FEEDING

7.1.1.1 Fuel Feed

Solid fuels like coal and petroleum coke are crushed in a crusher to below 10-mm size and then transported to the boiler. The above size is not necessarily firm. It could change depending on a number of factors including the operational experience of the specific plant. The crushed fuel is transported by a conveyor belt to bunkers from where it drops into a metering device like a screw or belt-weigh feeder.

The coal received from the feeder may be either sprayed from the top of the bed or pneumatically injected through the bottom of the bed.

The feeding system for a biomass fuel is a major challenge. Under-bed feeding systems are not as reliable as over-bed systems because the fuels, especially when wet, tend to plug. Special types of double-screw or ram-type feeders are more suitable for wet or sticky fuels. More details are given in Chapter 10.

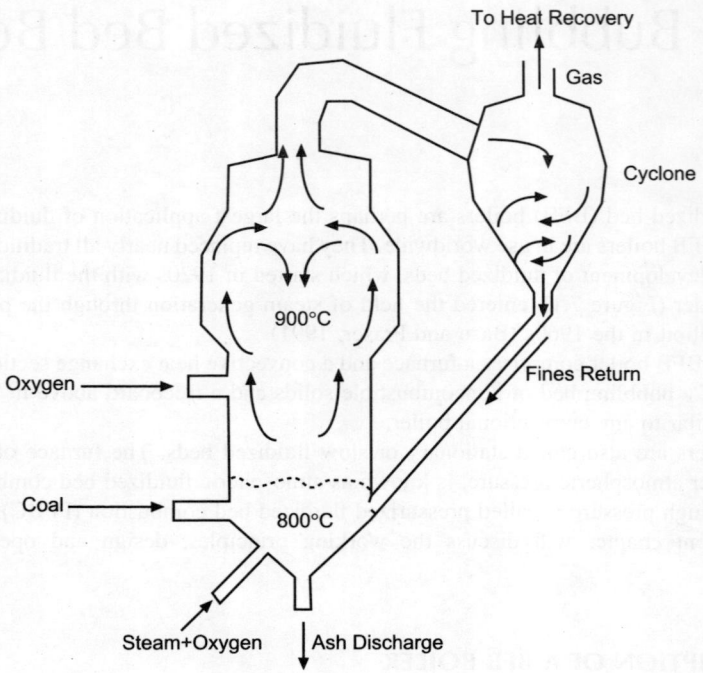

FIGURE 7.1 Winkler gasifier was one of the first applications of bubbling fluidized beds.

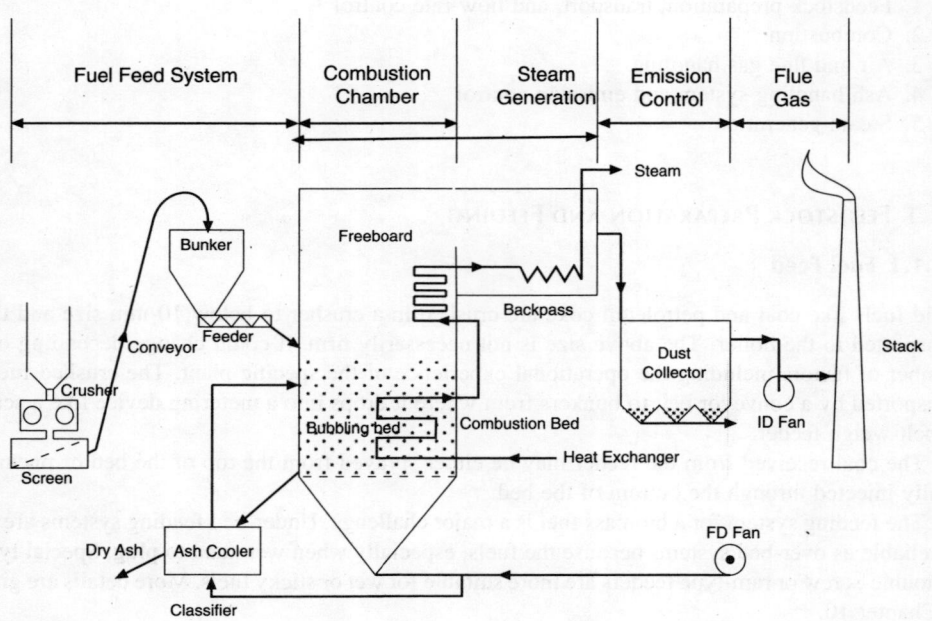

FIGURE 7.2 Schematic of a bubbling fluidized boiler system. The arrangement of steam water circuits in actual boilers is not necessarily as shown here.

Bubbling Fluidized Bed Boiler

7.1.1.2 Sorbent Feeding System

Boilers requiring sulfur capture must have a limestone crushing, conveying, and feeding facility. The sorbents are generally ground to finer sizes (<6 mm) and then conveyed pneumatically and injected into the bed.

7.1.2 AIR AND GAS HANDLING

7.1.2.1 Air Distribution

The combustion air is sometimes preheated in the air heater. The primary air enters the bed through a grate called a distributor, which helps distribute the air uniformly across the bed and prevent solids from dropping into the air plenum below. Design of this component is more critical to the operation of BFB boilers than that for CFB boiler. This topic is discussed further in Chapter 11.

The secondary air, if used, enters the bed through the side walls just above the bed level.

7.1.2.2 Flue Gas Path

Flue gas, generated in the bed, rises into the freeboard (Figure 7.2) and then exits the furnace from the top. The gas leaves the bed at bed temperature (800 to 900°C) and cools down to 600°C after losing heat to the superheater (SH) or reheater in the freeboard. After leaving the freeboard, the gas passes through the convective section, which may accommodate the bank tubes, superheater, reheater, economizer, and air heater as the design demands. Finally, the gas is discharged to the atmosphere through the stack.

7.1.3 ASH AND EMISSION SYSTEMS

Boilers create two waste products: flue gas and solid residues. The latter comprises ash from fuel and spent sorbents. The flue gas could potentially contribute significantly to air pollution. Therefore, these systems are designed with special care.

7.1.3.1 Ash Extraction

Coarse ash, generated during combustion, accumulates in the bed, while finer ash particles are entrained with the flue gas. Coarse ash particles are easily drained from the bed by opening the drain valve when needed. This ash sometimes passes through an air-cooled ash classifier (Figure 7.3) for heat recovery. This device also returns the finer fraction of drained solids back to the bed along with the air heated in it. This helps maintain a good particle size distribution in the bed. Unburnt carbon in this ash, drained through an ash classifier, is generally low.

An alternative design of ash extraction involves the use of a rotary drum cooler as shown in Figure 7.4. Here ash enters from the upper end. Water circulates through the drum wall, which cools the coarse ash tumbling in it. Cooled ash drops from the lower end. The water may be drawn from the feedwater heater system to return the waste heat to the boiler and thereby preserve the efficiency of the boiler. The ash is sufficiently cold for dry disposal.

The finer fraction of the ash, known as fly ash, is collected in the economizer hopper or a particulate collection device like an electrostatic precipitator (ESP) or bag-house. This ash may contain more than 3% unburnt carbon, making it unsuitable for its value-added use in cement industries. The fly ash can be collected by a dry-ash collection system while bed ash can be collected by either wet- or dry-ash systems.

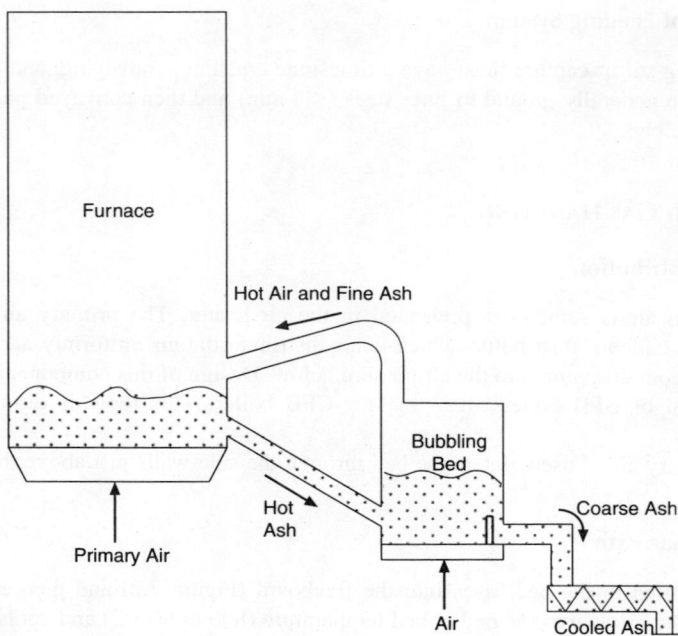

FIGURE 7.3 Fluidized bed ash classifier returns fines and drains coarse ash.

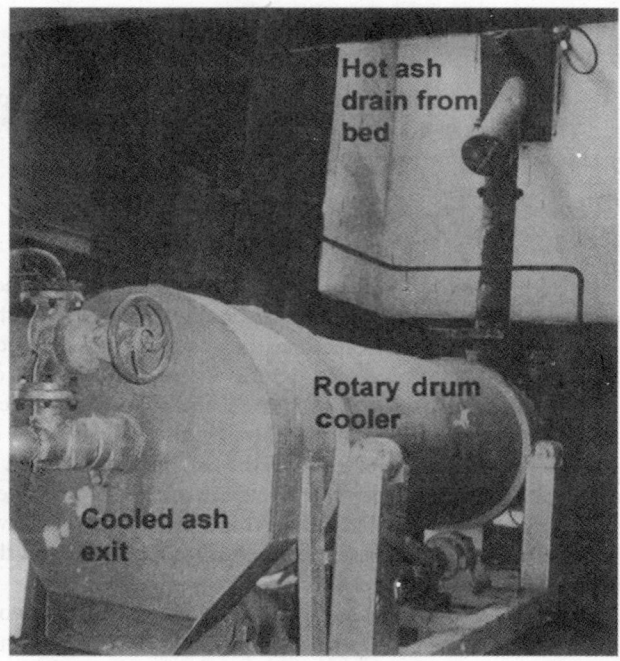

FIGURE 7.4 Rotary drum ash cooler receives hot ash from the bed above and cools it down by discharge from the front end.

7.1.3.2 Emission Control System

The low combustion temperature (800 to 900°C) creates favorable conditions for reductions in the emissions of NO_x and sulfur dioxide in a BFB boiler. Thus, the flue gas, leaving a BFB boiler with limestone injection, is relatively free from harmful gases like SO_2 and NO_x.

As a result, a well-operated fluidized bed boiler does not need post-combustion units like SCR or FGD for reduction of harmful gas emissions. However, for the control of mercury or for meeting exceptionally low SO_x emission standards some plants use a scrubber followed by a bag filter or ESP.

7.1.4 STEAM GENERATION SYSTEM

The feedwater is pumped to the drum through the economizer. From the drum, the hot water typically comes down to the lower header and then rises through the riser tubes, which could form the enclosure of the furnace. Evaporator tubes in some boilers pass through the bubbling bed to take advantage of its high heat-transfer coefficient. The superheater may be placed in the freeboard above the bed. Part of it may also be placed in the convective section downstream.

7.1.5 COMBUSTION CHAMBER

The furnace of a BFB boiler typically comprises a dense bubbling bed and a lean freeboard above it. These two parts constitute the furnace or combustion chamber. The energy released from the combustion of fuels is split between the fluidized bed and the freeboard approximately in the percentage ratio of 88:12 (Duqum et al., 1985). The temperature of the bubbling bed (also called dense bed) is generally maintained between 800 and 900°C by extracting appropriate amounts of heat.

The lower bed of a biomass-fired boiler is generally refractory-lined while that of a fossil fuel-fired boiler may be made of evaporator tubes. The latter type of boiler often has additional evaporator or superheater tubes immersed in the bed for effective heat absorption.

The height of the dense bed typically varies between 0.5 and 1.5 m and the average size of the solid particles (bed material) is around 1 mm and they may be sand, limestone, or ash.

The large mass of well-mixed hot solids allows BFB boilers to efficiently burn even hard-to-burn fuels. Thus a wide range of fuels like coal, paper sludge (Geisler et al., 1987), anthracite, paddy straw (Rajaram and Malliah, 1987) and fluid coke (Anthony et al., 1987) etc., can burn in a BFB boiler with minimum air pollution.

7.2 FEATURES OF BFB BOILERS

The BFB furnace operates, as the name suggests, in the bubbling fluidization regime. A discussion on the hydrodynamics of this regime is given in Chapter 2 (Figure 2.2). In BFB combustors the primary air should be adequate to fluidize most of the solids in the bed, but at a velocity less than the terminal velocity of the finer fraction of solids of the bed. Some designers take this velocity to be 3 to 4 times of the minimum fluidization velocity of average bed particle size.

7.2.1 ADVANTAGES

BFB boilers have nearly replaced chain-grate or fixed-grate boilers around the world. This popularity stems from the following positive features of BFB boilers:

1. A BFB boiler can be designed to burn any fuel including coal with ash as high as 60 to 70% or with volatiles less than 1%
2. Moving parts used by BFB boilers are less than those used in stoker-fired boilers

3. It does not need expensive fuel preparation plants like those required by pulverized coal (PC) boilers
4. It can meet moderately stringent gaseous emission standards without any expensive post-combustion scrubbing unit

7.2.2 LIMITATIONS

BFB boilers, however, suffer from some shortcomings as described below:

1. Erosion of in-bed tubes results in a rather poor availability especially when firing high-ash coal
2. A BFB boiler requires a much larger footprint than that required by a pulverized fuel-fired or circulating fluidized bed boiler. This limitation makes BFB firing unsuitable for retrofitting PC-fired boilers
3. BFB boilers require a rather large number of feed points limiting their capacity to small and medium-size boilers
4. Although they can be designed for any grade of fuel, BFB furnaces need modifications if they are required to burn off-designed fuels whether super or inferior to the designed fuel
5. Part-load operation of a BFB boiler is limited

7.3 THERMAL DESIGN OF BUBBLING FLUIDIZED BED BOILERS

The procedure for designing a BFB boiler is similar to that for a PC boiler except for the difference in heat-transfer mechanisms. A typical design starts with the stoichiometric calculation to obtain the air-fuel requirements. A heat balance analysis provides all components of the heat losses for boiler efficiency calculations. The thermal design provides the furnace dimensions and heating-surface areas of the evaporator, economizer, and superheater. The design steps include:

1. Overall heat balance of the entire boiler
2. Energy balance of the bed alone
3. Steam/water-side heat balance
4. Feed-rate of fuel and capacity of fans
5. Gas-side heat balance

This section will discuss the heat balance and the heat loads for the evaporator, economizer and superheater. It will also determine the total amounts of air required, flue gas generated and the fuel consumed.

7.3.1 OVERALL HEAT BALANCE OF THE BOILER

The combustion heat generated in the furnace is equal to the sum of all heat losses and the enthalpy gain of water/steam in the boiler. The heat balance of a BFB boiler is similar to that of any other boiler, except that its bed requires special consideration, as will be discussed in Section 7.3.3. The following section presents the overall heat balance of the boiler.

Heat losses are usually expressed as a fraction or percentage of the potential combustion heat. They are expressed in terms of the heat losses associated with the burning of 1 kg coal and its higher heating value (HHV).

The following components of the loss are discussed below:

1. Dry flue gas loss through stack
2. Moisture losses (in sorbent, fuels, air, and hydrogen burning)

Bubbling Fluidized Bed Boiler

3. Calcination losses
4. Sulfating credit — it is a negative loss
5. Unburnt carbon loss
6. Convection–radiation loss
7. Sensible heat loss in fly ash and bottom ash
8. Fan credit — it is also a negative loss
9. Unaccounted loss

7.3.1.1 Dry Flue Gas Heat Loss

When the temperature of the flue gas leaving the stack T_f, is higher than the ambient temperature T_a, an amount of heat, known as stack heat loss, is lost to the atmosphere. This loss is made up of two components: dry flue gas and moisture in it.

The dry flue gas loss can be calculated by Equation 7.1

$$L_{stack} = \frac{M_{df} C_f (T_f - T_a)}{HHV} \tag{7.1}$$

where: M_{df} is the dry flue gas mass per kg fuel burned.

7.3.1.2 Moisture Losses

7.3.1.2.1 Loss Due to Moisture in Air

The moisture, leaving the boiler above the ambient temperature, carries with it an amount of heat known as moisture loss. The moisture in the flue gas is made up of three components: moisture from air, coal, and hydrogen. The moisture losses for each of these three components are calculated separately.

Heat loss due to the moisture is relatively high because typical stack temperatures are above the vaporization point of water. Thus, the moisture carries with it the latent heat of vaporization, Q_{latent}, which cannot be recovered unless the flue gas is cooled below its vaporization temperature and the superheat, Q_{SH}. The moisture loss is defined as:

$$L_{m,air} = \frac{M_{wa}[EAC]X_m\{C_m(100 - T_a) + Q_{latent} + Q_{SH}\}}{HHV} \tag{7.2}$$

where: M_{da} is the stoichiometric amount of dry air required for 1 kg of fuel fired, X_m is the moisture fraction in the air, EAC is the excess air coefficient (1 + excess air fraction), C_m is the specific heat of water, and Q_{SH} is the enthalpy of superheating equal to $C_g(T_f - 100)$. Here, we assume that the saturation temperature of water is 100°C at stack pressure.

7.3.1.2.2 Moisture Loss from Coal

Coal may contain some moisture, H_2O, whose loss can be calculated by:

$$L_{m,fuel} = \frac{[H_2O][C_m(100 - T_a) + Q_{latent} + Q_{SH}]}{HHV} \tag{7.3}$$

7.3.1.2.3 Moisture Loss from Burning Hydrogen in Coal

When 2 kg of hydrogen is burned, 18 kg of water is generated. Thus, the moisture loss from the hydrogen content of the fuel, H is calculated by:

$$L_{m,h} = \frac{9[H][C_m(100 - T_a) + Q_{latent} + Q_{SH}]}{HHV} \tag{7.4}$$

7.3.1.3 Calcination Loss

When the boiler uses limestone to capture sulfur, two additional terms, calcination loss, and sulfation credit are considered in the heat balance. Both limestone and dolomite contain calcium carbonate, magnesium carbonate, and impurities. The amount of sorbent fed per unit mass of coal burned may be calculated from Equation 5.34. When exposed to the heat of the CFB furnace, both carbonates calcine to their respective oxides. The reaction is endothermic and absorbs heat from the combustor according to Equation A2.4 and Equation A2.5. The heat loss from calcination can be calculated from the following equations:

$$\text{Calcination Loss from } CaCO_3 = \frac{\text{Feed rate of } CaCO_3 \times 1830}{\text{Fuel feed rate} \times \text{HHV}}$$

$$\text{Calcination loss from } MgCO_3 = \frac{\text{Feed rate of } MgCO_3 \times 1183}{\text{Fuel feed rate} \times \text{HHV}}$$

(7.5a)

where: HHV is the higher heating value of the fuel in kJ/kg.

7.3.1.4 Sulfation Credit

The calcined limestone (CaO) reacts with sulfur dioxide producing calcium sulfate ($CaSO_4$) according to the following exothermic reaction:

$$SO_2 + CaO + \frac{1}{2}O_2 = CaSO_4 + 15,141 \text{ kJ/kg}$$

The resulting heat gain is:

$$\text{Heat gained from sulfation} = \frac{\text{kg of sulfur converted} \times 15,141}{\text{kg of fuel fed} \times \text{HHV}}$$

$$= \frac{E_{sor} S \, 15,141}{\text{HHV}}$$

(7.5b)

where: E_{sor} is the fraction of the sulfur content. S converted into $CaSO_4$.

This equation is unaffected by the calcium in the coal ash because sulfation from any other source produces the same amount of heat.

7.3.1.5 Unburnt Carbon Loss

The unburnt carbon loss, L_{uc} for BFBs is relatively high, in the range of 5 to 20%, depending upon the rank of coal. Low-rank fuel like lignite would have higher combustion efficiency than the high-rank anthracite. The bed ash may typically contain less than 3% unburnt carbon while the fly ash may have a much higher percentage.

7.3.1.6 Other Heat Losses (L_{others})

Other heat losses have to be estimated according to previous experience as they cannot be calculated directly. These losses, which include convective-radiation loss, ash sensitive heat loss, fan credit, and other unaccounted losses, are similar to those discussed in Section 8.5.2 for CFB boilers.

Bubbling Fluidized Bed Boiler

$$L_{others} = \text{convective radiation loss} + \text{ash sensitive heat loss} + \langle \text{fan credit} \rangle$$
$$+ \text{unaccounted losses}$$

No data on convection–radiation heat loss from fluidized bed boilers is available in the published literature. As the furnace is generally water-cooled, the convective-radiation heat loss through the exterior walls of a typical BFB boiler can be estimated using the graphs provided by the American Boiler Manufacturers Association (Singer, 1991) or ASME Performance test Code PTC-4.1.

7.3.2 Boiler Efficiency

Boiler efficiency is defined by the fraction of the total combustion heat released transferred to the water and steam. It is computed by subtracting all types of losses from the potential fuel energy input and then dividing it by the latter. This method is called heat-loss method. Alternately, it could be found directly dividing the enthalpy in the steam produced by the potential energy in the corresponding amount of fuel fired. However, owing to the practical difficulty in measuring, the heat-loss method is more frequently used for efficiency calculations.

The total heat loss is found by adding all the losses from above:

$$L_{total} = L_{stack} + L_{m,air} + L_{m,fuel} + L_{m,h} + L_{uc} + L_{others}$$

Using the information above, the boiler efficiency, η can be calculated by equation below:

$$\eta = (1 - L_{total}) \times 100\% \tag{7.6}$$

Example 7.1

A BFB boiler burns coal with a HHV of 14,654 kJ/kg.

Given:

- stoichiometric air required is 4.87 kg/kg
- excess air coefficient is 40%
- moisture in the air is 1.3%
- moisture in the coal is 10%
- hydrogen content of the coal is 2.1%
- dry flue gas mass is 7.2 kg/kg
- specific heat of flue gas 1.05 kJ/kg K
- flue gas temperature leaving stack is 160°C
- ambient temperature is 30°C
- specific heat of steam in flue gas is 1.902 kJ/kg K
- specific heat of water is 4.189 kJ/kg K
- latent heat of vaporization is 2256 kJ/kg
- convective-radiation loss is 0.8%
- unburned carbon loss is 5.5%
- sensitive heat loss in ash is 2.0%
- unaccounted loss is 1.5%
- fan credit is -1.0%.

Find:

- The boiler efficiency.

Solution

The dry flue gas loss is found from Equation 7.1:

$$L_{stack} = 7.2 \times 1.05(160 - 30)/14{,}654 = 6.7\%$$

Using Equation 7.2 we find the moisture loss due to water in air:

$$L_{m,air} = 4.87(1 + 0.4)1.3\%\{4.189(100 - 30) + 2256$$
$$+ 1.902(160 - 100)\}/14{,}654 = 1.6\%$$

The moisture loss from coal Equation 7.3 is:

$$L_{m,fuel} = 10\%\{4.189(100 - 30) + 2256 + 1.902(160 - 100)\}/14{,}654$$
$$= 1.8\%$$

The moisture loss from burning hydrogen in coal Equation 7.4 is:

$$L_{m,h} = 9 \times 2.1\%\{4.189(100 - 30) + 2256 + 1.902(160 - 100)\}/14{,}654$$
$$= 3.4\%$$

The unburned carbon loss is given:

$$L_{uc} = 5.5\%$$

The convective–radiative loss is given:

$$L_{cr} = 0.8\%$$

The sensitive heat loss in ash is given:

$$L_{ash} = 2.0\%$$

The fan credit is given:

$$L_{fan} = -1.0\%$$

The unaccounted loss is given:

$$L_{other} = 1.5\%$$

The total heat loss Equation 7.6 is:

$$L_{total} = L_{stack} + L_{m,air} + L_{m,fuel} + L_{m,h} + L_{uc} + L_{others}$$
$$= 6.7\% + 1.6\% + 1.8\% + 3.4\% + 5.5\% + (0.8\% + 2\% - 1\% + 1.5\%)$$
$$= 22.3\%$$

The boiler efficiency is:

$$\eta = 100 - L_{total} = (100 - 22.3)\% = 77.7\%$$

The block diagram given below illustrates the distribution of heat in the boiler, including the above losses.

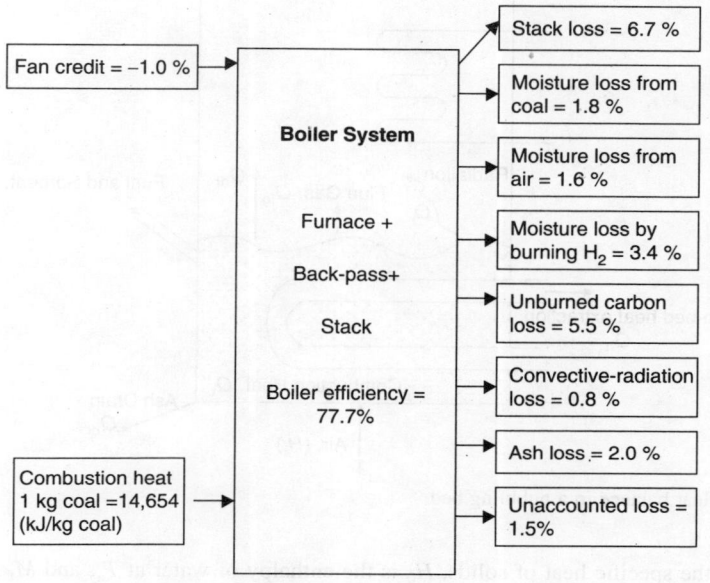

7.3.3 ENERGY BALANCE IN DENSE BUBBLING BED

The heat balance of the dense bed is a critical aspect of the design of BFB boilers. It determines the operating temperature of the bed. In a BFB, the furnace or bed temperature is controlled within a specified range of about 800 to 900°C through extraction of heat by heat-absorbing tubes and heat-carrying flue gases leaving the bed (Figure 7.5).

The combustion of fuel is rarely complete in the dense bubbling bed. Thus, the combustion heat is partially released in the bed, while some combustibles leave the bed to burn in the freeboard. The heat released in bed Q_i, can be found by:

$$Q_i = m_c X_b \text{HHV} \tag{7.7}$$

where: m_c is the fuel feed into the bed, X_b is the fraction of combustion taking place in the bed and HHV is the higher heating value of the fuel.

The enthalpy brought into the bed by the primary air H_i, depends on the mass of primary air m_a, and its preheat, T_i.

$$H_i = m_a C_{\text{air}}(T_i - T_a) \tag{7.8}$$

where: C_{air} is the specific heat of air at temperature T_i.

The energy brought into the bed by fuel and sorbents H_{fi}, includes the heat of the solids as well as the heat of the external moisture carried by them. The surface moisture M_f, of a fuel, stacked outside in the open atmosphere, can be substantial at times.

$$H_{fi} = (m_c + m_s)C_p T_a + (m_c M_f + m_s M_s)H_0 \tag{7.9}$$

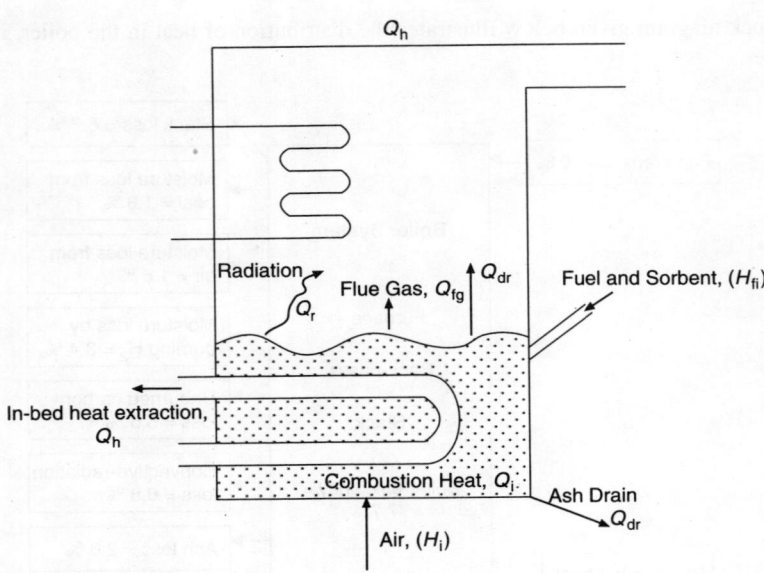

FIGURE 7.5 Heat balance in a bubbling bed.

where: C_p is the specific heat of solids, H_0 is the enthalpy of water at T_a, and M_f and M_s are the fractions of moisture in fuel and sorbents, respectively.

$$\text{Total heat input} = Q_i + H_i + H_{fi} \quad (7.10)$$

The heat absorbed Q_h, by in-bed tubes depends on the bed temperature T_b, average tube-wall temperature T_w, exposed heat transfer surface area A_s, and the $(-)h$, and is defined by $Q_h = hA_s(T_b - T_w)$.

Heat radiated from bed surface Q_r, is a major source of heat loss from the bed and is directly proportional to the bed surface area A_b, exposed to the freeboard.

$$Q_r = \sigma e_b A_b[(T_b + 273)^4 - (T_{fb} + 273)^4] \quad (7.11)$$

where e_b is the effective emissivity of the bed surface, σ is the Stefan–Boltzman constant, and T_{fb} is the freeboard temperature.

Heat loss through the bed drain is made up of sensible heat loss due to ash $(m_c X_{ash} x_d C_p T_b)$ and that due to sorbent, $(m_s x_d C_p T_b)$ for boilers with limestone injection.

$$Q_{dr} = (m_c X_{ash} x_d + m_s x_d) C_p T_b \quad (7.12)$$

where: X_{ash} is the ash fraction in fuel and x_d is the fraction of the total fuel ash or the spent sorbent present in the bed drain, m_s.

The flue gas carries the remaining ash (fly ash). Thus, the total heat loss in flue gas Q_{fg}, is made up of the enthalpies of the gas (including water vapor) and the fly ash:

$$Q_{fg} = m_c M_{flue} C_f T_b + m_c M_f H_{Tb} + (1 - x_d)(X_{ash} m_c + m_s) C_p T_b \quad (7.13)$$

where: H_{Tb} is the enthalpy of steam at T_b and G_g is the amount of flue gas produced per unit mass of fuel burned.

$$\text{Total heat leaving the bed} = Q_{fg} + Q_{dr} + Q_r + Q_h \quad (7.14)$$

Bubbling Fluidized Bed Boiler

The total heat input is equal to the heat output under a steady state. From Equation 7.10 and Equation 7.14 we get:

Total heat entering bed = total heat leaving bed

$$Q_i + H_i + H_{fi} = Q_{fg} + Q_{dr} + Q_r + Q_h \quad (7.15)$$

From Equation 7.15 one can solve for T_b. It is apparent from here that these equations will always yield a positive value of the bed temperature T_b, however, small it may be. If it is below the ignition temperature, the fuel will not ignite, but if it is above the softening temperature, the bed will agglomerate. This determines the operating or fuel flexibility limits of a BFB boiler.

7.3.4 Heat Balance of Boiler Components

A typical boiler comprises three heat exchangers: economizer, evaporator, and superheater (Figure 7.6). Large utility boilers often use additional heat exchangers like reheaters and air preheaters. The economizer preheats the cold water close to its saturation temperature. In the evaporator, the water is heated to saturation and the saturated water turns into saturated steam by absorbing the combustion heat. In rare exceptions, some evaporation may also be allowed in the economizer. The saturated steam is superheated by absorbing further heat in the superheater.

The primary goal of the thermal design is to determine the heating surface areas of the economizer, evaporator and superheater, which are dictated by their respective heat loads and heat-transfer coefficients. Steam conditions specify their heat loads.

The following four parameters of the water/steam are needed to obtain the heat loads or duties of the individual heat exchangers:

1. Steam temperature at the outlet of final superheater, $T_{sh,outlet}$
2. Steam pressure at the outlet of the final superheater $P_{sh,outlet}$
3. Feedwater temperature at the inlet of the economizer $T_{econ,inlet}$
4. Required steam mass flow rate G_s

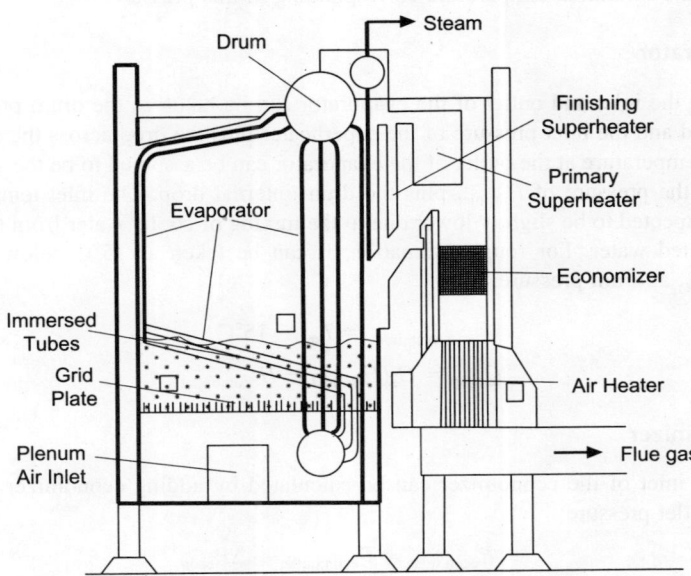

FIGURE 7.6 A typical BFB boiler comprises an economizer, evaporator, boiler drum, superheater, and an air heater.

If the steam cycle design requires reheating between two stages of the steam turbine, the pressure and temperature of the steam at the inlet of the reheater should be provided.

The pressure and temperature define the enthalpy H_{en}, of water and steam in heat exchangers.

$$H_{en} = f(P, T) \tag{7.16}$$

The enthalpy of steam or water is generally found from standard steam tables. After knowing the enthalpies at the inlet (H_{inlet}) and outlet (H_{outlet}) of the heat exchangers and the mass flow rate G_s, the heat load can be obtained from the general heat balance equation for a heat exchanger:

$$Q_{load} = G_s(H_{outlet} - H_{inlet}) \tag{7.17}$$

The calculation procedure could start with the superheater because the pressure of the steam is generally given and the pressure of the water in the economizer can be calculated considering the pressure losses through the heat exchangers in between.

7.3.4.1 Superheater

The superheater heat load is written as:

$$Q_{sh} = G_s(H_{sh,outlet} - H_{sh,inlet}) \tag{7.18}$$

The steam enthalpy at the outlet of the superheater is determined by knowing the pressure and temperature of the steam at the outlet of the final superheater. The pressure at the inlet of the superheater can be calculated by adding the pressure drop in the superheater to the designed outlet pressure: $P_{sh,inlet} = P_{sh,outlet} + \Delta P_{sh}$.

During the initial thermal design, when the type of superheater has not been decided, the pressure drop, ΔP_{sh} is estimated as 10% of the steam pressure at the outlet of superheater or obtained from experience or from the manufacturer of the superheater.

Steam at the inlet of the superheater is saturated at the pressure $P_{sh,inlet}$. Therefore, its temperature is the saturation temperature corresponding to this pressure.

7.3.4.2 Evaporator

The pressure at the inlet and outlet of the evaporator can be taken as the drum pressure plus the hydrostatic head and the inlet pressure of the superheater plus the drop across the drum internals, respectively. Temperature at the outlet of the evaporator can be assumed to be the saturated steam temperature at the pressure of $P_{sh,inlet}$ plus the drum internal drop. The inlet temperature in the evaporator is expected to be slightly lower due to the mixing of cooler water from the economizer with the saturated water. For rough estimation, it can be taken as 15°C below the saturation temperature T_{sat}, at drum pressure:

$$T_{evap,inlet} = T_{sat} - 15°C$$

$$Q_{evap} = G_s(H_{sh,inlet} - H_{econ,outlet}) \tag{7.19}$$

7.3.4.3 Economizer

Pressure at the inlet of the economizer can be calculated by adding economizer pressure drop, ΔP_{econ} to its outlet pressure:

$$P_{econ,inlet} = P_{econ,outlet} + \Delta P_{econ}$$

The economizer heat load is written as:

$$Q_{econ} = G_s(H_{econ,outlet} - H_{econ,inlet}) \tag{7.20}$$

Bubbling Fluidized Bed Boiler

The pressure at the economizer outlet can be assumed to be the drum pressure plus the hydrostatic head above its inlet and the friction in piping. The temperature at the economizer inlet is the same as the feedwater temperature, and the outlet temperature can be assumed as the temperature at the inlet of the evaporator.

The total heat load of the boiler is found by adding above terms:

$$Q_{steam} = Q_{econ} + Q_{evap} + Q_{sh} \qquad (7.21)$$

Actual heat load of the boiler will be slightly higher than this because the heat loss due to boiler blow-down, desuperheating has not been considered here.

Example 7.2

A BFB boiler generates steam at the rate of 32 t/h. The feedwater temperature is 30°C, the steam temperature at the outlet of final superheater is 270°C and the steam pressure at the outlet of final superheater is 25 bar.
Find the heat duties of the economizer, evaporator, and superheater.

Solution

A property table is constructed as below for the steam at different states in this boiler

Parameter	Property	Symbol	Equation	Value	Units
Main steam	Flow rate	G_s	Given	8.9	(kg/sec)
	Pressure	$P_{sh,outlet}$	Given	25	(bar)
	Temperature	$T_{sh,outlet}$	Given	270	(C)
	Enthalpy	$H_{sh,outlet}$	From steam table	2934	(kJ/kg)
Saturated steam in drum	Flow rate	G_s	Given	8.9	(kg/sec)
	Pressure	$P_{sh,inlet}$	$= 25 + 10\% \times 25$	27.5	(bar)
	Temperature	$T_{sh,inlet}$	Saturated temperature at 27.5 bar	229	(C)
	Enthalpy	$H_{sh,inlet}$	Saturated steam enthalpy, from steam table	2801.9	(kJ/kg)
Saturated water	Flow rate	G_s	Given	8.9	(kg/s)
	Pressure	P_{drum}	$= P_{sh,inlet}$	27.5	(bar)
	Temperature	T_{drum}	Saturated temperature at 27.5 bar	229	(C)
	Enthalpy	$H_{w,sat}$	Saturated water enthalpy, from steam table	985.9	(kJ/kg)
Water at outlet of economizer	Flow rate	G_s	Given	8.9	(kg/s)
	Pressure	$P_{econ,outlet}$	$= P_{drum}$	27.5	(bar)
	Temperature	$T_{econ,outlet}$	$= T_{drum} - 15$	214	(C)
	Enthalpy	$H_{econ,outlet}$	From steam table	916.4	(kJ/kg)
Feedwater	Flow rate	G_s	Given	8.9	(kg/s)
	Pressure	$P_{econ,inlet}$	$= 27.5 + 10\% \times 27.5$ (round to)	31	(bar)
	Temperature	$T_{econ,inlet}$	Given	30	(C)
	Enthalpy	$H_{econ,inlet}$	From steam table	128.5	(kJ/kg)
Heat load	Superheater	Q_{sh}	$= 8.9 \times (2934 - 2801.9)$	1175.7	(kW)
	Evaporator	Q_{evap}	$= 8.9 \times (2801.9 - 916.4)$	16,780.9	(kW)
	Economizer	Q_{econ}	$= 8.9 \times (916.4 - 129)$	7007.8	(kW)
Total heat		Q_{steam}	$= Q_{econ} + Q_{evap} + Q_{sh}$	24,964	(kW)

7.3.5 Mass Flow Rates of Fuel and Air

This part of the thermal design determines the capacities of the coal-feeding system and the air/flue gas-handling system of the boiler.

The fuel consumption rate can be calculated by the required combustion heat release rate, which in turn can be determined by the total heat load of the boiler and the boiler efficiency.

$$Q_{comb} = \frac{Q_{steam}}{\eta} \quad (7.22)$$

$$G_{fuel} = \frac{Q_{comb}}{HHV} \quad (7.23)$$

where: η is the thermal efficiency of the boiler and G_{fuel} is mass flow rate of fuel (kg/s).

The mass flow rate of air can be obtained by air/fuel ratio M_{wa}, and fuel mass flow rate. The air/fuel ratio can be obtained from the stoichiometric calculation.

$$G_{air} = G_{fuel} M_{wa} \quad (7.24)$$

G_{air} — air mass flow rate (kg/sec), M_{wa} — wet air required for burning fuel (kg air/kg fuel).

The flue gas mass flow rate can be obtained by flue gas/fuel ratio and fuel-mass flow rate. The flue gas/fuel ratio can be obtained from the stoichiometric calculations:

$$G_{flue} = M_{flue} G_{fuel} \quad (7.25)$$

G_{flue} — flue gas mass flow rate (kg/sec), M_{flue} — flue gas generated per kg fuel burned (kg/kg fuel).

7.3.6 Gas-Side Heat Balance

Flue gas temperatures at different points in the furnace and in the back-pass (convective section) help calculate the heat available for absorption in different sections and finally the temperature of the preheated air entering the furnace.

The flue gas temperature can be calculated by knowing the specific heat, heat load and mass flow-rate of the flue gas. If Q is the heat duty of the specific heating section one gets:

$$Q = G_{flue} C_f (T_{after} - T_{before}) - \text{surface losses} \quad (7.26)$$

T_{after} is the flue gas temperature at the exit of the heat exchanger (°C), T_{before} is the flue gas temperature at the entrance of the heat exchanger (°C), C_f is the specific heat of flue gas averaged over the temperature range of T_{after} and T_{before}. Table A4.5 gives specific heats of gas constituents at different temperatures.

7.3.6.1 Furnace Exit Gas Temperature

Flue gas temperature at the exit of the furnace, T_{fur} is an important parameter chosen by the designer. For example, the flue gas temperature at the furnace exit of a PC boiler is around 1100°C, while in a BFB boiler it is kept in the range of 800 to 900°C.

If one assumes that all evaporative surfaces are located in the furnace and all superheater surfaces are in the back-pass from the above equations we can write:

$$Q_{comb} + G_{air} C_{air}(T_{air} - T_a) = Q_{comb}(L_{uc} + L_{cr}) + Q_{evap} + G_{flue} C_f(T_{fur} - T_a) \quad (7.27)$$

Bubbling Fluidized Bed Boiler

where: L_{uc} and L_{cr} are the fractions of combustion heat lost as unburnt heat loss and through convection–radiation from the exterior of the furnace, respectively.

Example 7.3

Given:

- Combined heat duty of all surfaces in the furnace of a BFB boiler is 16,978 kW
- Potential combustion heat release rate is 30,394 kW excluding the moisture losses
- Ambient temperature is 30°C and the temperature at the exit of furnace is 850°C
- Unburned carbon loss is 1.5%; the convective-radiation heat loss is 0.8%; heat carried away by all types of ash is neglected
- Flue gas mass flow rate is 15.81 kg/sec, and the air mass flow rate is 14.33 kg/sec
- Average specific heat of flue gas is 1.134 kJ/kg K and the specific heat of air is 1.015 kJ/kg K.

Determine the air preheat temperature for the given boiler.

Solution

Using Equation 7.27 we get:

$$30,394 + 14.33 \times 1.015(T_{air} - 30) = 30,394(0.8\% + 1.5\%) + 16,978$$
$$+ 15.81 \times 1.134(850 - 30)$$

Solving it one gets:

$$T_{air} = 167°C$$

The air should be heated to 167°C in the air heater before it enters the furnace.

7.3.6.2 Gas Temperature after Superheater

If one assumes that the heat loss from the exterior of the furnace is negligible, the heat absorbed by the superheater would be equal to that lost by the flue gas:

$$Q_{sh} = G_{flue} C_f (T_{fur} - T_{sh}) \tag{7.28}$$

where: T_{sh} is the temperature of the flue gas leaving the superheater and T_{fur} is the temperature of leaving the furnace and entering the superheater.

7.3.6.3 Gas Temperature after Economizer

Similarly, the flue gas temperature after the economizer is:

$$Q_{econ} = G_{flue} C_f (T_{sh} - T_{econ}) \tag{7.29}$$

T_{econ} is the gas temperature of the economizer.

7.3.6.4 Heat Transfer through Air Heater

Air temperatures, before and after the air heater govern its heat duty. Assuming that air enters the air heater at ambient temperature T_a, and using Equation 7.26, the air heater duty can be written as:

$$Q_{\text{air heater}} = G_{\text{air}} C_{\text{air}} (T_{\text{air}} - T_a) = G_{\text{flue}} C_f (T_{\text{econ}} - T_{\text{stack}}) + \text{surface losses} \quad (7.30)$$

The air temperature after the air heater T_{air}, is to be specified to determine the flue gas temperature leaving the air heater.

7.4 COMBUSTION IN BUBBLING FLUIDIZED BED

In a BFB boiler, the fuel and oxygen undergo an exothermic reaction (combustion) in the furnace. The rate of combustion is governed by the kinetics of combustion, mass transfer, and heat-transfer processes. This section describes the combustion process from both design and operating perspectives of a BFB boiler furnace. The discussion is divided into two main sections:

- Coal combustion process
- Minimization of combustion losses

7.4.1 COAL COMBUSTION IN BFB BOILERS

Before it is fed into a BFB furnace coal is crushed to sizes typically smaller than 10 mm. Each coal particle undergoes the following sequence of events in the furnace as shown in Figure 7.7:

1. Drying
2. Devolatilization and volatile combustion
3. Swelling and primary fragmentation
4. Char combustion and attrition

Char particles may burn under any of the three combustion regimes discussed in Section 4.1.4.2 in Chapter 4 on Combustion.

Char combustion may have two different products, CO and CO_2. Their reaction equations are,

$$C + \tfrac{1}{2} O_2 = CO \qquad CO + \tfrac{1}{2} O_2 = CO_2 \quad (7.31)$$

It was shown in Chapter 4 that the relative amounts of CO and CO_2 produced by the combustion of carbon depend on the temperature. Table 7.1 gives values of this ratio at some typical temperatures.

When the bed temperature increases, the temperature of the carbon surface also increases, which in turn increases the production of carbon monoxide. The heat generated through the carbon monoxide reaction (Equation 7.31) is only a quarter of that of carbon dioxide generation. When a large amount of carbon monoxide is generated during combustion instead of carbon dioxide, much heat is lost due to incomplete combustion.

In a BFB combustor, an excess air coefficient of 1.2 to 1.3 is used to help complete the combustion. Here, combustion occurs in both the fluidized bed and the freeboard. The split of heat released between the bed and the freeboard depends on several factors including the fuel feeding arrangement. For example, if the fuel is fed from underneath the bed, the ratio would be higher. In a typical case the bed/freeboard ratio is 88:12 (Duqum et al., 1985).

During the combustion of a coarse char particle in a bubbling bed, a large amount of fines is produced through attrition. These fine char particles are typically smaller than 100 μm and are

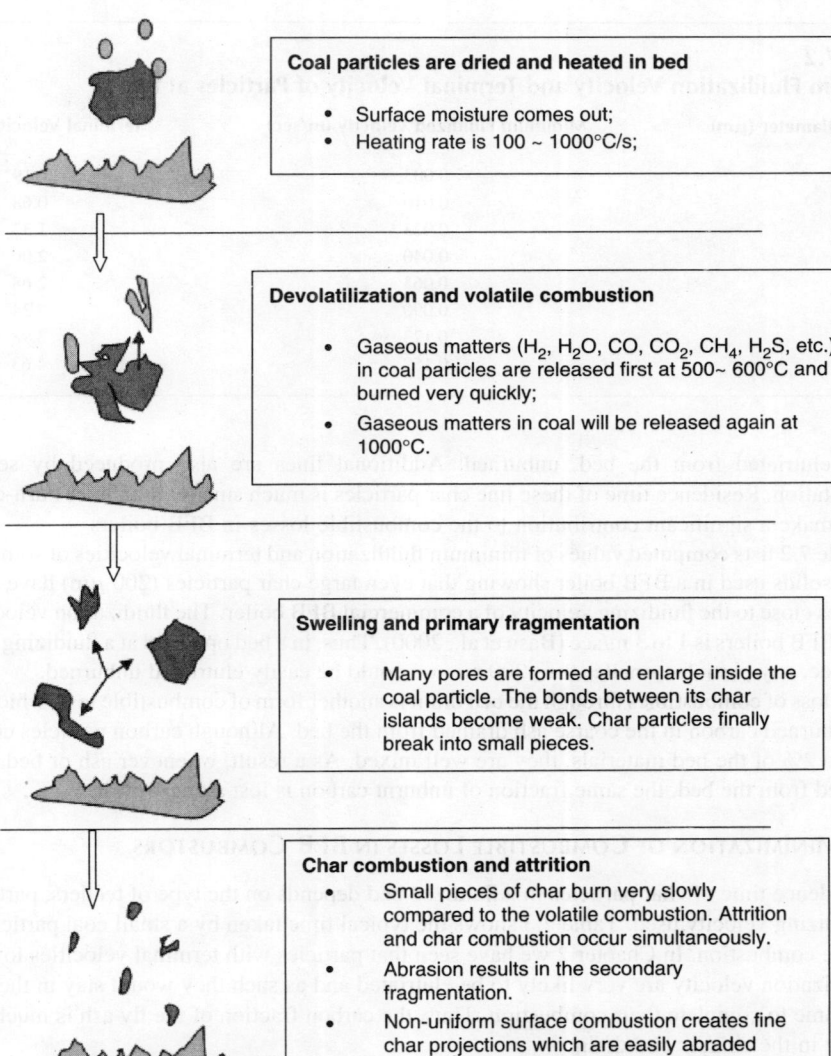

FIGURE 7.7 Coal particles in different combustion stages.

TABLE 7.1
Ratio of Carbon Monoxide to Carbon Dioxide Produced at Different Surface Temperatures

Surface Temperature of Carbon (°C)	[CO]/[CO$_2$]
800	7.2
900	11.8
1000	17.9

TABLE 7.2
Minimum Fluidization Velocity and Terminal Velocity of Particles at 850°C

Particle Diameter (μm)	Minimum Fluidized Velocity (m/sec)	Terminal Velocity (m/sec)
50	0.003	0.19
100	0.010	0.68
150	0.023	1.33
200	0.040	2.00
250	0.063	2.68
300	0.090	3.34
350	0.123	3.98
400	0.159	4.63

readily elutriated from the bed, unburned. Additional fines are also produced by secondary fragmentation. Residence time of these fine char particles is much smaller than their burn-out time, so they make a significant contribution to the combustible losses in BFB boilers.

Table 7.2 lists computed values of minimum fluidization and terminal velocities of some typical sizes of solids used in a BFB boiler showing that even large char particles (200 μm) have terminal velocities close to the fluidizing velocity of a commercial BFB boiler. The fluidization velocity used in most BFB boilers is 1 to 3 m/sec (Basu et al., 2000). Thus, in a bed operated at a fluidizing velocity of 2 m/sec, any particles smaller than 200 micron could be easily elutriated unburned.

The loss of combustibles through the bed drain is another form of combustible loss, which comes from unburned carbon in the coarse ash drained from the bed. Although carbon particles constitute only 1 to 2% of the bed materials, they are well mixed. As a result, whenever ash or bed material is drained from the bed, the same fraction of unburnt carbon is lost along with it.

7.4.2 MINIMIZATION OF COMBUSTIBLE LOSSES IN BFB COMBUSTORS

The residence time of char particles in a fluidized bed depends on the type of feeders, particle size and fluidizing velocity used. Table 7.3 shows the typical time taken by a small coal particle for its complete combustion. In Chapter 2 we have seen that particles with terminal velocities lower than the fluidization velocity are very likely to be elutriated and as such they would stay in the bed too short a time to complete their combustion. Thus, the carbon fraction of the fly ash is much greater than that in the coarser bed drain.

The above consideration would favor a low fluidization velocity. However, if the operating velocity is close to the minimum fluidization velocity of the coarser fraction of the bed solids, the bed would run the risk of defluidization.

Fuels with high reactivity or high oxygen content similar to those shown in Table 7.4 may have lower combustible losses compared to those with lower reactivities.

TABLE 7.3
Typical Time Taken by Coal Particles (<0.2 mm) for Combustion

Residence time in the fluidized bed	10 (sec)
Heat-up to devolatilization	0.1 (sec)
Devolatilization	0.2 ~ 0.5 (sec)
Char combustion	50 ~ 150 (sec)

Source: Adapted from Keairns et al., 1984.

TABLE 7.4
Typical Analysis of a Fuel with High Reactivity and High Oxygen

Proximate analysis		Ultimate analysis	
Moisture	24.8%	Moisture	24.8%
Ash	8.0%	Ash	8.0%
Volatile Matter	30.2%	O	11.31%
Fixed Carbon	37.0%	C	51.1%
		H	3.5%
		S	0.65%
		N	0.64%
High Heating Value = 20,374 (kJ/kg)			

Source: Adapted from Duqum et al., 1985.

The following section discusses several options for reducing the combustible loss or carbon fraction in fly ash.

7.4.2.1 Tapered Beds

By enlarging the cross-sectional area of the bed, the superficial velocity can be reduced, thereby reducing the carbon carryover loss. In a typical bubbling bed, solids segregate and coarser particles tend to accumulate in the lower sections of the bed. A reduced fluidizing velocity could result in defluidization particularly in the lower section where coarser particles are dominant and the superficial velocity could fall below their minimum fluidization velocity. Finer particles are concentrated near the top where the operating superficial velocity exceeds their terminal velocity.

This problem can be avoided by using a tapered bed, which will yield a higher fluidization velocity in the lower bed allowing coarser particles to remain fluidized and a lower velocity in the upper bed, reducing the entrainment of finer particles. This combined with a freeboard of larger cross section can reduce the carbon in the fly ash (Figure 7.8).

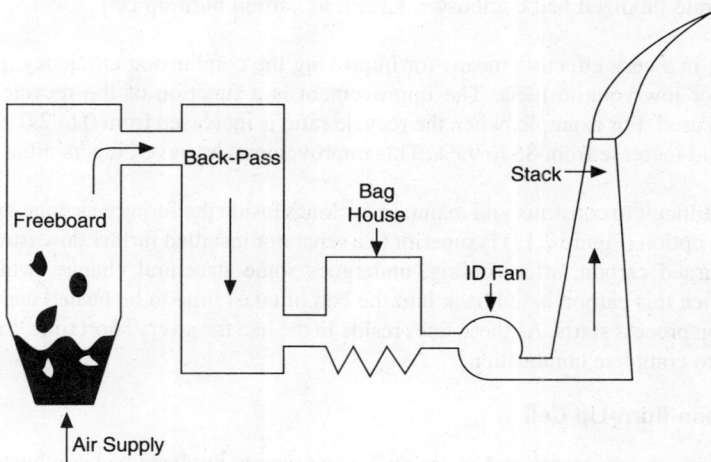

FIGURE 7.8 BFB boiler furnace with enlarging bed cross section.

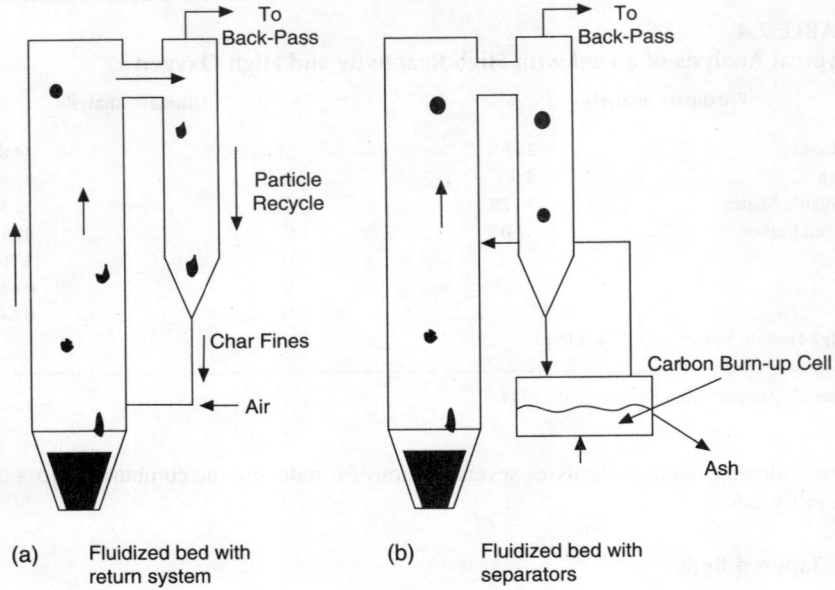

FIGURE 7.9 Fluidized bed with separators. Unburned carbon from fluidized combustion can be reduced by (a) refiring grits collected under back-pass, or (b) burning them in a separate low-velocity fluidized bed.

7.4.2.2 Recirculation

Recycling a part or the entire mass of entrained fly ash is another option for reducing the unburned carbon in the fly ash. Separators, installed beyond the exit or inside the furnace, could capture and recycle the particles back into either of the following two sections to burn the escaped carbon fines (Figure 7.9):

i) Main combustion bed
ii) A separate fluidized bed combustor, known as carbon burn-up cell

Recycling is a very effective means for improving the combustion efficiency, particularly for less reactive or low-volatile fuels. The improvement is a function of the recycle ratio (recycle rate/feed rate) used. For example, when the recycle ratio is increased from 0 to 2.0, the combustion efficiency could increase from 85 to 95%. This improvement, however, tapers off at higher recycle ratios.

It is very difficult to construct and maintain cyclones inside the furnace as done in FCC reactors. However, this option (Figure 7.13) is superior to a separator installed further downstream in a cooler section. Unburned carbon, after cooling, undergoes some structural change, which reduces its reactivity. When this carbon is fed back into the bed, it takes time to be heated and ignited before the combustion process starts. As these fines reside in the bed for a very short time, there may not be enough time to complete combustion.

7.4.2.3 Carbon Burn-Up Cell

The carbon burn-up cell, mentioned in option 2, is a separate fluidized bed combustor with its own air supply and start-up device (Figure 7.10) and is fed with carbon-bearing fly ash from the furnace. The carbon burns and the hot gas is fed back to the boiler. Being an independent unit, it has

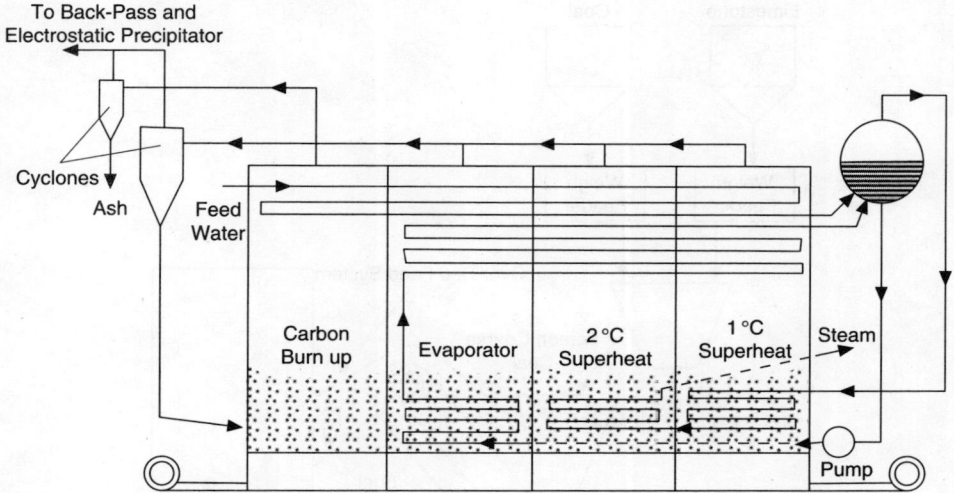

FIGURE 7.10 A separate fluidized bed (carbon burn-up cell) is used for a secondary combustion of unburned carbon fines.

the flexibility of being operated at a velocity most appropriate for the fine particles and at temperatures high enough to burn the less-reactive unburned fines. Carbon burn-up cells are necessarily large in cross section because of their lower operating velocity and are devoid of heat-absorbing surfaces. This makes them expensive. Also their operation at higher temperatures contributes to higher NO_x emissions.

7.4.2.4 Under-Bed Feed

The next option is to use an under-bed feeder instead of an over-bed one. The combustion in a BFB occurs mostly in the bed itself, rather than in the freeboard (Anthony et al., 1981). An under-bed feeder allows longer residence time for the finer fraction of the feed and, therefore, gives good combustion efficiency, especially for fine coal particles with low volatile content. Figure 7.11 shows a combined over-bed and under-bed feed system, where coarse particles are fed from the top and fines are injected from the bottom of the bed. For ratio of volatile matter to fixed carbon less than 1.0, the under-bed feed performs better than the over-bed feed.

Under-bed feeding will, of course, necessitate the use of a much larger number of feeders. Over-bed feeders, on the other hand, will work well for coarser coal with high volatiles. The hybrid system, shown in Figure 7.11, could sieve out the fines below 200 μm and inject them through a relatively small number of under-bed feeders while rest of the coal could be fed through the over-bed feeder to reduce the combustion losses.

7.4.2.5 Tall Freeboard

The next option is to increase the height of the freeboard, which will increase the residence time as well as reduce the carry-over of particles. The height of the freeboard should be, if possible, higher than the transport disengaging height. This is the most practical option and for this reason, very large BFB boilers usually demonstrate combustion efficiencies much higher than smaller units. However, a heavily cooled freeboard may not have this effect as that might quench the combustion. Furthermore some BFB boilers use a narrowed freeboard to improve the heat absorption. This adversely affects the effectiveness of the freeboard in improving the combustion.

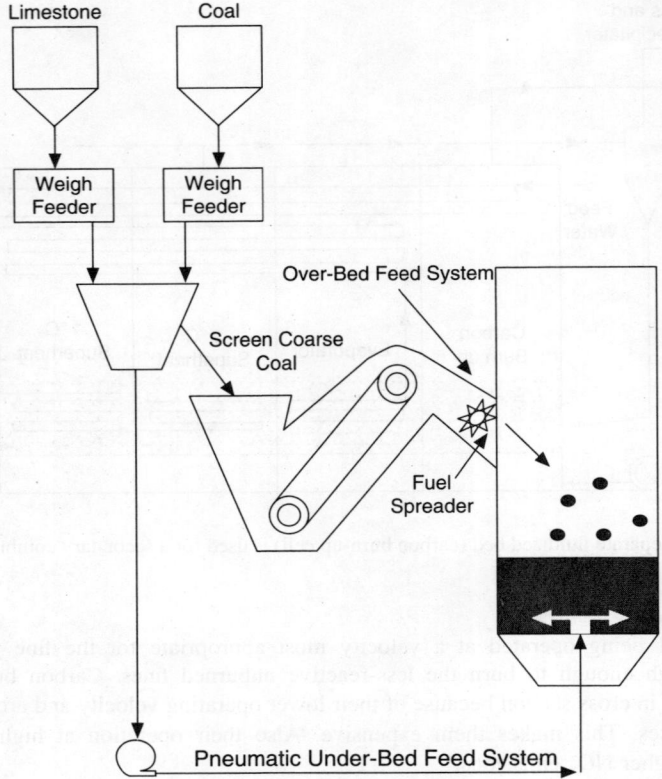

FIGURE 7.11 Combined under-bed and over-bed feeders for bubbling fluidized bed boilers.

7.4.2.6 Novel Design

A novel device like the creation of vortex in the freeboard (Greenfield Research Incorporated, 2004) could increase the residence time of char after it leaves the bed. Results for commercial units are, however, not available to judge the effectiveness of such techniques.

7.5 FURNACE DESIGN

This section presents some considerations and steps in the design of BFB boilers. The design of a BFB boiler would typically include the following aspects:

- Furnace design
- Distributor design
- Design of separators and return systems (if recycling is required)
- Determination of the sizes of water wall tubes, in-bed tubes and tube banks

It also includes the selection of auxiliary systems like:

- Steam and water feeding
- Air supply

Bubbling Fluidized Bed Boiler

- Fuel/sorbent preparation
- Fuel/sorbent feeding
- Waste solid handling

The design would specify the following operating conditions for the BFB boiler.

- Bed temperature
- Bed pressure
- Fluidizing or superficial velocity through the bed
- Excess air
- Fuel feed size distribution
- Sorbent feed size distribution, if sulfur capture is needed

7.5.1 Basic Design of Furnace

The furnace of a BFB boiler serves as both combustion chamber and heat exchanger. The furnace ensures generation of the required amount of energy with a minimum loss. It also ensures that the required thermal energy is transferred to water or steam to meet the specified furnace exit temperature. A typical furnace design will determine the following:

- Basic dimensions of the furnace such as height, width, and breadth
- Dimensions of combustor cells, if used
- Sizes of surface area for water wall and/or in-bed tubes
- Height of bed and freeboard

There is no unique and set design procedure. It varies from one designer to another. The following sections present one of the many approaches to the design of different components of a BFB boiler.

7.5.1.1 Cross-Sectional Area of Bed

The bed area can be determined from either of the following two considerations:

- Grate firing rate
- Fluidizing velocity in the bed

The grate firing rate is defined by the following equation:

$$q_g = \frac{Q_{comb}}{A_{bed}} \tag{7.32}$$

q_g — grate firing rate (MW/m^2), Q_{comb} — combustion heat rate (MW).

The grate firing rate can be determined by empirical values (or experience). A rule-of-thumb suggests the following equation (Basu et al., 2000):

$$q_g = 3.3 \frac{U_{27}}{\text{EAC}} \tag{7.33}$$

U_{27} is the fluidizing velocity referred to 27°C (m/sec), EAC is the excess air coefficient in fraction. The empirical value of the grate heat release rate for BFB boiler is given by Basu et al. (2000):

- Furnace with in-bed tubes, $q_g = 2 \text{ MW/m}^2$
- Furnace without in-bed tubes, $q_g = 1 \text{ MW/m}^2$

TABLE 7.5
Characteristics of Some Bubbling Fluidized Bed Boilers

Boilers	Capacity (Mwe)	Bed Area (m^2)	Bed Depth (m)	Heat Input (MWth)	Grate Release (MW/m^2)	Fluidizing Velocity (m/sec)	Fuel type	HHV (MJ/kg)
TVA	160	234.0	0.66	457.14	1.95	2.5	Bituminous	24–25
Black Dog	130	170	—	371.43	2.1	2 to 3	Coal	19.5–34.9
Wakamatsu	50	99.0	—	142.86	1.44	1.3	Coal	25.8
Shell	43MWth	23.6	1.22	43.46	1.84	2.74	Bituminous	—
Stork	90	61.0	1.05	100.00	1.64	1.6	Lignite	25

The cross-sectional area of the bed can also be determined by fluidizing velocity:

$$A_{bed} = \frac{G_g}{\rho_f U_f} \tag{7.34}$$

U_f — fluidization or superficial velocity through the bed (m/sec), ρ_f — density of flue gas (kg/m^3), G_g — flue gas flow rate (kg/sec).

Table 7.5 gives data on some large commercially operating BFB boilers. Grate release rate of these units are between 1.5 and 2.2 MW/m^2.

Example 7.4

Given:

- Flue gas mass flow rate: 15.8 kg/sec
- Primary air is 60% of the combustion air and is fed under the distributor
- Molecular mass of flue gas, n: 29.5 kg/kmol
- Bed temperature: 850°C
- Bed pressure is 1.013 bar
- Mean diameter of particles in bed: 1.2 mm
- Minimum fluidization velocity at bed temperature: 0.44 m/sec
- Fluidizing velocity in the bed should be 3 to 4 times the minimum fluidization velocity
- Gas constant: 8.314×10^{-2} m^3 bar/kmol K

Find: Cross-sectional area of bed.

Solution

- Using the equation of state the density of flue gas:

$$\rho_f = n\frac{P}{RT} = 29.5 \frac{1.013 \times 10^5}{8314(273+850)} = 0.32 \text{ kg/}m^3$$

- Choosing the fluidizing velocity 4 times that of U_{mf}:
  Superficial velocity: $U_f = 4 \times U_{mf} = 4 \times 0.44 = 1.76$ (m/sec)

- Since 60% of the air is used for fluidization, the air flow through the bed is $= 0.6 \times 15.8$ kg/sec
- Bed area: $A_{bed} = \dfrac{G_f}{\rho_f U_f} = \dfrac{15.8 \times 0.6}{0.32 \times 1.76} = 16.8 \text{ m}^2$

7.5.1.2 Bed Height

The bed height is limited by the volume of in-bed tube banks, residence time of air for combustion and/or desulfurization time. If in-bed tubes are used, the bed should be deep enough to immerse the entire tube bank. Furthermore, there should be a minimum distance of 100 to 300 mm between the distributor and the bottom of the in-bed tubes to avoid erosion due to air jets coming out of the grid nozzles.

When sulfur capture is required, the bed should be deep enough to provide the required gas residence time for sulfur capture. The minimum gas residence time for sulfur capture is 2/3 sec.

The bed height (or depth) is generally in the range of 0.4 to 1.5 m. It cannot be very high because the pressure drop through the bed increases linearly with height, greatly adding to the power consumption of the fluidizing air, and the bubble eruption size increases greatly, increasing the entrainment loss.

In some BFB boilers, the bed height is adjustable for part-load operation. When the bed height is increased, more rows of tubes are immersed in the bed (Figure 7.12). Thus, the tube surface exposed to the higher heat-transfer rate of the bed increases, increasing the heat transfer from the bed to the steam/water tube.

In the absence of in-bed tubes and sulfur capture requirements, the bed height is determined by combustion requirements. Fuel properties would affect the bed height. For example, when a hard-to-burn less-reactive fuel is burned, a shallow bed could become unstable below 50% loading of the boiler (Jai and Sang, 1987).

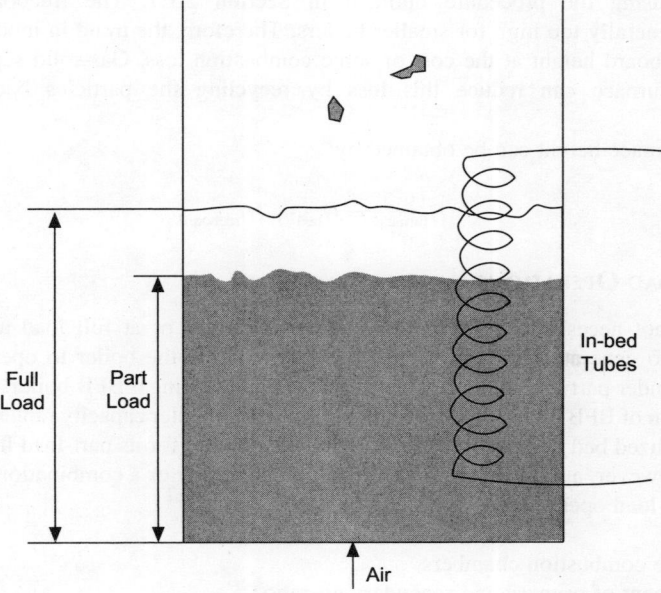

FIGURE 7.12 Adjustable bed height for different loads.

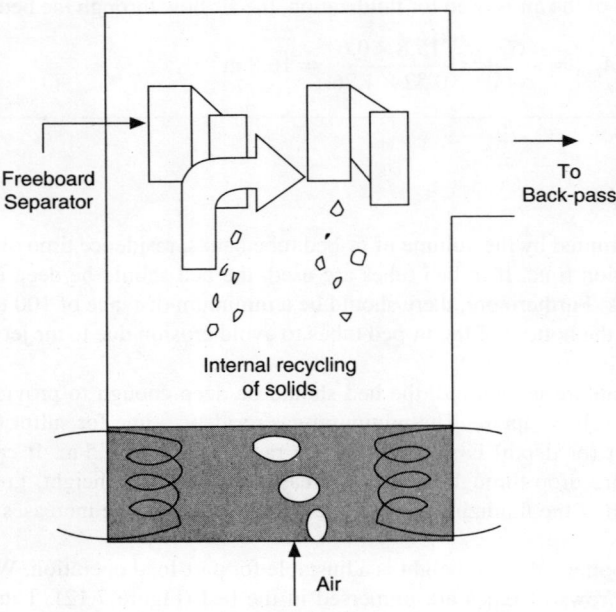

FIGURE 7.13 Bubbling fluidized bed boiler with separator.

7.5.1.3 Freeboard Height

The freeboard height should ideally be taller than the transport disengaging height, which can be calculated using the procedure outlined in Section 2.1.7. The freeboard height, thus calculated, is generally too high for smaller boilers. Therefore, the trend in modern designs is to shorten the freeboard height at the cost of some combustion loss. Gas-solid separators installed in the upper furnace can reduce this loss by recycling the particles back into the bed (Figure 7.13).

The total furnace height can be obtained by:

$$H_{\text{furnace}} = H_{\text{bed}} + H_{\text{freeboard}} \quad (7.35)$$

7.5.2 Part-Load Operations

A boiler does not necessarily operate in its design capacity or at full load all the time. It is often required to generate less than its full capacity, forcing the boiler to operate at part load. The operation under part load must be considered when designing a BFB boiler. Load control is a serious limitation of BFB boilers restricting their use to the smaller capacity range (10 to 25 MWe). Circulating fluidized bed boilers, on the other hand are popular for its part-load flexibilities among other things. However, a BFB boiler, if required, could use any or a combination of the following options for part-load operation:

- Multiple combustion chambers
- Adjustment of primary and secondary air ratio
- Adjustment of bed depth
- Adjustment of combustion temperature

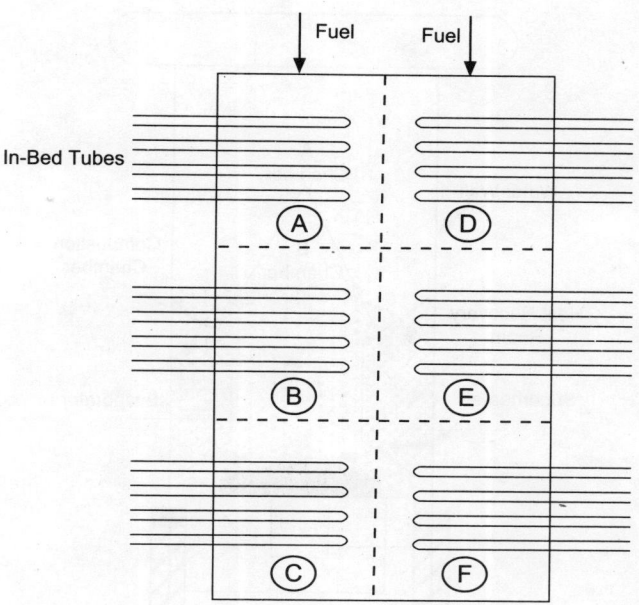

FIGURE 7.14 Plan view of a bed with three combustion zones.

7.5.2.1 Multiple Combustion Chambers

In multi-chamber designs (Figure 7.14), the furnace is divided into several chambers. When it is run under part load, the required number of chambers would be shut down with minimum heat generation (Makino and Seki, 1984) while the others could operate under full load. This technique works well on principle but accumulation of unburned carbon from neighboring chambers and defluidization are some of its practical problems.

In an alternative design, some chambers could be used exclusively for heat absorption while others are used for combustion as shown in Figure 7.15, where two chambers (A) in the wing absorb heat while the middle chamber (B) without heating surfaces is used for combustion. By regulating solid transfer between these chambers, one can control the heat absorption in the bed. A velocity differential between two chambers facilitates the solid transfer between them.

7.5.2.2 Combustion Temperature

The combustion temperature in the bed can be adjusted within a narrow range of 800 to 930°C by adjusting the fuel feed rate. This changes both heat-transfer coefficient and temperature difference between the bed and water and thus the heat absorption by water and steam.

7.5.2.3 Staged Combustion

In the staged-combustion design, no boiler tubes are generally located in the bed, allowing it to sustain the combustion even at low load. For this reason, instead of releasing all the combustion heat in the main bed, part of it is released in the freeboard and part in the bed. This way, neither the bed temperature is too high nor the freeboard temperature too low. Generally, the bed temperature should be kept at 850°C.

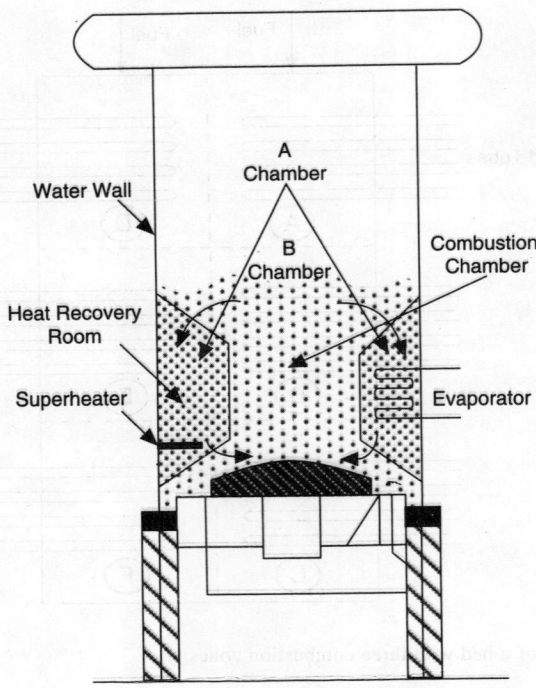

FIGURE 7.15 Bubbling fluidized bed boiler with two types of chambers with or without tubes.

Here, primary air is supplied through the distributor grid from the bottom of the combustor and secondary air is supplied above the bed. When the boiler runs under part load, the secondary air and fuel feeding rates are decreased.

In the case of over-bed feeding, volatiles from coarse fuel particles and some fine particles burn above the bed level using the secondary air.

7.5.2.4 Adjustable Bed Height

The load can be adjusted by changing the bed height as shown in Figure 7.12. It shows that the operator can change the heat-transfer surface area immersed by the fluidized bed materials by adjusting the bed height. There being a significant difference in the heat-transfer coefficient between tubes immersed in the bed and those outside the bed, the heat absorption in the bed can be changed by adjusting the bed depth. This could allow load changes of up to 5% per minute (Oka, 2004).

7.5.3 Other Operating Conditions

The operating parameters of BFB boilers include:

- Bed temperature and depth
- Superficial velocity through bed
- Excess air
- Fuel particle size distribution
- Sorbent particle size distribution

Bubbling Fluidized Bed Boiler

No general theory or equations are available to determine all these parameters. They are usually determined by experience and empirical design standards.

7.5.3.1 Bed Temperature

Bed temperature is normally kept around 850°C for optimal sulfur-capture performance. Low temperature adversely affects the combustion as it decreases the combustion rate and increases combustion losses. High bed temperature causes:

- Higher emissions of NO_x
- Higher erosions due to ash deformation points
- Higher thermal stresses on water tubes

Sometimes, when a less reactive, hard-to-burn fuel is fired, the combustion might become unstable at low temperatures and a high bed temperature is required.

An energy balance in the bed as shown in Figure 7.5 could give the steady-state temperature, T_b of the bed. It is apparent from this equation that by varying different parameters, one can control the bed temperature T_b. This has been studied in more detail in Section 7.3.3.

Combining Equation 7.7 to Equation 7.15, the energy balance is written as:

$$m_c X_b HHV + m_a C_{air}(T_i - T_a) + (m_c + m_s)C_p T_a + (m_c M_f + m_s M_s)H_0$$
$$= hA_s(T_b - T_g) + \sigma e_b A_b[(T_b + 273)^4 - (T_{fb} + 273)^4] + (m_c X_{ash} x_d + m_s x_d)C_p T_b$$
$$+ m_c G_g C_g T_b + m_c M_f H_{Tb} + (1 - x_d)(X_{ash} m_c + m_s)C_p T_b \qquad (7.36)$$

7.5.3.2 Bed Pressure

In practical operations, bed pressure is an important parameter to monitor. For atmospheric BFB boilers, the pressure at the exit of the furnace is kept slightly below the atmospheric pressure to reduce gas or dust leakage from the furnace. By adjusting the suction draft of the induced draft (ID) fans, the operator can control the pressure P_e, at the exit of furnace. The bed pressure P_b, immediately above the distributor can be calculated knowing the pressure drop across the bed and the freeboard:

$$P_b - P_s = \rho_p(1 - \varepsilon_b)H_{bed}g \qquad (7.37)$$

P_s — pressure at the interface between bed and freeboard (Pa), ε_b — voidage in the bed, H_{bed} — bed depth.

7.5.3.3 Fluidizing Velocity

The fluidization velocity controls the entire hydrodynamics of the bed including its heat transfer, entrainment of particles, defluidization, etc. Therefore, this value must be monitored and maintained carefully. One important difficulty in industrial units is the change in average bed particle size during extended operation. This, at times, leads to a reduction in heat absorption (due to a reduced heat transfer coefficient) and increased potential for defluidization. Nonuniform velocity distribution across a large bed is another problem.

Generally, the fluidization velocity is maintained at several times the minimum fluidization velocity of average bed particles with typical values being in the range of 1 to 2 m/sec. In special cases it might go up to 2.5 m/sec. Yet, at times the nonuniform velocity distribution around the bed may cause partial defluidization.

7.5.3.4 Particle Size

The particle size (typically less than 1mm) influences the bed hydrodynamics and combustion behavior to some extent, but exerts a dominant influence on the bed to tube heat transfer in a bubbling bed. The size distribution of the bed particles is governed by the complex interaction of the following parameters:

- size distribution of fuel ash particles as they enter through the feed
- fluidizing velocity
- attrition rate
- feed system: over-bed or under-bed
- bed drain rate and entrainment rate

One effective means of controlling the size distribution of particles in the bed is to use a fluidized bed ash classifier that indirectly classifies the bed materials by their size returning the desired fines sizes to the bed and draining the undesirable coarser fraction (Figure 7.3). Here, the solids drained from the bed are passed through a separate bed, fluidized by cold air. The superficial velocity through this bed is controlled to return the desired sizes of particles to the bed and drain the others. This equipment also cools the hot ash and is thus called an ash cooler. It would, however, work only when there is a net accumulation of solids as a result of combustion and entrainment.

7.5.3.5 Excess Air

As in any combustion device, excess air is necessary for BFB combustion. If only the theoretical amount of air is supplied, the combustion efficiency would be poor because of the poor distribution of air and fuel. To increase the oxygen concentration near coal particles, more air is needed. The value of excess air is around 30 to 40%, or 1.3 to 1.4 for BFBs (Basu et al., 2000). When the excess air is increased, the stack loss would be increased, which causes low boiler efficiency (see Section 7.3.1). Besides, more air would result in a low bed temperature. For this reason, sometimes a large amount of dilution air is supplied to control the bed temperature.

7.6 START-UP PROCEDURE AND CONTROL

The start-up procedure is an important aspect of the operation of a BFB boiler. Based on their respective experiences, different companies use different start-up procedures. It is not easy to comment which start-up procedure is the best, but generally, the whole process is divided into two stages — preparation and ignition.

Before start-up, the following preparations are necessary:

Standard safety examination: the standard safety check is similar to that of pulverized coal boilers. It includes examination of water/steam system, furnace, back-pass, fans, condensers, oil burner, etc

Bed preparation: bed preparation involves sieving the bed inventory from previous runs to remove the undesired over- or under-sizes. Depending upon the bed condition, fresh bed material of an appropriate size may be used. The bed materials should not be wet as it is hard to fluidize wet solids. The solids should be filled to the designed static height in the bed

Check air nozzles: operating the boiler in cold state can check air nozzles. First, a bed of 100 to 200 mm thick sand or ash (or other bed material) is built above the grid (distributor).

Then the superficial velocity is increased until the bed enters the bubbling region. The bed is watched from the top to see whether the bubbles are uniform. By this way, the plugged nozzles, if any, can be detected and they can be repaired or replaced before the bed is in operation.

7.6.1 Start-Up Procedure

In a typical start-up sequence the following steps may be followed:

1. Add bed materials by mixing them with about 5–10% coal by mass. Start the primary air fan to fluidize the bed evenly, just above the minimum bubbling velocity
2. Switch on the start-up burner fan. Check each system of the boiler
3. Switch on the burner or put a lighted object in the furnace (keeping it burning). When the flame turns orange, increase the primary air slowly. The bed should run evenly in the bubbling region by the end of this stage
4. When the bed temperature reaches 600 to 800°C, depending upon the manufacturer's recommendation, feed fuel into furnace. This temperature is an important parameter and should be fixed from experience. Stultz and Kitto (1992) give some values of minimum bed temperatures for starting the feed of different types of fuel, which are as shown in Table 7.6. A drop in the oxygen or increase in carbon dioxide concentration in the flue gas could be an indication of the onset of ignition and combustion of the fuel. Stop the fuel feed if the oxygen or CO_2 content in the flue gas suggest otherwise
5. The coal feed rate should increase gradually until the bed temperature is in the normal operating range. Shut off the burner; and stabilize the combustion by controlling the feed rate at the designed air flow rate

7.6.2 Control of BFB Boiler

In a power station, the control system is complex because the system controls a host of interconnected equipment including boilers, turbines, generators, etc. The present section briefly discusses the control strategy of the boiler with minimum details. This study will help develop a better understanding of the operation of BFB boilers and might help to improve their design.

Like any standard control problem, a BFB boiler control will have a set of operating variables. The operating variables are divided into two groups: inputs (independent variables) and state

TABLE 7.6
Minimum Bed Temperature for Fuel Feed

Fuel	Minimum Bed Temperature (°C)
Bituminous coal	480–510
Lignite	480
Anthracite	540–560
Mill rejects	550
Wet wood	650–670
Oil or Gas	760

Source: Adapted from Stultz and Kitto, 1992.

variables (dependent variables).

$$\begin{aligned} state\_1 &= function\_1(state\_1, state\_2, \ldots, input\_1, input\_2, \ldots); \\ state\_2 &= function\_2(state\_1, state\_2, \ldots, input\_1, input\_2, \ldots); \\ &\ldots \\ state\_N &= function\_N(state\_1, state\_2, \ldots input\_1, input\_2, \ldots); \end{aligned} \quad (7.38)$$

The state variables can be changed by changing the inputs. For example:
Bed temperature = f(coal flow rate, air flow rate, water/steam flow rate,...)
Bed temperature is a state variable as it can not be changed directly, but coal flow rate is an input, which can be changed directly. Bed temperature needs to be controlled within a small range. When it deviates from its normal range, some inputs should be adjusted to bring it back into the range.

The common state variables in BFB boilers are:

1. Steam pressure
2. Water height in drum
3. Furnace pressure
4. O_2 level, or combustible level
5. Superheater steam temperature
6. Bed temperature
7. SO_2 level

The common inputs in BFB boilers are:

1. Bed height
2. Feedwater flow rate
3. ID damper position
4. Airflow rate by FD damper
5. Spraywater flow rate
6. Coal flow rate
7. Sorbent (limestone) flow rate

A simple, ideal case would involve one input variable for each state variable as follows:

$$\begin{aligned} state\_1 &= f\_1(input\_1) \\ state\_2 &= f\_2(input\_2) \\ &\ldots \\ state\_7 &= f\_7(input\_7) \end{aligned} \quad (7.39)$$

This means that every state variable is function of a single input. In this case, when state_1 needs to be changed, simply adjusting the input_1 will work, provided an adjustment of input_1 does not affect other state variables. However, it may not be that simple in a BFB control system. A state variable may be affected by more than one input as in Equation 7.40. Thus, for operation, a simplified model needs to be developed as follows.

Bubbling Fluidized Bed Boiler

One solution for BFB boiler is following:

$$\text{state\_1} = f\_1(\text{input\_1}) + d\_1(\text{input\_2}, \text{input\_3}, \ldots);$$
$$\text{state\_2} = f\_2(\text{input\_2}) + d\_2(\text{input\_1}, \text{input\_3}, \ldots);$$
$$\ldots$$
$$\text{state\_N} = f\_N(\text{input\_7}) + d\_N(\text{input\_1}, \text{input\_2}, \ldots)$$

(7.40)

where: $f\_1(\cdot)$ means control function and $d\_1(\cdot)$ means disturbance.

State_1 is controlled by input_1, but is affected by other inputs. Thus, other inputs act as a disturbance to state_1. The key to do this is to find the strongest coupling input to this state variable. The other inputs would naturally have a far weaker coupling to the state variable.

By applying this theory to BFB boiler control, the following equations are developed;

$$\text{Steam pressure} = f\_1(\text{bed level}) + d\_1(\text{spray water, fuel flow});$$
$$\text{Drum level} = f\_2(\text{feed water flow}) + d\_2(\text{bed level});$$
$$\text{Furnace Pressure} = f\_3(\text{ID Fan}) + d\_3(\text{FD Fan});$$
$$O_2 \text{ level} = f\_4(\text{air flow}) + d\_4(\text{Bed level, fuel flow});$$
$$\text{SH temperature} = f\_5(\text{spray water}) + d\_5(\text{bed level, fuel flow});$$
$$\text{Bed temperature} = f\_6(\text{fuel flow}) + d\_6(\text{bed level, spray water});$$
$$SO_2 \text{ level} = f\_7(\text{sorbent flow}) + d\_7(\text{Fuel flow});$$

(7.41)

The whole relationship is summarized in Table 7.7, where it can be seen that there are seven control loops from the concept. All of these are close loops with feedback or feed-forward. When a disturbance crops up in a control loop, the equipment is designed to compensate its effect on the state variable in this loop.

The above seven control loops are divided into two groups: primary and secondary.

7.6.2.1 Primary Control Loop of BFB Boiler

Primary loop is the foundation of BFB control. Different power demands require different turbine throttle valve openings Figure 7.16, which will affect the steam pressure. By adjusting the bed

TABLE 7.7
Relationship among State Variables and Inputs for BFB Boilers

	Bed Level	Feedwater Flow	ID Fan	Air Flow by FD Fan	Spray Water	Fuel Flow	Sorbent Flow
Steam pressure	Control	—	—	—	Disturb	Disturb	—
Drum level	Disturb	Control	—	—	—	—	—
Furnace pressure	—	—	Control	Disturb	—	—	—
O_2 level	Disturb	—	—	Control	—	Disturb	—
SH temperature	Disturb	—	—	—	Control	Disturb	—
Bed temperature	Disturb	—	—	—	Disturb	Control	—
SO_2 level	—	—	—	—	—	Disturb	Control

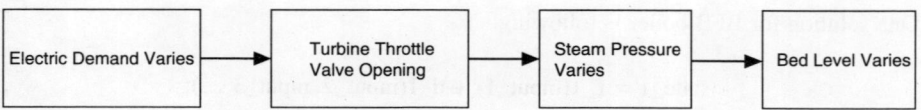

FIGURE 7.16 Primary control loop of a CFB boiler.

drain, the operator can control the bed level necessary to maintain the steam pressure under the changed load.

7.6.2.2 Secondary Control Loop of BFB Boiler

When the bed level is varied to maintain the superheater outlet steam pressure, many other state variables such as drum level, O_2 level, superheater outlet steam temperature and bed temperature also vary simultaneously. All of these state variables are controlled by six secondary loops. The details of these control loops are beyond the scope of this book.

When the adjustment of boiler load is not by bed level, another input needs to be found for this adjustment. The effect on control and disturbance, however, may not be the same as above. The introduction of the control theory of BFB boilers above provides an example. It may vary with the method to control boiler load, but the principles of control are similar to each other.

7.7 SOME OPERATIONAL PROBLEMS AND REMEDIES

Three common operating problems experienced by BFB boiler are:

1. High combustible loss
2. Erosion of refractory or tubes in furnace
3. Feed system malfunction

High combustible heat loss is a design issue. Once the superficial velocity, cyclone size and other parameters are frozen, the combustion efficiency is largely fixed for a specific fuel and its size distribution. Additional combustible losses could result only from improper operation.

Besides reducing the financial return of the plant operational problems influence the lifespan of boiler components. Erosion and feed system malfunction are two of the many operational problems of BFB boilers and are discussed below.

7.7.1 EROSION OF IN-BED COMPONENTS

Parts of the evaporator tube banks are often located in the bed as shown in Figure 7.17 to maximize heat absorption and maintain the bed temperature within desired limits. Unlike CFB boilers, the tubes and their supports in a BFB boiler are subjected to rather severe conditions as they suffer from impingement of hot solids and gas at the worst possible impingement angle. This causes severe erosion of the in-bed tubes.

Erosion also occurs on the caps of distributor nozzles, on which hot solids and air impinge while passing through. At times they are also eroded severely by particle-laden, high-velocity jets from neighboring nozzles. Following options are available for reduction of the erosion of boiler components in a BFB boiler:

FIGURE 7.17 In-bed tubes.

7.7.1.1 Bubble Size Reduction

Shang (1984) showed that in-bed erosion is mainly caused by gas bubbles, not by the abrasion of solids moving around them. The larger the bubble size, the more severe is the damage. An effort to reduce bubble size has been made using techniques such as reducing particle sizes, inducing bed vibration, and using pulse air supply.

7.7.1.2 Low Fluidizing Velocity

Bubbles are generated when the air supply exceeds that required for minimum fluidization. Any air, in excess of that required for minimum fluidization passes through bed as bubbles. Low superficial velocity can reduce the bubble size and decrease the erosion of in-bed tubes. Experimental data (Stringer, 1984) shows that when superficial velocity is lower than 2 m/sec, no erosion appears, but when it is above 2 m/sec, erosion is observed. It must be noted that there have been cases where BFB boilers operating even below 2 m/sec have experienced erosion. From this consideration, the preferred range of superficial velocity should be around 1 to 1.5 m/sec.

7.7.1.3 Vertical Tubes

Glicksman (1984) found that much of the particle movement near the surface of a tube is caused by bubbles generating at horizontal tubes rather than in the bed underneath. Thus, reducing the number of horizontal tubes may decrease the erosion. That would suggest using vertical tubes to replace horizontal tubes when designing the in-bed tube bank. This improvement has been applied to many BFB designs.

7.7.1.4 Fins

Using fins on the tubes is another popular anti-erosion measure. Usually, fins are located on the elbows of tubes and below horizontal tubes (Figure 7.18). Although this is the most effective means of erosion reduction, no universal design is available. For example, erosion-resistant fins that proved very successful in one boiler were found to be less than satisfactory in another. This is due

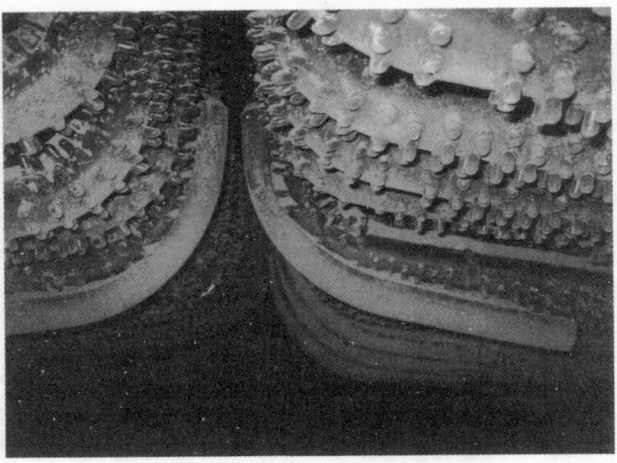

FIGURE 7.18 Shields are attached to the elbows of tubes.

to complex gas–solid motion in bubbling beds which depends on a large number of parameters, making it very difficult to predict their pattern precisely.

7.7.1.5 No In-Bed Tube

The above problem can be removed by avoiding the use of in-bed tubes altogether, as in pulverized coal boilers. The bed temperature in that case is controlled by either of the following two means.

(a) Using adequate excess air to carry the entire combustion heat
(b) Using substoichiometric air to limit the generation of combustion heat in the bed.

The latter option is more feasible for biomass than for fossil fuel. The first option is simple but uneconomic because it results in large flue gas loss. It is effective only in places where fuel cost is very low and combustion efficiency is less important than the boiler availability, like in combustion of waste coal.

The second option is very effective and is extensively used for biomass-fired BFB boilers, where a sizable amount of heat is released in the freeboard. BFB boilers firing very high-ash coal, could also use this option, but it cannot operate with moderate or high heat release in the bed. In that case, one needs to separate the combustion zone from the heat-transfer zone and arrange an efficient solid transfer between them. The heat-transfer zone may be kept in some form of moving bed region where solids move over the tube at a relatively low velocity like the multi-chamber bed shown in Figure 7.15.

Besides the above methods for the mitigation of the erosion problems one could use anti-erosion materials like hard-facing or harder materials. A sacrificial tube shield is another option for prolonging the life of eroding components (Figure 7.18).

7.8 DESIGN OF FEED SYSTEMS FOR BFB BED BOILER

The feed system for a BFB boiler is simple, but accounts for a significant part of the boiler's operational problems. Thus, a rational and reliable design of the fuel-feed system is essential for a BFB boiler. Chapter 10 discusses this aspect in detail.

NOMENCLATURE

A_b:	surface area of the bed exposed to the freeboard (m^2)
A_s:	surface area of tubes immersed in bed (m^2)
C_{air}:	specific heat of air (kJ/kg K)
C_f:	specific heat of dry flue gas (kJ/kg K)
C_g:	specific heat of gas or steam (kJ/kg K)
C_m:	specific heat of moisture/water (kJ/kg)
C_p:	specific heat of particle (kJ/kg K)
E_{sor}:	efficiency of sulfur capture
EAC:	excess air coefficient
e_b:	emissivity of bed
g:	gravity acceleration (m/sec^2)
G_{air}:	air mass flow rate (kg/sec)
G_{flue}:	flue gas mass flow rate (kg/sec)
G_g:	flue gas leaving the bed (kg/sec)
G_s:	mass flow rate of steam/water in a boiler (kg/sec)
h_{en}:	enthalpy of water/steam (kJ/kg)
h:	heat transfer coefficient (kW/m^2 K)
H_0:	enthalpy of water (kJ/kg)
H:	hydrogen content in coal (−)
HHV:	higher heating value of coal (kJ/kg)
H_{bed}:	height of expanded bed (m)
H_{en}:	enthalpy of water and steam in heat exchangers (kJ/kg)
H_{evap}:	heat transfer coefficient between bed and evaporator (kW/m^2 K)
H_{inlet}:	steam/water enthalpy at the inlet of a heat exchanger (kJ/kg)
H_{outlet}:	steam/water enthalpy at the outlet of a heat exchanger (kJ/kg)
$H_{freeboard}$:	height of freeboard (m)
$H_{furnace}$:	total height of furnace (m)
H_i:	enthalpy of primary air (kw)
H_{fl}:	enthalpy of fuel and its moisture (kw)
H_{Tb}:	enthalpy of steam at bed temperature (kJ/kg)
$H_{sh,inlet}$:	steam enthalpy at inlet of superheater (kJ/kg)
$H_{sh,outlet}$:	steam enthalpy at outlet of superheater (kJ/kg)
$h_{w,sat}$:	saturated water enthalpy (kJ/kg)
H_2O:	moisture content in coal (%)
L_{ash}:	heat loss from ash drain (%)
L_{cr}:	convective–radiative heat loss (%)
L_{fan}:	fan credit (%)
$L_{m,air}$:	heat loss moisture from air (%)
$L_{m,fuel}$:	moisture heat loss from coal (%)
$L_{m,h}$:	moisture loss from burning hydrogen in coal (%)
L_{others}:	sum of other heat losses except moisture losses and stack loss (%)
L_{stack}:	dry flue gas heat loss (%)
L_{total}:	total heat loss (%)
L_{uc}:	unburned carbon loss (%)
m_a:	primary air rate (kg/sec)
m_c:	fuel feed rate (kg/sec)
m_s:	sorbent feed rate (kg/sec)
M_f:	surface moisture fraction in coal (−)
M_s:	moisture fraction in sorbent (−)

M_{da}:	stoichiometric dry air for burning 1 kg coal
M_{df}:	dry flue gas produced by burning 1 kg coal
M_{flue}:	flue gas generated when 1 kg fuel is burned
M_{wa}:	wet air required for burning 1 kg fuel
n:	molecular mass of flue gas (kg/kmol)
P:	pressure of water/steam (Pa)
P_b:	pressure above distributor (Pa)
P_{drum}:	pressure in the drum (Pa)
P_e:	pressure at the exit of furnace (Pa)
$P_{econ,inlet}$:	pressure at the inlet of economizer (Pa)
$P_{econ,outlet}$:	pressure at the outlet of economizer (Pa)
ΔP_{econ}:	pressure drop in economizer (Pa)
P_s:	pressure between bed and freeboard (Pa)
P_{sh}:	superheater inlet pressure (Pa)
ΔP_{sh}:	pressure drop in the superheater (Pa)
$P_{sh,outlet}$:	steam pressure at outlet of final superheater (Pa)
q_g:	grate firing rate (MW/m^2)
$q_{in\text{-}bed}$:	heat transfer rate to in-bed tubes (kW)
q_{wall}:	radiative heat transfer rate to wall tubes in freeboard (kW)
$Q_{airheater}$:	heat duty of air heater (kW)
Q_{comb}:	combustion heat release rate (kW)
Q_{econ}:	heat duty of economizer (kW)
Q_{dr}:	bed drain loss (kW)
Q_{evap}:	heat duty of evaporator (kW)
Q_{fg}:	enthalpy of flue gas leaving bed (kW)
Q_h:	heat absorbed by in-bed tubes (kW)
Q_i:	heat released in bed (kW)
Q_{latent}:	latent heat of vaporization (kJ/kg)
Q_{load}:	heat load of a heat exchanger (kW)
Q_r:	radiative heat lost from bed surface (kW)
Q_{sh}:	heat load of superheater (kW)
Q_{SH}:	superheat enthalpy of moisture (kJ/kg)
Q_{steam}:	heat transferred to steam through economizer, evaporator and SH (kW)
R:	universal gas constant
S:	pitch of the tubes (mm); S=minimum (S_1, S_2).
T:	température (°C)
T_a:	ambient temperature (°C)
T_{after}:	flue gas temperature after passing the heat exchanger (°C)
T_{air}:	air temperature after air heater (°C)
T_b:	temperature of fluidized bed (°C)
T_{bed}:	fluidized bed temperature in the furnace (°C)
T_{before}:	flue gas temperature before passing through the heat exchanger (°C)
$T_{econ,inlet}$:	feedwater temperature at inlet of economizer (°C)
$T_{evap,inlet}$:	water temperature at the inlet of evaporator (°C)
T_f:	temperature of flue gas leaving the stack (°C)
T_{fb}:	freeboard temperature (°C)
T_{fur}:	furnace exit gas temperature (°C)
T_g:	average temperature of water in tube (°C)
T_{high}:	heat source temperature (°C)
T_i:	preheat of primary air (°C)
T_{sat}:	saturated temperature under the drum pressure (°C)

T_{sh}: flue gas temperature after leaving superheater (°C)
$T_{sh,outlet}$: steam temperature at outlet of final superheater (°C)
T_{stack}: stack temperature (°C)
Ts_{tube}: tube surface temperature, $\cong T_{sat}++30$, (°C)
T_w: temperature of tube surface (°C)
U_b: average bubble rise velocity (m/sec)
U_{mf}: minimum fluidization velocity of gas (m/sec)
U_f: superficial velocity through the Bubbling fluidized bed (m/sec)
U_{27}: superficial velocity referred to 27°C (m/sec)
X_{ash}: ash fraction in fuel ($-$)
x_d: fraction of total ash appearing as bed drain ($-$)
X_b: fraction of combustion in bed ($-$)
X_m: moisture content in supplied air ($-$)

Greek Symbols

ε: voidage of bed
ε_{mf}: voidage of bed at the minimum fluidization
η: boiler efficiency (%)
ρ_b: bulk density of bed (kg/m^3)
ρ_f: flue gas density (kg/m^3)
ρ_p: particle density (kg/m^3)
σ: Boltzman constant, 5.7×10^{-8}

REFERENCES

Anthony, E. J., Becker, H. A., Code, R. K., McCleave, R. W., and Stephenson, J. R., *Bubbling Fluidized Bed Combustion of Syncrude Coke*, Proceeding of Ninth International Conference on Fluidized Bed Combustion, Vol. 1, Mustonen, J. P., Ed., ASME, New York, pp. 322–329, 1987.

Anthony, E. J., Desai, D. L. and Friedrich, F. D., *Pilot-Scale Combustion Trials with Onakawana Lignite Phase II: Fluidized-Bed Combustor*, Canada Centre for Mineral and Energy Technology, May, 1981.

Basu, P., Kefa, C. and Jestin, L., *Boilers and Burners Design and Theory*, Springer, New York, 2000.

Basu, P. and Fraser, S. A., *Circulating Fluidized Bed Boilers Design and Operations*, Butterworth–Heinemann, London, 1991.

Duqum, J. N., Tang, J. T., Morris, T. A., Esakov, J. L., and Howe, W. C., *AFBC Performance Comparison for Under-Bed and Over-Bed Feed System*, Proceedings of Eighth International Conference on Fluidized-Bed Combustion, pp. 255–277, 1985.

Geisler, O. J., Heinrich, F. and Urban, U., *Operational Experience with a Fluidized Bed Boiler Burning Waste Material*, Proceeding of Ninth International Conference on Fluidized Bed Combustion, Vol. 1, Mustonen, J. P., Ed., ASME, New York, pp. 378–384, 1987.

Glicksman, L. R., Heat Transfer in Fluidized Bed Combustors, In *Fluidized Bed Boilers: Design and Application*, Basu, P., Ed., Pergamon Press, Toronto, pp. 63–100, 1984.

Greenfield Research Incorporated, *An innovative method for reduction in unburnt carbon in bubbling fluidized bed combustor*, Patent applied for, 2004.

Jai, S. C. and Sang, K. L., *Operating Experience in Fluidized bed Boilers with Deep/Shallow Bed in Korea*, Proceeding of Ninth International Conference on Fluidized Bed Combustion, Vol. 1, Mustonen, J. P., Ed., ASME, New York, pp. 424–429, 1987.

Keairns, D. L., Newby, R. A. and Ulerich, N. H., Fluidized-bed combustor design, In *Fluidized Bed Boilers: Design and Application*, Basu, P., Ed., Pergamon Press Canada Ltd, New York, 1984.

Keyes, M.A., Derreberry, B.C., and Kaya, A., Application of systems engineering to control fluidized-bed boilers, *Fluidized Bed Combustion and Applied Technology — The First International Symposium*, pp. III-147–III-167, 1984.

Keyes, M.A. and Kaya, A. Applications of systems engineering to control fluidized-bed boilers, *Fluidized Bed Combustion and Applied Technology — The First International Symposium*, Hemisphere Publishing Corp., New York, p. III-147, 1984.

Makino, K. and Seki, M., Experiences of IHI/FW 31.5-t/h commercial fluidized-bed boiler, *Fluidized Bed Combustion and Applied Technology — The First International Symposium*, Hemisphere Publishing Corp., New York, p. V-27, 1984.

Oka, S., *Fluidized Bed Combustion*, Marcel Dekker, New York, p. 439, 2004.

Plautz, D. A. and Johnstone, H. F., Heat and mass transfer in packed bed, *AIChE J.*, 1, 193–199, 1955.

Rajaram, S. and Malliah, K. T. U., *A 10 MWe Fluidized Bed Power Plant for Paddy Straw, Proceedings of Ninth International Conference on Fluidized-Bed Combustion*, Mustonen, J. P., Ed., p. 392–397, 1987.

Shang, J., An overview of fluidized-bed combustion boilers, In *Fluidized Bed Boilers: Design and Application*, Basu, P., Ed., Pergamon Press, New York, pp. 1–12, 1984.

Silvennoinen, J., *A New Method to Inhibit Bed Agglomeration Problems in Fluidized Bed Boilers, Proceedings of 17th International Conference on Fluidized-Bed Combustion*, Pisupati, S., Ed., ASME, New York, 2003, paper no FBC2003-081.

Singer, J.G. Combustion-Fossil Power, ABB Combustion Engineering, Connecticut, pp. 6–11, 1991.

Stultz, S.C. and Kitto, J.B., Steam — its generation and use Babcock & Wilcox, 40th edition, pp. 16–17, 1992.

Stringer, J., Materials selection in atmospheric fluidized bed combustion systems, In *Fluidized Bed Boilers: Design and Application*, Basu, P., Ed., Pergamon Press, New York, pp. 155–180, 1984.

Verhoeff, F., *Design and Operation of the 115 T/H FBC-Boiler for AKZO-Holland, Proceeding of 9th International Conference on Fluidized Bed Combustion*, Vol. 1, Mustonen, J. P., Ed., ASME, New York, pp. 61–69, 1987.

Werther, J., Saengen, M., Hartge, E.-U., Ogada, T., and Siagi, Z., Combustion of agricultural residues, *Prog. Energy Combust. Sci.*, 26, 1–27, 2000.

Yagi, S. and Kunii, D., Studies on Heat Transfer from Wall to Fluid in Packed Bed, *AIChE J.*, 6, 97–104, 1960.

8 Circulating Fluidized Bed Boiler

The circulating fluidized bed (CFB) boiler is a member of the fluidized bed boiler family. It has gained popularity, especially in the electric power-generation market, for its several practical advantages (see Chapter 1), such as efficient operation and minimum effect on the environment. Although it entered the market only in the 1980s, CFB technology is well beyond its initial stage of development. The technology has matured through successful operation in hundreds of units of capacities ranging from 1 MWe to 340 MWe (until 2005). The problems of the first generation have been solved and CFB is now considered to be a mature technology for atmospheric-pressure units. Its design methodology, however, is not as well-established as that of pulverized coal-fired boilers. Many aspects of its design are still based on rules of thumb. The present chapter describes different aspects of the circulating fluidized bed boiler including a brief outline of a design approach.

8.1 GENERAL ARRANGEMENT

A CFB boiler may be divided into two sections: the CFB loop and the convective or back-pass section of the boiler (Figure 8.1). The CFB loop consists of the following items making up the external solid recirculation system.

1. Furnace or CFB riser
2. Gas–solid separation (cyclone)
3. Solid recycle system (loop-seal)
4. External heat exchanger (optional)

Figure 8.1 shows the general arrangement of a typical CFB boiler without the external heat exchanger, while Figure 8.2 shows the same for one with the heat exchanger.

The back-pass is comprised of:

1. Superheater
2. Reheater
3. Economizer
4. Air heater

The following section describes the working of the boiler, tracing the path of air, gas, solids and water through it.

8.1.1 AIR SYSTEM

The air system is very important for the CFB boiler, as it consumes the greatest amount of power. A typical utility CFB boiler as shown in Figure 8.2 would use three types of fan/blowers:

1. Primary air fan
2. Secondary air fan
3. Loop-seal air fan or blower

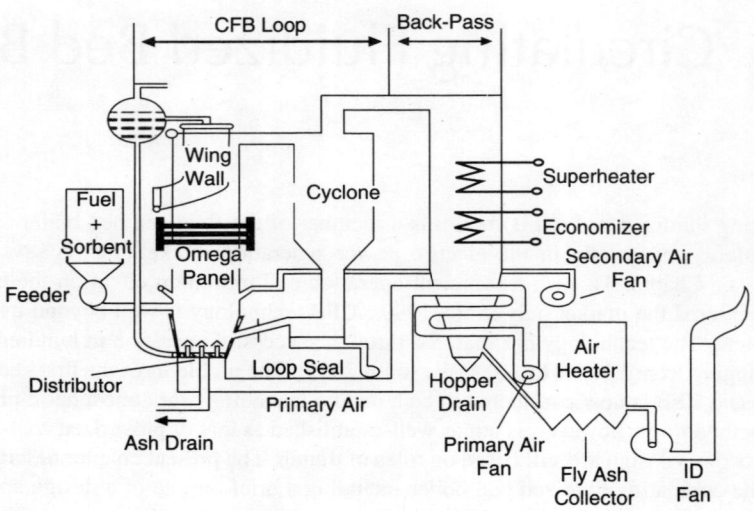

FIGURE 8.1 General arrangement of a typical circulating fluidized bed boiler.

The primary air fan delivers air at high pressure (10 to 20 kPa). This air is preheated in the air preheater of the boiler and then enters the furnace through the air distributor grate at the bottom of the furnace.

The secondary air fan delivers air, also preheated in the air preheater, at a relatively low pressure (5 to 15 kPa). It is then injected into the bed through a series of ports located around the periphery of the furnace and at a height above the lower tapered section of the bed. In some boilers, the secondary air provides air to the start-up burner as well as to the tertiary air at a still higher level,

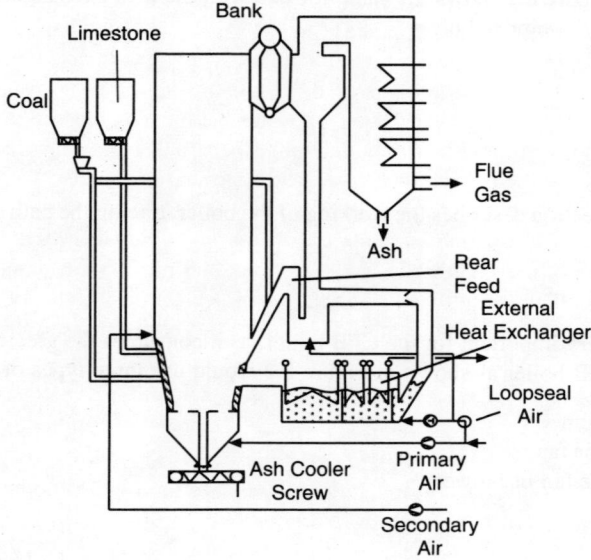

FIGURE 8.2 Air and feed circuit of a circulating fluidized bed boiler with an external heat exchanger.

if needed. The secondary air fan may also provide air to the fuel feeder to facilitate the smooth flow of fuel into the furnace.

Loop-seal blowers deliver the smallest quantity of air but at the highest pressure. This air directly enters the loop-seals through air distribution grids. Unlike primary and secondary air, the loop-seal air is not heated.

8.1.2 FLUE GAS STREAM

Generally, only one suction fan is used to handle the flue gas in a CFB boiler. This fan, called *induced draft* (ID) fan, creates suction in the system to draw flue gas from the boiler and through the dust control or any other gas emission-control equipment. The suction head of the ID fan is designed to have a balanced draft in the air/flue gas system with zero (or atmospheric) pressure at the mid or the top section of the furnace. This helps keep the boilerhouse clean and at the same time optimizes the power consumption by the ID fan.

8.1.3 SOLID STREAM

Fuel from the bunker drops on to a belt or some other type of feeder, which then feeds measured quantities of fuel into the fuel chute. In most large CFB boilers, the fuel chute feeds the fuel into the loop-seal's inclined pipe (Figure 8.2). Here, the fuel mixes with hot solids recirculating around the CFB loop, and therefore enters the bed better dispersed. Other boilers either take the fuel directly into the lower section of the bed through the front wall or use another conveyor to take it around the furnace for sidewall feeding.

The sorbent is generally finer than the fuel, so it is carried by conveying air and injected into the bed through several feed injection points. As sorbents react very slowly, the location of their feed points is not as critical as that for the fast-burning fuel.

The ash or spent sorbent is drained from the boiler through the following points:

1. Bed drain
2. Fly ash collection hopper under the fabric filter or electrostatic precipitator
3. Economizer or back-pass hopper

In some cases, ash is also drained partially from the external heat exchanger. In the case of a coarse bed drain, the ash is cooled by air or water before it is disposed of. The fly ash, being relatively cold, can be disposed of without cooling. Its particles are generally smaller than 100 μm with a mean size of 30 μm and are, therefore, easily carried pneumatically into a fly ash silo, where they are hauled away by truck or rail wagon as necessary.

The mixture of fuel, ash, and sorbents circulate around the CFB loop. Particles, coarser than the cyclone cut-off size, are captured in the cyclone and recycled near the base of the furnace. Finer solid residues like ash or spent sorbents, generated during combustion and desulfurization, escape through the cyclones. These are collected by the fabric filter or electrostatic precipitator located further downstream.

8.1.4 WATER–STEAM CIRCUIT

Figure 8.3 shows the water–steam flow circuit through a typical CFB boiler. Here, one can detect the following heat transfer surfaces in the boiler:

- Economizer in the back-pass
- Evaporator in the furnace wall

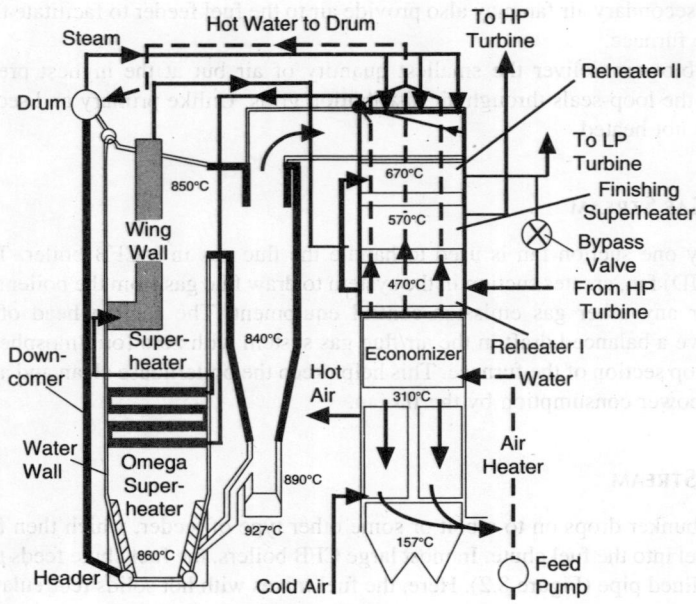

FIGURE 8.3 Water and steam circuit of a circulating fluidized bed boiler without an external heat exchanger.

- Superheaters in both the back-pass and furnace
- Reheaters in both the back-pass and furnace

A CFB boiler could locate parts of the superheater and reheater in an external heat exchanger as shown in Figure 8.2.

8.1.4.1 Economizer

The boiler feed pump feeds the water into the economizer located in the back-pass or convective section of the boiler (Figure 8.3). The economizer is a conventional shell-tube heat exchanger that uses the waste heat of the flue gas to preheat water. The water is forced through the economizer to flow directly to the drum. Water enters the cooler section and leaves from the hotter upper section of the economizer making it a counterflow heat exchanger. The temperature of the water leaving the economizer is generally kept at least 28°C below the saturation temperature of the water to ensure good circulation. Some high-performance boilers allow steam formation, but considering the possibility of nonuniform flow distribution between tubes, flow instabilities and other factors the rise in enthalpy in the economizer should be guided by the following equation (Stultz and Kitto, 1992):

$$H_2 - H_1 \geq \frac{2}{3}[H_f - H_1], \tag{8.1}$$

where H_1, H_2, and H_f are enthalpies of water entering the economizer, leaving the economizer and at saturated condition at the economizer outlet pressure, respectively.

The water velocity through the economizer is typically in the range of 600 to 800 kg/m²s and gas velocity is in the range of 7 to 15 m/s.

8.1.4.2 Evaporators

In a typical subcritical boiler, the water flows down large-diameter unheated pipes (known as *downcomers*) into distributing manifolds called *headers*. The header distributes water amongst vertical tubes rising along the walls of the furnace. Water rises through these tubes and hence they are called *riser*, or *water wall* tubes. To make an airtight enclosure around the furnace, these tubes are generally welded together by means of fins between them in the form of panels.

As the water rises up the tubes it absorbs heat from the furnace, converting part of it into steam. Hot water, carrying steam bubbles, leaves the top of the water wall panels and is collected in headers, which in turn carry it to the steam drum. Steam is separated from the water in the drum, which mixes with fresh water from the economizer and flows down through the downcomer and into the riser for heating again.

Sometimes four walls of the furnace cannot provide sufficient surface area to carry the entire evaporative load of the boiler. Additional surfaces are provided in the form of wing walls in the furnace (Figure 8.3) or in the form of bank tubes downstream of the furnace to take this load.

8.1.4.3 Superheaters and Reheaters

Figure 8.3 shows the arrangement of reheaters and superheaters in a typical CFB boiler. Saturated steam from the drum flows through a set of tube panels forming the walls of the back-pass. Then it goes to the *omega superheater* panels inside the furnace. These tubes are formed from a special steel section that, when joined, gives a flat vertical surface to minimize the erosion potential. The partially-heated steam then rises up through wing wall tubes as shown (Figure 8.3) and passes through the final superheater located in the back pass. Such a complex back-and-forth tube arrangement helps minimize the cost of tubes while minimizing any risk of tube overheating. Steam temperature can be controlled by spraying water into the steam at appropriate locations.

Low pressure steam enters the reheater section immediately upstream of the economizer (Figure 8.3). It then passes through the final reheater section upstream of the final superheater. One may use a bypass valve between the entry and exit of the reheater section to control the steam temperature.

8.2 TYPES OF CFB BOILERS

Numerous designs of CFB boilers are available in the market, some of which are more common than others. The following are four major types of CFB boiler designs:

1. Boilers with vertical, hot cyclones with or without in-furnace heating surfaces (Figure 8.1)
2. Boilers as above, with bubbling fluidized bed heat exchanger parallel in the CFB loop (Figure 8.2)
3. Boilers with impact or inertial-type separators (Figure 8.4)
4. Boilers with vertical, noncircular, cooled cyclones (Figure 8.5)

8.2.1 BOILERS WITHOUT BUBBLING BED HEAT EXCHANGERS

This is the most popular type and belongs to the first generation CFB boilers that entered the market in the 1980s. The furnace is connected by way of an expansion joint to a thick, refractory-lined, vertical, hot cyclone, which feeds the collected solids to a loop-seal. The loop-seal returns the solids to the furnace. Several expansion joints are used at different sections to compensate for the differential expansion between the cooled furnace and uncooled cyclone-loop-seal circuit as shown by Figure 9.14 in Chapter 9. Following types of in-furnace surfaces are used if needed to meet the demand for required furnace heat absorption:

- Wing wall (also called platen) (Figure 8.1)
- Omega tube panel (Figure 8.3)
- Division wall

258   Combustion and Gasification in Fluidized Beds

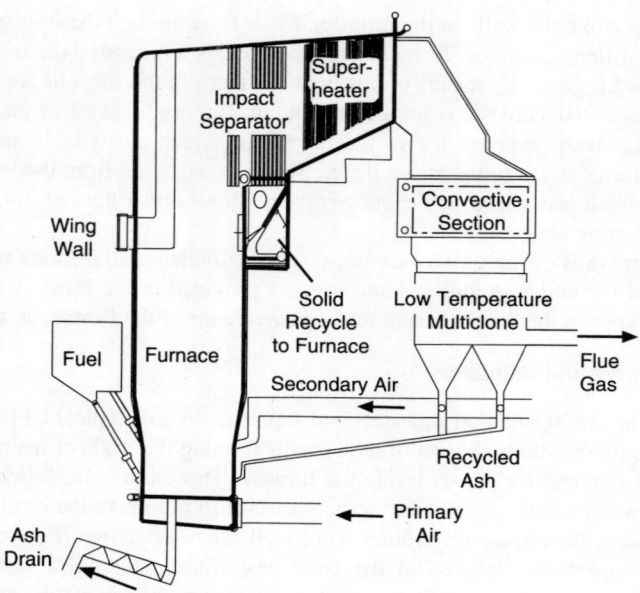

FIGURE 8.4 Arrangement of a CFB boiler with impact separators.

8.2.2 Boilers with Bubbling Fluidized Bed Heat Exchanger

The flue gas needs to be cooled down to the required temperature (800 to 900°C) before it leaves the CFB loop. In large boilers (>100 MWe) the furnace walls alone cannot absorb this heat, so additional surfaces like wing walls are required. Such surfaces do not give the flexibility of control

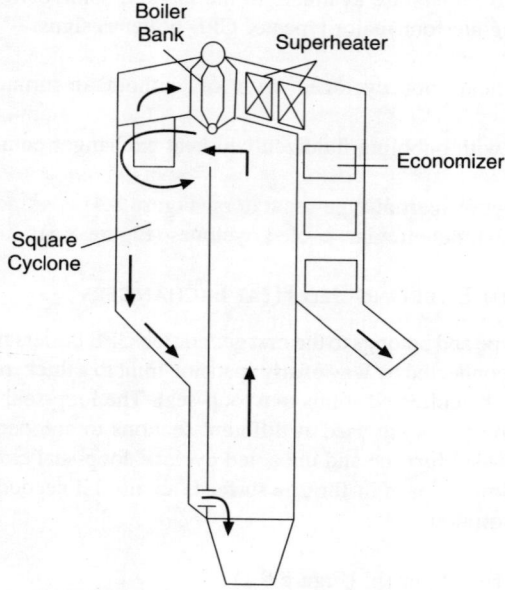

FIGURE 8.5 CFB boiler with a vertical noncircular cyclone.

of heat absorption, which may be required for partial load operation or for burning alternative types of fuel. For this reason, a bubbling fluidized bed heat exchanger as shown in Figure 8.2 is used in the CFB loop in this type of boiler. It is placed in parallel to the solid recycle line between the loop-seal and furnace. A part of the solid stream from the loop-seal is diverted through the bubbling fluidized bed heat exchanger. Boiler heat-absorbing tubes are located in the fluidized bed to absorb heat from the hot solids circulating through it. By regulating the amount of solids diverted through it, solid flow through the heat exchanger is easily controlled. Two type bubbling fluidized beds are used:

1. External heat exchanger located outside the furnace (Figure 8.2)
2. Internal heat exchanger located in the furnace.

8.2.3 Boilers with Inertial or Impact Separators

In order to avoid the high cost of hot cyclones an alternative type of gas–solid separator is used by this type of CFB boiler as shown in Figure 8.4. Here, the solids are separated through impact against a row of U-shaped flow barriers. Such separators are located partially in the furnace and partially outside it. They are not as efficient as centrifugal-type cyclones, so an additional multiclone or other type of gas–solid separator is required downstream of the back-pass. Solids from these separators are also recycled to the furnace. Compactness is a major feature of such boilers.

8.2.4 Boilers with Vertical, Noncircular Cyclones

This type of boiler is also known as *compact* design. Here, a geometric-shaped (square or octagonal) separator chamber is formed by boiler tubes covered with a thin refractory (Figure 8.5). Circular gas exits are located on the roof of these chambers. Gas–solid suspension from the furnace is made to enter the separator chamber through tangential entry points. Such entries create horizontal vortices, which separate the solids in the chamber and allow the gas to leave from the top.

8.2.5 Other Types

In addition to the above, many of other types of CFB boilers are available in the market and are generally used in smaller-sized units. An important type is the innovative Cymic© design shown in Figure 8.6. Here the gas–solid separator and the standpipe are located in the centre of the furnace, with risers around it. Gas–solid suspension enters the central cyclone through a number of tangential vanes, forming a vortex. The solids drop into the central standpipe while gas leaves from the top. The collected solids move to the riser through openings at the bottom of the standpipe as shown in Figure 8.6. This design is very compact and needs less refractory because it makes greatest use of heating surfaces. Large boilers can be built with multiple central tubes in a rectangular riser chamber.

8.3 NON-CFB SOLID CIRCULATION BOILERS

CFB boilers involve circulation of solids at a sufficiently high rate to create the *fast fluidized* bed condition in the furnace. Most of the CFB's positive features stem from this special hydrodynamic condition. A number of alternative designs involving solid circulation are available where the solid circulation rate is not sufficiently high to create the fast fluidized condition, but they enjoy some other benefits such as low cost. The present section describes a few such boilers, which are not CFB boilers but still use circulation of solids.

8.3.1 Circo-Fluid Boilers

This boiler is essentially a bubbling fluidized bed with high level of solid circulation (Figure 8.7). The furnace walls of the boiler are formed by water-cooled membrane tubes The lower section of

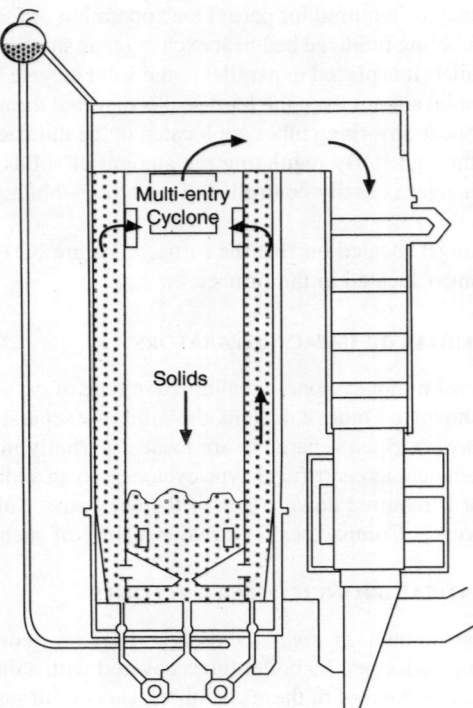

FIGURE 8.6 A novel design of CFB boiler with central multientry cyclone.

the furnace includes a bubbling fluidized bed with a freeboard above it. One part of the convective heating surfaces is located in the upper part of the furnace and the remaining parts are arranged in the back-pass. The first and the second pass are connected by a cyclone separator at a flue gas temperature 300 to 500°C. The solids captured are reintroduced to the combustion chamber by the siphon seal as shown in Figure 8.7.

8.3.2 Low-Solid Circulation Boiler

This is essentially a BFB boiler with recycling of fly ash. Figure 8.8 shows a sketch of a low-solid circulation boiler. The lower section of the boiler operates essentially in the bubbling fluidization regime. Several U-beam separators, located above, capture the solids leaving the bed. A special solid recycle valve (H-valve) recycles the solids at a controlled rate. From Table 8.1, we observe that the combustion efficiency improves with increased solid recycle rate (Yang et al, 1996). A major attraction of this boiler is its low power consumption due to its relatively dilute bed. A number of such boilers are being used in China.

8.3.3 Circulating Moving Bed Boiler

This is a relatively new but promising concept, where combustion and heat transfer are decoupled, allowing them to be independently optimized. The process is shown schematically in Figure 8.9. Coal burns in a bubbling fluidized bed at a relatively high temperature ($\sim 1100°C$), releasing hot gas into the freeboard above it (Jukola et al., 2005). Heat-carrying particles (like bauxite) rain from the

Circulating Fluidized Bed Boiler

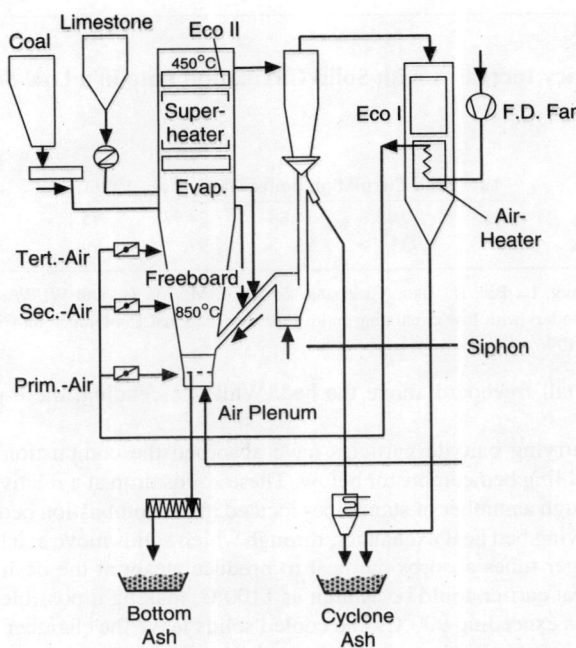

FIGURE 8.7 Circo-fluid boiler is a non-CFB type of boiler involving solid recirculation.

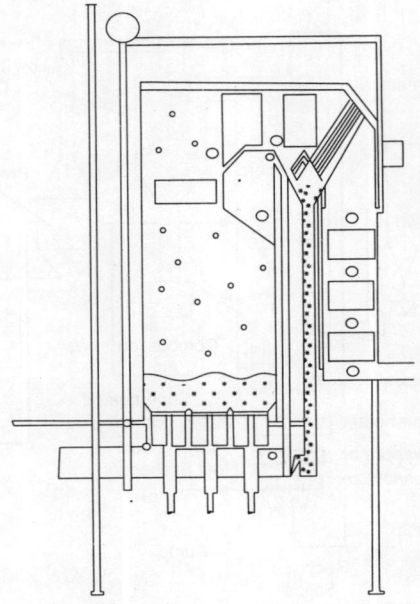

FIGURE 8.8 General arrangement of a low-solid circulation boiler.

TABLE 8.1
Combustion Efficiency Increases with Solid Circulation Rate in a Low-Solid Circulation Boiler

	Low-Solid Circulation Boiler (Yang et al., 1996)					Bubbling bed boiler (Tang and Engstrom, 1987)		
Circulation rate/Feed rate	7.1	10.48	14.88	24.98	75.1	0	1.0	1.6
Combustion efficiency %	90	93	95	97	99	85	96.5	98.5

Source: Adapted from Yang, L., Bie, R., Bao, Y., Zhang, Z., Zhao, M., Lu, H., and Wi, W., Design and operation of circulating fluidized bed boilers with low circulating ratio, Fifth International Conference on Circulating Fluidized Beds, Beijing, Paper no. CG10, 1996.

top of the relatively tall freeboard above the bed. While descending, these particles absorb heat from the flue gas.

Once the heat-carrying bauxite particles have absorbed the combustion heat they collect at the bottom of the bubbling bed combustor below. These solids drop at a relatively low velocity into a lower chamber through a number of standpipes located in the combustion bed. The lower chamber is a counter-flow, moving bed heat exchanger, through which solids move at a low velocity while an array of heat-exchanger tubes absorbs the heat to produce steam at the desired temperature. The temperature of the heat carrier could be as high as 1100°C, making it possible to produce steam or hot air at temperatures exceeding 900°C. The cooled solids leave the chamber from the bottom and are lifted to the top to start the cycle again. The flue gas leaving the freeboard passes through a low-temperature cyclone to remove any unburned fuel and fine dust, which are returned to the bed. The remaining heat in the flue gas is used to preheat the combustion air in an air preheater.

Combustion air is added in stages as shown in Figure 8.9 for better NOx control. A back-end desulfurizer unit is used for SO_2 control, the lime for which is calcined in the combustor.

This process has the potential for producing less expensive heat and being a better-performing boiler. It could also be used to capture CO_2 using the *chemical looping* concept.

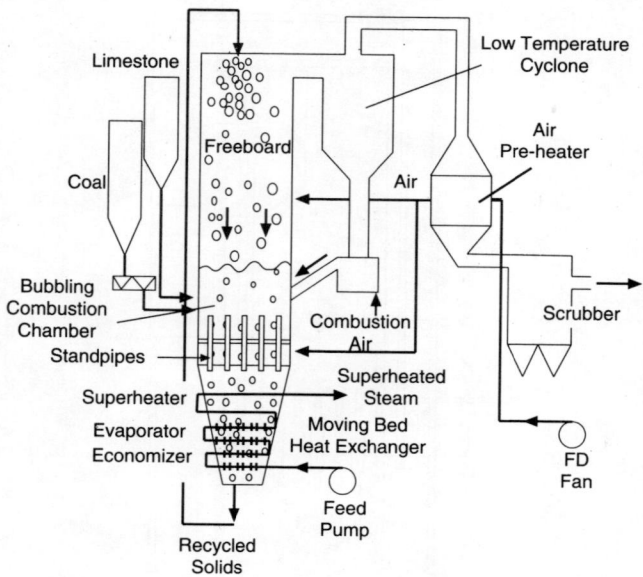

FIGURE 8.9 Schematic arrangements of a novel circulating moving bed boilers.

Circulating Fluidized Bed Boiler 263

8.4 COMBUSTION IN A CIRCULATING FLUIDIZED BED FURNACE

A detailed description of the basics of the combustion of fuel is presented in Chapter 4. Thus, the present section will describe only how the combustion of fuel could affect the furnace of a CFB boiler.

Although CFB boilers are fuel-flexible, the fuel composition exerts some effect on the furnace size for optimum performance. The effect of fuel type is, however, not as dominant as in the case of pulverized coal-fired boilers. An investigation into the effect of fuel types on the commercial design of CFB furnaces using commercial software (CFBCAD)$^{©}$ shows that the lower heating value (LHV) of coal is the main parameter influencing furnace size (Lafanechere et al., 1995a; 1995b). Figure 8.10 shows that as the design fuel moves from low-volatile bituminous to lignite, the furnace cross-sectional area increases by approximately 200% to keep the furnace velocity constant. The figure also shows how the optimum length/width ratio of the furnace would change along with the cyclone arrangements.

Fuel feed points are chosen to ensure a uniform distribution of burning particles. A more reactive fuel with higher-volatile content burns very rapidly near the feed port. Thus, it would require more feed points than a less reactive fuel. Table 10.4 presents design characteristics of several commercial CFB boilers. It shows that the bed area served by a feeder varies. The general rule is that more reactive bituminous coal and coal/peat need less feeders per unit bed area. For example such low-rank fuels use one feeder per 14 to 17 m^2 bed area while high-rank or less reactive petroleum coke and anthracite require only one feed point per 20 to 26 m^2 bed area. Some newer plants like the 262-MWe CFB plant at Turrow with eight points uses one feed point to serve about 27 m^2 grid area or generate about 93 MW heat. This is likely to give rise to nonuniform distribution of oxygen.

The combustion temperature of a CFB boiler is considerably lower than that of a conventional boiler. It should ideally be within the range of 800 to 900°C. High-sulfur fuels such as petroleum coke and some types of coal must be burned at around 850°C for optimum sulfur capture, while low-sulfur, low-reactivity coal such as anthracite culm should be burned at a higher temperature and/or with excess air for good combustion efficiency. Besides improved combustion efficiency, high temperatures also help reduce the emission of the greenhouse gas N$_2$O. Thus, the fuel characteristics set the optimum combustion temperature, which will in turn fix the amount of energy leaving the furnace. For example, a low-grade fuel will carry a high percentage of the generated

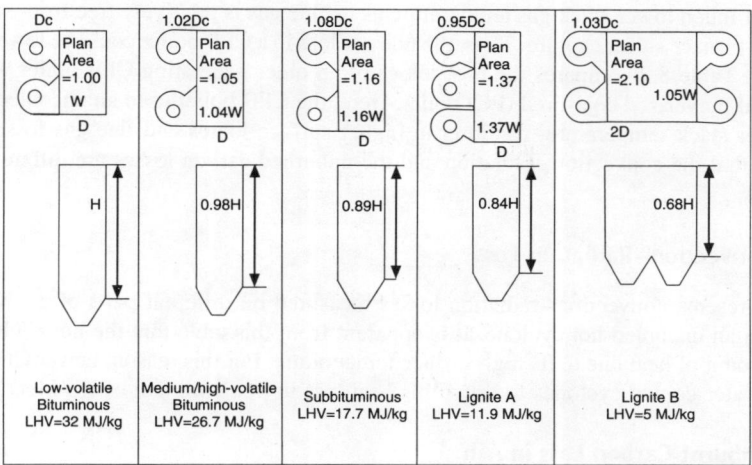

FIGURE 8.10 Influence of fuel type on the optimum size of CFB boilers.

heat out of the furnace. Less heat will thus be required to be absorbed in the CFB loop, and more heat will be absorbed in the convective section of the boiler.

8.5 DESIGN OF CFB BOILERS

Followings are some major steps involved in the design of CFB boilers:

1. Combustion calculations
2. Heat and mass balance
3. Furnace design
4. Heat absorption
5. Mechanical component design
6. Design for combustion and emission performance

The present section will discuss the above steps in sequence and will finally illustrate the design methodology through an example.

8.5.1 COMBUSTION CALCULATIONS

Combustion calculations (also known as stoichiometric calculations) provide the basic quantities of mass of gas and solids on which subsequent designs are based. It could determine the amount of air required to burn a unit mass of fuel and the amount of flue gas produced. Combustion calculation also helps specify the capacities of equipment, such as fans, feeders, and ash-handling systems. The equations for stoichiometric calculations and a detailed discussion are given in Appendix 2.

8.5.2 HEAT BALANCE

Only a fraction of combustion heat is utilized for the generation of steam. This fraction or the thermal efficiency of the boiler is calculated by carrying out a heat balance around the boiler. The heat balance of CFB boilers, though similar to that of other types of boilers, is not exactly the same. It requires some special considerations.

In a conventional pulverized coal-fired boiler the flue gas temperature is restricted by the dew point of the sulfur dioxide in the flue gas. A fluidized bed boiler, on the other hand can be designed for much lower stack gas temperature as its flue gas is relatively free from sulfur dioxide due to its in-furnace sulfur capture. Thus, the flue gas loss in a CFB boiler could be lower than that in a PC boiler. Table 8.2 compares the heat losses of an older generation CFB boiler with that of a conventional pulverized coal-fired (PC) boiler. Here, the CFB boiler used an uncooled hot cyclone and a higher stack temperature, resulting in higher surface losses and flue gas losses. Table 8.2 also shows that the convection–radiation and the unburned carbon losses are different in PC and CFB boilers.

8.5.2.1 Convection–Radiation Loss

Table 8.3 presents convection–radiation losses measured on different parts of a 110-MWe CFB boiler using an uncooled hot cyclone. It is apparent from this table that the hot cyclone loses the greatest amount of heat due to its high surface temperature. For this reason, newer CFB boilers use steam- or water-cooled cyclones to bring this heat loss in line with that of PC-fired boilers.

8.5.2.2 Unburnt Carbon Loss in Ash

This constitutes the combustible loss, which is a function of a number of design and operating factors. Combustible losses can be in the form of unburned hydrocarbons and carbon monoxide.

TABLE 8.2
Comparison of Heat Losses in a Typical Pulverized Coal-Fired Boiler with that in an Older Generation Circulating Fluidized Bed Boiler

Items	PC (%)	CFB (%) Optimistic	CFB (%) Pessimistic
Moisture in limestone	—	0.06	0.10
Calcination	—	1.02	1.69
Sulfation credit	—	⟨−1.60⟩	⟨−1.60⟩
Unburned carbon	0.25	0.50	2.00
Heat in dry flue gas	5.28	5.57	5.60
Moisture in fuel	1.03	1.03	1.03
Moisture from H_2 burning	4.16	4.19	4.19
Radiation and convection	0.03	0.30	0.80
Moisture in air	0.13	0.14	0.14
Sensible heat in boiler ash	0.03	0.09	0.76
Bottom ash	0.05	—	—
Fan power credit	⟨−0.25⟩	−0.75	⟨−0.40⟩
Pulverized credit	⟨−0.20⟩	—	—
Total loss	10.81	10.55	14.31

Source: Adapted from Gould, G. L., and McComas, M. W., *Power*, January, 39–40, 1987.

However, unburned solid carbon represents the major fraction. Available data from commercial CFB plants estimate that unburned carbon in the bottom ash is about 0.2 to 4.0% and 4.0 to 9.0% in fly ash (Lee et al., 1999). Taller furnaces with efficient cyclones usually have lower combustible losses. The unburned carbon loss U_{cl}, in percentage can be calculated as:

$$U_{cl} = \frac{X_c \times W_a \times 32790 \times 100}{HHV} \tag{8.2}$$

where X_c is the fractional carbon (not as carbonate) in the solid waste, and W_a is the ash per unit mass of the fuel feed.

TABLE 8.3
Heat Loss from a 110-MWe Circulating Fluidized Bed Boiler for an Average Boiler Building Temperature of 27°C

Components	Heat Loss (kW)	Area (m²)	Heat Flux (W/m²)	Temperatures Surface	Air (°C)
Cyclone and loop-seal	482	1241	374	89	44
Combustor	394	1804	212	78	43
Drum and piping	173	1355	123	67	42
Primary air duct	134	1610	80	56	36
Air heater	11	1408	76	51	36
Convection pass	98	862	109	63	44
Secondary air duct	47	537	84	58	39

Total loss = 0.44% of a fuel heat input of 327 MW.
Source: Recalculated from Jones, P. A., Hellen, T. J. and Friedman, *Proceedings of the 10th FBC Conference*, Manaker, A., Ed., ASME, New York, pp. 1047–1052, 1989.

8.5.3 MASS BALANCE

The mass balance specifies important items such as the division of fuel ash and spent limestone between the particulate collectors and bed drain. This also requires special attention in a CFB boiler, especially with sulfur-capture capability.

Fuel and sorbents are the two principal types of solids fed into a CFB furnace. Part of the fuel appears as the gaseous products SO_2 and CO_2. The remainder of the fuel appears as a waste product at various locations in the bed. Sometimes supplemental bed materials are added to maintain the solid balance. The input and output of solid streams in a CFB boiler are listed below:

Solid Input	Solid Output
Fuel	Drain from bed
Sorbent	Drain from back-pass
Make-up bed material	Drain from external heat exchanger
	Drain from bag-house/precipitator
	Solids leaving through stack
	Boiler/economizer/air heater hopper dust fallout

Knowledge of each component of the input and output solid stream is necessary for a proper design of the solid feed/drainage system and hopper/discharge chute. An overdesign of the ash discharge system results in waste of space and higher capital cost. An underdesign leads to overloading, and perhaps total failure of the solid handling system. It is thus important to know how the total solid waste produced in the combustor is distributed amongst different solid streams.

8.5.4 FLOW SPLIT

8.5.4.1 Division of Ash between Bottom Ash and Fly Ash

The ash and spent sorbents released in the furnace are extracted partially from the bed and partially from the flue gas. The size of the bag-house or electrostatic precipitator is specified on the basis of the expected solid load and particle size distribution. Similarly, the bed drains are designed according to the expected load of solids passing through them. The split of coal ash depends on the nature of ash, their attrition behavior in the bed and the cyclone performance. Coal ash could reside in the coal matrix in two forms, extraneous or intrinsic.

The intrinsic ash is extremely fine and is dispersed throughout the coal matrix. The extraneous ash includes the rock or discrete ash particles embedded in the coal structure. When coal burns down or fragments, the discrete ash particles are exposed and dislodged from the burning face of the char. The intrinsic ash, being fine, is progressively released as the coal burns and usually escapes the furnace to be collected in the bag-house. The extraneous ash, being large, generally concentrates near the bottom of the furnace and comes out as bed drain.

The mechanical attrition of sorbents and extraneous ash is a major contributor to the fine solids escaping the furnace. Hence, the attrition properties of the bed solids and the nature of the ash in the fuel influence the split of solids between the bed drain and the bag-house. The split is also governed by the performance of the cyclone or any other gas–solid separation device used.

General rule calls for 20 to 80% of the solid waste pass through the bed drain. Boilers firing high-ash coal without sorbent injection have higher bed drain. For lignite and subbituminous coals, the amount of ash retained in the bed as bottom ash is substantially smaller. About 30 to 80% passes through the bag-house (Gould and McComas, 1987). For the purpose of conservative design, the collection equipment for fine dusts may be designed for the disposal of 80% of the solid wastes. Similarly, the bed drain may be designed for 50% of the total wastes. These norms can be fine-tuned once more information on the ash split is available.

TABLE 8.4
Distribution of Primary and Secondary Air at Different Loads in an Anthracite-Fired 200-MWe CFB Boiler

Load	100%	75%	50%	30%
Coal feed rate (kg/s)	27.3	20.7	14.5	7.9
Primary air (m³/s)	87.2	76.3	65.5	68.3
Secondary air (m³/s)	53.1	29.7	29.7	29.7
Loop-seal + EHE (m³/s)	9.14	9.14	9.14	7.34

8.5.4.2 Division of Combustion Air between Primary and Secondary Air

In a CFB combustor, a significant part of the combustion air (40 to 60%) is injected into the bed as secondary air to reduce the NOx emission (see Section 5.3). To limit the formation of NOx, especially for highly volatile fuels, the lower zone of a CFB combustor is operated in a substoichiometric condition, meaning that the primary air must be less than the stoichiometric (or theoretical) amount. Table 8.4 gives the ratio of primary to secondary air used by a 200-MWe CFB boiler at different loads.

8.5.5 CONTROL OF BED INVENTORY

Maintenance of a stable inventory of solids is very important for the stable operation of a CFB boiler. The bed inventory needs to be maintained in terms of both quantity and quality. Quantity means the total mass of solids in the furnace while quality implies the size distribution of particles in the bed. This problem is especially acute in boilers without limestone feed and/or that are firing low-ash fuels like biomass. The following section discusses these two operational aspects of a CFB boiler.

8.5.5.1 Maintenance of Bed Inventory

Sometimes, the escape of solids from the cyclone exceeds the feed of ash and sorbent to the furnace, leading to a net depletion of solid inventory in the furnace. Following two options are available to maintain the inventory:

- Fly ash reinjection
- Addition of make-up bed materials

The fly ash reinjection system continuously recycles to the furnace part of the solids collected in the bag-house (or electrostatic precipitator). A reinjection system, besides maintaining the bed inventory improves combustion efficiency and sulfur capture to some extent and allows the use of a slightly less efficient cyclone, but it does increase the erosion potential of the back-pass tubes due to a higher dust loading.

Silica sand or similar bed materials are added, as required, to the combustor to maintain its inventory. This need is greatest with very low-ash, low-sulfur coal or nonash fuels such as wood.

8.5.5.2 Bed Quality

The correct amount and size of bed materials is crucial for maintaining both hydrodynamic and heat transfer characteristics of a CFB boiler. The size distribution of the particles in the furnace depends, among other things, on the following parameters:

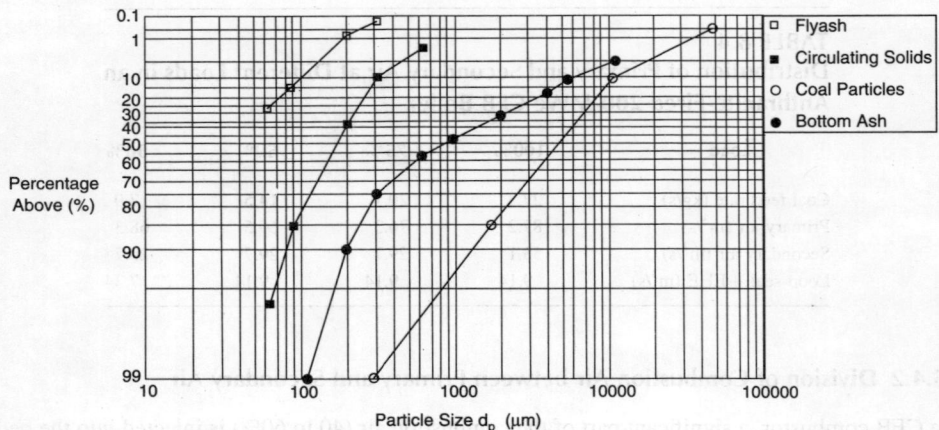

FIGURE 8.11 Size distributions of coal particles and different solid streams in a large CFB boiler.

- Feed size distribution and its friability
- Sorbent size distribution and its friability
- Grade efficiency of gas–solid separator

Figure 8.11, taken from a large commercial plant, shows typical distributions of particle sizes of fuel and sorbent and resulting solids circulating through the furnace and those in the bed drain and fly ash. Figure 8.12 shows the effect of particle size on the axial distribution of the bed density.

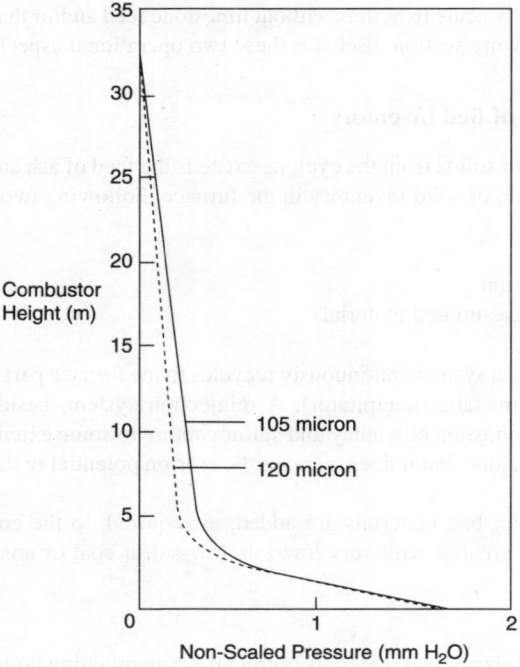

FIGURE 8.12 A change in the average particle size changed bed density profile of a CFB furnace.

Circulating Fluidized Bed Boiler

In this case, the design of the cyclone was changed, which in turn reduced the fraction of fine solids in the furnace. In another case, cyclone modification in a 450 t/h CFB boiler reduced the d_{50} particle size of the furnace from 180 to 80 μm (Lalak et. al., 2003). This increased the average suspension density in the upper regions because finer bed solids result in higher bed densities. Higher suspension density gives a higher heat-transfer coefficient (see Section 6.2.2) in the upper furnace, even though the overall pressure drop or bed inventory in the entire furnace is the same.

One important means for maintaining the bed quality is the use of an ash classifier like the one shown in Figure 8.13. A fluidized bed cooler is used in some boilers to drain the coarse particles and to return fines to the bed. The ratio of particles above and below the design size decreases with increasing classifier velocity. Thus, by varying this velocity, the operator can change the size distribution of solids in the bed.

The control of the inventory of fine bed solids W_{bi}, of a specific size i, is guided by the loss through entrainment, bottom ash, and reduction to finer sizes, and gain due to fresh ash feed and reduction of coarser sizes.

$$\frac{dW_{bi}}{dt} = G_{i,\text{input}} + W_{>bi,\text{decrepitation}} - G_{i,\text{bottomash}} - W_{<bi,\text{decrepitation}} - G_{i,\text{cyclone}} + W_{i,\text{classifier}} \quad (8.3)$$

where $G_{i,\text{input}}$, $G_{i,\text{bottomash}}$, and $G_{i,\text{cyclone}}$ are the mass flow rates of solids of size d_i, entering the bed, leaving through the bed drain, and cyclone respectively. $W_{i,\text{classifier}}$ is the mass of particles of size d_i returned to the bed. $W_{>bi}$ is the amount of solid entering size i through the reduction of larger sizes, and $W_{>bi}$ is the amount of solid leaving size i through reduction of particles of size i. For stable operation of the furnace, the rate of change of bed solids of size d_i must be zero:

$$\frac{dW_{bi}}{dt} = 0$$

Both the grade efficiency of the cyclone and the design of the bed drain have major influences on control of the bed inventory and its size distribution.

There is considerable segregation of particle size in a CFB furnace especially when no sorbents are used. Coarser particles tend to congregate near the bottom of the furnace, while finer particles are entrained out of the cyclone as fly ash. The captured part of the entrained bed solids circulate through the bed and around the primary loop.

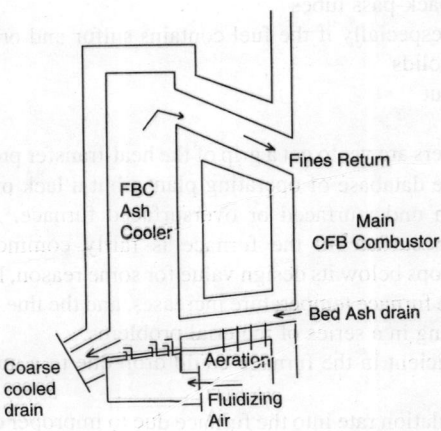

FIGURE 8.13 Ash classifier-cum-cooler for cooling and extracting coarse ash particles from the bed and returning the fines to the bed.

8.5.5.3 Effect of Feed on Bed Quality

The size distribution of the fuel and limestone fed into the bed could greatly affect the bed quality under certain situations. For example, with a low-ash coal (ash 25 to 30%), the feed size is less important from a heat-transfer point of view, just because the composition of the bed would be governed by the size of limestone. With high-ash coal, the feed size should be reasonably small so that the size of bed ash is not too high. Also coarser high-ash coal results in greater bottom ash production and higher heat losses (lower boiler efficiency), because of the higher temperature at which it leaves the system (unless there is a fluidized cooler to recover the heat into the system). This results in some negative effect on the boiler efficiency.

The particle size of the feed does not necessarily directly determine the size distribution of the bed solids. The size distribution of the ash in the feed (or primary ash size) matters most. A fine coal with large extraneous ash content may give bed materials coarser than that from a coarser coal with large intrinsic ash content. This may be measured easily by burning a sample of the fuel in a furnace and measuring size distribution.

The initial limestone sizing is much more important, especially with low-ash fuels like coal, wood, and sludge. Limestone and its reaction products contribute significantly to the quantity of ash circulating in the furnace. Loop-seal surging may occur due to excessive feed of limestone required for whatever reasons.

8.5.6 SOME OPERATING ISSUES

Like any equipment CFB boiler also experience some operating problems. The following are two important problems faced by many operating units.

8.5.6.1 High Furnace Temperature

Although the recommended furnace temperature of a CFB boiler is 800 to 900°C, and it is designed to operate in this range, a very large number of boiler furnaces operate at temperatures well in excess of 900°C. This results in a series of problems including the following:

1. Large increase in limestone consumption if sulfur capture is required
2. Increased fly ash burden and ash disposal cost
3. Higher erosion of back-pass tubes
4. Corrosion of tubes especially if the fuel contains sulfur and or chlorine
5. Agglomeration of solids
6. Drop in boiler output

CFB boiler manufacturers are yet to get a grip of the heat-transfer process in CFB furnaces. They make use of their extensive database of operating plants, but a lack of good understanding of the process often results in an undersurfaced or oversurfaced furnace. Alteration of the originally manufactured heat-transfer surface in the furnace is fairly common. When the heat-transfer coefficient in the furnace drops below its design value for some reason, less steam is generated in the boiler. As a result of this the furnace temperature increases, and the flue gas temperature entering the back-pass increases, resulting in a series of material problems.

The heat-transfer coefficient in the furnace could drop due to a number of factors:

1. A drop in solid circulation rate into the furnace due to improper operation of the loop-seal
2. Increased coarser particles in the bed solids circulating due to coarser fuel or sorbent feed
3. A reduction in cyclone grade efficiency for finer solids

4. A drop in bed inventory
5. Too optimistic design values of the heat-transfer coefficient

The following options are available for the mitigation of high furnace temperature problems:

1. Addition of a proper amount of heating surface areas in the furnace
2. Cyclone modification to enhance the collection efficiency of finer particles
3. Improved design of the loop-seal
4. Change grinding system to get a better size distribution of fuel or limestone

It was described earlier in Chapter 2, that finer particles move more easily to the upper furnace than coarser particles would. Thus for a given overall bed pressure, a furnace can have a leaner upper bed and a denser lower bed depending upon the size distribution of the bed solids. Werther (2005) showed by taking bed ash samples from an operating CFB boiler that a certain fraction of bed solids has a terminal velocity below the operating velocity in the CFB riser.

8.5.6.2 Agglomeration of Bed Materials

Agglomeration of bed materials could create a number of problems including flooding of the cyclone. This could happen if agglomerated particles prevent the free flow of solids through the loop-seal, forcing them to accumulate in the standpipe and the cyclone, leading to complete choking.

Petroleum coke-fired boilers see this problem more often than do coal-fired units. For this reason, many plants are forced to blend a minimum amount of coal with the cheaper petcoke fuels. The exact mechanism for agglomeration is not fully understood at the moment.

This problem can be avoided to some extent by designing the loop-seal liberally without leaving a potential dead spot in the EHE or loop-seal where particle mobility may be restricted.

Detailed measurements (Werther, 2005) in some commercial CFB boilers detected unusually high temperatures ($>1000°C$) at a certain height in the furnace. In another case a large increase in the temperature of solids across the cyclone has been observed, implying a large amount of combustion in hot cyclones. Both conditions could give rise to the agglomeration of bed solids in the circulating solids, which could plug the loop-seal.

8.6 FURNACE DESIGN

The following section discusses major considerations in the determination of basic sizes of a CFB boiler furnace, which include:

1. Furnace cross section
2. Furnace height
3. Furnace openings

Both the ash and moisture content of the fuel have major effects on the mechanical design of the boiler. Figure 8.14 and Figure 8.15 illustrate the effect of ash and moisture content of the fuel, respectively, on the solid and gas flow rates for a given thermal output of the boiler.

Figure 8.14 implies a minor effect of the ash content of fuel on the flue gas volume and hence on the major dimensions of the furnace, cyclone, convective pass, dust collection systems, forced-draft (FD), and induced-draft (ID) fans. However, the handling system for fuel and ash must conform to the requirement of additional solids loading for high-ash fuels.

The moisture content of fuel increases the flow rate of both the flue gas and the fuel. The higher gas volume per unit heat input for a high-moisture fuel will require the size of the furnace

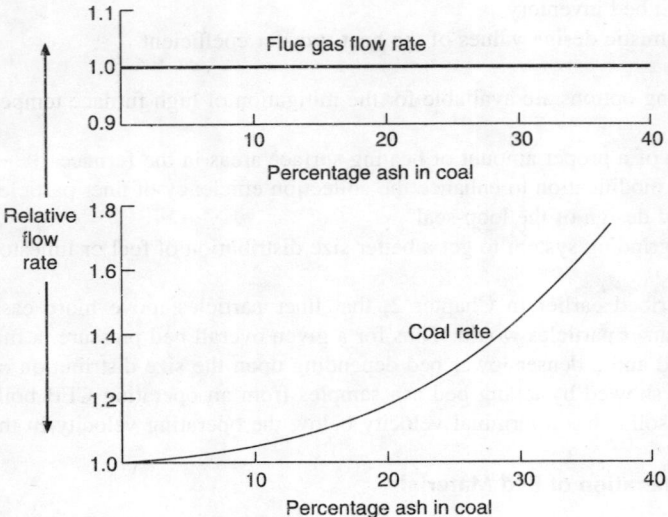

FIGURE 8.14 Effect of ash content of coal on flow rate.

and hot cyclone to be greater. The bed area will also need enlargement if it is designed on the basis of the superficial velocity. In addition, the flow area through convective tubes in the back-pass must be increased to maintain a safe gas velocity in this pass. This reduces the fuel flexibility of the boiler.

The sulfur content in the coal and the limestone characteristics may not affect the furnace dimensions, but the limestone bunker, feeder, and solid-waste disposal system will be influenced by these characteristics. In addition to the ash and moisture content of the coal, its other

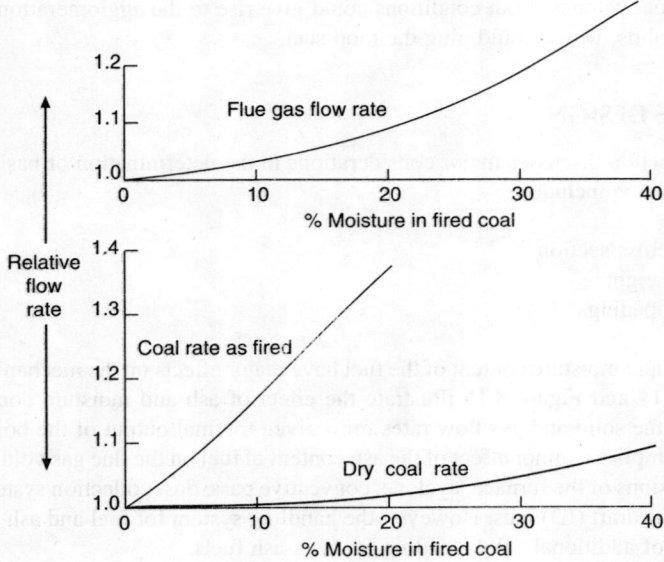

FIGURE 8.15 Effect of coal moisture on flow rates.

Circulating Fluidized Bed Boiler

татLE 8.5
Effect of Coal Properties on the Design and Performance of a CFB Boiler

Coal Property	Design Parameter	Affected Performance
Friability	Cyclone grade efficiency	Boiler efficiency, carbon carryover
Reactivity	Airflow distribution, feed point allocation	Boiler efficiency, carbon carryover, CO emission
Inherent vs. extraneous ash	Ash removal, ash split, heat absorption surface design	Ash carryover, bed drain, dust collector loading
Ash chemistry	Ash removal, back-pass flow area, heating surface design	Bed agglomeration, tube fouling
Moisture	Heat absorption, dimensional requirements, capacity of cyclone and downstream equipment	Thermal efficiency, excess air
Heating value	Dimensional requirements	Capacity, thermal efficiency
Sulfur content	Sorbent handling equipment	Emission, bed drain requirement

properties, such as reactivity, mechanical attrition, ash properties, sulfur content, and heating value also have an important influence on the overall design and performance of a CFB boiler. The effects of various properties of the fuel on the performance of the boiler are illustrated in Table 8.5. The most important parameter affecting the size of a CFB boiler is the LHV as seen in Figure 8.10.

8.6.1 Furnace Cross-Section

Unlike pulverized coal-fired boilers, the cross-sectional average velocity of a CFB boiler needs to be checked carefully. To derive the maximum benefit of the *fast fluidized bed* process, the bed needs to be operated within certain limits of velocity and circulation rates, as discussed in Chapter 2. The following section describes how different operating parameters affect the furnace cross section. Table 8.6 presents the grate heat release rates of some large commercial CFB boilers along with the type of fuel fired. Although the grate heat release depends to some degree on the fuel properties and the manufacturer's preference, the grate heat release rate is now generally of the order of 3.0 to 4.5 MW/m^2 in the upper section of the bed. The volumetric heat release rate is not a major design criterion in a CFB boiler. Table 8.6, however, computes this and it is found to favor the range of 0.08 to 0.15 MW/m^3. The danger of erosion and the high fan-power requirement may be the reason for such a low grate heat release rate. The general norm for the avoidance of furnace erosion is the use of a fluidization velocity that does not exceed 5 m/s.

8.6.2 Width and Breadth Ratio

For a given cross-sectional area the furnace can have different shapes or breadth-to-width aspect ratios. CFB boiler furnaces generally have a rectangular cross section and are made of water wall panels. Three major considerations may influence the aspect ratio:

1. Heating surface necessary in the furnace
2. Secondary air penetration into the furnace
3. Solid feeding/lateral dispersion

TABLE 8.6
Volumetric and Grate Heat Release Rates of a Number of Commercial CFB Boilers

Plant	Electrical Output MWe	Upper Bed Area (m^2)	Furnace Height (m)	Grate Release Rate ($MWth/m^2$)	Volume Release Rate (MW/m^3)	Fuel
Sheng Li Pan, China	26 Mwth	8	16.5	4.08	0.25	Stone coal
Henan Pingdingshan, China	12	13	20.0	2.99	0.15	Subbituminous coal
Jonesboro and West Endfield, USA	25	18	24.0	3.92	0.16	Waste wood and chips
Stadwerke Pforzheim, Germany	25	23	25.0	3.05	0.12	Hardwood
Bayer, Germany	35	22	31.0	4.52	0.15	Hard coal + brown coal
Sinarmas Pulp, India	40	43	34.0	2.65	0.08	Coal
GEW Koln, Germany	75	53	31.0	4.04	0.13	Brown coal
Duisberg, Germany	96	64	30.5	4.31	0.14	Bituminous coal
Nisco Co-Generation, USA	100	78	34.1	3.66	0.11	PetCoke + bituminous coal
Fenyi	100	95	35.7	3.00	0.08	Bituminous
Northampton, USA	110	90	30.4	3.49	0.11	Culm-anthracite
GIPCL, India	125	81	30.0	4.39	0.15	Lignite
Akrimota, India	125	104	37.9	3.45	0.09	Lignite
Emile Huchet, France	125	95	33.0	3.78	0.11	Coal/slurry
Point Aconi, Canada	190	135	30.5	4.02	0.13	Subbituminous coal, PetCoke
Tonghae, Korea	200	133	32.0	4.30	0.13	Anthracite
Turrow, Poland	235	209	42.0	3.21	0.08	Brown coal
Alholmens, Finland	240	206	40.5	3.34	0.08	Bark, wood residue, peat
Turrow, Poland	262	217	42.0	3.43	0.08	Brown coal
Provence, France	250	170	37.0	4.20	0.11	Subbituminous coal
Jacksonville, USA	300	171	34.7	5.00	0.14	Coal, PetCoke

Furnace heat input, where unavailable, is found by assuming generation to heat input ratio as 0.35. Grate heat release rate is based on the cross-sectional area of the upper bed.

Circulating Fluidized Bed Boiler

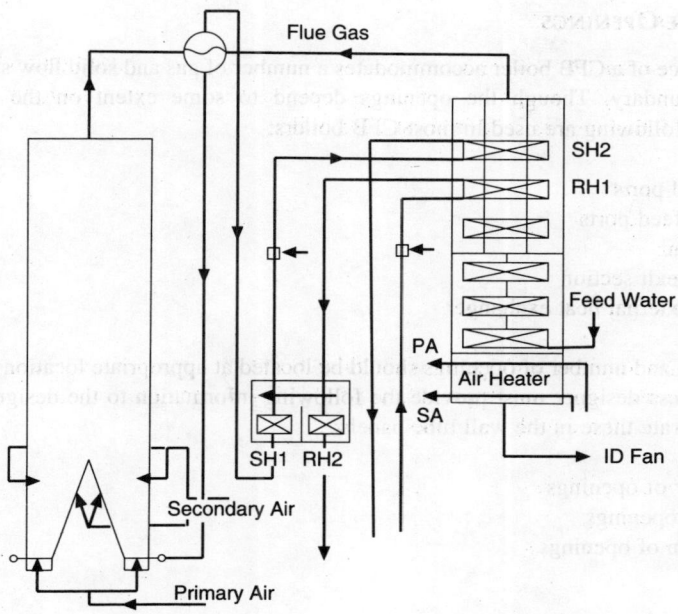

FIGURE 8.16 Pant-leg design of the lower section allows secondary air injection into the interior of a very large CFB boiler.

The width (or depth) of the furnace should not be so great as to result in a poor penetration of the secondary air into the furnace and nonuniform dispersal of volatile matter. Though no set rule is available at present, operating units may provide a good guide to the selection of this ratio. The depth of a single-bed furnace does not usually exceed 7.0 m. This may also be the current upper limit of furnace width from the secondary air injection consideration. For very large boilers of 200 MWe or above, this limitation would make the furnace uneconomically narrow. To avoid this problem, *pant-leg* designs are used for the lower furnace. Here like a pair of trousers, two narrow legs meet into one common section (Figure 8.16). Such designs also allow the feeding of fuels from the centre aisle promoting good mixing in the case of very large boilers.

8.6.2.1 Design of the Lower Furnace

Primary air enters through the floor of the lower furnace to fluidize the bed. To keep the bed moderately fluidized, even under a low load, most designers use a narrow cross section. This helps maintain similar superficial velocities above and below the secondary air level under all operating conditions, and thus minimizes the risk of agglomeration under a low load. In addition, the walls are tapered upwards. In most fluidized bed boilers there is a segregation of particles, especially near the grid, where coarser particles tend to accumulate. The downward narrowing (Figure 8.16) helps to increase the superficial velocity near the grid, minimizing the chance of segregation and consequent clinker formation. The taper angle is decided from experience.

The other important consideration is the minimum pressure drop required for stable fluidization, especially under part-load conditions. This needs to be reflected in the design of the grate or distributor.

For reasons explained later, the walls of the lower furnace are lined with refractory. Some designers use an expanded lower section instead of a contracted one to allow bubbling fluidization in the lower furnace, but this design is falling from favor.

8.6.3 FURNACE OPENINGS

The lower furnace of a CFB boiler accommodates a number of gas and solid flow streams crossing the furnace boundary. Though the openings depend to some extent on the manufacturer's preference, the following are used in most CFB boilers:

1. Fuel feed ports
2. Sorbent feed ports
3. Bed drain
4. Furnace exit section
5. Exit to external heat exchanger

Proper sizes and number of openings should be located at appropriate locations on the furnace walls. The process designer must provide the following information to the designers of pressure parts to incorporate these in the wall tube panels:

- Number of openings
- Size of openings
- Location of openings

8.6.3.1 Fuel Feed Ports

CFB boilers require considerably fewer feed points than bubbling bed boilers due to the higher lateral mixing capability and deeper beds of the former type. This aspect is discussed in detail in Section 10.4.1

8.6.3.2 Location of Primary and Secondary Air Ports

The primary air is generally fed through the bottom of the air distributor supporting the bed. It has to overcome the resistance of the air distributor and that of the lower section of the bed, which is the densest (100 to 200 kg/m^3). Primary air thus needs a high-head fan. However, one can reduce the auxiliary power consumption of the boiler by injecting part of the primary air above the distributor, and the other part, which is just adequate to keep the bed materials above minimum bubbling fluidized state ($U > U_{mf}$), through the distributor. The first part of the primary air is injected into the bed through a set of low-resistance air nozzles set on the furnace walls around the distributor. The design of a CFB boiler calls for substoichiometric conditions below the secondary air injection level. This part is refractory-lined to prevent corrosion and heat loss.

The secondary air is usually fed in above the refractory-lined lower section of the bed. The addition of secondary air greatly increases the superficial gas velocity through the upper furnace. The bed density is, thus, much lower above this level resulting in a reduced level of back-mixing of solids than that in the denser bed below the secondary air level. A deeper lower bed offers a longer residence time of carbon fines for better combustion, but at the expense of a higher head for the primary air fan.

A complete understanding of the hydrodynamics of mixing of the secondary air jet with the fast bed is still lacking. The optimum number of secondary air ports for a given furnace geometry is difficult to assess. However, the secondary air ports may be conservatively located along the wider side of the furnace cross section, assuming that the secondary air will penetrate the depth of the furnace within a reasonable height. The penetration depth is a function of the furnace height. For furnace heights in the range of 15 to 20 m, the depth does not usually exceed 5 m, but for 20 to 30 m it can be as deep as 7.5 m. For this reason, large beds use the pant-leg design shown in Figure 8.16.

Circulating Fluidized Bed Boiler

8.6.3.3 Solid Entry for Circulating Solids

Typical dust loading of the flue gas exiting the CFB boiler furnace is approximately 7.5 kg solid/kg gas (Mathew et al., 1983) or lower. This ratio depends on the hydrodynamic condition and bed density at the top of the bed, which can be a low as 1 to 4 kg/m^3, but is rarely in excess of 30 kg/kg of air (suspension density 10 kg/m^3) (Herbertz et al., 1989). Solids are separated from the flue gas in a cyclone or impact separator. Collected solids are returned to the furnace through a solid entry port. To extend the residence time of carbon and unreacted sorbent particles in the furnace, the solid entry port is located below the secondary air port, i.e., into the lower refractory-lined section of the furnace. The density of this zone of the furnace being much greater than that in the upper section provides a greater refluxing and hence longer residence time.

8.6.3.4 Solid Drain Valve

The bed drain, which extracts solids from the lowest section of the bed, serves two purposes:

1. Maintain the required inventory of solids in the bed
2. Maintain the size distribution of solids

To perform the first task, a mechanical (or nonmechanical) valve receives a signal from a sensor, measuring pressure drop across the bed, which opens or closes the valve as necessary to maintain the required bed inventory.

The drain valve should ideally be located at the lowest point in the bed, preferably through the distributor. The drainpipe needs to be designed such that it ensures a free passage of solids through it. Sometimes, bed drains are provided on the furnace walls very close to the distributor level.

8.6.3.5 Height of the Primary Zone

The purpose of the primary zone in the furnace is to gasify and pyrolyze fresh coals. It also serves as a thermal storage device. A large volume of this zone helps in the completion of pyrolysis of the coal and helps to increase the thermal stability of the bed under varying loads. Earlier designs used primary zones as deep as 7 m, but the present trend is towards a shorter height, with 3 to 6 m being fairly common. Since the primary zone is much denser than the secondary, its depth has a direct bearing on the fan power. To reduce the fan power requirement, designs had to compromise the virtues of a deeper primary zone. The primary zone, however, cannot be made too shallow because it needs room for entry points of many solid and airflow streams. The possibility of corrosion under an oxidizing–reducing environment and the thermal storage feature of the primary zone of the CFB combustor prevent it from containing any heat-transfer surfaces.

8.7 DESIGN OF HEATING SURFACES

The first step in the design of heating surfaces is determining the heat duties of different components of the boiler heating surfaces. A typical boiler would use the following four types of heating surfaces:

- Economizer
- Evaporator
- Superheater
- Reheater (for reheat boilers)

Their locations in a CFB boiler are shown in Figure 8.3. Heat duty of these elements depends on the designed steam parameter of the boiler. It is best illustrated by an example (Figure 8.17), which shows how the relative heat duty of different boiler elements changes with steam pressure. As the

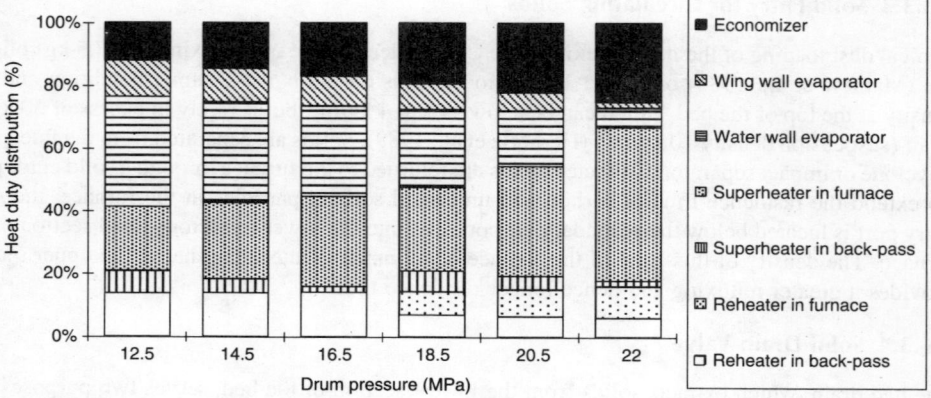

FIGURE 8.17 Variations in heat duty distribution with drum pressure for boilers without external heat exchangers.

steam pressure increases, the heat duty of the evaporator decreases and that of superheater increases. At low pressure the evaporator duty is so high that a water wall or wing wall alone cannot absorb the required amount of heat. So, a separate heating surface, called *bank tubes*, is needed.

After the heat duties of individual elements (economizer, evaporator, superheater, and reheater) are determined by the steam table, their disposition can be determined. From the viewpoint of heat absorption, a CFB boiler may be divided into two regions, the CFB loop and back-pass, (Figure 8.1).

Primary Loop

The CFB loop includes the furnace, cyclone/impact separator, loop-seal, and external heat exchanger as shown in Figure 8.1 and Figure 8.2.

Secondary Loop or Back-pass

The back-pass is the section of gas pass between the exit of the cyclone/impact separator and the exit of the air heater.

The furnace usually accommodates:

- Evaporator tubes
- Parts of the superheater
- Parts of or the entire reheater

The economizer is normally located in the back-pass between the superheater and the air heater. Evaporator tubes may form the walls of the furnace and those of the back-pass. Parts of it may also be located in the external heat exchanger. Sometimes, the superheater tubes also form parts of the back-pass enclosure.

The disposition of the reheater and superheater tubes in the furnace, back-pass, and external heat exchanger is the designer's choice. This choice is, however, influenced by the type of fuel, as shown below. Some designs also use a steam-cooled cyclone, as shown in Figure 12.16.

8.7.1 Effect of Fuel Type on Furnace Heat Load

The energy balance in the furnace is greatly influenced by the type of fuel fired because a specific fraction of the heat input to the furnace must be extracted in the CFB loop (furnace and return leg) to maintain a given bed temperature, and that fraction depends on the fuel. Table 8.7 presents

TABLE 8.7
Fraction of the Heating Value of the Fuel Contained in the Flue Gas at 850°C

Fuel	Volatile matter (%)	Fixed carbon (%)	H_2O (%)	ASH (%)	S (%)	HHV (kJ/kg)	Flue-Gas Enthalpy at 850°C (kJ)/fuel fired (kJ)
Wood waste	—	45.0	2.0	—	0.01	9883	0.562
Anthracite culm	7	45.3	20.1	27.6	0.56	17482	0.447
Anthracite	7	60.8	16.1	16.1	0.66	22855	0.436
Lignite	28.2	30.8	34.8	6.2	0.7	16770	0.431
No. 6 oil	—	0.3	0.07	—	0.45	42119	0.417
Petroleum coke	15	74.3	10.0	0.3	4.7	32564	0.403
Bituminous coal	34	52	7.5	6.5	1.5	30238	0.400
Natural gas	—	—	—	—	0.35	49767	0.391

Source: Recalculated from Beissenger, H., Darling, S., Plass, L., and Wechsler, A., *Proceedings of eighth International Conference on Fluidized Bed Combustion*, DOE/METC-856021, Morgantown, pp. 619–632, 1985.

the enthalpy of the flue gas leaving the furnace at 850°C for a number of fuels to show how the enthalpy of the residual gas leaving the furnace changes with the fuel composition.

Table 8.7 shows that the flue gas from low-grade fuels, such as wood waste, carries nearly 60% of the heat input into the convective pass, whereas high-grade fuels, like bituminous coal carry only about 40% of the combustion heat into the convective section.

Thus, the heat duties of the primary CFB loop, as well as those of the convective section, vary depending on the type of fuel. However, this variation may be taken care of to some extent by varying bed density through adjustment of:

- Primary/secondary air ratio
- Bed inventory
- Circulation rate
- Bed temperature
- Excess air

As discussed earlier, the extent to which these changes can be made is limited by considerations of emission control and combustion efficiency.

TABLE 8.8
Heat Duty of the Fluid Bed Heat Exchanger as a Function of the Heat Absorbed in the CFB Loop

Fuel	Plant	Plant capacity MW	FBHE duty (% of primary CFB loop duty)
Ruhr coal	Flensburg	108	63
Bituminous	Duisberg	208	63
Lignite	Ione	56	0
Anthracite culm	Scott Paper	203	25
Bituminous	Chatham	69	39
Bituminous	Bayer	98	70

Source: Adapted from Beissenger, H., Darling, S., Plass, L., and Wechsler, A., *Proceedings of Eighth International Conference on Fluidized Bed Combustion*, DOE/METC-856021, Morgantown, pp. 619–632, 1985.

8.7.2 Heat Absorption in the External Heat Exchanger

Some CFB boilers employ external heat exchangers to absorb a fluctuation in heat duty while leaving the bed temperature and excess air relatively unaffected. The amount of heat extraction depends on the fuel fired, as shown by examples on Table 8.8. Heat transfer to the fluidized bed external heat exchanger is discussed in detail in Section 6.3 of Chapter 6.

The capacity of the boiler and the quantity of fuel burned dictates whether a fluid bed heat exchanger or extra combustor surface is needed, in addition to the water wall, to maintain the furnace temperature at an optimum level. The following conditions require the provision of an external fluid bed heat exchanger:

Total heat duty of (superheater + reheater)

> maximum heat absorption by (superheater + reheater) in back-pass and furnace (8.4)

If m_s is the mass of solids escaping through the cyclone per unit mass of fuel fired, the enthalpy loss due to these solids is $m_s C_a (T_{ce} - T_0)$. The total amount of heat leaving the CFB loop Q_y, is the sum of the loss due to the flue gas Q_{fg}, and that due to the solids.

$$Q_y = m_s C_a (T_{ce} - T_0) + Q_{fg} \tag{8.5}$$

Here, T_{ce} is the cyclone exit temperature. The amount of solids m_s, leaving the cyclone with the flue gas depends on the recycle rate and the grade efficiency of the cyclone. About 30 to 80% of the total solid fed to the bed escapes through the cyclone. For stable operation, this amount can never exceed the ash and sorbent feed to the combustor. This stipulation may give an estimate of the dust loading in the back-pass.

The heat available for absorption in the external heat exchanger H_{ehe}, may be found by subtracting from heat input, $(HHV + m_a C_a T_0)$ the heat going to the back-pass Q_y, the heat loss through the bed drain $[m_d C_{ps}(T_b - T_0)]$, and the heat absorbed in the furnace and cooled cyclone Q_f, on unit-mass-of-fuel-fired basis.

$$H_{ehe} = HHV + m_a C_p T_0 - m_d C_a (T_b - T_0) - Q_y - Q_f \tag{8.6}$$

To calculate the surface area of the external heat exchanger one can use the equations given in Section 6.3 in Chapter 6.

The overall heat balance dictates the heat absorption inside the furnace Q_f, and heat extracted in the external heat exchanger H_{ehe}, though the ratio depends to some degree on the designer's choice.

8.7.3 Heat Absorption in the Furnace and Rest of the Boiler

A number of other practical considerations in the design of the heat distribution system around a CFB boiler (Gottung and Darling, 1989) are necessary. For example, owing to some heat loss:

Total heat required for evaporation

> maximum heat absorbed in (economizer + evaporator) (8.7)

Maximum heat absorbed by economizer = $M_w(H_s - H_w)$ (8.8) where M_w is the feedwater flow rate, H_s is the enthalpy of feedwater at $(T_{sat} - \Delta T_w)°C$, and H_w is the enthalpy of feedwater entering the economizer. We assume that feedwater enters the drum at ΔT_w below its saturation temperature at drum pressure.

$$\text{Total flue gas enthalpy available for absorption in back-pass} = W_c(H_b - H_e) \tag{8.9}$$

where W_c is the flow rate of flue gas produced (kg/s), H_b is the enthalpy of the flue gas entering the back-pass (kJ/kg), and H_e is the enthalpy flue gas in the back-pass after the air heater.

$$\text{Maximum heat absorbed in the back-pass by the superheater/reheater} = W_c(H_b - H_a) \tag{8.10}$$

The flue gas, entering the economizer, cannot be cooled below the saturation temperature of water at the economizer exit. So the flue gas must leave the economizer at ΔT_g above the saturation temperature. Thus, the enthalpy of flue gas H_a, leaving the economizer is taken at $(T_{sat} + \Delta T_g)°C$.

The flue gas generally enters the air heater at a certain temperature ($\sim 38°C$) above the feedwater inlet temperature.

$$\text{Minimum thermal duty of air heater} = W_c(H_{38} - H_e) \tag{8.11}$$

where H_{38} is the enthalpy of feedwater at $38°C$ above its temperature at the inlet of the economizer.

$$\text{Maximum heat absorption by the water wall} = U_h A_w(T_b - T_{sat}) \tag{8.12}$$

where U_h is the overall heat-transfer coefficient between the furnace and the water in the evaporator tubes, A_w is the water wall area, and T_b is the furnace temperature. The heat-transfer coefficient inside the riser tubes remains fairly constant but that outside it varies.

Chapter 3 showed that the convective heat transfer varies along the furnace height due to the axial variation of the suspension density. This variation is strongly influenced by the hydrodynamics of the bed. The convective heat-transfer coefficients along the height of a fast bed can be fitted by an equation of the following form:

$$h_c(y) = a(y - z_0)^{-n} \tag{8.13}$$

where a and n are fitting parameters. The height of the section y, is measured from the distributor plate. It is greater than z_0, the height of the secondary air port or the uncooled lower section of the furnace.

The effect of different variables on the heat-transfer coefficient is discussed in Chapter 6. The heat-transfer coefficient is calculated by knowing the approximate metal temperature, tube emissivity, furnace temperature, and bed density. The above relation (Equation 8.13) is subject to the following assumptions:

1. Radiative component of heat transfer is independent of suspension density
2. No axial temperature gradient exists in the furnace

The water inside the furnace enclosure tubes remains at saturated temperature. Thus, the wall temperature of the tubes is approximated as $(T_{sat} + 25)°C$.

The furnace of a CFB boiler is usually constructed out of wall tube panels. If tubes with outer diameter D are at a pitch p, the ratio of the actual area and projected area of the wall would be:

$$A_x = \frac{\left(\dfrac{\pi D}{2} + p - D\right)}{p} \tag{8.14}$$

The lower section of the furnace is lined with refractory. For a first approximation, if one neglects heat absorption in this zone, the wall area of the rest of the furnace where heat is absorbed can be found as follows:

$$\text{Total projected wall area} = 2(W + B)L \tag{8.15}$$

where W, B, and L are the width, breadth, and height of the water-cooled section of the furnace, respectively.

Assuming the roof and exit area of the furnace to be equal the total heat-transferring area of the furnace enclosure S, is approximated as:

$$S = 2A_x(W + B)L, \tag{8.16}$$

Since boiling heat-transfer coefficients are an order of magnitude higher than the furnace side heat-transfer coefficients, the latter will control the overall heat transfer. Furthermore, the evaporator tubes carry boiling water, whose temperature may be assumed to be the saturation temperature of the water at the drum pressure. This may be well above the superheater outlet pressure. The average heat-transfer coefficient can be found by integrating the local coefficient over the height of the furnace:

$$U_c = \int_{z_0}^{z} \frac{a \, dy}{(z - z_0)(y - z_0)^n} = \frac{a}{(1 - n)(z - z_0)^n} \tag{8.17}$$

Thus, the convective component of the furnace heat absorption Q_c is:

$$Q_c = S \, U_c (T_b - T_{\text{wall}}) \tag{8.18}$$

where S is the effective surface area in the furnace.

This equation gives the convective heat transfer. The radiative heat-transfer coefficient is not a strong function of bed density in a typical CFB boiler. Thus, unless there is a large temperature difference along the bed, the average bed temperature can be used to find the radiative heat transfer by:

$$Q_r = 2 \, (W + L) \, L \cdot h_r (T_b - T_{\text{wall}}) \tag{8.19}$$

A furnace heat balance should be carried out to check the adequacy of wall heat absorption. The heat contained in flue gas leaving the furnace Q_y, is a function of fuel composition, excess air, and cyclone exit temperature T_{ce}. Use of this approach is illustrated by the example given below.

An alternative simpler approach, based on experimental data, for the determination of furnace heat absorption is given in Equation 8.31:

Example 8.1

> Thermal capacity of the boiler is 406 MW. The flue gas temperature is 143°C, and the sulfur capture target is 90%, with calcium to sulfur molar ratio of 2.0. The average temperature of all ashes leaving the boiler is 220°C. Ash contains about 1.5% carbon. Other data are as follows:
> *Steam condition*: 130 bar, 588°C; saturation temperature, 363°C; Feedwater temperature, 240°C.
> *Bituminous coal*: $M_f = 7.0\%$, C = 56.59%, H = 4.21%, N = 0.9%, Cl = 0.4%, S = 4.99%, O = 5.69%, ASH = 20.6%, HHV = 24657 kJ/kg.
> *Limestone*: $X_{CaCO_3} = 90\%$, $X_{MgCO_3} = 9.0\%$, $X_{inert} = 1.0\%$;
> Combustion efficiency = 98%.
> Find the principal dimensions of the furnace.

Circulating Fluidized Bed Boiler

Solution

1. Heat duties of different boiler elements as found using a steam table are as follows

Economizer	= 70.0 MW
Evaporator	= 168.0 MW
Superheater	= 113.0 MW
Reheater	= 54 MW
Total thermal output of the boiler	= 406 MW

Stoichiometry

The stoichiometric calculations for the above coal and limestone are carried out in Appendix 2; so we take the relevant values directly from there.

2. Combustion (stoichiometric) calculation (Example A2.2) gives

Excess air		= 20%
Theoretical air		= 8.05 kg/kg
Total dry air	= 8.05 × (1 + 0.2)	= 9.66 kg/kg
Total wet air	= 9.66 × (1 + 0.013)	= 9.78 kg/kg
Flue gas mass		= 10.71 kg/kg
Primary air is assumed to be 78% of theoretical air		
Primary air	= 0.78 × 8.05	= 6.28 kg/kg
Limestone required/coal feed from Example A2.2		= 0.319 kg/kg
Solid waste produced (from Example A2.2)		= 0.502 kg/kg

3. Heat balance on the basis of higher heating value of coal, 24657 kJ/kg

We use 20°C as the base temperature

Enthalpy of water at 20°C	4.186 × 20 = 83.7 kJ/kg
Enthalpy of water vapor at 143°C	2762 - 83.7 = 2678 kJ/kg
Loss due to moisture in sorbent	[0.319 × 0.07 × 2678 × 100]/24657 = 0.242%
Loss due to moisture in fuel	[1 × 0.07 × 2678 × 100]/24657 = 0.76%
Calcination loss $CaCO_3$ (Equation 7.5a)	[(0.9 × 0.319) × 1830 × 100]/24657 = 2.13%
Calcination loss $MgCO_3$	[0.09 × 0.319 × 1183 × 100]/24657 = 0.14%
Sulfation credit (Equation 7.5b)	[0.9 × 0.0499 × 15141 × 100]/24657 = <-2.75%>
Unburned carbon	[0.15 × 0.502 × 32790 × 100]/24657 = 1.00%
Heat in dry flue gas	10.71 × 1.03 × (143 - 20) × 100/24657 = 5.5%
Moisture in hydrogen	9 × 0.0421 × 2678 × 100/24657 = 4.12%
Moisture in air	0.013 × 8.05 × 1.2 × 2678 × 100/24657 = 1.36%
Radiation and convection (assumed from Table 8.2)	= 0.5%
Ash loss	0.502 × 0.2 × 4.18 × 220 × 100/24657 = 0.37%
FD fan credit (assumed 1%)	= <-1.0>
Total loss is found through algebraic addition of all losses	= 12.37%

4. Fuel heat input

$$Q_i = \frac{406}{1 - 0.1237} = 462 \text{ MW}$$

5. Mass balance

Coal feed required	$m_c = Q_i/HHV$	$46200/24657 = 18.74$ kg/s
Total airflow	$m_c \times M_{wa}$	$18.7 \times 9.78 = 183.3$ kg/s
Coal ash	$M_a = ASH \times m_c$	$0.206 \times 18.74 = 3.86$ kg/s
Fly ash (10% of total)	$0.01 \times M_a$	$0.1 \times 3.86 = 0.39$ kg/s
Sorbent feed		$18.74 \times 0.319 = 5.84$ kg/s
Total ash		$0.502 \times 18.74 = 9.77$ kg/s
Using the results of stoichiometric calculations of all		
Primary air		$6.28 \times 18.74 = 117.7$ kg/s
Total air		$9.78\ 18.74 = 183.2$ kg/s
Fuel gas wt		$10.70 \times 18.74 = 200.7$ kg/s

6. Furnace cross section

Heat release chosen from Table 8.6	$= 3.51$ MW/m^2
Bed upper cross section	$462/3.51 = 132$ m^2
Density of flue gas at 850°C	$= 0.326$ kg/m^3
Fluidizing velocity at 850°C	$200.7/(0.326 \times 132) = 4.67$ m/s
Check it for fast fluidization (refer to Chapter 2)	
Primary area required to maintain same velocity	$(117.7/183.2) \times 132 = 84.8$ m^2
Shape of cross section(refer to Section 6.3.2), width = $\frac{1}{2}$ length	
Width	$(132/2)^{0.5} = 8.1$ m
This is too wide and > 7.5m. So we limit to 7.5m. So the new value	
Width	$= 7.5$ m
Length	$132/7.5 = 17.6$ m

7. Furnace height
   (a) Required gas residence time for sulfur capture is assumed = 3.5 s
       Assuming that the core velocity may exceed the average by 50%,
       Maximum core velocity $1.5 \times 4.67 = 7.00$ m/s
       Height of the furnace $7.0 \times 3.5 = 24.5$ m
   (b) The adequacy of the height for sulfur capture may be further checked using Equation 5.34 of Chapter 5. However, it is most important to check if this height would allow absorption of the heat necessary to maintain the furnace at 850°C. Flue gas leaving the furnace at this temperature carries 40% heat (Table 8.7). So the heat to be absorbed by furnace wall for a bituminous coal is: $100 - 40 = 60\%$
       Heat to be absorbed $= 0.6 \times 462 = 277.2$ MW

The heat-transfer coefficient in the furnace can be estimated using information from Chapter 6. However, the average heat-transfer coefficients in typical CFB boilers are in the range of 130 to 200 W/m^{2}K. For these reasons, we chose a value of 0.175 kW/m^{2}K

Wall temperature $= T_{sat} + 25$	$363 + 25$	$= 388$°C
Evaporator duty	168 MW $\times$ 1000	$= 168000$ kW
Total evaporator area required	$168000/[(850-388) \times 0.175]$	$= 2078$ m^2
For 2 in pipe(3 in pitch, actual/project area	$(3.14 \times 2/2 + 1)/3$	$= 1.38$
Projected area required	$2078/1.38$	$= 1505$ m^2
Roof area = furnace cross section		$= 132$ m^2
Wall area required	$1505 - 132$	$= 1373$ m^2
Opening area in the wall is assumed ($\approx 30\%$) of bed cross section		
So, the wall area unavailable for heat transfer	0.3×132	$= 39.6$ m^2
The wall area required	$1373 + 39.6$	$= 1412.6$ m^2
Height required	$1412.6.[2 \times (7.5 + 17.6)]$	$= 28.1$ m

Circulating Fluidized Bed Boiler

The new furnace height, 28.1 m, is greater than 24.5 m calculated earlier. Thus, we make the furnace 28.1 m tall above the refractory zone. This will provide a gas residence time greater than 3.5 s. Therefore, the design is safe from both sulfur capture and heat-transfer standpoints. Now, the furnace height needs to be checked to ensure that this will allow adequate height of the standpipe in the cyclone for the smooth flow of solids through the loop-seal. Since we did not design the cyclone, we will ignore this part in the present exercise.

Additional superheater tubes and reheater tubes inside the furnace, in the form of panels or protected cross-tubes, can absorb the remainder of the heat. For rigorous design, one can check the furnace height using the sulfur-capture and combustion-performance models.

8.7.4 Heat Balance around a CFB Loop

CFB boilers operate in a closed loop. A major part of the solids, leaving the furnace, returns back to the base but at a different temperature, depending upon the heat exchange with different elements around the CFB loop. This is shown diagrammatically in Figure 8.18. Thus, the temperature of the gas–solid leaving the furnace depends upon its temperature while reentering the furnace. For this reason, an iterative solution of a complete energy and mass balance around the entire CFB loop is necessary for proper design of the CFB boiler. The following section presents that analysis.

The analysis neglects any gas entry into the standpipe from the loop-seal.

8.7.4.1 Mass Balance

Section A (lower bed)

$$\text{Solid}: \quad M_c + G_d + G_s = G_u + (1-x)M_a + xM_a \tag{8.20}$$

Since the solids circulating (G_d, G_s, and G_u) are considerably larger than the amount of solid fuel gasified we neglect that from the mass balance.

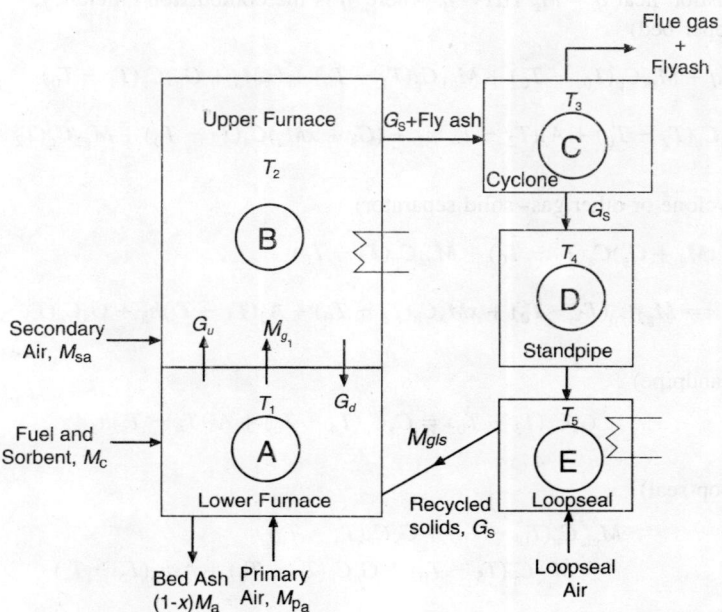

FIGURE 8.18 Heat and mass balance around a CFB loop.

$$\text{Gas}: \quad M_{pa} + M_{gLs} = M_{g1}$$

Section B (upper bed)

$$\text{Solid}: \quad xM_a + G_u = G_s + G_d + xM_a \quad \text{Gas}: \quad M_{g1} + M_{sa} = M_{g2} \quad (8.21)$$

Section C (cyclone or other gas–solid separator)

$$\text{Solid}: \quad xM_a + G_s = G_s + xM_a \quad \text{Gas}: \quad M_{g2} = M_{g2} \quad (8.22)$$

Section D (standpipe)

Solid: $G_s = G_s$ Gas: we assume that there is no exchange of gas in the standpipe (8.23)

Section E (loop-seal)

$$\text{Solid}: \quad G_s = G_s \quad \text{Gas}: \quad M_{gls} = M_{gls} \quad (8.24)$$

8.7.4.2 Energy Balance

Section A (lower bed)

$$\begin{aligned}
& M_c C_c (T_c - T_0) + \alpha q + M_{pa} C_p (T_{0p} - T_0) + G_s C_a (T_5 - T_0) + M_{gls} C_g (T_5 - T_0) \\
& + G_d C_a (T_2 - T_0) = (1 - x) M_a C_a (T_1 - T_0) + S(T_1 - T_w) U_h \\
& \qquad\qquad + (xM_a + G_u) C_a (T_1 - T_0) + M_{g1} C_g (T_1 - T_0)
\end{aligned} \quad (8.25)$$

where combustion heat $q = M_c$ HHV η, where η is the combustion efficiency.

Section B (upper bed)

$$\begin{aligned}
& (1 - \alpha) q + M_{sa} C_p (T_{0s} - T_0) + M_{g1} C_g (T_1 - T_0) + (xM_a + G_u) C_a (T_1 - T_0) \\
& = G_d C_a (T_2 - T_0) + A_2 (T_2 - T_w) h_2 + (G_s + xM_a) C_a (T_2 - T_0) + M_{g2} C_g (T_2 - T_0)
\end{aligned} \quad (8.26)$$

Section C (cyclone or other gas–solid separator)

$$\begin{aligned}
& (xM_a + G_s) C_a (T_2 - T_0) + M_{g2} C_p (T_2 - T_0) \\
& = M_{g2} C_p (T_3 - T_0) + xM_a C_a (T_3 - T_0) + A_3 (T_3 - T_s) h_3 + G_s C_a (T_3 - T_0)
\end{aligned} \quad (8.27)$$

Section D (standpipe)

$$G_s C_a (T_3 - T_0) = G_s C_a (T_4 - T_0) + A_4 (T_4 - T_s) h_4 \quad (8.28)$$

Section E (loop-seal)

$$\begin{aligned}
& M_{gls} C_p (T_{0l} - T_0) + G_s C_a (T_4 - T_0) \\
& = M_{gls} C_p (T_5 - T_0) + G_s C_a (T_5 - T_0) + A_5 h_5 (T_5 - T_s)
\end{aligned} \quad (8.29)$$

The last term includes heat losses from exterior walls as well as any heat absorbed by air or steam from the heat exchanger. Above equations are to be solved by iteration.

TABLE 8.9
Comparison of Auxiliary Power Consumption in the Boiler Island of a Typical 210-MWe Coal-Fired Thermal Power Plant

Equipment	PC (%)	CFB (%)
PA fan	0.46	1.75
SA fan	0.28	0.58
Ball Mills	1.53	—
Coal feed	—	0.05
Others including loop-seal blower	1.17	1.52
Total boiler island consumption	3.44	3.90

Figures are percentage of the net power output.
Source: Calculated from Rajaram, S., *Chemical Engineering Science*, 54, 5568, 1999. With permission.

8.8 AUXILIARY POWER CONSUMPTION

Auxiliary power consumption of a CFB boiler is one of its major shortcomings. It is slightly higher than that of a PC boiler or BFB boiler. The high-head primary air fan is the main power consumer in a CFB boiler as are the pulverizing mills for a PC boiler. Elimination of mills slightly offsets the high power consumption of PA fans, but is not sufficient to offset it entirely. BFB boilers do not need such high-head fans. Total power consumption is slightly higher in the case of a CFB boiler. Table 8.9 shows a comparison of power consumption of different elements of a basic PC boiler with that of a basic CFB boiler without any provision for sulfur capture. As one can see the difference is only 0.5%.

If sulfur capture is required, the power consumption in the FGD in a PC boiler will out pace the extra power consumption for sorbent handling in a CFB boiler.

Power consumption in the entire plant including the boiler island varies between 7.5 to 11%. Beyond the boiler island most equipment in a typical thermal power plant is similar. Thus the difference in total plant auxiliary load between a PC- and a CFB-based power plant is not much different. Newer generation CFB boilers are less power-hungry than older generations.

8.9 CONTROL OF CFB BOILERS

The control system of a CFB boiler is not very different from that of a pulverized coal-fired (PC) boiler except that its combustion control is much simpler. In a PC-fired boiler the control system has to supervise several burners, detect the presence of flame and guard against flame-out among other things. In a CFB boiler there is no flame. Thus, there is no need for sophisticated flame detection and monitoring system. The combustion of fuel in the bed is easily detected by thermocouples measuring temperatures. The large inventory of hot solids makes any response to a fuel interruption or excess fuel sufficiently slow to allow the operator or the automatic control system adequate time to take corrective measures. The bed inventory, sensed by differential pressure drop across the bed, is one unique control requirement of the CFB boiler. Distributed control systems (DCS) and programmable logic control (PLC) are used for control, display, alarm, and operator interface functions.

8.9.1 PART-LOAD OPERATION

CFB boilers can adequately respond to variation in load demands. Typically, they can handle load changes of 2 to 4% (of full load) per minute in the 100 to 50% and 1 to 2%/min in the 50 to 30% load range without any problem. In most cases, the boiler is not the limiting factor.

The allowable rate of change of turbine metal temperature restricts the pace of load change (Lyons, 1993).

A CFB boiler can reduce its output to 30% of its maximum continuous rating (MCR) without firing any auxiliary fuel such as oil. This greatly reduces the consumption of expensive auxiliary fuel in CFB-fired power plants. However, at 30% load, the furnace would operate in bubbling bed mode.

The load on a CFB boiler is controlled by its unique heat-transfer characteristics. Here, a change in the bed density can easily change the bed-to-wall heat-transfer coefficient. The heat absorbed Q_a, by water/steam in a fluidized bed furnace may be written as:

$$Q_a = A_2(h_c + h_r)(T_b - T_{\text{wall}}), \qquad (8.30)$$

where A_2 is the boiler surface areas exposed to the upper furnace, T_b is the bed temperature, and T_{wall} is the tube wall temperature. Heat absorbed in the insulated lower furnace is relatively low.

The average convective component h_c is dependent on the square root of average bed density, ρ_b (Glicksman, 1997). The height average radiative heat-transfer coefficient h_r, may be expressed as $h_r = K\rho_b^{n_1} T_b^{n_2}$. So, Equation 8.30 may be written as:

$$Q_a = A_2(K_c \rho_b^{0.5} + K_r \rho_b^{n_1} T_b^{n_2})(T_b - T_{\text{wall}}) \qquad (8.31)$$

where n_1, n_2, K_c, and K_r are empirical constants derived from experimental data.

The *constant* depends on a number of factors. The bed density ρ_b, refers to the average density of the bed above the primary combustion zone, where most heating surfaces are located. It is determined from the pressure drop measured across the upper section of the furnace.

To reduce the load on the boiler, for example, the operator could reduce the primary airflow. This will increase the density of the lower bed and reduce the density of the upper bed ρ_b, which according to Equation 8.31 influences Q_a, the heat absorption in the upper furnace. The fuel feed rate would of course change correspondingly. The bed temperature is another parameter that can be adjusted within a certain range to control the load. Table 8.4 illustrates with an example from a 200-MWe CFB boiler how the load can be controlled by adjusting air flow rates.

Example 8.2

> Find the average heat-transfer coefficient over the upper section of a furnace that runs from 3 m to 25 m above the grate of a furnace. The axial variation in the bed density was fitted by the equation $\rho_b = 214/z^{1.105}$ kg/m^3, where z is the height above the distributor. Assume the bed temperature to be uniform at 850°C. The convective and radiative heat-transfer coefficients measured at a bed density of 15 kg/m^3 were observed to be 0.098 and 0.126 kW/m^{2}K, respectively.

Solution

No substantial effect of bed density on the radiative component of heat transfer is known at the moment. Thus, this is taken to be independent of the bed density. However, when the average bed density is so low that the emissivity of the bed is less than 0.6 to 0.8, i.e., Equation 8.31 cannot be used. For the calculation of radiation from the entire bed in such a situation, the effect of bed density on the emission needs to be considered. For a first approximation, we assume here that the radiative heat transfer is not strongly affected by the bed density. So we can write Equation 8.31 as:

$$h = Q_a/[A \times (T_b - T_s)] = (\text{constant} \times \rho_b^{0.5} + h_r)$$

The convective heat-transfer coefficient is 0.098 kW/m²K. So we can find the *constant* of Equation 8.31 as:

$$\text{Constant} = 0.098/(15^{0.5}) = 0.025$$

By substituting the values of the *constant*, h_r and ρ_b one gets:

$$h_c = 0.025 \left(\frac{214}{z^{1.105}} \right)^{0.5} = \frac{0.365}{z^{0.552}}$$

Average convective coefficient

$$h_c = \frac{1}{(25-3)} \int_3^{25} 0.025 \left(\frac{214}{z^{1.105}} \right)^{0.5} dz = \frac{1}{22} \int_3^{25} \frac{0.365}{z^{0.552}} dz = 0.096 \text{ kW/m}^2 \text{ K}$$

The radiative heat transfer is assumed to be uniform throughout the height. So the average total heat-transfer coefficient is

$$h = 0.096 + 0.126 = 0.222 \text{kW/m}^2 \text{ K}$$

8.9.2 LOAD CONTROL OPTIONS

The steam demand on the boiler is rarely constant and remains at its design value. A boiler is often required to generate less steam to meet reduced load demand. Followings are two basic approaches to the load control:

1. Furnace heat absorption control
2. External heat absorption control

8.9.2.1 Furnace Heat Absorption Control

Figure 8.19 explains qualitatively the heat absorption in a CFB boiler under part-load conditions. The bed density in the lower section of the boiler is very high, and that in the upper section is very dilute, making the particle convective heat transfer-dominant in the lower section and radiation-dominant higher up. Under full load, both convective and radiative components of heat transfer will control the heat absorption in the bulk of the furnace. To respond to a partial load (70%, for example), the bed density in the upper or the cooled section of the furnace could be reduced through a reduction in the primary air making the radiation dominant over larger part of the furnace. The convective component will still play some role. To reduce the load to about 40%, the primary air needs to be reduced such that the bed will operate like a bubbling fluidized bed. The solid concentration above the refractory walls would approach that of the freeboard of a bubbling bed. The heat transfer in the dilute zone will be low and mainly radiative.

8.9.2.2 External Heat Absorption Control

One could control the load by controlling the heat absorption by the circulating solids elsewhere in the CFB loop. Figure 8.20 shows two schemes for load control. The scheme in Figure 8.20a uses a CFB loop with an external fluidized bed heat exchanger where 20 to 60% of the total heat (excluding the boiler back-pass) is absorbed. The heat absorbed in this bubbling bed heat exchanger is controlled by changing the temperature of the bubbling bed by adjustments of solid flow through

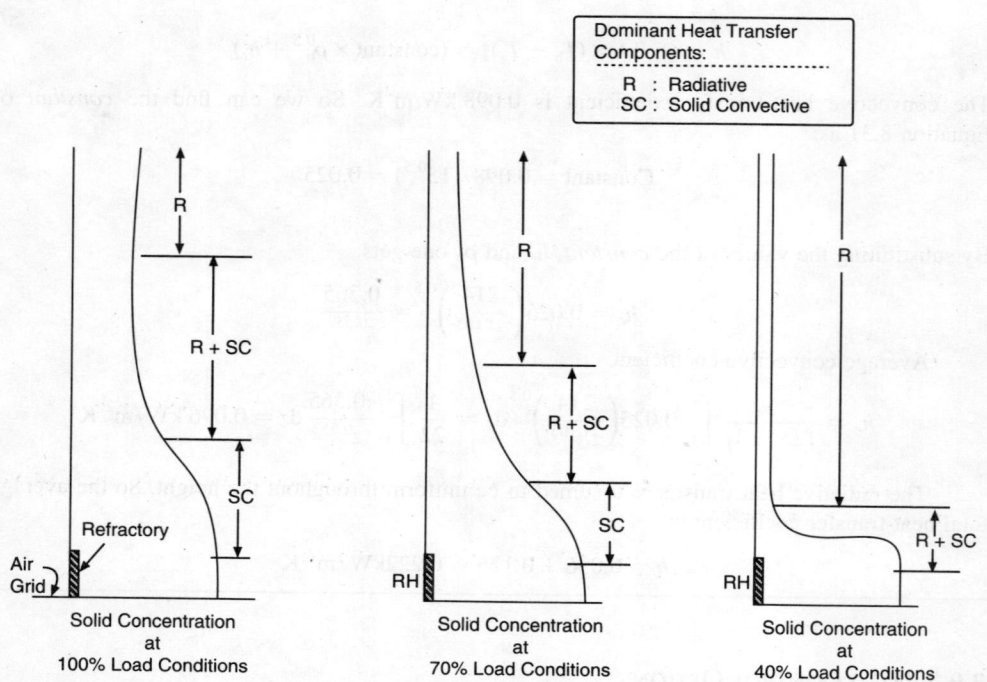

FIGURE 8.19 Heat absorption distribution in the furnace of a CFB boiler at different loads (R-radiative, SC-solid convective heat transfer).

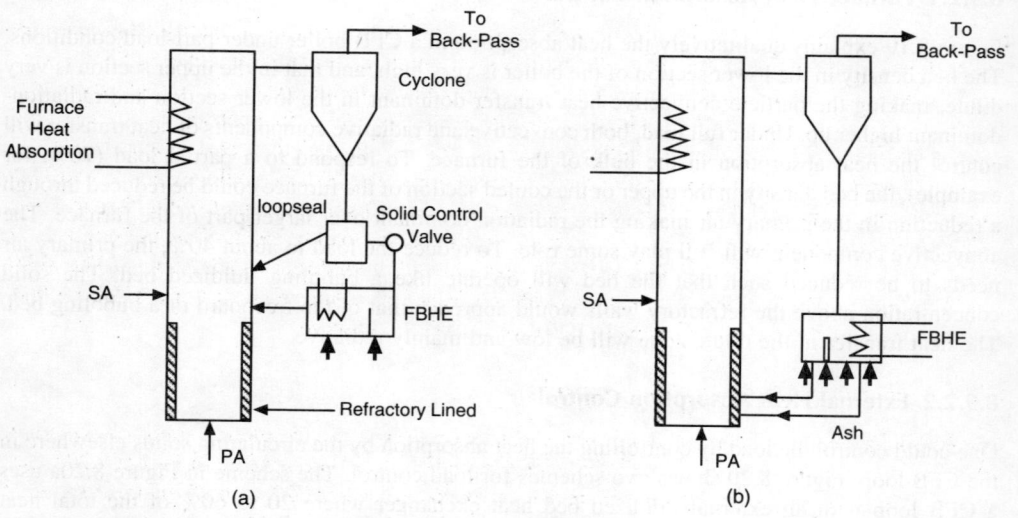

FIGURE 8.20 Different schemes for load control of a CFB boiler.

Circulating Fluidized Bed Boiler

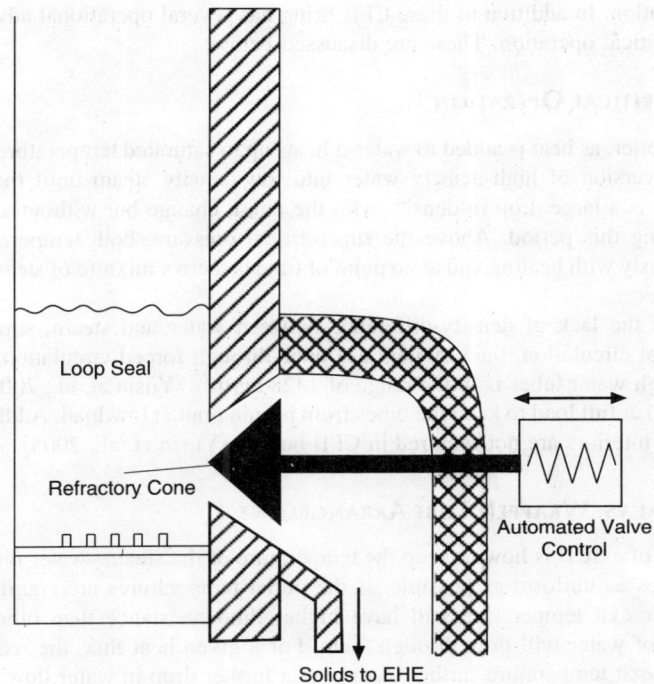

FIGURE 8.21 Solid control valve for diversion of recirculating solids through the external heat exchanger.

it. This is done by controlling the amount of solids diverted from the loop-seal. Here, only part of the recycled solids pass through the external fluidized bed. A schematic of the mechanical valve used for control of the diversion of solids is shown in Figure 8.21.

In another version of load control (Figure 8.20b), the entire circulating solids are passed through a bubbling fluidized bed external heat exchanger located at the base of the cyclone dip leg. The bubbling bed is divided into two sections, one of which contains heat-exchange surfaces. Two separate streams return solids to the CFB furnace; one stream passes through the heat exchanger and another bypasses it. The split of solid flow between two L-valves is adjusted by controlling airflow to the fluidized bed according to the heat to be extracted in the external heat exchanger.

The above adjustments have limits. Thus, in commercial boilers, the overall gas velocity and bed temperatures are also adjusted to some degree. Under part load, the gas flow rate through the back-pass of the boiler also reduces. Thus, there is an automatic reduction in heat absorption by the heat transfer elements in the back-pass. When further fine-tuning of the heat absorption is necessary for controlling the temperature of superheated and reheated steam, an additional control loop in the back-pass or the injection of water to superheated steam may be used.

8.10 SUPERCRITICAL CFB BOILER

Supercritical boilers (SCB) operate at pressures above the critical pressure of 221.2 bar, where the overall efficiency of the steam cycle is relatively high. Supercritical boilers consume less steel due to their use of smaller diameter tubes and the absence of large steam drum. In the bidding process of a 460-MWe supercritical project (including SCR and FGD) at Lagisza, Poland, the CFB firing option was found to be 20% cheaper in capital cost and 0.3% higher in net efficiency than the competing PC option. In addition to these CFB firing has several operational advantages over PC firing for supercritical operation. These are discussed below.

8.10.1 SUPERCRITICAL OPERATION

In a subcritical boiler, as heat is added to water it heats up to saturated temperature. Further heating causes slow conversion of high-density water into low-density steam until the entire water is converted. There is a large drop in density with the phase change but without any change in the temperature during this period. Above the supercritical pressure, both temperature and density change continuously with heating and at no point of time is there a mixture of steam and water as in a subcritical boiler.

Because of the lack of density-difference between water and steam, supercritical boilers cannot use natural circulation. Such boilers use once-through forced circulation instead. Typical water flow through water tubes is in the range of 1428 kg/m^2s (Yusin et. al., 2005) to 700 kg/m^2s (Lundqvist, 2003) at full load to keep the tubes from burning out at low load. Additional provisions of rifling of tube interiors are not required in CFB boilers (Yusin et. al., 2005).

8.10.2 VERTICAL VS. WRAPPED TUBE ARRANGEMENT

A major concern of a SCB is how to keep the temperature of the steam–water mixture exiting the parallel wall tubes as uniform as possible. If the outlet temperatures are significantly different, tubes with higher exit temperature will have higher fluid resistance than others, and hence a reduced amount of water will flow through them. For a given heat flux, the reduced water flow will increase the exit temperature further leading to a further drop in water flow. Thus, through a spiral effect the water flow through the tube could drop, leading up to total dry-out. Thus, it is vitally important that the heat absorbed by all parallel tubes is similar. For this reason PC boilers, which have a wide distribution of heat flux around their perimeter, use parallel tubes wrapped around the furnace periphery. Such an arrangement is not very convenient from the standpoint of mechanical strength because the entire load of the water wall is borne by the fin weld rather than the tubes themselves.

A CFB boiler is relatively free from this problem because of its relatively uniform heat flux distribution around the furnace perimeter. Heat-transfer coefficients measured in a 135-MWe CFB (Figure 6.20) show that the lateral variation is within 6%. (Yusin et al., 2005). Lundqvist (2003) found through a numerical simulation what would be the lateral distribution of heat flux on

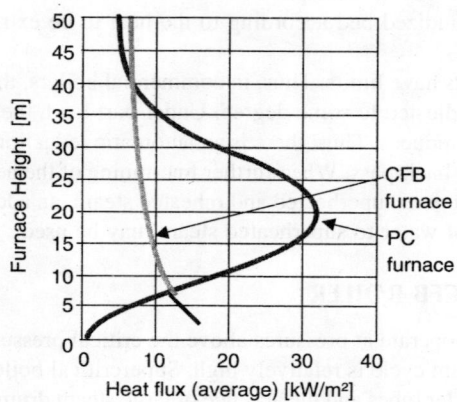

FIGURE 8.22 Heat flux in a PC boiler is much greater than that in a CFB boiler.

Circulating Fluidized Bed Boiler

TABLE 8.10
Effect of Steam Parameter on the Cost of Electricity Generation from a 600-MWe Coal-Fired Power Plant Using CFB Firing

Steam Parameter	Change in Boiler Cost (%)	Change in Investment Cost €/kW (%)	Efficiency increase (%)	Change in Cost of electricity (%)
265 bar, 535°C/571°C	Reference	Reference	Reference	Reference
270 bar, 583°C/601°C	+4.6	+0.2	1.9	−0.7
290 bar, 603°C/621°C	+9.5	+1.6	3.5	−0.7

Adapted from Laffont, P., Barthelemy, J., Scarlin, B., and Kervenec, C., A clean and efficient supercritical circulating fluidised bed power plant Alstom Power, PowerGen-Europe, www.siemenswestinghouse.com, 2003. With permission.

the furnace walls if one of the feed points trips in a 500-MWe CFB boiler leaving other feeders around the wall to take up the slack. The resulting lateral variation in the heat flux was within 3 to 5%, which could result in a evaporator exit temperature of less than 35°C.

8.10.3 LOW HEAT FLUX

Figure 8.22 compares the axial heat flux in a PC furnace with that in a CFB furnace. The average heat flux in a CFB boiler is lower than that in PC boiler, making its furnace tubes safer and more resilient to operational upsets. Furthermore, the peak heat flux in a PC boiler is significantly higher than that of CFB boiler. Furthermore the peak in a PC boiler occurs near the most vulnerable point where the transition from water to steam takes place. In a CFB boiler the peak heat flux takes place where water is just heated. Thus the CFB boiler tubes are safer, even without the use of flow equalization orifices in the tubes.

In today's deregulated fuel market, a fuel-flexible supercritical CFB boiler is much more attractive than a supercritical PC boiler. Since it does not use a drum, SCBs enjoy the cost savings and their furnace length is not limited by the drum length.

8.10.4 COST ADVANTAGE

Studies carried out by Ahlstom (Laffont et al., 2003) shows that, for an increase in steam pressure by 90 bar and 20 K increase in steam temperature, the cycle efficiency of a steam power plant could increase by 2.2% and 1.0% (with reference to a subcritical cycle), respectively.

Most subcritical steam power plants operate with net efficiency of about 35%. An improvement in the cycle efficiency by few percentage points without any major increase in the investment cost can greatly reduce the emission of CO_2 and the cost of electricity. Above studies found that even after allowing for somewhat higher investment cost for the boiler there is a net reduction in the cost of electricity even before taking into account credit for CO_2 reduction (Table 8.10).

NOMENCLATURE

a:	coefficients of Equation 8.13
A_1:	total heat-transfer surface in lower furnace, m^2
A_2:	total heat-transfer surface in upper furnace, m^2
A_3:	total heat-transfer surface in cyclone, m^2

A_4: total heat-transfer surface in standpipe, m^2
A_5: total heat-transfer surface in loop-seal/ EHE, m^2
A_w: furnace heat-transfer surface area, m^2
A_x: ratio of actual to projected wall area of boiler's water wall
ASH: mass fraction of ash in coal
B: breadth of furnace, m
Ca/S: calcium to sulfur molar ratio
C_a: specific heat of ash, kJ/kg K
C_c: specific heat of fuel and sorbent, kJ/kg K
C_g: specific heat of gas at furnace temperature, kJ/kg K
C_p: specific heat of air, kJ/kg K
D: diameter of water wall evaporator tubes, m
G_d: downwards solids flow rate in furnace, kg/s
G_i: mass flow rate of solids of size i, kg/s
G_s: external solids recirculation rate, kg/s
G_u: upward solids flow rate in furnace, kg/s
h_1: overall heat-loss coefficient at the lower furnace, kW/m^{2}K
h_2: overall heat-transfer coefficient at the upper furnace, kW/m^{2}K
h_3: heat-absorption coefficient at the cyclone, kW/m^{2}K
h_4: heat-loss coefficient at the standpipe, kW/m^{2}K
h_5: heat-loss coefficient at the loop-seal, kW/m^{2}K
$h_c(y)$: convective heat-transfer coefficient at a height y, kW/m^{2}K
h_r: furnace averaged radiative heat-transfer coefficient, kW/m^{2}K
h_c: furnace averaged convective heat-transfer coefficient, kW/m^{2}K
H, O, S, C, N, Cl: mass fractions of hydrogen, oxygen, sulfur, carbon, nitrogen, and chlorine, respectively, in coal
HHV: higher heating value of coal, kJ/kg
H_a: total enthalpy of flue gas entering economizer at $T_{sat} + 38°C$, kJ/s
H_b: total enthalpy of flue gas entering back-pass, kJ/s
H_{ehe}: heat absorbed in heating surfaces in EHE per unit mass of fuel-fired basis, kJ/kg
H_e: total enthalpy of flue gas leaving the back-pass, kJ/s
H_f: saturation enthalpy at economizer outlet pressure, kJ/kg
H_s: enthalpy of feedwater at $(T_{sat} - 10)°C$, kJ/kg
H_w: enthalpy of feedwater entering the economizer, kJ/kg
H_1, H_2: enthalpies of water entering and leaving economizer, kJ/kg
H_{38}: enthalpy of water at 38°C above feedwater inlet temperature, kJ/kg
L: height of the furnace above the uncooled section, m
K_c, K_r: empirical constants in Equation 8.31
m_a: mass of air per unit mass of fuel fired, kg/kg fuel
m_c: feed rate of fuel, kg/s
m_d: solid drained from bed per unit mass of fuel fired, kg/kg fuel
m_i: mass flow rate of bed solids of size i, kg/s
m_s: mass of solid entering back-pass per unit mass of fuel fired, kg/kg fuel
M_a: ash flow rate, kg/s
M_c: combined feed rate of fuel and sorbent, kg/s
M_f: moisture content of coal, kg/kg fuel
M_{gls}: loop-seal air flow rate, kg/s
M_{g1}: gas leaving lower furnace, kg/s
M_{g2}: gas leaving upper furnace, kg/s

M_{pa}:	primary air flow rate, kg/s
M_{sa}:	secondary air flow rate, kg/s
M_{wa}:	amount of wet air feed per unit fuel fired
M_w:	flow rate of feed water, kg/s
n:	exponent in Equation 8.13 and Equation 8.17
n_1, n_2:	experimentally derived exponents of Equation 8.31
p:	pitch of tubes, m
q:	total combustion heat released, kW
Q_a:	heat absorbed in the furnace, kW
Q_c:	convective component of heat absorption by the wall, kW
Q_i:	heat input from fuel, kW
Q_r:	radiative component of heat absorption by the wall, kW
Q_f:	heat absorbed in furnace and cyclone per unit mass of fuel fired, kg/kg fuel
Q_{fg}:	enthalpy of flue gas leaving CFB loop (per unit mass of fuel fired), kJ/kg fuel
Q_y:	enthalpy of gas–solid suspension entering the back-pass, kJ/kg fuel
S:	total surface area of furnace enclosures, m^2
t:	time, s
T_b:	bed temperature, °C
T_0:	ambient temperature, °C
T_{0p}:	primary air temperature, °C
T_{0s}:	secondary air temperature, °C
T_1:	temperature at the lower furnace, °C
T_2:	temperature at the upper furnace, °C
T_3:	temperature at the cyclone, °C
T_4:	temperature in the standpipe, °C
T_5:	temperature at the loop-seal, °C
T_s:	exterior wall temperature of loop-seal, standpipe or cyclone, °C
T_c:	temperature of fuel and sorbents, °C
$T_{ce}, T_{sat}, T_f, T_{wall}$:	temperatures of cyclone exit gas, saturated steam at drum pressure, feedwater and wall, respectively, °C
T_{ol}:	loop-seal air temperature, °C
T_{sat}:	saturation temperature of water in rise, °C
T_{wall}:	wall temperature of the furnace wall, °C
U_h:	overall heat-transfer coefficient in furnace, kW/m^{2}K
U_c:	height-averaged convective heat-transfer coefficient in furnace, kW/m^{2}K
U_{cl}:	unburned carbon loss, %
W:	width of the furnace, m
W_a:	mass of solid residue produced per unit mass of coal burned
W_{bi}:	mass of bed solids of size i, kg
$W_{i,\ classifier}$:	mass of bed solids of size i returned to bed by classifier, kg/s
W_c:	total mass of flue gas due to combustion, kg/s
y:	distance above the uncooled section of the furnace, m
x:	fraction of fuel ash appearing as fly ash
X_c:	fractional carbon in solid wastes
X_{inert}:	inert fraction in limestone
X_{CaCO_3}, X_{MgCO_3}:	mass fraction of calcium and magnesium carbonate in limestone
X_{loss}:	total heat loss in the boiler in fraction
y:	running height in the furnace above the distributor, m
z:	height in bed above the distributor, m
z_0:	height of the uncooled section above the distributor, m

GREEK SYMBOLS

α: fraction of combustion heat released in the lower bed
η: combustion efficiency in fraction
$\rho_b(y)$: bed density at a height y, kg/m^3
ρ_b: average suspension density in the furnace of a CFB, kg/m^3

REFERENCES

Beissenger, H., Darling, S., Plass, L., and Wechsler, A., *Burning multiple fuels and following load in the lurgi/ce circulating fluidized bed boiler*, Proceedings of the Eighth International Conference on Fluidized Bed Combustion, DOE/METC-856921, Morgantown, pp. 619–632, 1985.

CFBCAD® — An expert system for design and evaluation of circulating fluidized bed boiler, Greenfield Research Inc. Box 25018, Halifax, Canada B3M 4H4.

Glicksman, L. R., Heat transfer in circulating fluidized beds, In *Circulating Fluidized Beds*, Grace, J. R., Avidan, A. A. and Knowlton, T. M., Eds., Chapman & Hall, London, pp. 261–310, 1997.

Gottung, E. J. and Darling, S. L., Design considerations for CFB steam generators, In *Proceedings of the 10th International Conference on Fluidized Bed Combustion*, Manaker, A., Ed., ASME, New York, pp. 617–623, 1989.

Gould, G. L. and McComas, M. W., Know how efficiencies vary among fluidized bed boilers, *Power*, January, 39–40, 1987.

Herbertz, H., Lienhard, H., Barnie, H. E., and Hansen, P. L., Effects of fuel quality on solids management in circulating fluidized bed boilers, In *Proceedings of the 10th International Conference on Fluidized Bed Combustion*, Manaker, A., Ed., ASME, New York, pp. 1–7, 1989.

Jones, P. A., Hellen, T. J., and Friedman, M. A., Determination of heat loss due to surface radiation and convection from the CFB boiler at NUCLA, *Proceedings of the 10th FBC Conference*, Manaker, A., Ed., ASME, New York, pp. 1047–1052, 1989.

Jukola, G., Liljedahl, G., Nskala, N., Morin, J., and Andrus, H., An Ahlstom's vision of future CFB technology based power plant concept, In *Proceedings of the18th International Conference on Fluidized Bed Combustion*, Lia, J., Ed., ASME, New York, 2005, paper FBC2005-78104.

Lafanechere, L., Basu, P., and Jestin, L., Effects of fuel parameters on the size and configuration of circulating fluidized bed boilers, *J. Inst. Energy*, UK, 68(December), 184–192, 1995a.

Lafanechere, L., Basu, P., and Jestin, L., Effect of steam parameters on the size and configuration of circulating fluidized bed boilers, In *Proceedings of the 13th International Conference on Fluidized Bed Combustion*, Heinschel, K., Ed., ASME, New York, pp. 38–54, 1995b.

Laffont, P., Barthelemy, J., Scarlin, B., and Kervenec, C., A clean and efficient supercritical circulating fluidized bed power plant Alstom Power, PowerGen-Europe, www.siemenswestinghouse.com, 2003.

Lalak, I., Seeber, J., Kluger, F., and Krupka, S., Operational experience with high efficiency cyclone comparison between boiler A and B in Zeran power plant — Warsaw Poland, In *Proceedings of the 17th International Fluidized Bed Combustion Conference*, ASME, Pisupati, S., Ed., 2003, Jacksonville, Paper no FBC2003-146.

Lee, J. M., Kim, J. S., Kim, J. J., and Ji, P. S., *Status of the 200 MWe Tonghae CFB boiler after cyclone modification*, 13th US-Korea Joint Workshop on Energy & Environment, Reno, p. 199, 1999.

Lundqvist, R. G., Designing large-scale circulating fluidized bed boilers, *VGB PowerTech*, 83, 10, 41–47, 2003.

Lyons, C., Current comparison of coal-fired technologies to meet today's environmental requirements, Power-Gen Americas'93, 1993.

Mathew, F. T., Payne, H. M., Wechsler, A. T., Saunders, W. H., Berman, P. A., and Dille, J. C., Design & Assessment of a PCFB Boiler, EPRI Report CS 3206, 1983.

Stultz, S. C. and Kitto, J. B., *Steam — its generation and use*, 40th ed., Babcock & Wilcox, Barberton, 1992, pp. 19–23.

Tang, J. and Engstrom, F., Technical assessment on the Ahlstrom pyroflow circulating and conventional bubbling fluidized bed combustion systems, In *Proceedings of the ninth International Conference on Fluidized Bed Combustion*, Mustonen, J. P., Ed., ASME, New York, pp. 38–54, 1987.

Werther, J., Fluid dynamics temperature and concentration fields in large-scale CFB combustors, In *Circulating Fluidized Bed Technology VIII*, Cen, K., Ed., International Academic Publishers, Beijing, pp. 1–25, 2005.

Yang, L., Bie, R., Bao, Y., Zhang, Z., Zhao, M., Lu, H., and Wi, W., *Design and operation of circulating fluidized bed boilers with low circulating ratio, Fifth International Conference on Circulating Fluidized Beds, Beijing*, 1996, Paper no. CG10.

Yusin, W., Junfu, L., Jiansheng, Z., Guangxi, Y., and Long, Y., Conceptual design of an 800 MWe supercritical pressure circulating fluidized bed boiler and its performance prediction, In *Circulating Fluidized Bed Technology VIII*, Cen, K., Ed., International Academic Publishers, Beijing, pp. 529–536, 2005.

9 Material Issues

Fluidized bed boilers have acquired sufficient operating experience to be called a matured technology. Innovative modifications of different components of the boilers largely addressed the problems affecting materials and performance. Still, some problems remain, forcing occasional shutdowns. Most forced outages or shutdowns of a boiler are related to materials, operation, or design issues. Some fluidized bed boilers of older design have experienced availability as low as 70% due to failures of their refractory parts or erosion of pressure and nonpressure parts. Failure of a critical component results in shutdown of the entire power plant and requires immediate attention, making material issues the most immediate concern for the plant operators. Thus, a good selection of materials and the understanding of their behavior in a fluidized bed environment are critical to the operators.

The present chapter discusses the behavior of materials in fluidized bed boilers, and how to reduce their failure through good design and operation practices.

9.1 MATERIAL SELECTION CRITERIA

Main construction materials used in fluidized bed boilers are:

- Steel
- Refractory and insulations
- Special materials such as expansion joints

A designer selects the optimum material keeping the following requirements in mind:

- Structural requirements
- Heat transfer duty
- Erosion potential
- Corrosion potential

These, of course, must be met simultaneously in a manner such that, with limited repairs and replacement, the boiler has an effective service life of 30 years or longer.

9.1.1 STRUCTURAL REQUIREMENTS

The most important requirement of any material used in a boiler is to bear the imposed structural loads continuously throughout its life. These normally include:

- Gravitational loads
- Hydrodynamic loads imposed by static and dynamic heads of gases and solids. These loads could be oscillatory, especially in case of circulating fluidized beds
- Structural loads transmitted by other boiler components
- Miscellaneous loads, including earthquake reactions, hydrostatic testing and all other types of operational events

Internal working pressure and temperature of the water and steam circuits are two major criteria for the selection of pressure parts materials of a fluidized bed boiler. The allowable stress in the candidate materials depends on the operating metal temperature as shown in Figure 9.1 for some commonly used steels.

9.1.2 HEAT-TRANSFER DUTY

The primary function of a boiler tube is to transfer heat from the furnace or flue gas to the steam or water inside the tube. Though a fluidized furnace, like any other furnace, offers a rather hostile environment, the walls of modern boilers are still made of water/steam-cooled tubes to minimize the use of expensive and high-maintenance refractory. Uncooled surfaces made of refractory increase the start-up or cooling-down time of a boiler and also add to the load on the boiler structure. This makes it imperative to avoid refractory wherever possible.

In a bubbling fluidized bed, where horizontal or slightly inclined tubes are used, care must be taken not to exceed the *critical heat flux* to these tubes, because it might bring about the phenomenon of *departure from nucleate boiling* (DNB). Under nucleate boiling, water boils inside the tubes, continuously releasing bubbles on the surface. This keeps the tube wall sufficiently cool, within 20 to 30°C of the saturated temperature of the water. When the boiling process moves from the nucleate to film boiling, a steam blanket covers the tube wall resulting in a sharp drop in its heat-transfer coefficient. This pushes the temperature of the tube wall closer to the temperature of the furnace. As one can see from Figure 9.1, for most steels the allowable stress drops sharply once the metal

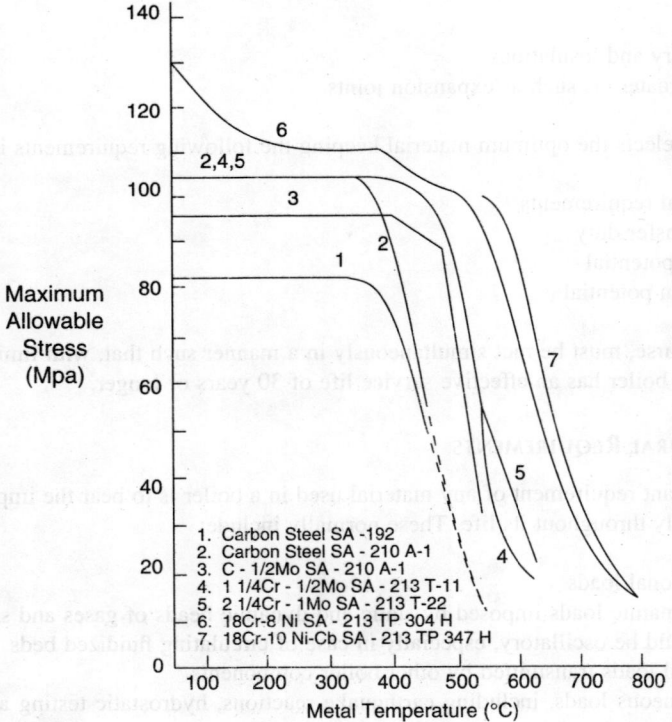

FIGURE 9.1 Maximum allowable stress for various carbon steels and alloys, and its dependence on the operating metal temperature.

temperature exceeds a certain value. Thus, an excursion in the tube wall temperature severely weakens the wall, which may finally rupture. High heat flux could also produce deposits, leading to caustic gouging.

Several precautions, such as selection of a conservative heat flux, higher water velocities inside the tubes, or use of rifled tubes (Singer, 1991) could avoid the potential occurrence of DNB.

9.2 EROSION POTENTIAL

The most significant material problem facing the designer and operator of a fluidized bed boiler relates to tube wastage by erosion (Larson et al., 1985). The solid concentration within a fluidized bed combustor is much higher that that in other types of boilers, and as such it remains a challenge to protect its tubes from erosion. Table 9.1 presents the range of solid densities and velocities found in various zones of different types of boilers.

Erosion is a function of density, velocity, particle characteristics, and trajectory of solids. The solid impingement velocity greatly affects the rate of erosion, and as such is one of the primary considerations in the design of convective passes of a boiler (Stringer and Wright, 1989). In dense zones, such as the lower combustor of a circulating fluidized bed (CFB) or bubbling bed, the erosion is also dependent on the motion of the particles relative to the exposed surface and gas flow (Stringer, 1987).

Erosion in a CFB boiler is primarily due to the flow of solids down the water walls of the combustor. The resulting erosion is therefore greatly influenced by the geometry of the combustor wall. The areas subjected to serious erosions are normally associated with abrupt changes in the downward solids flow, which could be caused by:

1. Water wall refractory interface
2. Protruding instrumentation
3. Irregularities and weld defects in the membrane wall

9.2.1 BASICS OF EROSION

A brief review of the erosion process might help form a better understanding of the erosion mechanism in fluidized bed boilers. Erosion is a process of wear in which materials are removed from a solid surface through impingement of solids on it. Its mechanism is different from that of abrasion.

The process of erosion is similar to that of metal cutting. Figure 9.2 explains the process. A particle impacts a surface at an angle α with a velocity V_p. The impact force has two components: force normal to the surface causing deformation of the material and force parallel to the surface, which removes the material from the surface.

TABLE 9.1
Typical Solids Density and Gas Velocities for Various Boiler Types

Boiler Zone	Solids Density (kg/m³)	Gas Velocity (m/sec)
CFB lower combustor	100–200	2–6[a]
CFB upper combustor	50–5	4.5–7
CFB convection pass	4	12–16
Bubbling fluidized bed	300–200	1–2.5
Pulverized coal convection pass	<2	20–25
Gas-fired convection pass	0	30+

[a] Some designs deliberately maintain a bubbling bed in the lower combustor.

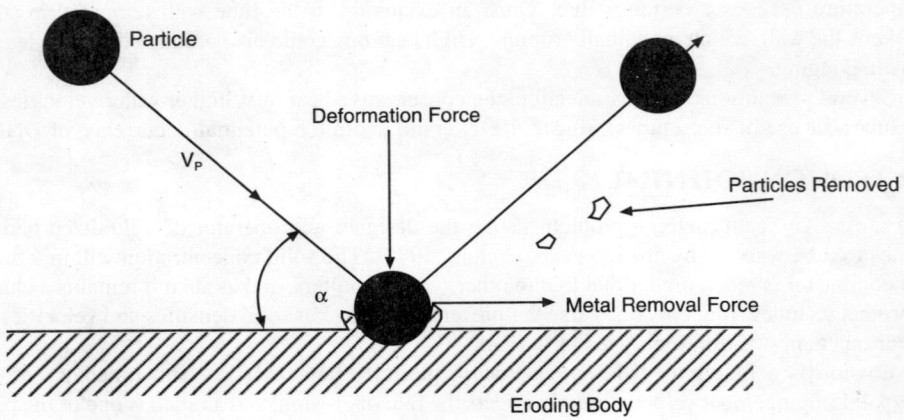

FIGURE 9.2 An eroding particle hits a target metal surface causing deformation and removes the deformed piece.

Using the metal cutting theory, Finnie (1960) developed a model for surface erosion. Chinese Boiler Thermal Standards (1973) developed the following equation by following this model and using a large amount of experimental data. The maximum erosion rate in the convective section is written as (Basu et al., 1999):

$$E_{max} = aM\mu k_\mu \tau (k_v V_g)^{3.3} R_{90}^{2/3} \left(\frac{1}{2.85 k_p}\right)^{3.3} \left(\frac{S_1 - d}{S_1}\right)^2 C_t \eta \qquad (9.1)$$

where

k_μ and k_v are coefficients to account for the nonuniformity of the fly ash density and gas velocity field, temperature, respectively. For double-pass arrangement, $k_v = 1.25$ and $k_\mu = 1.2$. When the gas turns 180° in front of the tube banks, $k_v = k_\mu = 1.6$.
τ is the time of erosion of the tubes (h).
μ is the density of the fly ash in the gas in the calculation section of the tube banks (g/m³).
k_p is the ratio of gas velocities at designed and average running load, respectively.
R_{90} is the percentage of the fly ash smaller than 90 μm.
M is the abrasion-resisting coefficient of the tube material. ($M = 1$ for carbon steel tube, $M = 0.7$ for steel with alloy).
V_g is the gas velocity in the narrowest section of the tube (m/sec).
a is the erosion coefficient of fly ash in the gas, mm sec³/(g h). It depends of the type of coal, but a default value of $a = 14 \times 10^{-9}$ mm sec³/(g h) may be used if appropriate test data is not available.
C_t factor taking account of strike efficiency.
η strike efficiency.

Erosion is affected by the following three major factors:

1. Flow and environmental conditions
2. Particle properties
3. Nature of target materials

A brief discussion of these factors is given below.

9.2.1.1 Flow and Environmental Conditions

The following flow and environment related conditions affect the erosion rate:

9.2.1.1.1 Angle of Impingement

The most important factor influencing erosion is the angle of impingement of solids, α on the target (Figure 9.2). Being minimal at $\alpha = 0$, erosion increases with the angle of attack and reaches a maximum at around 45°. Beyond this angle the erosion decreases again, especially for ductile materials, reaching a minimum value at 90°. If the material is brittle, an impact at this angle may crack it. It is now apparent that erosion is minimal on vertical tubes in a CFB boiler, where solids flow parallel to the wall.

The erosion rate of ductile materials (e.g., steel) increases with the angle of attack. The rate peaks at around 45° and then declines beyond this angle of attack. For brittle materials like refractories the erosion rate increases continuously. In fact, above 45° the rate of rise in erosion increases with the angle of attack (BEI, 1992a, 1992b).

9.2.1.1.2 Particle Velocity (or Momentum)

Tubes in the back-pass of a boiler are eroded primarily due to the impact of fine particles carried by the flue gas. The erosion rate, E is given as (BEI, 1992a):

$$E = \frac{K I_a R}{T_g} \left(\frac{\text{ASH}}{\text{C}} \right) V_g^n \text{ nm/h} \tag{9.2}$$

K is a constant dependent upon tube material.
I_a is ash abrasiveness factor.
T_g is flue gas temperature, K.
V_g is the gas velocity between tubes, m/sec.
ASH/C is the ash to carbon ratio in fuel.
n is the exponent 2.3 to 2.7 for ductile materials and 2 to 4 for brittle materials (Basu et al., 1999). Raask (1985) took this to be 3.3.

Operating temperature and pressure do not exert much influence on erosion, especially in fluidized bed boilers where the temperature is rarely high enough to make the boiler tube walls soft. The erosion rate is directly proportional to the number of particles striking a wall as shown by Equation 9.3.

A corrosive environment may increase the erosion rate due to the simultaneous occurrence of erosion and corrosion.

9.2.1.2 Particle Properties

9.2.1.2.1 Mechanical Properties

Mechanical and physical properties of the impacted surfaces have a major influence on erosion. For example, a soft material could have a considerably lower erosion rate than a hard material. The powder hardness (Vickers Hardness, for example) alone does not define the erosion; the abrasiveness of the particle also plays a major role. Coal ash contains pyrite as well as quartz silica. The quartz (1081 kg/mm^2) is much more abrasive than pyrite (650 kg/mm^2). For this reason the overall abrasiveness I_c, of coal mineral is expressed in terms of the concentration of quartz C_q, and pyrite C_{py}, as (DOE, 1992):

$$I_c = [C_q + (0.2 - 0.5)C_{py}]I_q \qquad (9.3)$$

where I_q is the abrasiveness of quartz.

The crystalline structure of the quartz particles is also important. For example, the alpha form of quartz is significantly more abrasive than other forms in coal. Angular quartz particles exhibit higher erosion rates. In fluidized bed boilers, the combustion temperature being low, the silica in ash remains as silicon dioxide instead of being converted into alpha quartz. Furthermore, sometimes alkali-induced low-melting point eutectics result in a coating around the ash particles making them more rounded and hence less erosive.

Particle density may also have some effect on the erosion. For example, a denser particle would cause greater erosion due its larger momentum.

9.2.1.2.2 Equivalent Size and Sphericity

Particle size (>5 μm) has a major influence on the erosion, where the erosion rate increases as the size increases. Nonspherical angular particles cause greater erosion than rounded ones (Wang et al., 1992).

9.2.1.3 Nature of Target Materials

Material properties such as surface hardness, impact strength and ductility are important factors influencing the erosion. The crystallographic structure rather than the Vickers hardness affects the erosion of surface coatings.

9.2.2 Types of Erosion

The physical mechanism of erosion is of two types:

1. Two body (Figure 9.3a)
2. Three body (Figure 9.3b)

In two-body erosion, solids impact the wall and scour it using their own momentum, as shown in Figure 9.3. In three-body erosion, solids moving along the wall are impacted by particle clusters, which use the former solids as an abrading agent to erode the surface (Figure 9.3).

9.2.3 Erosion in Bubbling Fluidized Beds

Banks of water tubes are commonly located in the dense bed of BFB boilers (Figure 9.4) to maintain the bed temperature at desired levels by taking advantage of the high heat-transfer coefficient of the bubbling bed.

The tubes immersed in the bubbling bed are subjected to severe conditions such as contact with oxidizing–reducing gas pockets, impingement of hot solids and gas, causing both corrosion and erosion of the tubes. Erosion also occurs on distributor caps, which also come into contact with hot solids and high-velocity air jets, making erosion one of the major operational problems of BFB boilers.

Erosion of tubes immersed in bubbling beds is caused mainly by gas bubbles. The solids carried in the wake of the bubbles hit the tubes with the velocity of the bubbles, which could at times attain very high values. Therefore, a reduction in the bubble velocity might help reduce the erosion. We have seen in Chapter 2 that the bubble velocity is a function of the bubble size. Thus, efforts to reduce bubble size could be made by reducing bed height, lowering gas velocity, reducing particle size or employing novel techniques like bed vibration and pulse air supply.

Material Issues 305

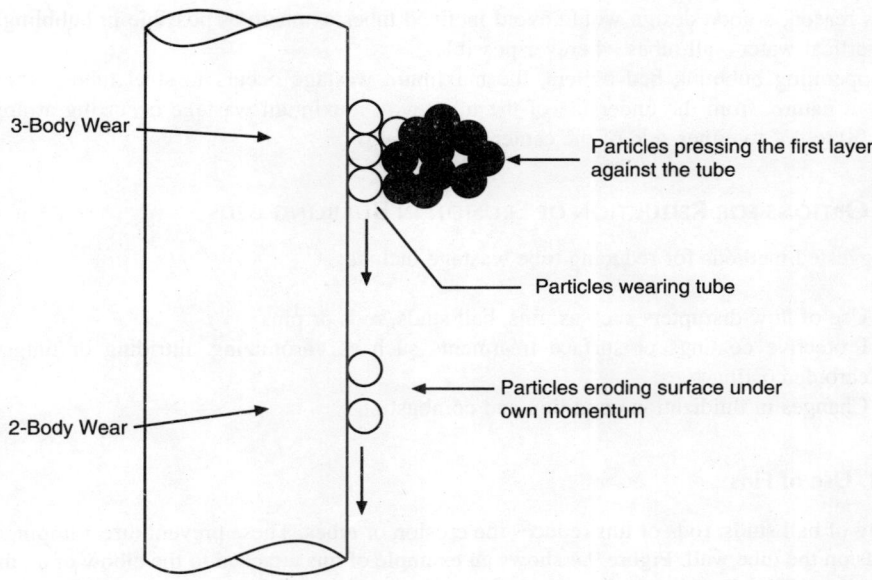

FIGURE 9.3 Simplified scheme of two-body and three-body erosion.

Bubbles are generated when the fluidizing air velocity exceeds that required for minimum fluidization (for Group B and D particles). The air in excess of minimum fluidization passes through bed as bubbles. Thus, low superficial velocity can reduce the bubble size and decrease in-bed erosion. In BFB boilers, which use coarse Geldart D particles in the bed, the erosion is low to moderate when superficial velocity is less than 2 m/sec (Stringer, 1984), but above 2 m/sec, the rate of erosion is high. Thus, for this bubbling bed boiler, the fluidizing velocity should be around 1.0 to 1.5 m/sec.

Inclined evaporative tubes are frequently used in the bed to ensure proper steam-water circulation, but are subjected to severe erosion. Buchanan (1995) found a wall thickness loss of 4.8 mm per side after 6100 service hours of a 2 in. diameter tube. For vertical wall tubes it was only 1.7 mm.

The particle motion near the surface of a tube is caused by bubbles generated at the tube rather than in the bed below. This may explain why the 45 t/h fluidized bed equipment (FBC) boiler at Georgetown had more erosion in the middle of the tube banks rather than in the lowest tubes.

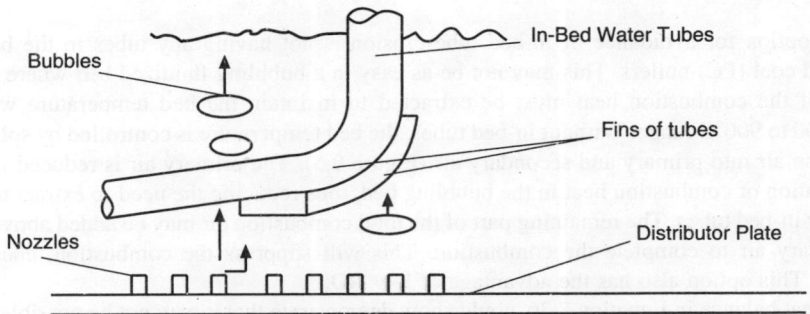

FIGURE 9.4 Use of fins on tubes to reduce erosion of tubes in a bubbling bed.

For this reason, a good design would avoid inclined tubes as much as possible in bubbling beds, using vertical water wall tubes wherever possible.

In operating bubbling bed boilers, the maximum wastage occurs in steel tubes, which are ductile in nature, from the underside of the tube, with maximum wastage occurring at approximately 30 to 45° to either side of the center line.

9.2.4 Options for Reduction of Erosion in Bubbling Beds

The suggested methods for reducing tube wastage include:

1. Use of flow disrupters such as, fins, ball studs, rods or pins
2. Protective coatings or surface treatments such as chromizing, nitriding or tungsten carbide coating
3. Changes in fluidization condition and combustion

9.2.4.1 Use of Fins

Welding of ball studs, rods or fins reduces the erosion of tubes. These prevent direct impingement of solids on the tube wall. Figure 9.4 shows an example of fins attached to the elbow of a tube in a bubbling bed. The elbows are more prone to erosion as this orientation, allowing 45° attacks from rising bubbles. Figure 7.18 shows a photograph of a tube bank with studs welded on for erosion protection. It shows that fins are welded on both the top and bottom of the tube to shield it from falling, as well as rising, particles.

9.2.4.2 Use of Coatings

It may not be possible to weld fins to all tube configurations and even doing so may not give the desired service life. In such cases, hard coatings may be applied to the surfaces of the tubes. Certain coatings applied by the hypersonic velocity oxygen fuel (HVOF) process (Buchanan, 1995) were found best for erosion protection, and reduced the erosion by a factor of five on inclined tubes and twelve on vertical tubes (HH-15 88%WC-12CO coating). Wang and Lee (1997) also found that ceramic Cr_3C_2 coating (HVOF) exhibits the highest erosion-corrosion resistance due to its favorable composition and fine structure.

The metallography and structure, rather than the hardness of the coating material determines its erosion resistance.

9.2.4.3 Avoidance of In-Bed Tubes

The best option for avoidance of in-bed tube erosion is not having any tubes in the bed, as in pulverized coal (PC) boilers. This may not be as easy in a bubbling fluidized bed where a certain fraction of the combustion heat must be extracted to maintain the bed temperature within the desired 800 to 900°C range. Without in-bed tubes, the bed temperature is controlled by splitting the combustion air into primary and secondary air (Figure 9.5). The primary air is reduced to restrict the generation of combustion heat in the bubbling bed, thus reducing the need to extract heat from the bed by in-bed tubes. The remaining part of the total combustion air may be added above the bed as secondary air to complete the combustion. This will suppress the combustion, encouraging pyrolysis. This option also has the advantage of low NO_x.

The heat balance in Equation 7.36 could show demonstrate that it may not be possible to avoid in-bed heat extraction for all fuels. However, for highly volatile, high-moisture fuels like biomass, or high-ash fuels like washery rejects, it is often possible to avoid the use of in-bed tubes.

Material Issues

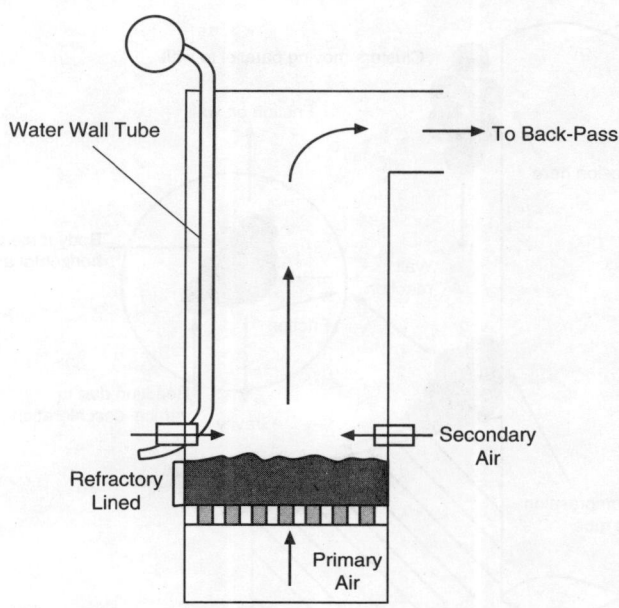

FIGURE 9.5 Use of secondary air in the absence of in-bed tubes.

9.2.5 EROSION IN CFB BOILERS

The following section describes the typical areas of a CFB boiler that may experience erosion.

9.2.5.1 Erosion in Combustor

In CFB boilers, solids flow down the vertical furnace walls as shown in Figure 9.6. Solids travel down the wall within a relatively narrow annular region at a typical velocity of 1 to 2 m/sec for small unit to about 6 to 7m/sec for very large boilers. As the angle of attack is zero, the two-body erosion on vertical walls is practically zero. Since there is no change in momentum in the horizontal direction, there is no external lateral force that could give rise to three-body erosion, but whenever there is a change in momentum severe erosion is observed. The following section discusses some of the areas of erosion.

9.2.5.2 Erosion of Water–Wall Interface

If there is any projection on the wall, the solids are forced to change direction, resulting in a lateral change in momentum. This gives rise to the three-body erosion, which is evident from the wear patterns observed on combustor/refractory interface (Figure 9.6). Small projections on the vertical wall are, therefore, easily eroded away. Figure 9.7 shows an example of how a hemispherical butt weld on tubes is worn down into a smooth shape.

The most common place of erosion is the water wall/refractory interface. A sudden change in particle momentum makes this site most vulnerable for erosion. Figure 9.8 shows a schematic of the refractory/tube interface showing the erosion pattern.

Weld overlays, as shown in later on Figure 9.20, have been the most commonly-used erosion protection at the refractory interface with the lower bed. It provides a sacrificial wear surface that can be replaced when it is worn out without affecting the parent tube. The life of the weld overlay depends upon the severity of erosion. Flame spraying has been tried to protect the tubes near

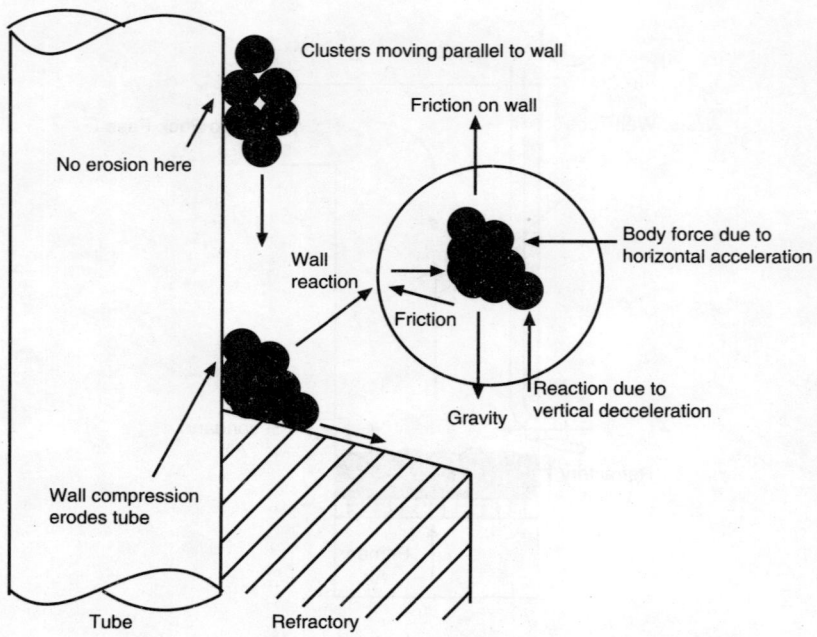

FIGURE 9.6 A change in momentum of vertically flowing solids would cause severe erosion in a CFB boiler.

the refractory interface, but it has not been very successful. Hard spray material generally cracks and peels off from the surface whereas soft materials wear very quickly.

The best means for avoiding erosion is to avoid the change in horizontal momentum of the solids near a tube wall. Figure 9.9 shows a modified design, where tubes are bent away from the location of momentum change to prevent erosion. This is the only long-term solution and is now widely used in new CFB boilers.

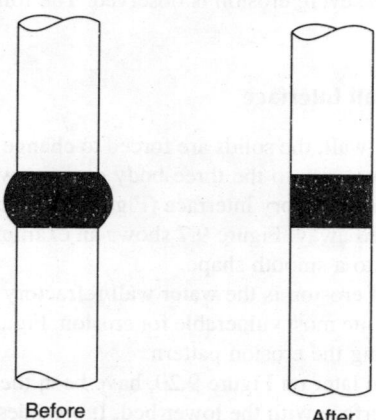

FIGURE 9.7 Typical erosion of butt welds on a tube.

Material Issues

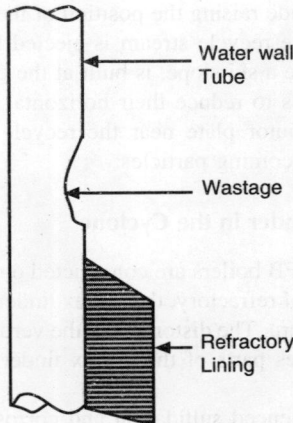

FIGURE 9.8 Metal loss at the water wall/refractory interface.

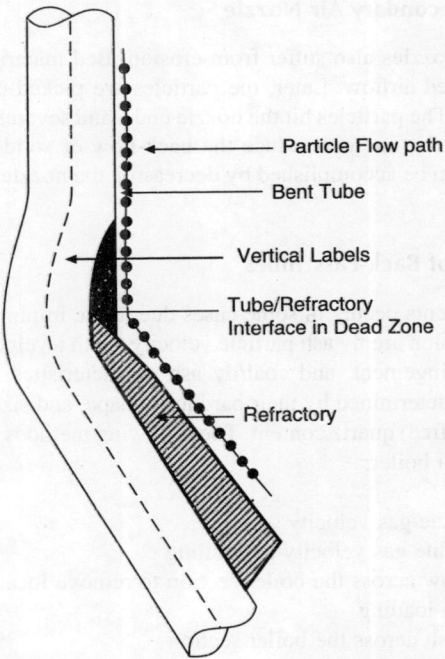

FIGURE 9.9 Erosion at the refractory/tube interface can be reduced by bending the tube away from the erosion prone zone.

9.2.5.3 Erosion of Distributor Plates

The distributor plate is sometimes damaged due to the erosive cutting of air jets coming out of its nozzles. The immediate vicinity of the inlet of recycled solids is one vulnerable spot of the distributor plate. Here, a concentrated stream of recycled solids from the cyclone impinges at a relatively high velocity on to the distributor plate. A significant component of the impinging velocity is parallel to the plate, which erodes the bed nozzles nearby.

Possible remedial actions include raising the position of the inlet relative to the air distributor or changing the angle at which the recycle stream is ejected from the loop seal, among others. Sometimes a small upward lip, like a ski-slope, is built at the exit of the recycle duct and bottom of furnace walls force the particles to reduce their horizontal momentum. An increased airflow through the nozzles in the distributor plate near the recycle exit may also provide increased upward flow to help deflect the incoming particles.

9.2.5.4 Erosion of the Vortex Finder in the Cyclone

Most vortex finders operating in CFB boilers are constructed out of metallic plates. Unlike the rest of the cyclone, which is made out of refractory, the vortex finder is made from stainless steel plates of thickness varying from 6 to 12 mm. The distortion of the vertical plates due to high temperatures is common. This distortion exposes parts of the vortex finder to the path of the exit gas–solid stream, causing severe erosion.

Some boilers have also experienced sulfidation and corrosion of the plates due to reducing conditions. The solution to this problem has been to replace the plates by high temperature-resistant materials such as Incoloy and Hestalloy.

9.2.5.5 Erosion of the Secondary Air Nozzle

The lower secondary air nozzles also suffer from erosion. Bed materials enter the nozzles during shutdown or during reduced airflow. Later, the particles are picked up by the secondary air and blown through the nozzle. The particles hit the nozzle ends, and severely erode them. By increasing the air velocity in the nozzle, one can reduce the back-flow of solids into the nozzle and hence reduce the erosion. This can be accomplished by decreasing the nozzle diameter and increasing the air flow rate.

9.2.5.6 Fly Ash Erosion of Back-Pass Tubes

Erosion of back-pass elements occurs in some cases due to the impingement of fly ash. Important variables affecting this erosion are fly ash particle velocity (with a velocity exponent between 2 and 4) and size, angle of impingement, and coal/fly ash characteristics. The erosive characteristics of the fly ash particles are determined by their hardness, shape, and size. The erosiveness of the fly ash is usually based on its (free) quartz content. The following methods are used for reducing fly ash erosion at any location in a boiler:

1. reducing the bulk flue gas velocity
2. reducing the local flue gas velocity by baffling
3. level out the gas flow across the boiler section to remove localized turbulent regions
4. reducing the fly ash loading
5. spread out the fly ash across the boiler section
6. decreasing the erosivity of fly ash

9.2.5.7 Thermocouple and Temperature Elements

In the case of large openings for instruments, the water wall tubes are bent out of the vertical plane as shown in Figure 9.10, creating a new site for erosion. The tubes above the opening experience little or no erosion, whereas the return bends below the opening are worn significantly.

Thermocouple locations on the combustor wall are also subjected to erosion as thermocouples project into the furnace, disturbing the solid flow in their vicinity. Thermocouple probes penetrate sufficiently far into the fluid stream to allow accurate assessment of the fluid temperature as shown in Figure 9.11. These probes are the source of erosion for the adjacent water wall tubes. The probes

Material Issues   311

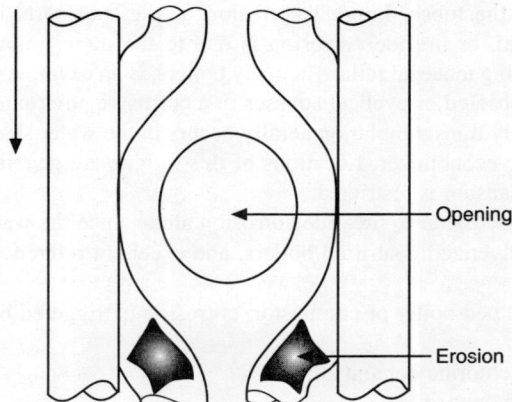

FIGURE 9.10 Erosion of boiler tubes below an opening on the wall.

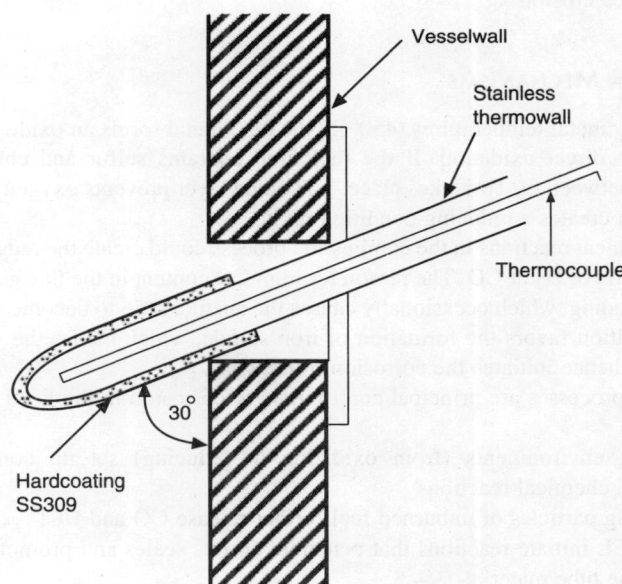

FIGURE 9.11 Thermocouple penetrating a CFB water wall needs an erosion protection sheath for temperature measurements.

are therefore inserted at an angle of 30° to prevent erosion, and protected by a shield preferably made of 309 SS steel.

9.3 CORROSION POTENTIALS

Corrosion, slagging, scaling, deposition, and fouling are some of the major reasons for failure of pressure parts in a utility boiler. These phenomena cause both failure of tubes and increased fuel consumption through a decrease in heat-transfer efficiency. Corrosion can occur either on the outer or the inner surface of the tubes. Internal corrosion of pressure parts is mostly due to water chemistry, while external, or fireside corrosion is due to the direct impact of fuel combustion. Corrosion fatigue, a leading material failure in many boilers, is an example of water-side corrosion. It is the failure due to repeated or cyclical stresses in a corrosive environment. Corrosion fatigue, whose cracks are typically transgranular, generally occurs in the water side of the boiler, either in the water walls or in the economizer. Locations of this fatigue are generally associated with the areas where thermal expansion is restricted.

Here, we limit the discussion to fire-side corrosion alone since the water-side chemistry is no different from that in pulverized coal-fired boilers, and excellent references are available on that topic.

In a typical fluidized bed boiler or combustor, corrosion is triggered by:

- Fuels with high chlorine content
- Poor combustion control
- High gas temperature
- High metal temperature
- Local reducing atmosphere
- Tube surface erosion

9.3.1 Corrosion Mechanisms

At typical operating metal temperatures (450°C), the tube metal forms an oxide layer that protects the tube wall from direct oxidation. If the fuel fired contains sulfur and chlorine, a complex chemical reaction between the two takes place. The oxide layer prevents oxygen from reaching the tube surface, which creates a reducing condition on it.

Undesired chemical reactions in the combustion process could create the reducing environment through the formation of H_2 or CO. The reason for high CO content in the flue gas in many cases is the unstable fuel feeding, which occasionally causes the air/fuel ratio to become substoichiometric. The reducing condition favors the formation of iron sulfide, which makes the scales on the tube nonprotective, and hence initiates the corrosion process.

The following processes are principal contributors to corrosion in fluidized bed boilers:

- Fluctuating environments (from oxidizing to reducing) set up complicated non-equilibrium chemical reactions
- Smouldering particles of unburned fuel further release CO and HCl
- CO and HCL initiate reactions that penetrate porous scales and promote intergranular attack of the tube material
- Sintered ash deposits prevent the access of oxygen to the tube surface, further encouraging the reducing conditions in the porous scales

9.3.2 Examples of Fire-Side Corrosion

This type of corrosion occurs on the outside of the tubes, which are exposed to the hot gas or fluidized solids.

9.3.2.1 Furnace Wall Fire-Side Corrosion

Fire-side corrosion of furnace wall tubes is caused primarily by localized reducing conditions and high heat fluxes. An example of tube rupture due to corrosion is shown in Figure 9.12.

FIGURE 9.12 Furnace wall tube failure by fire-side corrosion.

9.3.2.2 Fire-Side Corrosion Fatigue

Damage initiation and propagation result from corrosion in combination with thermal fatigue. Outer tube surfaces experience thermal fatigue stress cycles, which can occur from normal shedding of slag, or from cyclic operation of the boiler. Thermal cycling, in addition to subjecting the material to cyclic stress, can initiate cracking of the less elastic external tube scales and expose the tube base material to repeated corrosion. The indicator of this phenomenon is the tubes developing a series of cracks that initiate on the outside surface and propagate into the tube wall.

9.3.2.3 Fire-Side Ash Corrosion of Superheater

Fire-side ash corrosion is a function of the ash characteristics of the fuel and the boiler design. Damage occurs when certain constituents of the coal ash remain on the superheater tube surfaces in a molten and highly corrosive state.

9.3.2.4 Fire-Side Corrosion of Superheater

Combustion of biomass and recycled fuels is known to increase the risk of fire-side corrosion of the superheaters located in the flue gas-pass beyond the solids separator. Three main types of corrosion have been reported to take place on the heat-transfer surfaces:

1. Oxidation
2. Sulfidation
3. Chlorination

Chlorine content of the deposits on the tubes often is the primary cause of metal loss. The evaluation of corrosion risk is often done by examining the fuel characteristics, but the role of bed material and combustor type must not be neglected.

9.3.3 CHEMISTRY OF FIRE-SIDE CORROSION OF HEATING SURFACES

9.3.3.1 Sulfur Corrosion

Sulfur corrosion, often referred as sulfidation, is a very common phenomenon in coal combustion, where the sulfur in fuel reacts with oxygen in the combustion air forming SO_2 and SO_3. The SO_2

and SO_3 react with calcium in the bed material, forming $CaSO_4$ along with other sulfides and sulfates. The sulfides and sulfates formed may deposit on a superheater surface, initiating corrosion at the interface between the metal oxide and the deposit.

Sulfidation does not normally create a protective layer as oxidation does. The formed layer has a more pronounced defect structure, which can be scaled off more easily, leaving the base metal vulnerable to further oxidation and sulfidation. The sulfur corrosion may be due to either sulfate or sulfide.

9.3.3.1.1 Sulfate Corrosion

Sulfate corrosion can take place in an oxidizing environment when the SO_3 concentration is high. Sulfate deposit in a fluidized bed boiler is possible only when the fuel contains sodium or potassium. Sodium and potassium react with SO_2 in flue gas to produce Na_2SO_4 and K_2SO_4 (Figure 9.13), which vaporize at a high temperature (>900°C). For this reason, a fluidized bed boiler, which generally operates within 800 to 900°C, does not see much of this corrosion, but if the furnace exceeds 900°C there is a rapid rise in the corrosion of downstream tubes.

The dew point of these products is approximately 877°C. When the gaseous Na_2SO_4 and K_2SO_4 reach the relatively cold tube surfaces, they condense on the oxide film of the tube walls. The deposition rate of these compounds increases greatly when the chlorine content of the fuel is above 0.5%. The K_2SO_4 and Na_2SO_4 condense on heating surfaces that absorb Fe_2O_3, forming compound sulfates, $(Na, K)(Fe, Al)(SO_4)_3$:

$$3Na_2SO_4 + Fe_2O_3 \rightarrow 2Na_3Fe(SO_4)_3$$
$$3K_2SO_4 + Fe_2O_3 \rightarrow 2K_3Fe(SO_4)_3$$
(9.4)

The sulfate compounds $K_2Fe(SO_4)_3$ and $Na_3Fe(SO_4)_3$ cannot form a stable protective layer on the tubes as does Fe_2O_3. When the thickness of the sulfate deposit on the tube increases, the surface

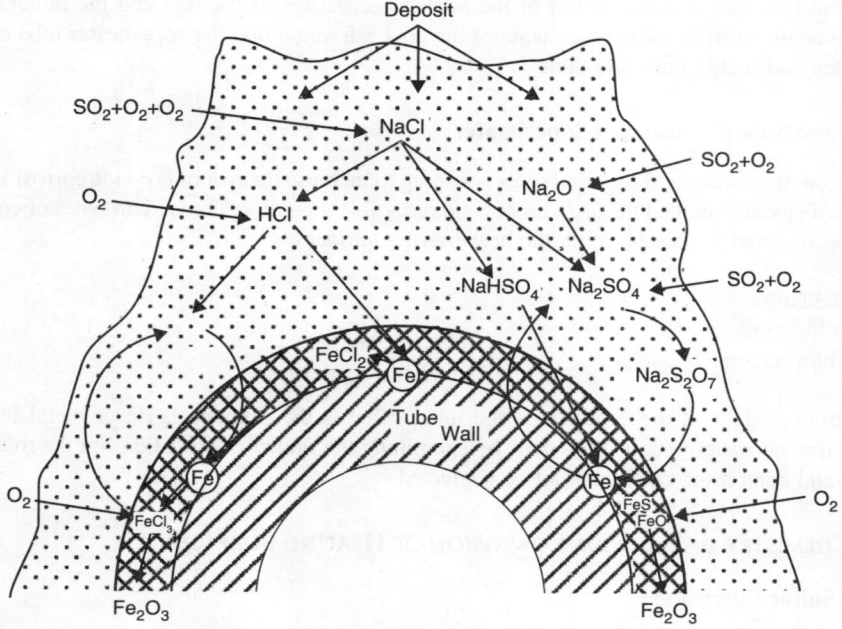

FIGURE 9.13 Chlorine and sulfur corrosion mechanisms (cyclic corrosion).

temperature of the tube rises to its melting point. The Fe_2O_3 protective membrane is dissolved by compound sulfate, which causes corrosion of tube walls.

Another possibility is pyrosulfate melting corrosion of alkaline metal:

$$FeS_2 \rightarrow FeS + [S] \tag{9.5}$$

When alkaline pyrosulfate exists in an adhesive layer, it melts at normal wall temperature because its melting point is relatively low, making the melting sulfate corrosion rate faster. Tests show that the corrosion rate of a melting sulfate ash deposit layer on a metal wall surface is much faster than in the gas phase.

The sulfur in fossil fuels produces sulfur dioxide in the course of combustion, but only a small part of this may be converted into sulfur trioxide. In the case of a CFB combustor, the sulfur dioxide is captured in the furnace by the injection of limestone. Thus, the amount of sulfur trioxide formation is much lower than in a conventional, PC-fired boiler.

9.3.3.1.2 Sulfide Corrosion

Sulfide corrosion occurs in reducing environments. Sulfides are formed from alkali metals and sulfur oxides in the presence of CO.

Another main factor leading to the corrosion of water wall tubes is hydrogen sulfide in the flue gas. The hydrogen sulfide reacts with the metal of the heating surface to produce iron sulfide, which then forms iron oxide. Unlike iron oxide, iron sulfide and ferric trioxide are porous and therefore do not protect the tube from corrosion:

$$\begin{aligned} H_2S + Fe \rightarrow FeS + H \\ H_2S + FeO \rightarrow FeS + H_2O \end{aligned} \tag{9.6}$$

In a CFB combustor, the only reducing zone is in the bed below the secondary air level. Thus, the water wall tubes in this zone are covered with refractory. Though there have been reports regarding failure of the refractory, no failures of pressure parts in this zone have been reported to date.

Another area of hydrogen sulfide attack in CFB boilers is the grid nozzle. Corrosion of grid nozzles has been reported in a number of CFB boilers.

9.3.4 CHLORINE CORROSION

Chlorine-induced corrosion is common in boilers, incinerators or gasifiers firing chlorine-containing fuels like biomass, high-chlorine coals and waste fuels. Figure 9.13 shows a schematic for chlorine induced corrosion.

Chlorine corrosion can take place as gas-phase corrosion in which the HCl in the flue gas, produced in the combustion of high-chlorine fuels, comes in contact with unprotected metal iron of the tube wall to consume the iron oxide and produce iron chloride (Figure 9.13). This process is demonstrated below:

$$\begin{aligned} Fe + 2HCl &\rightarrow FeCl_2 + H_2 \\ FeO + 2HCl &\rightarrow FeCl_2 + H_2O \\ Fe_2O_3 + 2HCl + CO &\rightarrow FeO + FeCl_2 + H_2O + CO_2 \\ Fe_2O_3 + 2HCl + CO &\rightarrow FeO + FeCl_2 + H_2O + CO_2 \end{aligned} \tag{9.7}$$

Owing to its low evaporating temperature, ferric chloride ($FeCl_2$) volatilizes as soon as it is formed corroding away the pipe.

At the same time, due to the destruction of the oxide film, hydrogen sulfide, formed in other combustion or gasification reactions, reaches the surface of the metal, accelerating the rate of metal wastage through corrosion.

These reactions take place in the region between the deposit and the metal oxide. The chlorine corrosion rate depends on the chlorine concentration in the flue gas, tube metal temperature, and tube steel composition. At a metal temperature of around 500°C, a temperature increase of 15°C approximately doubles the chloride corrosion rate.

Chlorine corrosion is often accelerated by alkaline components in the fuel. At low temperatures, chlorine corrosion can take place as hydrochloric acid corrosion, but in the case of superheater corrosion, the mechanism is initiated when the fuel contains sufficient amounts of chlorine and the superheater tube temperature is sufficiently high for chlorides to form molten eutectics. If sulfur-containing components are present in the fuel, chlorine corrosion can cause very high metal loss rates.

There is a direct correlation between the chlorine content of the fuel and the corrosion rate. The corrosion rate r, of austenitic tube depends on the metal temperature T_{metal}, temperature T_g, and the chlorine content of fuel Cl. Based on results from a large number of PC plants the following dependency has been suggested (BEI, 1992a):

$$r \alpha (T_{metal} - C)^p (T_g - 273)^q (Cl - D) \quad \text{nm/h} \tag{9.8}$$

where C and D are constants and p and q are exponents.

9.3.5 VANADIUM CORROSION

Vanadium (especially vanadium pentoxide) causes a rapid increase in the rate of metal loss in a sulfur corrosion environment. The role of vanadium in this corrosion case is not fully understood. Some researchers claim that the vanadium pentoxide acts as a catalyst, while others think that the role of vanadium is more complex. However, significant concentrations of these salts require high amounts of vanadium in the fuel. Some fuels, such as petroleum cokes and refinery tars may contain sufficient amounts of vanadium for the corrosion attack to initiate.

When the high-temperature flue gas flows over the convection surface, the SO_2 converts into sulfur trioxide because of the catalytic actions of V_2O_5 and Fe_2O_3 in the ash deposit. Sulfur trioxide actively participates in corrosion reactions.

9.3.6 PREVENTION OF HIGH-TEMPERATURE CORROSION

The following measures can reduce the risk of fire-side corrosion in a boiler:

9.3.6.1 Low Oxygen Combustion

The use of low excess air can reduce the formation of sulfur trioxide and increase the formation of sulfur dioxide. The latter goes up the stack, while the former contributes to corrosion. The low oxygen combustion may reduce the amount of vanadium pentoxide formed in case of oil firing. The vanadium may instead be oxidized to less corrosive vanadium trioxide. However, the low excess air may lead to a loss in combustion efficiency.

9.3.6.2 Uniform Distribution of Combustion Air

In the case of CFB boilers, the combustion air is supplied above the bed by nozzles set all around the furnace. The air jets leaving the nozzles disperse into the bed at a finite rate. The uniform distribution of combustion air in the boiler is ensured by maintaining a good ratio of bed area to the number of nozzles. Maintaining the uniformity of temperature all around further reduces the chance of localized reducing zones.

9.3.6.3 Uniform Distribution of Fuel in the Combustion Chamber

The fuel is fed into a CFB boiler either above the bed or through the loop seal return leg. As the amount of fresh coal feed is very small compared to the mass of bed material going through the loop-seal, fuel fed into the loop-seal is automatically mixed into the bed. However, the coal fed from the side is not immediately mixed, leaving fuel-rich and fuel-lean areas around the furnace. The larger the number of feed points, the better the mixing. Therefore, this mixing can be ensured by maintaining a small ratio of bed area to the number of feed points.

9.3.6.4 Avoidance of Local Hot Spots

The higher the temperature of the heating surface, the greater is the potential for high-temperature corrosion. Thus, efforts should be made to avoid the high temperature spots in the bed.

9.3.6.5 Use of Additives to Prevent High-Temperature Corrosion

The addition of metal magnesium or dolomite can reduce the corrosion in oil-fired boilers. The magnesium sulfate or the mixture of magnesium oxide and vanadium oxide forms high-melting-point deposits on the tubes. Thus, the high-temperature corrosion due to melting of the deposit is reduced.

9.3.6.6 Control of the Flue Temperature at the Furnace Outlet

From a corrosion standpoint, the furnace heat release rate may be restricted. The furnace exit gas temperature should also be lower for coal with higher propensity for corrosion. Since the furnace exit gas temperature of a fluidized bed boiler is lower than that of a PC-fired boiler, this has very little significance in the present context.

9.3.6.7 Use of Corrosion-Resisting Alloys

High-chromium steel can be used in furnaces firing corrosive coals. This has superior anticorrosion properties. However, high-chromium steel is more expensive than carbon steel. Thus, a compromise between initial cost and service life expectation should be made while selecting the tube material. The maximum allowable stress at different metal temperatures for several types of steel and alloys are given in Figure 9.1.

9.3.6.8 Adoption of Flue Gas Recycling

Flue gas recycling may reduce the corrosion of high-temperature heating surfaces because flue gas recycling reduces the maximum furnace exit temperature as well as the sulfur trioxide content of the gas.

9.3.6.9 Avoidance of Simultaneous Occurrence of High Gas Temperature and High Wall Temperature

It is known that for a given wall temperature, the corrosion rate is higher for higher gas temperatures. Also, it is known that if the wall temperature is lower than 550°C the corrosion rate is greatly reduced. The boiler configuration (i.e., gas path design) should therefore be made such that the metal temperature and gas temperature combination does not give excessive corrosion.

9.3.7 Erosion–Corrosion

Corrosion and erosion could take place simultaneously in some places. In the simultaneous erosion–corrosion process, the damaged protective oxide layer, or the corrosion product formed, is removed from the tube surface due to particles colliding with the metal surface. The erosion rate depends on the particle velocity and characteristics, as well as on the metal surface properties. The erosion–corrosion is considered to be the main reason for superheater material wastage in the vicinity of soot-blowers. The erosion–corrosion rate in the furnace is often high, and for this reason the lower edges of the Omega panels used in CFB boilers have to be protected with shields.

9.3.8 Fouling and Deposit Formation

When solid fuel is combusted in a fluidized bed, some ash will be carried with the flue gas out of the furnace. Part of the ash deposited inside the furnace can melt and cause *slagging*. If the ash is deposited in the gas duct downstream of the furnace, the phenomenon is called *fouling*, which does not involve melting of the ash. Fouling could occur on the superheaters located outside the furnace. The primary concern of fouling is reduced heat-transfer rate in the superheater. If the ash contains the ingredients needed for corrosion, the previously mentioned corrosion can take place in the superheater tubes. The main initiators of a fire-side corrosion attack are the deposits formed on the surface of the superheater. In order to remove the deposits from the superheater tubes, different soot-removal systems are used, the most common being steam soot-blowing.

9.4 STEELS USED IN FLUIDIZED BED BOILERS

To be cost-competitive, a boiler design must utilize the lowest-cost materials that meet the service and performance requirements of the boiler. As with conventional boilers, the fluidized bed boiler designer cannot afford the luxury of overconservatism when selecting a material for a particular application. This could increase the cost beyond acceptable range. Optimistic choice of low cost materials, on the other hand, may lead to subsequent material changes in troublesome areas. Each fluidized bed boiler manufacturer optimizes material selection differently but follows these general application guidelines:

1. Low-carbon and alloy steels are used for pressure part heat transfer duty under oxidizing conditions as well as other structural duties.
2. Refractory materials are used for erosive and/or reducing conditions, including the lower combustor, cyclone, and smaller components, such as the loop seals and FBHE casing.
3. Expansion joints are used to allow the differential expansion between large boiler components, such as the cyclone and combustor.

The application of these materials in a typical CFB boiler is indicated in Figure 9.14.

9.4.1 Carbon and Alloy Steels

The most important use of steel in a fluidized bed boiler is in the high-pressure boiler tubes, including those in the evaporator, superheater, reheater and water wall surfaces. These tubes are arranged in a variety of complex configurations, including:

1. Membrane wall (water- or steam-cooled)
2. Convective tube bundles (superheater, reheater, and economizer)
3. Support tubes used to support tube bundles

Other, more specialized tube arrangements include:

1. Fluidized bed heat exchanger bundles
2. Suspended panels in the upper combustor

Material Issues 319

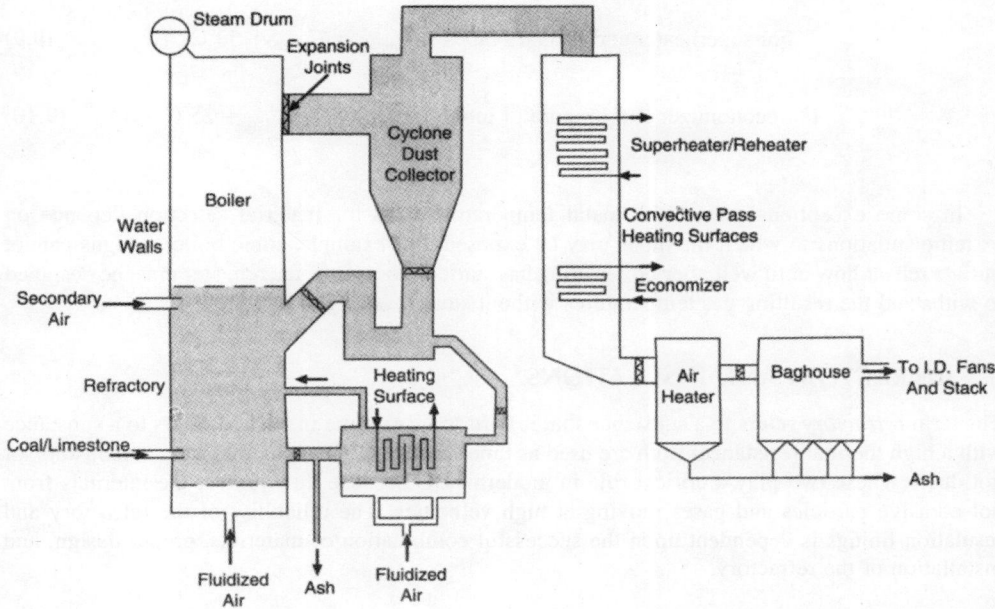

FIGURE 9.14 Application of steel, refractory, and expansion joints to a CFB boiler.

3. Planten tubes (wing walls) in the combustor
4. Water-cooled air plenum
5. Water-cooled cyclones

When selecting steel it is useful to note the function of different alloying materials. For example:

- Chromium improves corrosion resistance in general.
- Nickel is good against chlorine corrosion, but poor against sulfur corrosion.
- Aluminum is good against chlorine corrosion and creates good oxide protection, but is expensive.
- Silica is good against chlorine corrosion, but may be affected by alkali metals.

While selecting the steel, the primary consideration is the predicted metal temperature followed by the corrosion and erosion potential. Once the material is chosen, the tube wall thickness is calculated based upon the allowable stress at the predicted metal temperature plus allowances for the predicted erosion rate over the design service life.

In fluidized bed boilers applications, the majority of the pressure parts (tubes), including the evaporative and steam-cooled membrane wall, economizer surface, and low-temperature superheater, are made of low-carbon steel. More expensive alloys are used selectively in the higher metal temperature section of the superheater and/or reheater.

Owing to the high steam-side heat-transfer coefficient the tube metal temperature is closer to the steam temperature T_{steam}, and less dependent on the gas/solids temperature in the furnace. Thus, for determining the strength of the tube metal, its temperature T_{metal}, is assumed to be as follows:

For superheater and reheater tubes : $T_{\text{metal}} = T_{\text{steam}} + 50°C$ (9.9)

For economizer and evaporator tubes : $T_{\text{metal}} = T_{\text{saturated}} + 25°C$ (9.10)

In some exceptional cases, the metal temperature used for material selection depends on extreme situations to which the tubes may be exposed. For example, some boiler designs cannot initiate reheat flow until well after boiler firing has started. Therefore, the reheater must be designed to withstand the resulting gas temperatures without steam cooling at low loads.

9.5 REFRACTORY AND INSULATIONS

The term *refractory* refers to a substance that is hard to fuse, while insulation refers to a substance with a high thermal resistance. Both are used as inner linings of gasifiers, furnaces, combustors, or hot ducts. These two play a critical role in modern FBC because they protect the internals from hot abrasive particles and gases moving at high velocities. The reliability of the refractory and insulation linings is dependent upon the successful combination of materials, proper design, and installation of the refractory.

9.5.1 IMPORTANCE OF LINING

The inner lining of a furnace or hot duct serves two purposes: protection against erosion and protection against high temperatures. Refractory serves the first purpose and insulation serves the second.

9.5.1.1 Erosion Resistance

Erosion resistance is a critical criterion for refractory selection. Potential for severe erosion in the cyclones and transfer lines requires that special erosion-resistant materials be used in these areas. ASTM test C-704 is a commonly-used procedure for evaluating the relative erosion resistance of refractory materials.

9.5.1.2 Insulating Surfaces

Another important feature of the lining is to insulate against heat loss in areas where heat cannot be transferred to the water wall tubes. It is discussed in Section 9.5.4

9.5.2 PROPERTIES OF REFRACTORY

Principal qualities required in a refractory material are (BEI, 1992a, 1992b):

1. Resistance to the temperatures to which it is likely to be exposed
2. Resistance to any stress likely to be imposed by adjacent material
3. Resistance to any vibrations and mechanical blows that may occur
4. Resistance to the slagging action of the fuel
5. Uniform expansion and contraction properties
6. Resistance to environmental attack associated with oxidizing or reducing conditions

Important physical properties of the materials selected for the lining include erosion resistance, thermal conductivity, volume stability, and thermal expansion/shrinkage.

9.5.2.1 Physical and Mechanical Properties of Refractories

The physical and mechanical properties of refractory materials are as follows:

1. Refractoriness (pyrometric cone equivalent) and maximum continuous operating temperature
2. Changes in dimensions when heated
3. Strength of refractories (modulus of rupture, cold crushing strength)
4. Spalling resistance
5. Thermal conductivity

The thermal conductivity of refractory materials increases with increasing temperature. Figure 9.15 shows the variation of the thermal conductivities of several refractory materials with temperature.

9.5.3 TYPES OF REFRACTORY

Refractory materials may be divided into three major groups (BEI, 1992a, 199b):

1. Prefired shapes such as bricks, tiles, special shapes from fire clay, grog or silicon-carbide materials.

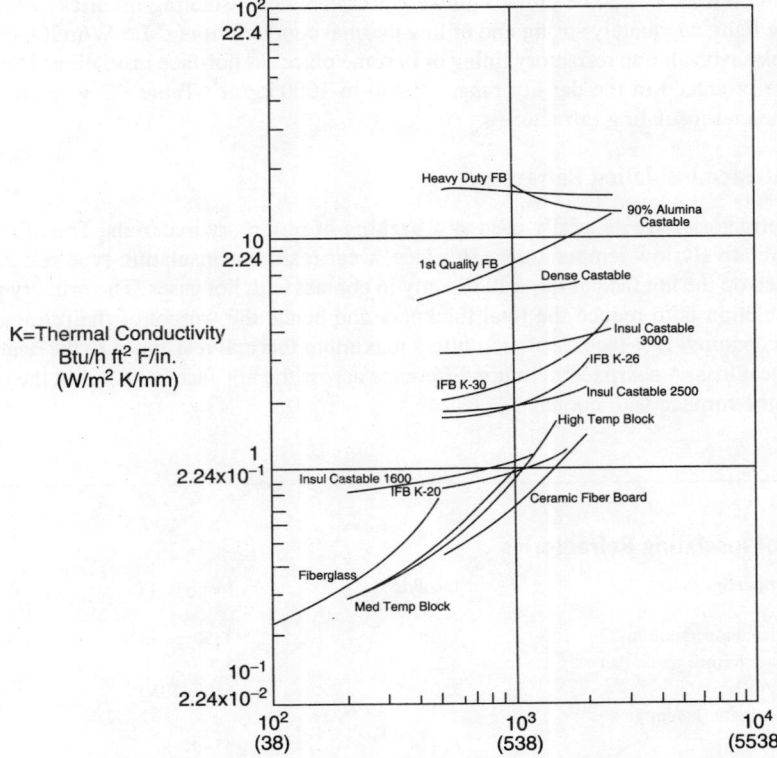

FIGURE 9.15 Thermal conductivity of various refractory materials.

2. Clay-based mouldable materials which can also be rammed into place. These are either alumino-silicate or silicon-carbide chemical-bonded refractory.
3. Castable refractories, which are used by casting or gunning. These are cast by mixing one of the following cements with aggregate or steel fiber.
   1. Portland cement
   2. High-alumina cement
   3. Low-iron high-alumina cement

For gunning one could use refractory cement, aggregate with certain amount of clay.

9.5.3.1 Physical Properties of Refractories

The most important physical or mechanical properties of refractories are follows:

1. Refractoriness and maximum continuous operating temperature
2. Changes in dimension when heated and shock resistance
3. Strength (modulus of rupture and cold crushing strength)
4. Spalling resistance
5. Thermal conductivity

9.5.4 TYPES OF INSULATION

The insulating refractory can be either lightweight castable or insulating firebrick. The insulating castables are light, adequately strong and of low thermal conductivity (<1.5 W/m K), thus serving as either back-up insulating refractory lining or in some places as hot-face insulation. The insulating castables are produced in the density range of 400 to 1600 kg/m^3. Table 9.2 gives properties of some commercial insulating refractories.

9.5.4.1 Hot-Face Insulating Refractory

Insulating refractories are generally used as a backing of refractory materials. Therefore, they are exposed to relatively low temperatures. Hot-face, wear-resistant, insulating-type refractories are, however, used on the hot face of the wall directly in contact with hot gases. The primary purpose of hot-face insulation is to reduce the total thickness and hence the weight of the refractory, in the interests of economy. Hot-face insulation offers maximum thermal resistance to the heat, reducing the overall heat loss. A sharp temperature difference across the hot-face insulator in the hotter zone helps make the furnace wall cooler.

TABLE 9.2
Properties of Insulating Refractories

Properties	Insulyte 7	Insulyte 11	Insulyte 13
Maximum service temperature in °C	1100	1350	1350
Chemical analysis ferrous oxide (%)	8.5	3.5	3.5
Dry density (kg/m^3)	750–850	1100–1250	1300–1450
Cold crushing strength (kg/cm^2)			
at 110°C	12–14	35–45	50–60
at 800°C	4–6	25–30	30–40
at 1100°C	6–8	25–30	30–40
at 1300°C		40–45	50–70

9.5.4.2 Fire Brick or Formed Refractory

The advantages of fire brick refractory are:

- Reduced surface penetration
- Resistance to chemical attack
- Limited surface wear
- Expansion characteristics tighten the lining during operation

It has some limitations as listed below:

- Labor intensive
- Longer installation schedule
- Difficult to repair

The common types of fire bricks refractory are as follows:

- Fire clay enhanced matrix — 42% Al_2O_3
- Andalusite — 59% Al_2O_3
- Phosphate bonded, burnt — 85% Al_2O
- Mulite bonded, tabular alumina — 91% Al_2O_3

They have crushing strengths varying from 500 to 1300 kg/cm^2.

9.5.4.3 Castables

Castables are cement-like substances that are formed on-site:

1. *Low-cement castables*. Properties of low-cement castables (Table 9.3) are low porosity and high strength
2. *Whytheat*. Whytheat (Table 9.4) is another type of castable used for hot-face, wear-resistant surfaces. This is a general-purpose, high-temperature castable used for service temperatures up to 1750°C and where erosion is less severe

TABLE 9.3
Properties of Low-Cement Castables

Properties	LC-45	LC-60	LC-90
Maximum service temperature in °C	1550	1600	1700
Chemical analysis			
Aluminum oxide	45	60	90
Ferrous oxide	0.8–1.0	1.5	0.8–1.0
Dry density (gm/cc)	2.3	2.6	3.0
Maximum grain size	6	6	6
Percentage of water required for casting	5.5–6	5.5–6.0	4.0–4.5
Cold crushing strength (kg/cm^2)			
at 110°C	12–14	35–45	90–100
at 800°C	4–6	25–30	60–70
at 1100°C	6–8	25–30	60–70
at 1300°C		40–45	70–90

TABLE 9.4
Properties of Whytheat Refractory

Properties	Whytheat A	Whytheat C
Maximum service temperature in °C	1750	1500
Chemical analysis		
Aluminum oxide	90	50
Ferrous oxide	0.8	1.5
Dry density (kg/m^3)	2800	2100
Maximum grain size	5	5
Linear change (%)	±1	±1
Cold crushing strength (kg/cm^2) at 110°C	550	350

9.5.4.4 Plastics

Plastics are dense, strong, erosion-resistant materials that can be rammed, vibrated or gunned into place and are used for patching existing units as well as in original installations. High-alumina plastics are in common use. Generally, plastics are phosphate-bonded, air-setting plastic masses, which were popular at one time, but have restricted use today due to their higher shrinkage characteristics.

Disadvantages of plastic refractories

- Installation for large, overhead areas requires forming
- Limited shelf-life
- Careful thermal capture required
- High-shrinkage on initial thermal cure

9.5.4.5 Ceramic Fiber

Ceramic fiber is often used in fluidized beds as expansion joint filler and as a *wrap* around metal projections that penetrate the lining to accommodate thermal expansion of the metal. It is used in the form of boards or blankets of various thicknesses.

9.5.5 ANCHORS

A refractory cannot hold to its enclosure by itself. Often, anchors are used to hold it to the metal enclosure of the refractory lining. The design of anchors is very important to the stability of the refractory, since they hold the lining to the wall in position. The anchors are generally V- or Y-shaped and are connected to a threaded stud, which is welded to the steel shell. The ends of the anchors are covered with plastic caps to allow the thermal expansion of the anchors as the temperature rises. Two common types of anchor materials are alloys and ceramics. Alloys commonly used are the 300-series stainless steel and Incoloy 800, which are fabricated from plate, rod, and bar stock or cast in foundry.

9.5.5.1 Alloy Anchors

Ductility, ease of manufacture, and low cost are some of the advantages of alloy anchors, whereas their limitations are reduced tensile strength and oxidation resistance at high temperatures. The tips of alloy anchors should be coated with combustible material to provide for expansion of the alloy inside the lining. Commonly used materials for anchors are: SS 304, SS 316, Incoloy-800HT, and Hayness-230.

Material Issues   325

9.5.5.2 Ceramic Anchors

Ceramic anchors are typically made of fire clay or high-alumina materials having a maximum diameter of 3 to 5 in.

9.5.6 DESIGN CONSIDERATIONS OF LINING

The design of the refractory lining of a fluidized bed boiler should meet both thermal and mechanical requirements.

9.5.6.1 Thermal Requirement

The lining of a furnace is designed for a specified heat loss Q. The thickness of the lining is calculated considering the heat conduction between the furnace and its exterior:

$$Q = A \frac{\Delta T}{R}$$

where ΔT is the temperature difference between the outside of the casing and the hot face, and A is the area. The thermal resistance R, is calculated as the sum of the various film resistances at the hot face of the refractory lining, the outside film resistance on the shell, and the conductive resistance of the refractory lining. The hot-face temperature of the lining is either calculated from the thermocouple fitted in the hot face to measure the flue gas temperature or taken as an input parameter at the design stage. The thickness is known from the conductive resistance of the lining.

9.5.6.2 Mechanical Requirements

Before calculating the final thickness of a lining, the mechanical requirements of the lining must be taken into consideration. The lining should be supported by a sufficient number of properly spaced anchors. The rule is to space the anchors at 2 to 3 times the lining thickness for flat or nearly flat surfaces.

To measure the erosion resistance of the hot-face lining, one could place a spinning disc of refractory in a fluidized bed of the actual abrasive media. The erosion is measured as a function of the refractory's loss in weight.

9.5.7 AREAS OF REFRACTORY USE IN FLUIDIZED BED BOILERS

Refractory linings are used in a fluidized bed boiler to minimize erosion of the walls and to insulate components which are uncooled. The refractory helps protect the steel and reduce surface losses from large cyclones. In a CFB boiler (Figure 9.14), refractories are also used in the lower combustor wall, combustor outlet roof, loop-seal return legs, loop-seals, start-up duct-burner and external fluidized bed heat exchangers. Table 9.5 gives a list of suggested refractory and insulating materials for different areas of circulating fluidized bed boilers. The following section discusses these application areas in detail.

9.5.7.1 Combustor

The lower combustion zone (furnace) in most CFB boilers operates at 800 to 900°C and is generally in the reducing (i.e., oxygen deficient) condition. Severe thermal shock and thermal cycling are common here. Due to the reducing atmosphere, refractories low in free iron or iron oxide are recommended for the lower combustor.

The lower furnace chamber is covered with refractory of 1.2- to 2.5-cm thickness. Thick linings have frequently failed due to excessive cracking followed by pinch spalling. When the lining goes through a thermal cycling, the crack opens up at high temperatures, trapping bed materials and then

TABLE 9.5
Suggested Refractory Materials for CFB Boilers

Location	Refractory Type	Remarks
Cyclone — inlet and combustion chamber roof	Insulating refractory with ceramic fiber blanket	LCC insulating brick Ceramic fiber blanket
Cyclone target zone	Multi-layer thick lining with dense hot-face refractory	Block insulation LCC insulating brick Fireclay brick/phosphate bonded, burnt 85% Al_2O_3
Cyclone cone section	Multi-layer thick lining, dense hot-face refractory with insulating materials	Block insulation Insulating brick Fire clay brick
Cyclone return leg and loop-seal	Dense-phase refractory with insulating bricks	Insulating brick Low-carbon castable phosphorous bonded high-alumina brick
Lower combustor bed	Multi-layer thick lining with insulating refractory of low FeO and carbon with fire clay bricks as hot face	Block insulation LC-60 Superduty fire clay brick/SiC-based castables
Expansion joint between loop-seal leg and furnace	Insulating ceramic block insulation, fire clay brick	Block insulation Fiber ceramic blanket and wool Fire clay brick
Vortex finder	Abrasion-resistant alloy	Incolloy and Hestalloy
Furnace wall	Double-layer insulation lining	Insulating brick Ceramic fiber blanket
Bypass (superheater, economizer section)	Insulating refractory along with firebrick (brick and ceramic fiber blanket)	Insulating brick Ceramic fiber blanket/wool

Location	Layers of Material	Layer Thickness (mm)	Max Temperature of Service
Lower furnace wall	Shell insulating castable dense erosion-resistant castable	75–150	93
Lower furnace with water wall	Tubes, anchors dense abrasion-resistant castable	25–50	
Water wall interface	Tube, thin SiC tiles or Gunmite on studs		
Cyclone	Shell calcium silicate brick insulating firebrick dense abrasion-resistant superduty/mullet brick		982–1093 Temperature fluctuation
Dip-leg, loop-seal	Shell vibration-cast insulating castable dense fused silica-based castable with stainless steel fibers and burn-out fibers		
Roof	Shell insulating layer dense castable with stainless steel fibers		870
	Water tube dense high-conductivity SiC-based	25–30	
External heat exchanger	Multi-layer castables fused silica-based material for division wall		

at low temperature tries to shrink, but is prevented from doing so by the trapped materials. Therefore, it may fail.

Cracks are also formed due to shrinkage upon drying, coupled with creep of the hot face under compressive stresses. Generally, a 25- to 100-mm layer of a dense, abrasion-resistant castable or plastic with a high thermal conductivity is used for water wall tubes. Silicon-carbide-based castables are also used in new boilers. For repairs, phosphate-bonded plastics could be used after removing the defective castable. If the feedstock or recycle ash is high in alkalines, phosphate-bonded plastic is a better choice for the original lining. SiC tile is also used in this area.

9.5.7.2 Cyclone Inlet and Combustion Chamber Roof

The combustor roof and cyclone inlet are subjected to impact by high-velocity flue gases containing large particle clusters at high temperatures (800 to 900°C). Both gas and bed materials change direction here, and thereby cause greater impact on the wall. Therefore, the tubes adjacent to the inlet on either side, above and below are protected with refractory as shown in Figure 9.16. The thickness of the total lining varies from 300 to 400 mm. In the cyclone inlet, a dense-phase castable is used on the hot face along with one or two layers of insulating material. If the roof is made of water wall tubes, a dense, high conductivity castable, 25 to 50 mm thick, is used on studded tubes.

9.5.7.3 Cyclone

The cyclone wall of a CFB boiler experiences the most severe conditions. Thermal cycling is rather common as the temperature varies between 850 and 950°C with little change in operating parameters. For this region, a dense-phase refractory along with an insulating back-up of 400 to 500 mm is used.

Generally, a multilayer brick lining with a calcium silicate block next to the shell, followed by insulating firebrick and a hot face of dense, abrasion-resistant, superduty or mullite brick is used to resist the erosion of the refractory lining. Repairs to these linings, where some bricks have spalled or fallen out, can be made with phosphate-bonded plastic.

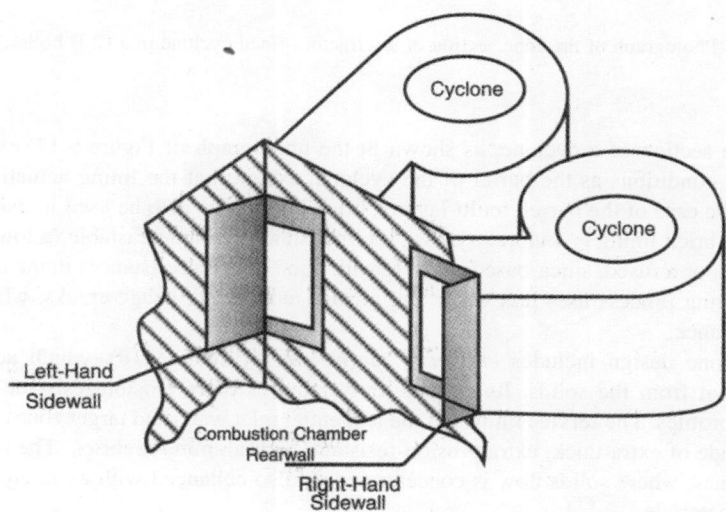

FIGURE 9.16 Protective refractory lining around the cyclone entry section protects the furnace tubes from erosion.

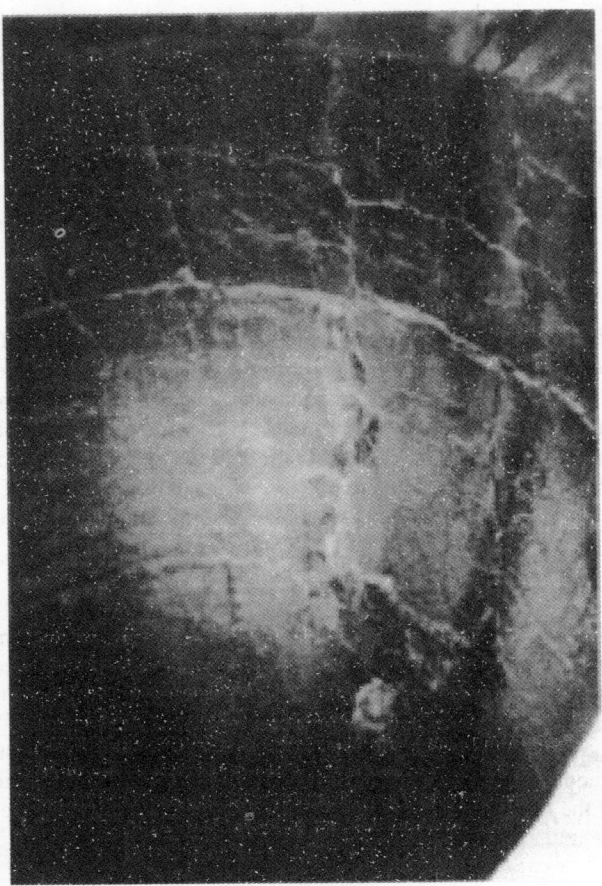

FIGURE 9.17 Photograph of the cone section of a refractory-lined cyclone in a CFB boiler.

The cone section of a cyclone, as shown in the photograph of Figure 9.17, experiences the same service conditions as the barrel of the cyclone, except that the lining actually rests on the shell. As in the case of the barrel, multi-layered brick linings can also be used in many cases. One alternative to brick lining is a vibratory-cast, thermal shock-resistant castable (a low- or ultra-low cement-based, or a fused, silica-based castable with good erosion resistance) in the cone area. The vibratory casting process uses less water and greatly reduces shrinkage cracks, while improving erosion resistance.

The cyclone design includes *enhanced target zones* (Figure 9.18), which accept the first and direct heat from the solids. Its extreme duty requires extra erosion-resistant materials and streamlined profiles. The service linings of the tangential inlet walls and target zones of the cyclone barrel are made of extra-thick, extra-erosion-resistant, high-alumina firebrick. The roof just above the target zones, where solids flow is concentrated, is also enhanced with extra-erosion-resistant, monolithic materials.

The balance components of the cyclone, hot ash ducts, and hot gas ducts are lined with high-purity block insulation, insulating firebricks and castables, erosion-resistant high-alumina bricks, and erosion-resistant refractory monolithics.

Material Issues   329

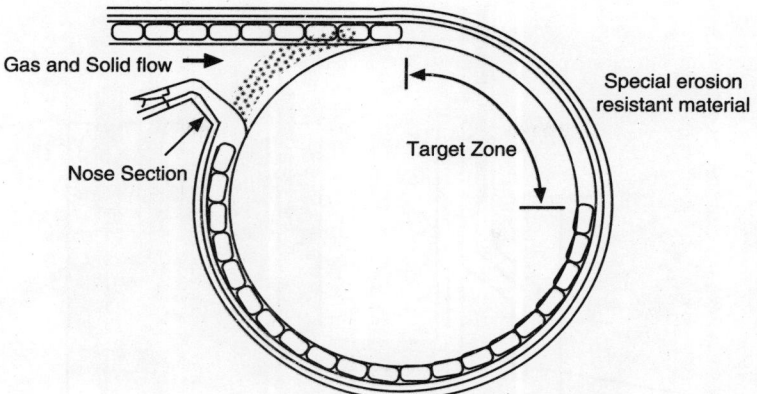

FIGURE 9.18 Cyclone target zone.

9.5.7.4 Cyclone Return Leg and Loop-Seal

Materials commonly used in cyclone return legs and loop-seals include vibration-cast linings of insulating castable covered by dense shock-resistant castable (fused silica-based) with stainless steel fiber and *burn-out* fibers. Dense, erosion-resistant brick with an insulating firebrick back-up lining of 400 to 500 mm thick is also used.

9.5.7.5 External Heat Exchanger and Ash Cooler

The external heat exchanger (EHE) undergoes moderate to severe service conditions. A properly designed multi-layer castable lining could give a reasonable service life. Fused silica-based materials are required for the division walls.

Two-component castables are mostly used in ash coolers. Other design options are multilayer brick or water walls with relatively thin lining (2 to 4 in.) of abrasion-resistant castable or plastic.

9.5.7.6 Wing Walls

Wing walls (superheat and/or evaporative) are normally installed in the upper part of the furnace. Refractory is applied at the bottom of the wing walls on the section of the tubes that run perpendicular to the flow of circulating bed material as shown in Figure 9.19.

9.5.7.7 Other Areas

The evaporator and superheater panels, located in the lower furnace, are covered with erosion-resistant refractory secured by studs. Spent ash return lines from the loop-seal back to the combustor could use phosphate-bonded plastic on thick two-component castable lining.

9.5.8 Failure Analysis

The most critical step in repairing a refractory lining is to determine the mode of failure. Primary causes of failure are:

- Lining disintegration/spalling
- Erosion
- Anchor failure

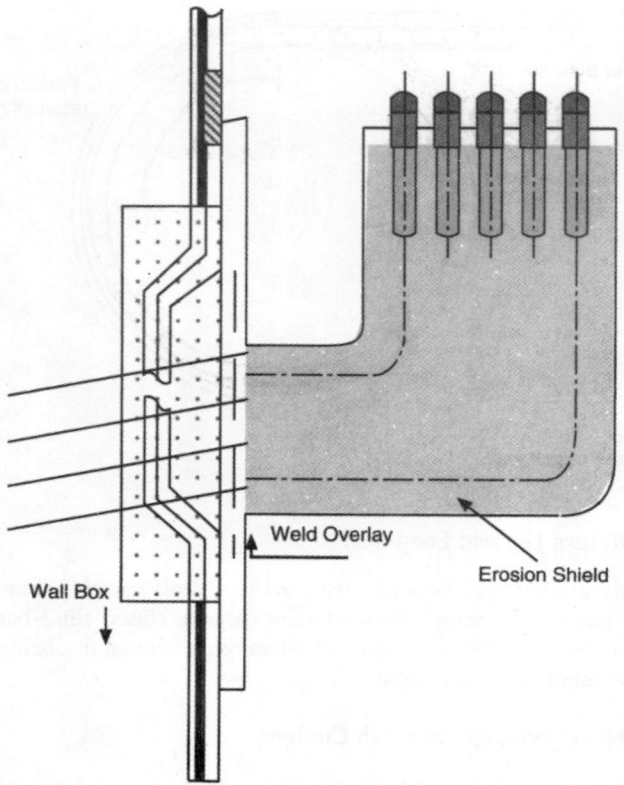

FIGURE 9.19 Lower section of a wing wall is covered with refractory.

9.5.8.1 Lining Disintegration

Lining disintegration is defined as loss of materials due to thermal or mechanical stresses. Thermal stresses are caused due to thermal cycling of the refractory lining, which suffers from the differential expansion coefficient of the binder and aggregates. Thermal cycling results in large cracks and spalling of the lining as small cracks fill with circulating solids and exert compressive stresses on the hot face. Thermal shock due to rapid temperature changes can cause spalling of the hot face on initial heat-up when the internal steam pressure exceeds the tensile strength of the material. Mechanical stresses are caused by the expansion of steel penetrations through the lining material unless an adequate expansion allowance is provided. In addition, the differential expansion between the steel shell and the refractory lining induces stresses on the lining material.

9.5.8.2 Erosion

Erosion is caused by the high velocity of the flue gases (proportional to the second to fourth power of the velocity) and the size of the particles that the flue gases carry. The impact zone of the cyclone, where the particles move to turn, is often the area of extreme erosion. Brick or SiC tiles are often the best materials for this application. Erosion-prone areas are visibly detected by rounded edges and smooth polished appearances. Another area where erosion is high is near the cyclone vortex finder. Erosion is also severe at the vertical refractory of combustor, where loss is caused by local eddies. Figure 9.20 shows erosion on the lower water wall that has been repaired by weld overlay.

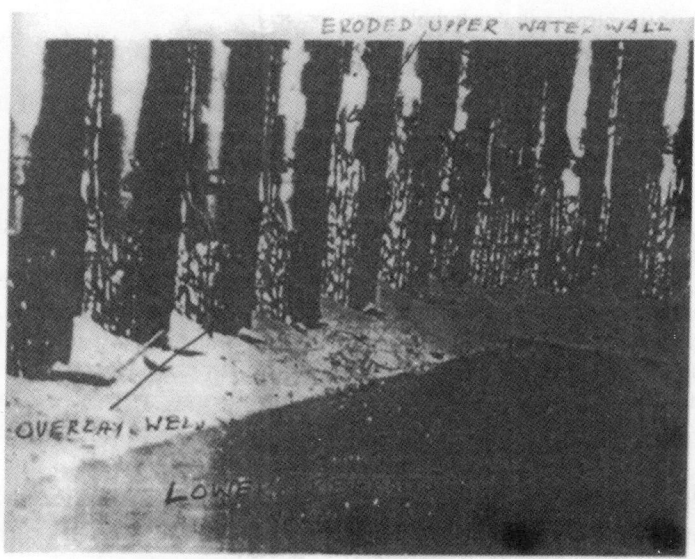

FIGURE 9.20 Photograph showing weld overlay applied to reduce erosion damage to vertical boiler tubes at the refractory interface.

9.5.9 REFRACTORY MAINTENANCE

The refractory should be inspected on an annual basis and repaired in the following critical places:

- Furnace walls and refractory partitions
- Grid nozzles and refractory
- Start-up burner hoods
- Water wall/refractory interface
- Loop-seal return legs

Attempts to repair the failed areas in dense-phase, hot-face lining by phosphate-bonded refractory have not proved very successful. For this reason, the fluidized bed boiler operators in many installations have shifted from monolithic lining to all-brick or brick plus insulating castable or insulating firebricks. Another good option is to use multi-layer brick lining with calcium silicate block next to the shell, followed by insulating firebricks and a hot face of dense, abrasion-resistant, superduty mullite bricks.

A steam- or water-cooled cyclone is a good option. In this case, the entire cyclone is made of water wall tubes, with 50 to 75 mm thick phosphate-bonded, dense-phase refractory on the studs fixed to the water wall tubes. This reduces thermal shock and spalling.

For anchors, materials like Incoloy 800 HT and Haynes Alloy 230 have substantially higher allowable stress levels at higher temperatures than 300 series stainless steels and these can be easily formed, machined, and welded.

9.5.9.1 Lower Combustor

The material problems encountered within the lower combustor include:

- Failure of the refractory layers in the lower dense bed

- Erosion of the air distributor bubble caps
- Erosion of the water wall/refractory interface

The refractory in the lower combustor is exposed to erosion by the dense bed and to chemical attack due to the prevailing reducing environment. In extreme cases, it is also subject to both rapid thermal and mechanical shock. These lead to premature thinning and perhaps total failure (Elsner and Friedman, 1989). Hard, high-density tiles, attached to the water wall in an overlapping pattern (similar to shingles on a roof) proved effective (Zychal and Darling, 1989), though its economic feasibility is under review.

Early CFB designs tended to utilize an air distribution nozzle designed for bubbling bed combustors, which require very accurate air distribution. These specialized nozzles, generally referred to as *bubble caps*, are subject to high temperatures (Elsner and Friedman, 1989) and severe erosion in the dense bed of the CFB combustor. Presently, Tuyer-type nozzles are used by taking special care that the air jet from one nozzle does not hit neighboring ones and that there is no backflow under low load.

9.5.9.2 Upper Combustor

The erosion of the water wall/refractory interface has been a major problem in most CFB boilers, as indicated previously. Sometimes horizontal shelves are welded to the tubes to protect the parent tube from erosion (Figure 9.21). Presence of the shelves at the water wall/refractory interface

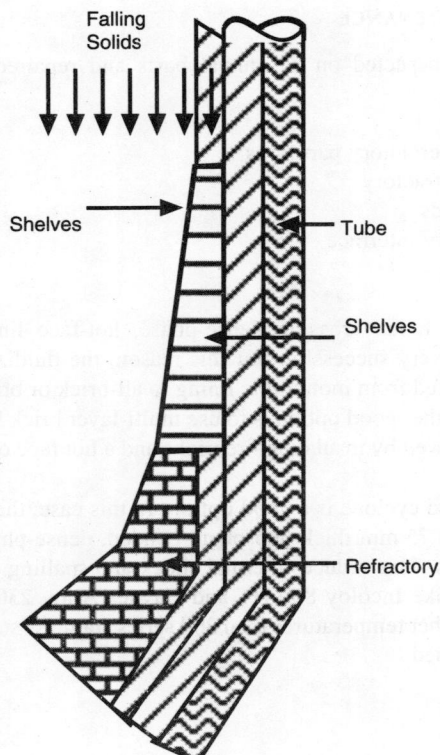

FIGURE 9.21 Horizontal shelves make tubes from erosion at the water wall/refractory interface.

Material Issues   333

disrupts the downward flow of solids resulting in erosion of the adjacent tube surfaces. Several solutions to this problem have been implemented with limited success, including:

- Minimizing the size of the refractory shelf
- Adding multiple steps to decrease the individual jetting effect (Slusser et al., 1989)
- Hard-facing the area prone to erosion
- Bending the tube/refractory interface backwards

9.6 EXPANSION JOINTS

The arrangement of a CFB boiler is very different from that of a pulverized coal-fired boiler because it uses a closed loop, part of which is water-cooled and the other part uncooled. Owing to their temperature difference these two legs of the CFB boiler expand differently creating a significant expansion problem.

In many CFB boilers, large refractory-lined cyclones sit between a water-cooled combustor and a steam/water-cooled convective pass as shown in Figure 9.14. Furthermore the bottom of the cyclone is connected to the bottom of the combustor. The cyclone's steel wall, which maintains an average temperature of the support elements of approximately 90°C, expands to a lesser degree than the combustor's wall tubes because the wall temperature of the tubes varies between 170 to 200°C. It is approximately equal to the saturated temperature of water in the drum plus 25°C. This difference in temperature can lead to a differential movement of up to 150 mm or more in three dimensions. This calls for multiple expansion joints in the CFB boiler. In addition, the need to positively seal all penetrations in the lower combustor due to high bed pressure requires a large number of smaller expansion joints for feeds such as coal, limestone, and other flow paths into the lower combustor. For example, the bottom of the cyclone and the top of the standpipe are connected through a bellows-type expansion joint to allow thermal expansion.

Although the temperature and pressure rating of these joints is not extremely high, the designer must consider the effect of bed material, which can infiltrate the joint. Packing of bed material into such joints can reduce the desired flexibility and lead to premature failure. The expansion joints normally use high-temperature, insulating stuffing inside specially designed interlocks on the hot side (Figure 9.22). To prevent gas leakage, a fiber-cloth sealant, capable of withstanding 250 to 300°C, is used on the cooler side.

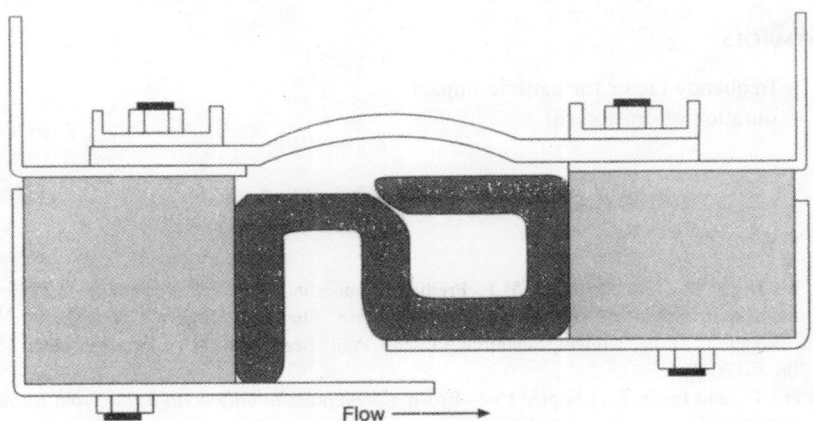

FIGURE 9.22 Nonmetallic expansion joint.

NOMENCLATURE

a: erosion coefficient of fly ash in gas (default value $= 14 \times 10^{-9}$) (mm sec^3/g h)
A: area (m^2)
ASH/C: ash to carbon ratio in fuel
C_{py}: concentration of pyrite in coal
C_q: concentration of quartz in coal
C_t: factor in Equation 9.1 to take account of changes in surface temperature
Cl: chlorine content in fuel
D: diameter of tube (mm)
E: erosion rate (nm/h)
E_{max}: maximum erosion rate in the convective section (mm/h)
I_a: ash abrasiveness factor
I_c: overall abrasiveness of coal mineral
I_q: abrasiveness of quartz
K: constant dependent upon tube material
k_p: ratio of gas velocity at design load to that at average running load
k_μ: coefficient to account for nonuniformity in fly ash density
k_v: coefficient to account for nonuniformity in gas velocity field
M: abrasion-resisting coefficient of the tube material
n: constant in Equation 9.2
p, q: exponents in Equation 9.8
Q: allowable heat loss between furnace and exterior (kW)
R: sum of resistances to heat transfer between furnace and exterior
R_{90}: percentage of fly ash smaller than 90 μm in diameter
R: corrosion rate (nm/h)
S_1: horizontal pitch of tubes (mm)
ΔT: temperature difference between outside of casing and hot face (°C)
T_g: flue gas temperature (K)
T_{metal}: temperature of tube metal (°C)
$T_{saturated}$: saturation temperature (°C)
T_{steam}: temperature of steam (°C)
V_g: gas velocity in narrowest section of tube (m/sec)
V_p: particle velocity (m/sec)

GREEK SYMBOLS

η: frequency factor for particle impact
τ: duration of erosion (h)

REFERENCES

Backman, R., Hupa, M., and Skrifvars, B.J., Predicting superheater deposit formation in boilers burning biomasses, In *Impact of Mineral Impurities in Solid Fuel Combustion*, Gupta, R. P., Wall, T. F. and Baxter, L., Eds., Kluwer Academic/Plenum Publishers, New York, pp. 405–416, 1999, ISBN 0-306-46126-9.

Basu, P., Kefa, C., and Jestin, L., Chapter 15 — Erosion prevention in boilers, In *Boilers and Burners Design and Theory*, Springer, New York, pp. 426–456, 1999.

BEI, *Boiler and Anciliary Plants, Modern Power Station Practice*, 3rd ed., Vol. B, Pergamon Press, London, 1992a, p. 14.

BEI, *Modern Power Station Practice*, 3rd ed., Vol. E, Pergamon Press, London, 1992b, pp. 431–438.
Buchanan, E. R., Evaluation of several coatings after extended FBC service, In *Proceedings of the 13th International Conference on Fluidized Bed Combustion*, Vol. I, Heinschel, K., Ed., ASME, New York, pp. 191–199, 1995.
Chinese Boiler Calculation Standards (1973), CBSC.
Crowley, M. S., Bakker, W. T., and Stallings, J. W., Refractory applications in circulating fluid bed combustors, *Proceedings of the 11th International Conference on Fluidized Bed Combustion*, Vol. 2, Anthony, E. J., Ed., ASME, New York, pp. 417–424, 1991.
DOE, US Department of Energy News Letter, *Materials and Components*, 98, June 1, pp. 1–5, 1992.
Elsner, R. W. and Friedman, M. A., CFB material related Issues at CUEA nucla station, In *Proceedings Workshop on Materials Issues in Circulating Fluidized Bed Combustors*, EPRI, 1989.
Finnie, I., Erosion of Surfaces by Solid Particles, *Wear*, 31960.
Jia-Sheng, L., Stevenson, D. A., and Stringer, J., The role of carburization in the in-bed corrosion of alloys in FBC, In *Proceedings of the Ninth International Conference on Fluidized Bed Combustion*, Vol. 2, Mustonen, J. P., Ed., ASME, New York, pp. 656–662, 1987.
Johnson, R. C. and Seshamani, M., Refractory lining system for circulating fluidized bed water cooled cyclones, In *Proceedings of the 10th International Conference on Fluidized Bed Combustion*, Vol. 1, Manaker, A. M., Ed., ASME, New York, pp. 419–426, 1989.
Kalmanovitch, D. P., Hajicek, D. R., and Mann, M. D., The effects of coal ash properties on FBC boiler tube corrosion, erosion and ash deposition, In *Proceedings of the 10th International Conference on Fluidized Bed Combustion*, Vol. 1, Manaker, A. M., Ed., ASME, New York, pp. 847–856, 1989.
Larson, J.W, Nack, H, Stringer, J., and Wright, I., Summary-Components — Materials, *Proceedings of the Eighth International Conference on Fluidized Bed Combustion*, US DOE/METC-85/6021, Conf-Vol II, pp. 711–713, 1985.
Raask, E., *Mineral Impurities in Coal Combustion*, Hemishpere Publications, USA, 1985.
Salam, T. F., Malik, S. R., and Gibbs, B. M., Characterising corrosion potential in fluidized bed combustors, In *Proceedings of the 10th International Conference on Fluidized Bed Combustion*, Vol. 1, Manaker, A. M., Ed., ASME, New York, pp. 575–582, 1989.
Salmenoja, K., Hupa, M., and Backman, R., Laboratory studies on the influence of gaseous HCl on fireside corrosion of superheaters, In *Impact of Mineral Impurities in Solid Fuel Combustion*, Gupta, R. P., Wall, T. F. and Baxter, L., Eds., Kluwer Academic/Plenum Publishers, New York, pp. 513–524, 1999b, ISBN 0-306-46126-9.
Singer, J. G., *Combustion-Fossil Power Systems*, Combustion Engineering Inc., Windsor, CT, 1991, Ch. 7–10.
Skrifvars, B.-J., Laurén, T., Backman, R., and Hupa, M., The role of alkali sulfates and chlorides in post cyclone deposits from circulating fluidized bed boilers firing biomass and coal, In *Impact of Mineral Impurities in Solid Fuel Combustion*, Gupta, R. P., Wall, T. F. and Baxter, L., Eds., Kluwer, New York, 1999.
Slusser, J. W., Bartlett, S. P., and Bixler, A. D., Materials Experience with the Stockton CFBC, In *Proceedings Workshop on Materials Issues in Circulating Fluidized Bed Combustors*, EPRI, 1989.
Stringer, J., Material selections in atmospheric pressure fluidized bed combustion systems, In *Fluidized Bed Boilers*, Basu, P., Ed., Pergamon Press, Toronto, pp. 155–180, 1984.
Stringer, J., Current information on metal wastage in fluidized bed combustors, In *Proceedings of the Ninth International Conference on Fluidized Bed Combustion*, Mustonen, J. P., Ed., ASME, New York, pp. 685–696, 1987.
Stringer, J. and Minchener, A. J., The Effect of Overall In-Bed Oxygen Concentration on Corrosion in Fluidized Bed Combustors, In *Proceedings of the Seventh International Conference on Fluidized Bed Combustion*, Vol. 2, US Department of Energy, Washington, DC, 1982, pp. 1010–1019.
Stringer, J., Minchener, A. J., Lloyd, D. M., Hoy, H. R., and Stoke, O., In-Bed Corrosion of Alloys in Atmospheric Fluidized Bed Combustors, In *Proceedings of the Sixth International Conference on Fluidized Bed Combustion*, Vol. 2, US Department of Energy, Washington, DC, 1980, pp. 4437–447.
Stringer, J. and Wright, I. G., Materials problems in circulating fluidized bed combustors, In *Proceedings Workshop on Materials Issues in Circulating Fluidized Bed Combustors*, EPRI, 1989.

Wang, B. Q. and Lee, S. W., Erosion resistance of cooled thermal spray coatings under simulated erosion conditions at Waterwalls in FBCs, In *Proceedings of the 14th International Conference on Fluidized*, Vol. I, Preto, F. D. S., Ed., ASME, New York, pp. 335–342, 1997.

Wang, B. Q., Geng, G. Q., Levy, A. V., and Mack, W., Erosivity of particles in circulating fluidized bed combustions. *Wear*, 152, 201, 1992.

Zychal, R. J. and Darling, S. L., A summary of combustion engineering; erosion/wastage experience in CFB boilers, In *Proceedings Workshop on Materials Issues in Circulating Fluidized Bed Combustors*, EPRI, 1989.

10 Solid Handling Systems for Fluidized Beds

The functionality of fluidized bed equipment is limited by its fuel feed system. A major fraction of the interruptions in the operation of a fluidized bed unit can be attributed to the breakdown of its feed system. This system is vulnerable, especially for biomass or when the solid fuel picks up an excessive amount of surface moisture from rain or snow while lying in an outdoor stockyard. A fluidized bed plant feeds the fuel into the furnace or gasifier directly after crushing without drying or pulverizing it as is done in a pulverized coal-fired (PC) boiler. Thus, a blockage could easily occur either in the fuel bunker or in the feeder itself due to the surface moisture of the fuel. Excessive fines in the fuel also reduce its mobility interrupting the boiler operation.

One of the attractive features of a fluidized bed boiler is that it can tolerate even several minutes of interruption in its fuel flow. This is a significant improvement over a PC boiler, which cannot tolerate a fuel interruption more than few seconds in duration. However, an interruption any longer would cool even a fluidized bed excessively, requiring a hot restart.

Besides the problem of interruption in operation, there is the issue of boiler performance. How the fuel or sorbent is fed into the furnace has a major influence on the combustion and sulfur capture efficiency. A good design of the fuel and sorbents feeding system is therefore essential for proper operation of a fluidized bed boiler. The present chapter describes different designs of fuel feed systems and their effects on the operation of the bubbling (BFB) or circulating (CFB) fluidized bed boilers and gasifiers.

10.1 SOLID HANDLING SYSTEMS

Fluidized bed boilers, gasifiers and incinerators use a wide range of feedstock and produce solid residues corresponding to the feedstock used. The following section gives a general discussion on solid handling systems for fossil fuels and waste fuels.

10.1.1 FOSSIL FUEL PLANTS

Solid handling is required in the following areas of a fluidized bed boiler or gasifier plant:

1. Fuel feed into the boiler
2. Sorbent feed into the boiler
3. Grit refiring
4. Ash handling for disposal or storage

Figure 10.1 shows the general arrangement of the solid handling system of a typical fluidized bed boiler plant firing coal. A gasification plant will have a very similar system.

The fuel is stored in large piles, generally in the open. When required, trucks or front-loaders transport the raw fuel into the crusher that crushes it to typically below 10 mm in size. The crushed fuel is then lifted to a vibratory screen (Figure 10.1) to remove the oversized particles and feed the undersized to the feed fuel bunker. The oversized may be returned to the crusher for further crushing. The feeder (screw, for example) draws measured amounts of fuel from the bunker and drops them into

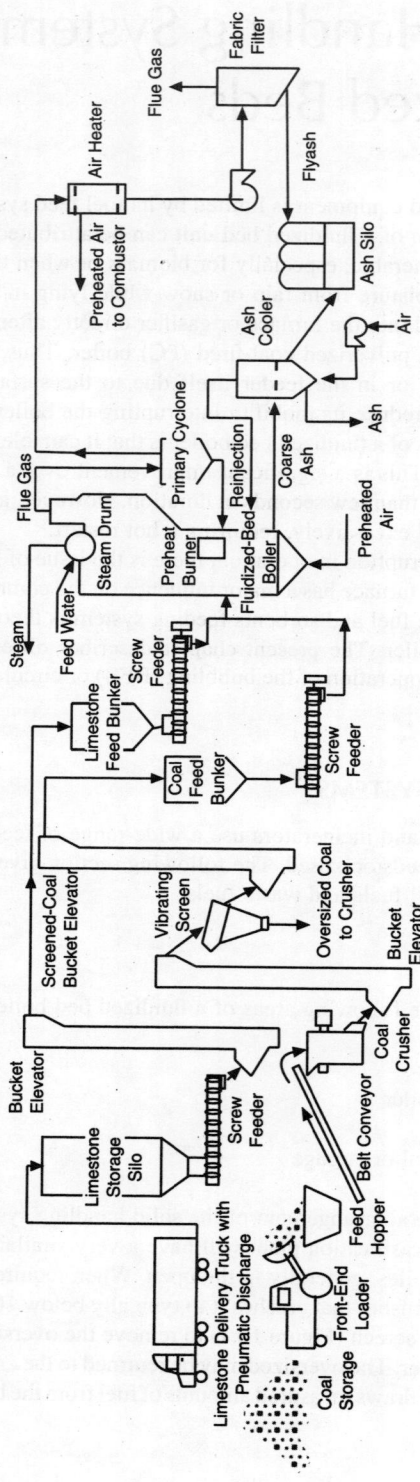

FIGURE 10.1 Solid handling system of a typical fluidized bed boiler plant firing solid fuels.

the feed line or chute, which in turn conveys the fuel to the furnace. The sorbent (limestone, for example) used for sulfur capture is handled in a similar way before it is fed into the boiler.

The ash and spent sorbents, generated in the furnace, are split into two streams: coarse and fine.

The finer fraction, known as *fly ash*, is collected in the primary collector (cyclone, for example) as well as in the final collector (fabric filter, for example). The solids collected in the primary collector are often reinjected into the bed to burn the unburned carbon and to utilize the unreacted lime in the spent sorbent. The fly ash collected in the final collector, is conveyed in ash silos for appropriate disposal.

Figure 10.2 shows additional details of the solid handling system of a bubbling fluidized bed boiler. This figure shows one of many types of systems in use. Crushed fuel is stored in hoppers, from where it drops into a feeder. The feeder feeds measured amounts of fuel into a fuel conveying system, comprising a high-pressure blower, a conveying pipe system, and a flow splitter for solids. The limestone is similarly fed into the flow stream and conveyed to the bed.

Part of the fly ash escaping the bed is captured underneath the back-pass. Fly ash often contains a large amount of unburned carbon (4 to 9%) and free lime (20 to 40%, as $Ca(OH)_2$) (Lee, 1997). Therefore, to improve the combustion efficiency and the utilization of sorbents, part of the collected fly ash, known as *grit*, is reinjected pneumatically into the bed. The remaining fly ash moves further downstream into particulate (dust) collection equipment such as a bag filter or electrostatic precipitators (ESP). Solids collected in these devices are fed into a pneumatic conveying line by means of a special valve. The solid is then conveyed to ash silos for final disposal by appropriate means.

The coarser fraction of the ash is generally drained from the bed into an ash cooler that cools it before disposal. In some cases the hot ash is dropped into a water trough and then is pumped as slurry. Unlike fly ash, the bed ash contains a considerably smaller amount (0.2 to 4%) of unburned carbon.

10.1.2 Municipal Solid Waste

Municipal solid waste (MSW) is a type of nonfossil fuel, which may be loosely termed *renewable* as its source is not finite like those for coal and other fossil fuels. The composition of MSW varies significantly depending on socioeconomic factors of the city and its geographical location.

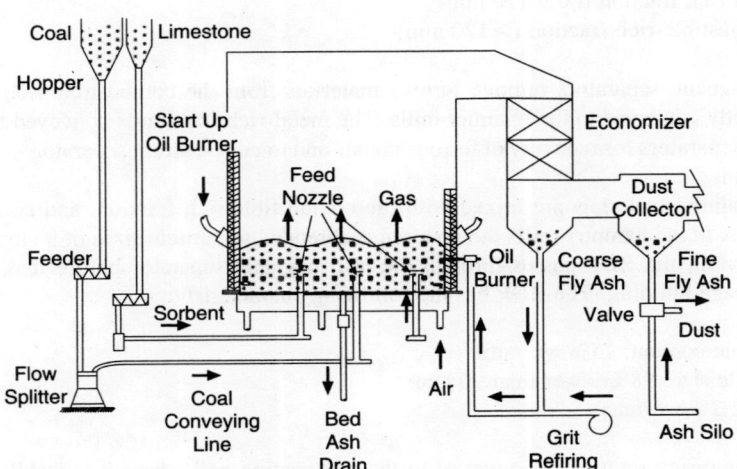

FIGURE 10.2 Coal and ash handling system of a bubbling fluidized bed boiler using under-bed pneumatic feed system.

TABLE 10.1
Typical Composition of Municipal Solid Wastes (MSW)

Material	% by wt
Food waste	20.0
Paper and cardboard	27.5
Plastic	13.5
Textiles	3.5
Metals	3.5
Wood	3.5
Yard waste	7.5
Glass	8.0
Screenings	5.0
Other	8.0
LHV	2200 kcal/kg

An example of the composition of MSW from a typical North American city is shown in Table 10.1 for illustration. Modern recycling systems could considerably reduce the paper and cardboard content of this MSW.

It is difficult to burn most MSW in its raw form as it contains a significant amount of noncombustibles, such as glass (Table 10.1). For this reason MSW is processed to produce a refuse derived fuel (RDF). In a typical plant, trucks deliver the waste to the tipping building. The waste is dumped in the waste pit, large enough to hold several days' supply. Bridge cranes, equipped with a grapple, handle incoming MSW for feeding into the MSW processing line, which is composed of a shredder, primary trommel, secondary trommel, magnetic separators, and hammer mill. After primary shredding, the MSW is sorted into three streams, as below, in order to obtain a quality fuel and to recover recyclable materials:

1. Organic-rich fraction (<60 mm)
2. Metal-rich fraction (60 to120 mm)
3. Combustible-rich fraction (>120 mm)

After magnetic separators remove ferrous materials from the combustible-rich fraction, the latter is directly conveyed to the hammer mills. The metal-rich fraction is conveyed to another set of magnetic separators for recovery of ferrous metals and to eddy-current separators for recovery of aluminum cans.

The remaining materials are mixed with the combustible-rich fraction, and conveyed to the hammer mills. In the hammer mills the material is shredded to particle sizes of 9 cm or less. After the hammer mill, the RDF passes through the last magnetic separator before it is conveyed to the RDF storage building. The RDF has the following characteristics:

1. Organic content: 15% wt. max
2. Particle size: 98% lower than 90 mm
3. Inerts: 2% wt. max

The organic-rich fraction is conveyed to the composting hall where it is stabilized using an aerobic fermentation process, which takes about 1 month. The process air for composting is partially taken from the MSW receiving and sorting building. A biofilter treats the air from composting and from MSW processing buildings. The stabilized product is sent to the secondary trommel where materials over 2 cm in size, consisting mainly of paper and plastic, are recovered and conveyed to the RDF stream. The remaining organic fraction can be refined using an air classifier for the separation of glass and other solid inerts.

The low-quality compost derived from stabilization of the MSW organic fraction can be directed to compost storage or to the RDF stream. The stabilized and cleaned product coming from the source-sorted organic fraction becomes quality compost after 2 months of composting. To minimize dust in the RDF storage areas, combustion air for the boiler is drawn from this section of the facility. Modern plants are designed for zero water discharge, where leachates from the compost are recycled in the compost process itself.

The RDF, produced this way, is fed into the boiler through suitable feeding devices. Further discussion on this is given later.

10.2 BIOMASS HANDLING SYSTEMS

One area of a biomass power plant that demands great attention is its fuel yard and fuel feed system. Most of the plants spend significant time and money, especially during their initial years of operation, solving problems such as fuel pile odors and heating, excessive equipment wear, fuel hang-ups and bottlenecks in the feed system, tramp metal separation problems, wide fluctuations in fuel moisture to the boiler or making changes in the fuel yard to respond to market opportunities. Thus, the design of biomass handling systems requires careful consideration.

Biomass can be loosely classified under two broad groups:

1. Harvested
2. Nonharvested

Harvested fuels include long and slender biomass like straw, grass and bagasse. They also carry considerable amounts of moisture. Examples of nonharvested fuels are wood chips, rice-husk, shells, bark and prunings. These fuels are not as long as harvested fuels and some of them are actually granular in shape.

There are two general types of biomass feeding system chosen depending upon the type of biomass being handled. For solid biofuels, two different ways of handling the feedstock during harvest have lead to the development of two different feeding systems, each optimized for one main fuel type.

10.2.1 FEEDING OF HARVESTED FUELS

These fuels are pressed into bales to facilitate transportation and handling. Figure 10.3 shows the handling system for straw bales. Bales are handled by crane with a number of grabs depending on the specific fuel consumption. The bales are brought to the boiler house from storage by chain conveyors.

Whole bales are fed into a bale shredder and rotary cutter chopper. This reduces the straw into sizes adequate for feeding into the fluidized bed gasifier or combustor. In the final leg, the chopped straw is fed into the furnace by one of several types of feeders. Figure 10.3 shows the use of a ram feeder, which would push the straw into the furnace. In some cases, the straw falls down into a double-screw stoker (Figure 10.4), which presses it into the furnace through a water-cooled tunnel.

10.2.2 FEEDING OF NONHARVESTED FUELS

Wood and by-products from food-processing industries are generally granular in shape. The majority of the by-products from food processing are delivered by trucks, boxcars or barges as bulk particles (nuts, shells, etc.). Wood, bark, and prunings are also delivered predominantly in bulk in

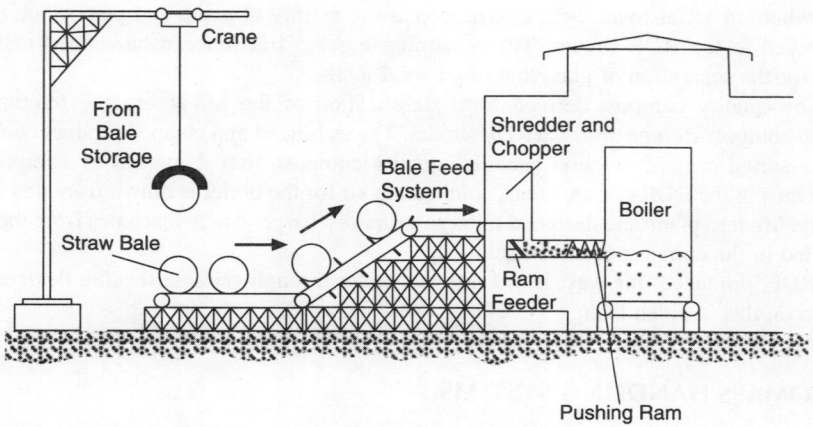

FIGURE 10.3 Feeding system for a harvested fuel in bales.

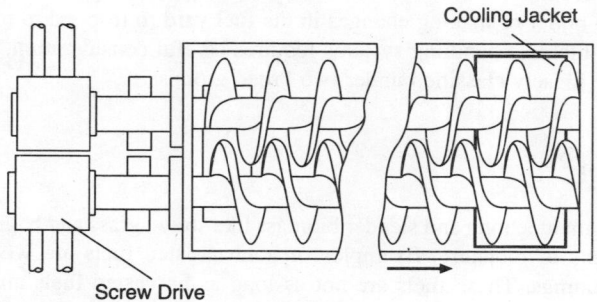

FIGURE 10.4 Double-axial screw feeder.

the form of chips. Different bulk-material handling equipment such as moving-bottom, scraping or belt conveyors are used to bring the fuel to the day-silo in the boiler house. Biomass such as wood chips or bark may not be of the right size when delivered to the plant, so, it needs to be shredded to the desired size in a wood chopper. Some fuels like rice-husk and coffee beans are of fixed granular size. As such, they do not need further chopping.

Speed-controlled feeders take the fuel from the silo and drop measured amounts into several conveyors. Each conveyor takes the fuel to an air-swept spout, which feeds it into the furnace. In case the moisture in the fuel is too high, one can use augurs to push the fuel into the furnace.

10.3 FEED SYSTEM FOR BUBBLING BEDS

This section discusses design principles of feed systems for bubbling fluidized bed boilers or gasifiers.

10.3.1 TYPES OF FEEDERS

Five main types of feeders are available for metering and feeding fuels into a bubbling bed:

1. Gravity chute
2. Screw conveyor

3. Spreader-stoker
4. Pneumatic injection
5. Moving-hole feeder

A brief discussion of each type is given below.

10.3.1.1 Gravity Chute

The gravity chute is a simple device, where coal particles are fed into the bed via a chute with the help of gravity (Figure 10.5). It requires the pressure in the furnace to be at least slightly lower than atmospheric pressure, otherwise, hot gas and fine particles would be blown back into the chute, creating operational hazards, and possible chocking of the feeder due to coking near its mouth.

In spite of the excellent mixing capability of fluidized beds, a fuel-rich zone is often created near the outlet of the chute feeder causing severe corrosion in that zone. Since the fuel is not well dispersed in the gravity chute feeder, much of the volatile matter is released near the outlet of the feeder, which causes a pressure surge and a reducing atmosphere. This pressure surge might blow back fine fuel particles into the chute while reducing conditions might encourage corrosion. An air jet could help disperse fine particles away from that zone. To avoid this problem to some extent, the chute can be extended into the furnace as shown in Figure 10.5. The extended part would, however, need insulation around it and some cooling air to avoid premature devolatilization.

The gravity feeder is not a metering device, meaning that it can neither control nor measure the feed rate of the fuel. A separate metering device like a screw is required upstream of the chute.

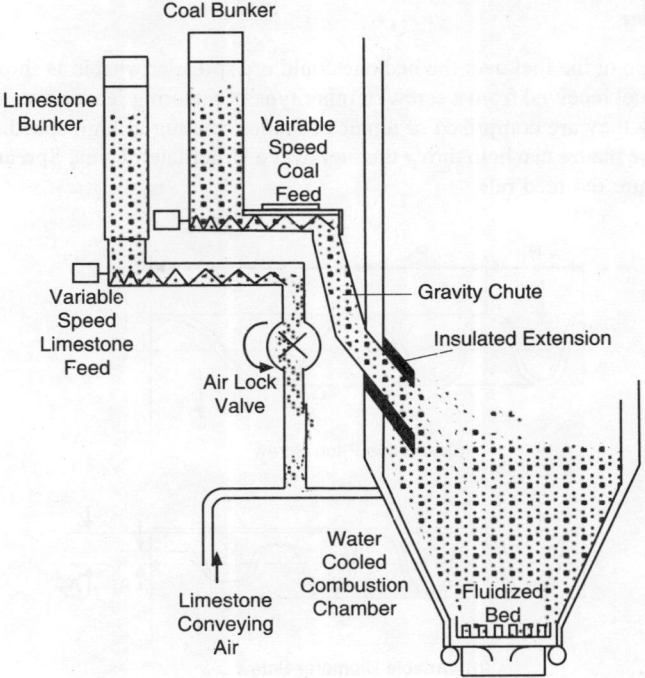

FIGURE 10.5 Coal metered by a screw is fed into the furnace through a gravity chute that feeds the coal from above the bed. The limestone is injected pneumatically into the bed.

10.3.1.2 Screw Conveyor

The screw conveyor is a positive displacement pump. It can move solid particles from a low-pressure zone to a high-pressure zone with a pressure seal. It can also measure the amount of fuel fed into the bed. By varying the RPM of its drive, a screw feeder can easily control the feed rate. Similar to the case of a gravity chute, the fuel from a screw feeder does not have any means for dispersion. An air dispersion jet, employed under the screw conveyor, could serve this purpose.

Plugging of the screw is a common problem. Solids within the screw flights are compressed as the move downstream. Sometimes the solid is packed so hard that it does not fall off the screw, but compacts against the sealed end of the trough carrying the screw. This often leads to jamming of the screw. Following design options may avoid the problem of jamming:

1. Variable-pitch screw (Figure 10.6a)
2. Variable diameter to avoid compression of fuels toward discharge end (Figure 10.6b)
3. Multiple screws (Figure 10.4)

Multiple screws are especially effective in handling biomass fuels. Figure 10.4 shows a feeder with two screws. Some feed systems use more than three or four screws.

The hopper outlet, connected to the inlet of the screw, needs careful design. A vertical wall of the hopper towards the discharge end of the screw is superior to the traditional inclined wall because it avoids formation of rat holes in the hopper as shown in Figure 10.7 for a belt feeder.

Each screw feeder typically serves only 3 m^2 of bed area. For this reason, it is only good for small-sized beds. Another major and very common operational problem arises when fuel contains high moisture. It has to be dried before it enters the conveyor otherwise it may plug the system. The screw feeders can serve pressurized boilers with pressures up to several atmospheres.

10.3.1.3 Spreader

For wide dispersion of the fuel over the bed one could use spreader wheels as shown in Figure 10.8. These throw the fuel received from a screw or other type of metering feeder over a large area of bed surface. Typically they are comprised of a pair of blades rotating at high speed. Slightly opposite orientation of these blades can help throw the fuel over a larger lateral area. Spreader wheels are not designed to measure the feed rate.

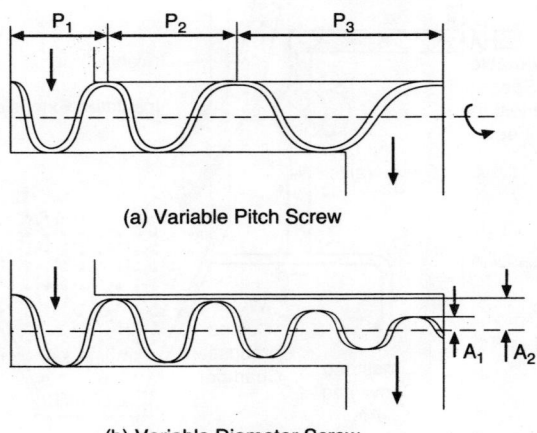

FIGURE 10.6 By changing the pitch or the diameter of the screw, one can avoid the compaction of solids that occur in screw feeders.

Solid Handling Systems for Fluidized Beds

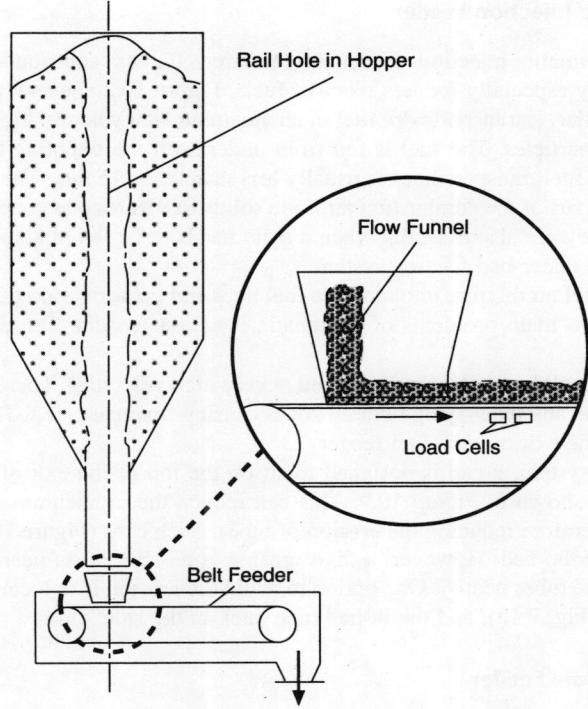

FIGURE 10.7 Gravimetric belt feeder showing the potential formation of flow funnel in the hopper.

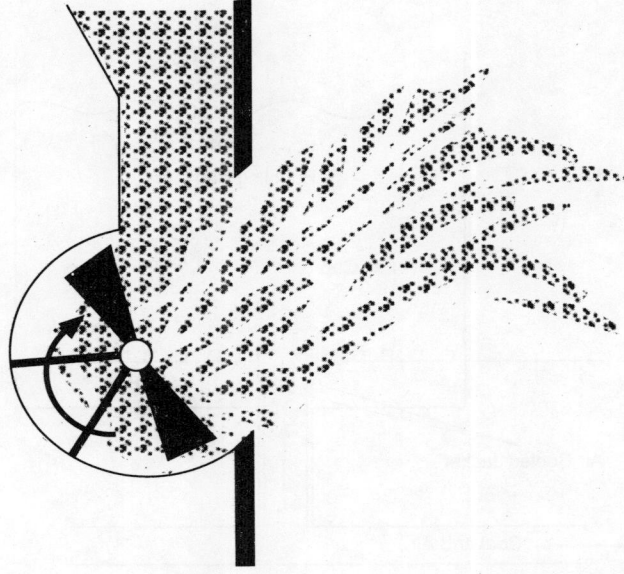

FIGURE 10.8 Fuel spreaders help spread the fuel by throwing it across over a large area.

10.3.1.4 Pneumatic Injection Feeder

The under-bed pneumatic injection feeder is preferred for its contribution towards higher combustion efficiency especially for less reactive fuels. Figure 10.2 shows a pneumatic injection system for a BFB boiler. It transports dry fuel in an air stream at a velocity higher than the settling velocity of the fuel particles. The fuel is fed from underneath the bubbling bed. The maximum velocity of air in the fuel transport lines is usually less than 11 to 15 m/s. The air for transporting coal also constitutes part of the combustion air. Fine solids like limestone generally use pneumatic injection feeders. They are also essential when a solid needs to be fed at a specific point inside a dense bed like in the under-bed feeding system.

Splitting of the fuel/air mixture into multiple fuel lines and the formation of high velocity air jet in the bed are the two main problems of pneumatic injection feeders. The design of splitters is discussed later.

Air jets, carrying solid particles, enter the bed at very high velocities. These particles with high momentum, impact on any tubes lying in these zones causing severe erosion. Thus a highly erosive zone could form at the exit of each fuel feeder.

To improve the system, a cap is designed to sit on the top of the exit of each nozzle of the pneumatic feeder as shown in Figure 10.9. This can reduce the momentum of the jets breaking into the freeboard therefore reducing the erosion of tubes. Such caps (Figure 10.9) also help better distribute the fuel in the bed. However, a high erosion zone still exists near each feeder outlet, which might erode the tubes nearby. One option to avoid the erosion is to locate all feed nozzles in the center zone (see Fig. 7.15), and the in-bed tube bank in the side zones.

10.3.1.5 Moving-Hole Feeder

This feeder is particularly useful for materials that are not free-flowing and could experience excessive packing in the hopper and screw feeder, such as:

1. Biomass
2. Solids with flakes

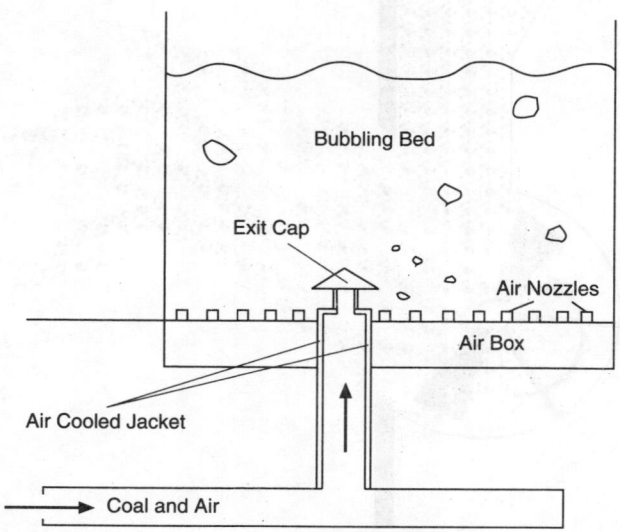

FIGURE 10.9 Air jets, which could erode tubes, can be dissipated by caps on top of pneumatic feed nozzles.

The moving bed feeder avoids the problem of solid compaction typically developed in screw feeders. To avoid formation of rat holes, the hopper uses vertical walls. A moving-hole feeder essentially consists of slots that traverse back and forth, without any friction between the stored material and the feeder deck (www.kamengo.com). The solids immediately above the hole naturally drop through the slot or hole into a trough located below the feeder. The feed rate is governed by the size of the hole, and the speed of the hole.

10.3.1.6 Fuel Auger

In this system, a metering device like a screw is used to meter fuels such as hog-fuel and feeds it on to a fuel belt, which carries it to the boiler front where the fuel-stream is divided into several fractional capacity fuel trains. Each train consists of a surge bin with a metering bottom and a fuel auger to deliver the fuel into the furnace. The auger is of cantilevered design, driven at constant speed through a gear reducer. The shaft bearing is located away from the heat of the furnace. Cooling air could cool the auger's inner trough and help propel the fuel toward the near bed region.

10.3.1.7 Ram Feeder for RDF

A ram feeder is essentially a hydraulic pusher. Some fuels such as RDF are too fibrous or sticky to be handled by any of the above feeders. A ram-type feeder can be effective in forcing the fuel into the boiler. A fuel auger, as described above, can convey solids like RDF into a hopper, at the bottom of which sits a ram feeder. The pneumatic ram of the feeder pushes the RDF on to a sloped apron-type feeder that feeds the fuel chute. From the fuel chute, the RDF drops into the fuel spout, where sweep air is used to transport the fuel into the furnace. This air also prevents any back flow of hot gases. The RDF, stored in the inlet hopper also provides a seal against a positive furnace pressure.

Ram feeders are supplied with variable-speed drivers. So, they can feed controlled amounts of fuel into the system.

10.3.2 Mode of Feeding in Bubbling Beds

The feed system of a bubbling fluidized bed (BFB) boiler performs the following two tasks:

- Crush coarse or lump fuel particles into desired sizes (0 to 12 mm).
- Feed the crushed particles into the bed evenly.

Design requirements for the feed system of a BFB boiler include:

- High combustion efficiency and low unburned carbon loss
- High reliability
- Simple design for easy maintenance and low cost
- Maximum residence time of feed particles in the combustion zone

Bubbling fluidized bed boilers use two types of feed systems:

1. Over-bed system (Figure 10.5)
2. Under-bed system (Figure 10.2)

Table 10.2 compares main features of these two types of systems. As one can see, the over-bed feeder can handle coarser particles while the under-bed feeder needs finer particles with less moisture. An under-bed feed system consists of crushers, bunkers, gravimetric feeders, air pumps, a splitter, and small fuel transporting lines.

TABLE 10.2
Comparison of Process Parameters of Over-Bed and Under-Bed Feeding Systems

Fuel	Over-Bed Feeder	Under-Bed Feeder
Top size	25–50 mm	10 mm
Surface moisture	No limit	<6%
Fines	<25% — 16 mesh	<20% — 30 mesh

An over-bed feed system consists of crushers, bunkers, gravimetric feeders, small storage bins, a belt conveyor, and spreaders.

In under-bed feed systems, coal or other fuel particles are crushed into sizes smaller than 8 to 10 mm. The crushed coal typically has 50% of the particles smaller than 1 to 2 mm in diameter. A metering feeder such as a screw feeder drops measured amount of the fuel into a simple pneumatic system (Fuller–Kinyon pumps for example), which mixes the fuel with air. The air pump (or blower) transports the fuel particles pneumatically into the combustor, injecting them from the bottom of the bed. Flow-stream splitters are used to split the fuel flow stream into the desired number of feed points. Coal nozzles, mounted on the grid plate of the bed, are used to inject the entering fuel into the bed.

In an over-bed feed system, coal particles are crushed to sizes less than 20 mm. This size is much coarser than the 1 to 2 mm particles used in an under-bed feeding system. In a typical system the fuel passes through a bunker, gravimetric feeders and a belt conveyor (Figure 10.7), and then drops into a feed hopper. From this hopper, fuel particles are spread into combustor by a set of rotary spreaders set above the bed surface (Figure 10.8).

The design of a feed system would generally involve the following steps:

1. Choose the type of feeder: over-bed or under-bed.
2. Determine the average size of fuel particles.
3. If under-bed is used, determine how many feeder points should be used for a specified combustor.
4. If over-bed is used, choose the number of spreaders to be used.

10.3.2.1 Over-Bed Feed System

The over-bed feed system is simple, reliable and economical, but results in a relatively low combustion efficiency, especially for less reactive fuels like anthracite and petroleum coke.

With an over-bed feed system, the top size of coal particles can be greater than that in under-bed feed systems making the coal preparation simpler and less expensive. The feed could, however, contain a large amount of fines whose terminal velocity is higher than the superficial velocity in the freeboard. When a particle's terminal velocity is lower than the superficial velocity of the boiler, the particle will be elutriated before it burns out completely, resulting in a combustion loss. Unburned carbon makes up most of the combustion loss in BFB combustors.

When the fuel is very reactive (as with high oxygen content), the over-bed feed system is more suitable for bubbling bed boilers, as its combustion efficiency can reach up to 97% without a recycle system (Duqum et al., 1985). For such highly reactive fuels there is no major difference in combustion efficiency between over-bed and under-bed systems. The burning rate of such fuels is so high that most of their fine particles burn out before they leave the freeboard. Most biomass-fired BFB boilers use over-bed feed systems because biomass is generally more reactive than solid fuels like anthracite. An example of a highly reactive coal is presented in Table 7.4.

While firing ordinary bituminous coals, the over-bed feed system would typically have a combustion efficiency 4 to 12% lower than that expected in a under-bed system.

Over-bed systems are less expensive than under-bed systems, because the latter require additional components like the pneumatic transport system, fuel splitter, and multiple feed nozzles. The over-bed feed system also uses coarser coal. The coarser particles would, in theory, stay in the bed far longer than would finer ones, however, the coarse particles would, however, undergo fragmentation when thrown into the bed. Additionally, while burning in the bed, the coarse particles are subjected to combustion-assisted attrition, producing fines. These fines have a much shorter residence time in the combustion zone than would be expected for coarser parent particles.

A smaller number of feed points is an important characteristic of over-bed feed systems that throw fuels over the bed surface by using spreaders. Coarser coal particles reach further deep into the furnace while finer particles drop closer to the feeder. Thus, the bed receives a nonuniform size distribution of coal. The maximum throwing distance of a typical spreader is around 4.5 m. The location of spreaders is dependent on the dimensions of the bubbling bed. When the width (distance between side walls) is less than the depth (distance between front and rear wall), spreaders are located on the sidewalls. When the depth is less than the width, spreaders are located on the front wall. When both width and depth are greater than 4.5 m, they can be located on both sidewalls.

Sometimes air assists the throw of coal by spreaders. When the combustor is small, the transporting air could be about 10% of the total air, which results in poor air supply to the dense bed. However, when the combustor is very large, the amount of air through the spreaders is a negligible portion of the total air, and hence does not affect the combustion in the dense bed.

Over-bed feed systems are more suitable for high reactivity coals which burn out rapidly in the bed. Otherwise, it is better to use an under-bed feed system.

10.3.2.2 Under-Bed Feed System

The under-bed feed system is expensive, complicated, and less reliable (especially with moist fuels), but it helps achieve high combustion efficiency.

The spacing between feed points greatly affects the combustion. At a given operating condition, an increase in feed point spacing from 1 to 4 m^2 per feed point could significantly drop the combustion efficiency (Duqum et al., 1985).

Bed height can affect the spacing of under-bed feeders (Figure 10.10). One could increase the spacing without sacrificing combustion efficiency only if it is compensated by a corresponding increase in the bed height. A decrease in bed height must be matched with decreased feed point spacing, otherwise the combustion efficiency could drop. For example, in one specific BFB boiler the feeder spacing was 1 m^2/feed point. Its combustion efficiency was 97% when the bed height was 1.22 m, but the combustion efficiency dropped to 94% when the feeder spacing was increased to 4 m^2/feed point though the bed height was increased to 1.78 m.

The feeder spacing also affects the optimum size of coal. For example, when the spacing is 1.0 m^2/feeder, the optimum size of coal is 1.6 mm, but when the spacing is 4 m^2/feeder, the optimum fuel size of is 1.8 mm (Duqum et al., 1985).

There is an optimum size of fuel for a given feeder and operating condition. When fuel size is either greater or lesser than this size, the combustion efficiency will drop. Larger coal particles have a slower burning rate for a given mass of carbon, and a larger carbon fraction in the bed for a given heat generation. This could increase the carbon loss through bed drain and fly ash. On the other hand, finer coal particles will have a faster burning rate on a unit mass basis, but they are more likely to be entrained adding to the carryover carbon loss as was observed by Duqum et al. (1985) in the 20-MWe BFB boiler at TVA.

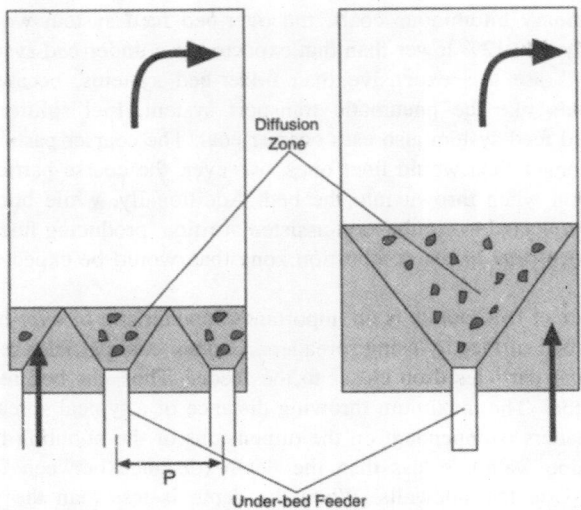

FIGURE 10.10 Effect of bed height and feed point spacing on fuel dispersion in a bubbling fluidized bed.

In the case of under-bed feeding, a feed-point spacing of 1 m²/point is a reasonable choice. This will help combustion efficiency and reduce operating costs. For very large boilers, the spacing may be between 1 to 4 m²/point. When the spacing is increased, the bed height has to be increased correspondingly. Also, the freeboard height should be greater than the transport disengaging height (TDH). This is very important, especially when wide feeder spacing is used.

10.3.3 Feed Point Allocation

The excellent solid–solid mixing in a fluidized bed helps disperse the fuel over an area of 1 to 2 m². As such, a single feeder is adequate for a small boiler having a bed less than 2 m² in cross section. Larger beds would, however, need multiple feeders. The number of feeders required for a given bed depends on several factors like quality of fuel, type of feeding system used, amount of heat input, and bed area. A highly reactive fuel with high volatiles would need more feeders because such fuels burn relatively fast, while a less reactive fuel like anthracite requires less feeders. Such fuels burn slowly and thus get adequate time to be dispersed over a larger area before significant combustion of these fuels starts. Table 10.3 lists the characteristics of a number of commercial bubbling

TABLE 10.3
Feed Points for Some Commercial Bubbling Fluidized Bed Boilers

Boilers	Boiler Rating, (MWe)	Bed Area (m²)	Feed Points	MWth/ Feed	m²/ Feeder	Feed Type	Fuel Type	HHV of Fuel (MJ/kg)
Shell	43 MWth	23.6	2	21.7	11.8	OB	Bituminous	
Black Dog	130	93.44	12	31.0	7.8	OB	Bituminous	19.5–34.9
TVA	160	234	120	3.8	2.0	UB	Bituminous	24–25
Wakamatsu	50	99	86	1.7	1.2	UB	Bituminous	25.8
Stork	90	61	36	2.8	1.7	UB	Lignite	25

OB — over-bed spreader feeder; UB — under-bed pneumatic feed.

fluidized bed boilers. Here, we note that the heat release rate per feeder q_{feed}, is about 1.5 to 4 MW for under-bed feeders and 20 to 30 MW for over-bed feed systems.

The number of feeders N_f, can be calculated from:

$$N_f = \text{Integer}\left(\frac{Q_{comb}}{q_{feed}}\right) \quad (10.1)$$

where q_{feed} may be taken either from experience or chosen from Table 10.3.

10.3.3.1 Redundancy

Industrial designs often provide redundancy. For example, if a boiler needs two over-bed feeders, the designers may provide three feeders each with 50% capacity of the design feed rate. This way, if one feeder is out of service, the boiler could still maintain the full load with two operational feeders. Thus, depending on the capacity and reliability required of the plant, feeders in addition to that calculated from Equation 10.1 are provided for.

Example 10.1

Given
- Superficial velocity of air through the bed is 2.0 m/s
- Primary air fraction is 60%
- Stoichiometric air is 4.87 kg/kg coal
- Excess air coefficient is 1.4; (moisture is negligible)
- Bed temperature is 900°C
- Density of air 0.3 kg/m³ at 900°C
- Air temperature in fuel transporting line is 45°C; air density 1.11 kg/m³
- Outside diameter of fuel line is 115 mm, and thickness is 5 mm
- Coal/air ratio in weight in fuel line is 1.3
- Velocity of air in the lines is 12 m/s

Find:
- The number of under-bed feed points per square meter.

Solution

We base the calculation on a bed cross-sectional area of 1 m².
Amount of primary air through the bottom of a 1 m² bed = 0.30 (kg/m³) × 1 (m²) × 2.0 (m/s) = 0.60 kg/s.
This constitutes 60% of the total air.
Therefore, the total air mass flow rate = 0.60/60% = 1.00 kg/s.
Total air per kg coal = 4.87 (kg/kg coal) × 1.4 = 6.82 kg/kg coal.
Coal mass flow rate for the 1 m² of bed = 1.00 (kg/s)/6.82(kg/ kg coal) = 0.15 kg/s.
Air mass flow through the fuel lines = 0.15 (kg/s)/1.3 = 0.11 kg/s.
Sum of cross-sectional area of fuel lines/m² bed = 0.11 (kg/s)/1.11 (kg/m³)/12 (m/s) = 0.0085 m².
Cross-section area of fuel line = π × (0.115 − 2 × 0.005)²/4 = 0.0087 m².
The number of fuel lines (feeders) = 0.0085 (m²)/0.0087 (m²) = 0.97 (points/m²) ≅ 1 point/m².

10.4 FEED SYSTEM FOR CIRCULATING FLUIDIZED BEDS

Circulating fluidized bed (CFB) boilers also use crushed coal. Feed size for CFB boilers is, however, slightly smaller than that required for BFB boilers. The fuel is generally specified to be below 6–10 mm.

10.4.1 FUEL FEED PORTS

An important characteristic of CFB boilers is that they require considerably fewer feed points than used by bubbling bed boilers. This is a result of the higher lateral mixing capability and deeper beds of CFB boilers.

No theoretical method for calculating the number of feed points per unit cross section is yet available in the published literature. However, it is known to be a function of the fuel characteristics (char reactivity, volatile yield) and the degree of lateral mixing in the specific design of the furnace. Table 10.4 lists thermal input per unit feed point and bed area served by each feed point for a number of commercially operating CFB boilers. This provides a rough guide for the selection of appropriate numbers of feed points. Tang and Engstrom (1987) suggested one feed point per 9 to 27 m^2 of bed area. The values of bed area per feed point used in operating units (Table 10.4) are in line with their suggestion.

TABLE 10.4
The Feed Point Allocation of Some Commercial Circulating Fluidized Bed Boilers

Plant	Capacity MWe	Feed Points	MWth/ Feed	Bed area (m^2/feed)	Fuel
Henan Pingdingshan, China	12	1	34.3	13.4	Subbituminous coal
Kanoria Chemicals, India	25	2	35.7	9.9	Coal/slurry
Sogma, Spain	25	3	23.8	11.3	Municipal solid waste
Stadwerke Pforzheim, Germany	25	2	35.7	11.7	Hardwood
Bayer, Germany	35	2	50.0	11.1	Hard coal + 20% brown coal
Stadwerke Saarbrucken, Germany	40	4	28.6	9.5	Hard coal
Sinarmas Pulp, India	40	2	57.1	21.6	Hard coal + 20% brown coal
GEW Koln, Germany	75	4	53.6	13.3	Brown coal
Duisberg Co-Generation, Germany	96	4	68.6	15.9	Bituminous coal
Nisco Co-Generation, USA	100	4	71.4	19.5	PetCoke + bituminous coal
Goaba, China	100	2	142.9	44.5	Anthracite coal
Colorado-Ute, USA	110	6	52.4	16.3	Bituminous coal
Northampton, USA	110	4	78.6	22.5	Culm-anthracite waste
Colver, USA	110	4	78.6	22.5	Bituminous gob
GIPCL, India	125	2	178.6	40.7	Lignite
Akrimota TPS, India	125	4	89.3	25.9	Lignite
Emile Huchet Power Stn., France	125	6	59.5	15.8	Coal/slurry
Point Aconi, Canada	190	4	135.7	33.8	Subbit. coal
Tonghae, Korea	200	4	142.9	33.3	Anthracite coal
Turow, Poland	235	8	83.9	26.2	Brown coal
EC Zeran Plant, Polant	235	4	167.9	22.9	Coal
Alholmens. Finland	240	12	57.1	17.1	Bark, wood residue, peat
Turow, Poland	235	8	92.9	26.1	Brown coal
Provence, France	250	4	178.6	21.3	Subbituminous coal
Jacksonville, USA	300	4	214.3	42.8	PetCoke coal
Fenyi, China	100	4	214.3	23.8	Bituminous

Solid Handling Systems for Fluidized Beds

Though coal is normally fed by gravity, the feed line should have a pressure higher than that in the furnace to prevent blow-back of hot gases from the combustor. The feed point should be located within the refractory-lined substoichiometric lower zone of the furnace, as far below the secondary air entry point as possible to allow the longest possible residence time for the fine fresh coal particles. In some boilers, coal is fed into the loop-seal where it is heated and partially devolatilized before entering the bed. Since the fuel travels with a large volume of hot bed solids it has an opportunity to arrive at the bed in a well-mixed state. The practice of feeding fuel into the loop-seal is especially advantageous for high-moisture or sticky fuels (Tang and Engstrom, 1987).

10.4.2 LIMESTONE FEED SYSTEM

The location of limestone feed points is less critical than that of the coal feed points, because the reaction rate of limestone is much slower than that of coal combustion. Limestone is smaller in quantity and finer in size and thus is suitable for pneumatic injection into the bed. It is sometimes injected into the recycled solids to help them mix better with bed materials.

Limestone is fed either pneumatically or mechanically. A pneumatic system feeds limestone directly into the furnace through openings in the front and rear wall of the furnace, while the mechanical system involves dropping a metered amount of limestone into the fuel feed hopper by means of a rotary feeder. It then slides under gravity into the furnace along with the fuel.

10.5 DESIGN OF PNEUMATIC TRANSPORT LINES FOR SOLIDS

This section presents a simple approach to the design of transport lines, which can be used in the solid-handling system of a fluidized bed gasifier or boiler.

The basic objective of pneumatic transport is to convey as much solid as possible with minimum pressure drop while avoiding any chance of settling or plugging. A mixture of gas and solids transported through a tube could demonstrate one of the three behaviors shown in Figure 10.11. When the solid/gas ratio is very low and air velocity is high, the mixture travels as a lean phase, in which the solid is uniformly dispersed (Figure 10.11a). If the gas velocity is decreased while the solid feed rate is held constant, there could be segregation of solids in the tube. Some solids may settle on the tube floor, while some may be transported (Figure 10.11b). This is called *saltation*. If the solid feed rate is above a critical level, the solids could flow like plugs (Figure 10.11c). This is called *dense-phase flow*, which might give pressure pulsation in the system.

Table 10.5 provides design values of safe density and safe velocity for transport lines carrying coal and/or ash. It is apparent from the table that horizontal pipes need higher velocity and have a lower solid-carrying capability than vertical pipes. A lower gas velocity, but higher solid loading gives rise to the settling of solids on the floor of the pipe or sometimes formation of solid plugs or dunes. Either of these two conditions should be avoided because the pressure drop through the pipe increases dramatically as soon as settling of solids takes place in the pipe. Thus, it is important to ensure that the gas velocity is well above the saltation velocity for the system.

10.5.1 PRESSURE DROP IN A PNEUMATIC TRANSPORT LINE

The pressure drop $(P_1 - P_2)$, through a pneumatic transport line with diameter d_t, and length L, inclined at an angle θ, with the horizontal, may be expressed as the sum of three components: static head, kinetic energy of solids, and the friction resistance of the mixture (Kunii and Levenspiel, 1991):

$$P_1 - P_2 = [\rho_p(1 - \varepsilon_g) + \rho_g \varepsilon_g]\left[g \sin\theta + \frac{2f^1 U^2}{d_t} \right] L + \frac{G_s^2}{\rho_p(1 - \varepsilon_g)} \qquad (10.2)$$

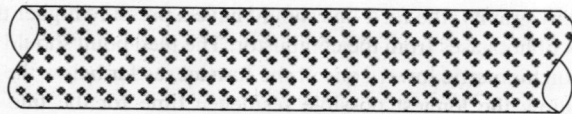 

(a) Lean Phase
(Dilute or Stream Flow)

18<30m/sec
High Velocity

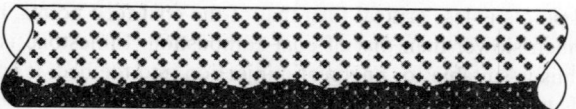 

(b) Medium Phase

10<20m/sec
Low Velocity

(c) Dense Phase
(Plug Flow)

2<10m/sec

FIGURE 10.11 Effect of conveying velocity on the solid transportation.

TABLE 10.5
Safe Values for Pneumatic Transport of Solids

Solid	Size (μm)	Average Bulk Density (kg/m³)	Minimum Safe Air Velocity (m/s)		Maximum Safe Density for Flow (m/s)	
			Horizontal	Vertical	Horizontal	Vertical
Ash	90% < 150	720	4.6	1.5	160	480
Coal	100% < 480	560	4.6	1.5	110	320
	75% < 78					
	100% < 6400	720	12.2	9.2	16	24
	100% < 12.7	720	15.3	12.2	12	16
Silica	95% < 105	800–960	6.1	1.5	80	320

Source: From Kunii, D. and Levenspiel, O., *Fluidization Engineering*, Butterworths Heinemann, Stoneham, pp. 385–388, 1991. With permission.

TABLE 10.6
Values of Coefficient f_s for Horizontal

Solid	Size (mm)	Pipe Diameter (mm)	Air Velocity (m/s)			
			10	15	20	30
Coal	0–1	25.4	—	0.0014	0.0011	0.0011
Sand	0.8–1.4	44.4	0.005	0.0045	—	—

where ρ_p is the density of particles, ρ_g is the density of gas, ε_g is the voidage in the mixture, U is the gas velocity, u_s is the solid velocity, and G_s is the solid flux.

The friction coefficient f_s, is a function of air transport velocity and size of particles. Table 10.6 gives some values of the friction coefficient.

For bends in the line, one can use the expression given by Kunii and Levenspiel (1991):

$$\Delta P = 2f_b[\rho_p(1 - \varepsilon_g) + \rho_g\varepsilon_g]U^2 \quad (10.3)$$

The bend coefficient f_b, depends on the radio of bend radius r, and pipe diameter d_t. Table 10.7 presents some values of the bend coefficient for a few ratios of bend radius and pipe diameter.

10.5.2 Design of a Pneumatic Injection System

In a pneumatic transport system, the coal/air weight ratio in the transporting lines should be chosen to avoid saltation, and at the same time avoid use of excessive transport air. For a typical BFB boiler, a coal/air mass ratio of around 1.3:1 is acceptable. Following engineering data based on operating experience could help design of a pneumatic transporting line.

Typical sizes of the fuel line diameter are 115 mm O.D to 168 mm O.D. The fuel pipe may project about 200 mm above the distributor plate. An air jacket, covering the projected portion of the feed tube, may prevent a coal particle from heating to its devolatilization point. A typical thickness of the air-cooled jacket is 25 mm. The cooling air comes from the wind-box underneath the distributor plate. To prevent direct impact on the bed tubes, the gap between the top of the injection nozzle's cap to the bottom of bed tube should not be less than 130 mm.

10.5.3 Fuel Splitter

Fuel splitting or division of a gas–solid suspension into a large number of equal flow streams, requires major consideration. This task is considerably more difficult than that of dividing a gas stream or liquid stream. Here, high-pressure air conveys the fuel pneumatically through tubes to individual coal injection nozzles in the bed. The air pressure in the transport line has to be increased to acquire a good distribution of the solid fuel among the downstream fuel lines. This necessitates the use of special devices, as shown in Figure 10.12a and b for splitting a gas–solid stream into multiple streams.

TABLE 10.7
Values of Bend Coefficients in Pneumatic Transport

r/d_t	2	4	6
f_b	0.375	0.188	0.125

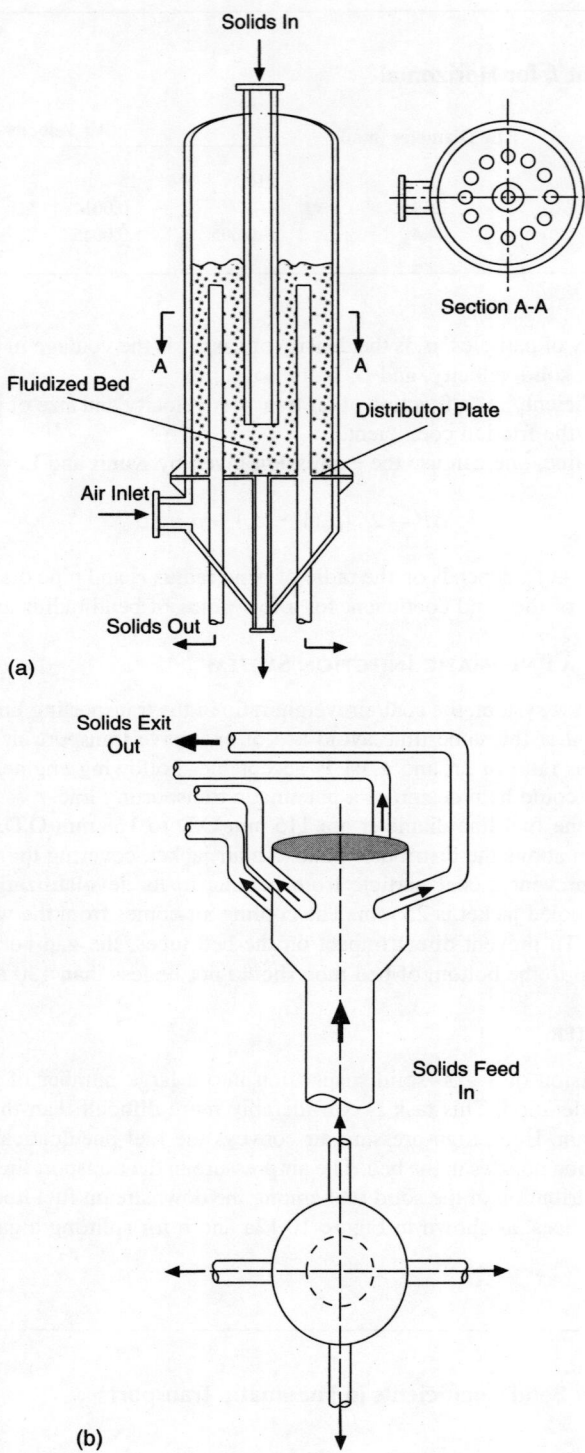

FIGURE 10.12 (a) Fluidized bed gas–solid flow splitter. (b) Pneumatic gas–solid flow splitter.

Two types of splitters are shown in Figure 10.12. The first (Figure 10.12a) is a small fluidized bed with multiple drains, where solids enter through one line. Since the freeboard is sealed, the fluidizing air must pass through the multiple drain pipes carrying solids through them.

Figure 10.12b shows another type of splitter, which is a simple gas manifold. This type of splitter can experience severe erosion, particularly while using erosive feedstock.

NOMENCLATURE

d_t: Diameter of transport line, m
f_b: Bend coefficient
f_s: Solid friction coefficient
f_g: Gas friction coefficient
G_s: Solids mass flux, kg/m^2 s
g: Acceleration of gravity, 9.81 m/s^2
L: Length of transport line
N_f: Number of feeders
$P_1 - P_2$: Pressure drop across the transport line, Pa
Q_{comb}: Combustion heat release, MW
q_{feed}: Combustion heat release per feed point specified, MW
U: Superficial gas velocity in pipe, m/s
u_s: Velocity of solids

GREEK SYMBOLS

ε_g: Gas void fraction in pipe
ρ_p, ρ_g: Density of the solid and gas respectively, kg/m^3
θ: Inclination of transport line with horizontal, degree

REFERENCES

Duqum, J. N., Tang, J. T., Morris, T. A., Esakov, J. L., and Howe, W. C., *AFBC Performance Comparison for Under-bed and Over-bed Feed System*, Proceedings of the Eighth International Conference on Fluidized-Bed Combustion, 1985, pp. 255–277.

Kunii, D. and Levenspiel, O., *Fluidization Engineering*, Butterworths Heinemann, Stoneham, 1991, pp. 385–388.

Lee, Y. Y., Design considerations for a CFB boilers, In *Circulating Fluidized Beds*, Grace, J. R, Avidan, A. A, and Knowlton, T. M., Eds., Chapman & Hall, London, pp. 430, 1997.

Tang, J. and Engstrom, F., Technical Assessment on the Ahlstrom Pyroflow Circulating and Conventional Bubbling Fluidized Bed Combustion Systems, In *Proceeding of the Ninth International Conference on Fluidized Bed Combustion*, Mustonen, J. P., Ed., ASME, New York, pp. 38–54, 1987.

11 Air Distribution Grate

The air distribution grate, also called the *distributor plate*, supports the bed materials and homogeneously distributes the fluidizing gas into the bed of solids. This is an important aspect of the design of the fluidized bed because beyond the distributor plate there is no other physical means for influencing the distribution of air through the solids. Nonuniform distribution of air may result in a number of problems; from reduced performance of the combustor or gasifier to complete collapse of the bed due to agglomeration.

One of the most common problems with fluidized bed boilers has been with fluidizing nozzles. Problems include high maintenance due to erosion, plugging, and back-sifting. Nozzle plugging also contributes to general unit performance problems. Serious grid nozzle plugging could result in frequent cleaning, high bed temperatures, excessive limestone and ammonia consumption, and/or high excess air requirements.

The fluidizing air has to lift a large mass of solids against gravity and overcome the resistance of the distributor plate. Thus, it consumes a sizeable amount of energy and adds to the auxiliary power consumption of the fluidized bed boiler or gasifier. Therefore, the auxiliary power consumption of a fluidized bed is generally higher than that of an entrained bed, fixed bed or transport bed. Auxiliary power consumption is a major issue in a power plant as it directly reduces the efficiency of the plant. For example, in a circulating fluidized bed (CFB) boiler plant the total auxiliary power consumption is about 7.5 to 11% of the total power generation compared to about 9% for a pulverized coal boiler plant. In a bubbling bed boiler it is slightly lower than that in CFB boilers, but still higher than that of stoker-fired boilers. A good rational design of the distributor can greatly contribute to the reduction in the auxiliary power consumption.

In addition, most fluidized beds have to deal with the problems of nonuniform fluidization and back-sifting of solids. A proper design of the grid plate is, therefore, very important from an operational as well as an economic standpoint.

The present chapter discusses the basics of the operation of air distribution grids as well as general design methodology.

11.1 DISTRIBUTOR PLATES

In a bubbling fluidized bed (BFB) the grid plate is used in the furnace alone, but in a circulating fluidized bed, distributor plates are required in three different locations: the lower furnace, the loop-seal, and the external heat exchanger. Owing to its relatively low fluidization velocity (0.5 to 1.7 m/sec), a bubbling fluidized bed needs a much larger grate area for a given energy output. The distributor plate required for this type of bed is therefore much larger than that required by a circulating fluidized bed that operates with a higher fluidizing velocity of 4 to 6 m/sec.

The primary function of a distributor plate is to distribute the fluidizing air uniformly across the cross section of the bed. This uniformity should be maintained under all operating conditions of the boiler. For a large boiler furnace this could be a serious problem.

A BFB boiler or gasifier uses a relatively coarse particle and low fluidizing velocity. If the fluidizing air above its grid is not uniformly distributed, the air velocity in one region could drop below the minimum fluidizing velocity of the coarser fraction of the bed materials. The air will thus fail to fluidize the coarser particles making that section partially defluidized. Fuel particles burning in that defluidized section will fail to dissipate their heat and lead to a local hot spot. Since the burning rate of most fuel particles is higher at higher temperatures, heat generation in that region will rise faster than its dissipation through solids mixing. As a result the temperature of the particles in this region will rise above their softening temperature allowing the particles to stick to each other. This further restricts the mobility and consequently the heat dissipation from that zone. Thus the region quickly turns into clinker, which grows throughout the bed.

Distributor plates can be broadly classified into three groups:

1. *Porous* and *straight-hole orifice* type plates generally use punched or drilled vertical holes through a plate or sintered plates. It is also called a *plate-type distributor*. Such plates are not commonly used in industrial boilers or gasifiers (Figure 11.1a)
2. *Nozzle*-type or *bubble cap*-type uses nozzles, which distribute air into the bed through horizontal vertically, or downward holes. (Figure 11.1b)
3. *Sparge pipe*-type which comprises of air-carrying tubes with holes. These are introduced directly into the fluidizing bed without a grid plate or a plenum box below it (Figure 11.1c)

Three generic types of distributors are shown in Figure 11.1. In addition to these, many other novel designs are in use with varying degrees of success. For example, the *bent-tube nozzle* was developed and used by one CFB boiler manufacturer to prevent the back-flow of solids, but is not used any longer. The following section elaborates each of the above three types of distributors.

11.1.1 PLATE-TYPE DISTRIBUTOR

11.1.1.1 Perforated Plates

Plate-type distributors are the simplest type to manufacture. They are popular amongst researchers, but they are not used widely in the industry. Figure 11.2 shows some types of perforated plate-type distributors. The plate (Figure 11.2a) is made of a fine wire mesh sandwiched by two perforated plates. The load of the bed materials is borne by the thicker plates while the wire mesh provides the necessary protection against any leakage of solids during shut-down.

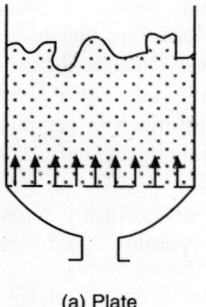
(a) Plate

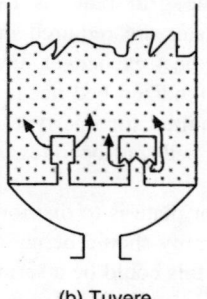
(b) Tuyere

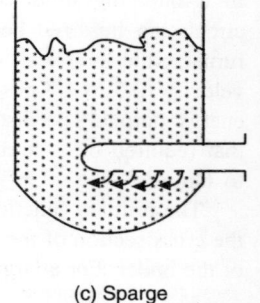
(c) Sparge

FIGURE 11.1 Three generic forms of grid design.

Air Distribution Grate

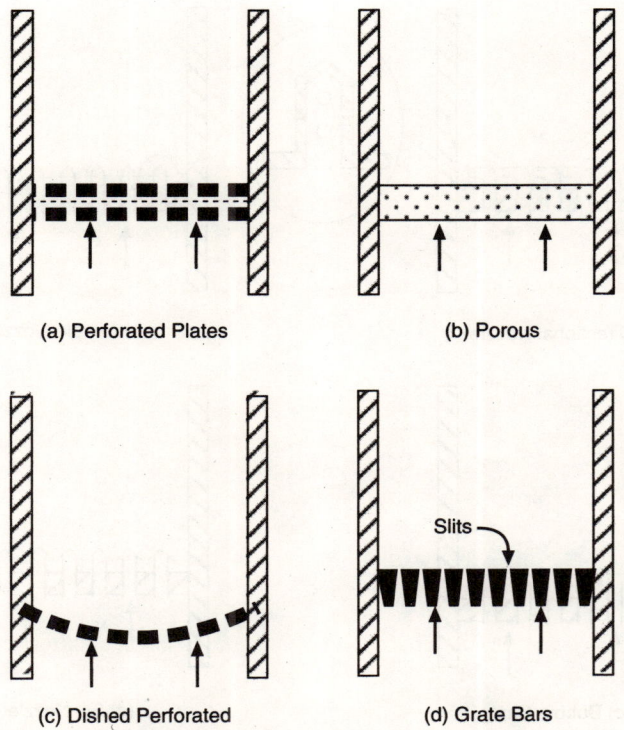

FIGURE 11.2 Plate and grate distributors are relatively inexpensive and easy to construct.

Figure 11.2c shows a bent perforated plate for a specific application, where it was necessary to induce a circulation of solids within the fluidized bed by inducing higher fluidizing velocities on two edges and lower velocity at the center. By using a bowl-shaped grid plate (edges raised by 0.2 m), the CFB boiler at Bewag increased airflow near the walls and thereby reduced the erosion of the wall and sintering near coal feed points (Werther, 2005).

Slot-type grid plates (Figure 11.2d) are formed by a series of slotted bars.

11.1.1.2 Porous Plates

Porous plates can be made of the following materials:

- Synthetic
- Ceramic
- Sponge metal

Such plates are available in various porosities and thicknesses. Porous plates cannot provide good mechanical strength at acceptable cost, but they provide total protection against back-sifting of bed materials. Another important feature of the porous plate is its pressure drop vs. velocity characteristic. In all other types, the pressure drop across the plate is proportional to the square of flow through the plate, but in this case they are directly proportional. Thus, under low airflow rates porous plates are not exposed to potential misdistribution of air as other types would be due to drastic reduction in the plate pressure drop at low boiler load.

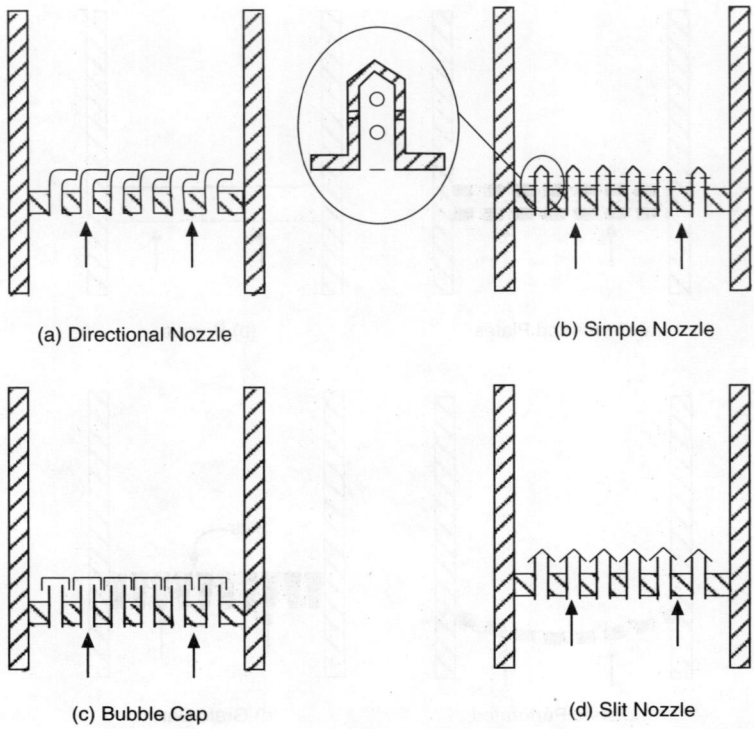

FIGURE 11.3 Examples of tuyer or nozzle-type distributors.

11.1.2 NOZZLE-TYPE DISTRIBUTOR

This is the most commonly-used distributor in energy industries with numerous designs of nozzles available in the market. Figure 11.3 shows only a few types of nozzles or *tuyers*. Figure 11.3a shows an example of nozzles oriented in a specific direction to facilitate flow of coarser bed solids towards the bed drain. Figure 11.4 shows how these nozzles could be arranged to facilitate solid flow in the desired direction.

Directional nozzles (Figure 11.3a) can be cast instead of made from welded tubes. In such cases, nozzle openings can be made noncircular to reduce the risk of solid back flow. Details of one such nozzle is shown in Figure 11.5.

These nozzles being large in diameter may suffer from back-flow of solids unless their openings are designed specifically to reduce the back-flow. Figure 11.6 shows an example of bent tube nozzles, which are designed to minimize back-flow of solids through the nozzle.

Upright tubes with holes drilled into their sides (Figure 11.3b) are the most common type of nozzles. These are sometimes cast into special shapes instead of being manufactured from tubes. The bubble caps shown in Figure 11.3c are more popular in chemical than in energy industries. The slit type nozzle (Figure 11.3d) uses an inclined roof to avoid the dead solids, which is a problem with the bubble cap. Another popular nozzle, known as arrowhead nozzle, is made of two long sloped holes with inclined roofs similar to this type.

In each of the above types, a smaller opening at the entry of the nozzle can be employed instead of one at the exit to control the plate pressure drop as shown in Figure 11.6. This will avoid the high jet velocity (20 to 60 m/sec) at the exit orifice, and yet obtain the desired pressure drop across the distributor.

FIGURE 11.4 Directional nozzles are set in square arrangements leaving sufficient space between nozzle rows.

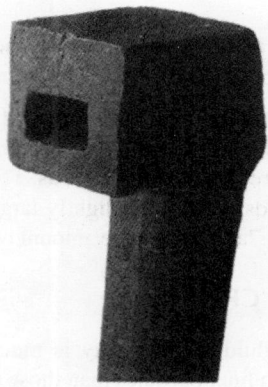

FIGURE 11.5 A directional nozzle.

The nozzles are generally supported on a grid plate which is made of either a water-cooled membrane wall or a refractory-coated steel plate. The water-cooled membrane grid is effective in handling very hot air sometimes used for start-up.

11.1.3 Sparge Pipe Distributor

Sometimes it is impractical to use an air box under the bed-support plate due to space limitations or when an explosive/inflammable mixture of gases is introduced to the reactor. In this case, the gas is carried by pipes and distributed directly into the bed as shown in Figure 11.7. One can maintain gas velocity in the nozzle well in excess of flame velocity to guard against any possible blow-back. Since the volume of gas in the pipe grid is considerably smaller than that in a large air box, the chance of explosion is also greatly reduced. Such situations are common in the case of gas-fired boilers. An alternative means for avoiding the risk of explosion is to inject gas directly into the air nozzles of the bed as shown in Figure 11.7.

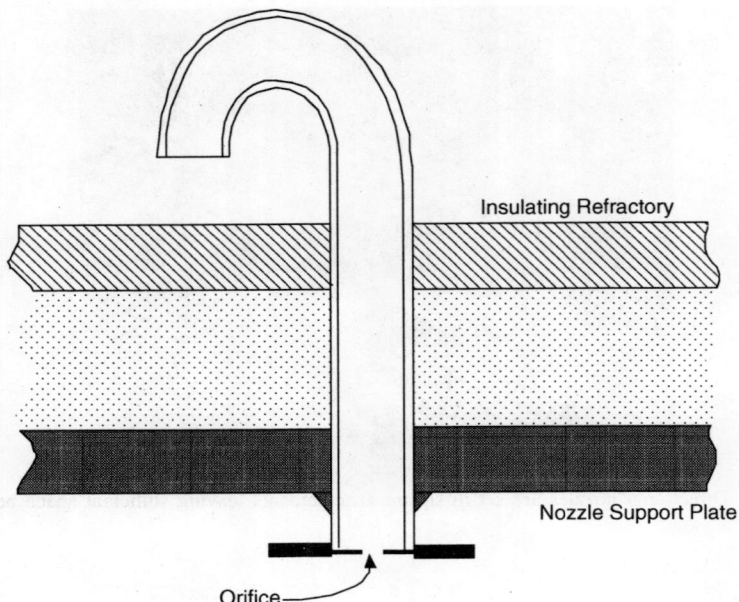

FIGURE 11.6 A downward nozzle with an inflow orifice to reduce the velocity of the gas exiting into the bed.

In a sparge pipe distributor, air enters through a tube inserted into the fluidized bed (Figure 11.7a). Holes or orifices can be drilled directly into the pipes, but that might lead to solids entering the pipe during shut down even if they are oriented downwards. To avoid this one could use shrouds as shown in Figure 11.7a. These shrouds, which are slightly larger than the holes, can be located in alternate pitch as shown in Figure 11.7a to reduce the amount of dead space above the sparge pipes.

11.1.4 Distributor Grids for CFB

In a circulating fluidized bed, the fluidizing velocity is much higher than that in the bubbling system. Also, CFB systems use much finer particles than those used by bubbling beds. The *fluidization number*, defined as the ratio of operating to the minimum fluidizing velocities of average particles, is much higher in a CFB system compared to that in a BFB. Thus, even if there is some nonuniformity in the air distribution across the bed, chances of any part getting defluidized is very low in a CFB riser.

Thus, a high pressure drop across the grid plate may not be needed in a CFB riser to ensure uniform airflow across it. A sufficiently high orifice velocity and a pressure drop just adequate to avoid back-shifting could meet its requirement. In fact, some nonuniform airflow distribution at the level of the grid may actually be advantageous in enhancing lateral solid distribution through the creation of a solid-circulation vortex. A large CFB boiler in Germany improved its performance by deliberately having nonuniform fluidizing air distribution in the lower furnace (Werther, 2005).

The height of a CFB riser is typically in the range of 10 to 40 m while its width is only 2 to 8 m. This allows adequate time to the solids to be laterally mixed over the riser height.

The potential of sinter formation due to nonuniform fluidization is relatively low in a CFB. The high shear force in the high-velocity CFB riser tends to break up any sintered mass if formed.

The current engineering practice for air-distributor plates in CFB calls for the use of a high pressure drop grid to provide uniform fluidization. Because of this, the fan power consumption of a typical CFB boiler is about 2.3% compared to only 0.75% for a PC boiler (Table 8.9).

Air Distribution Grate

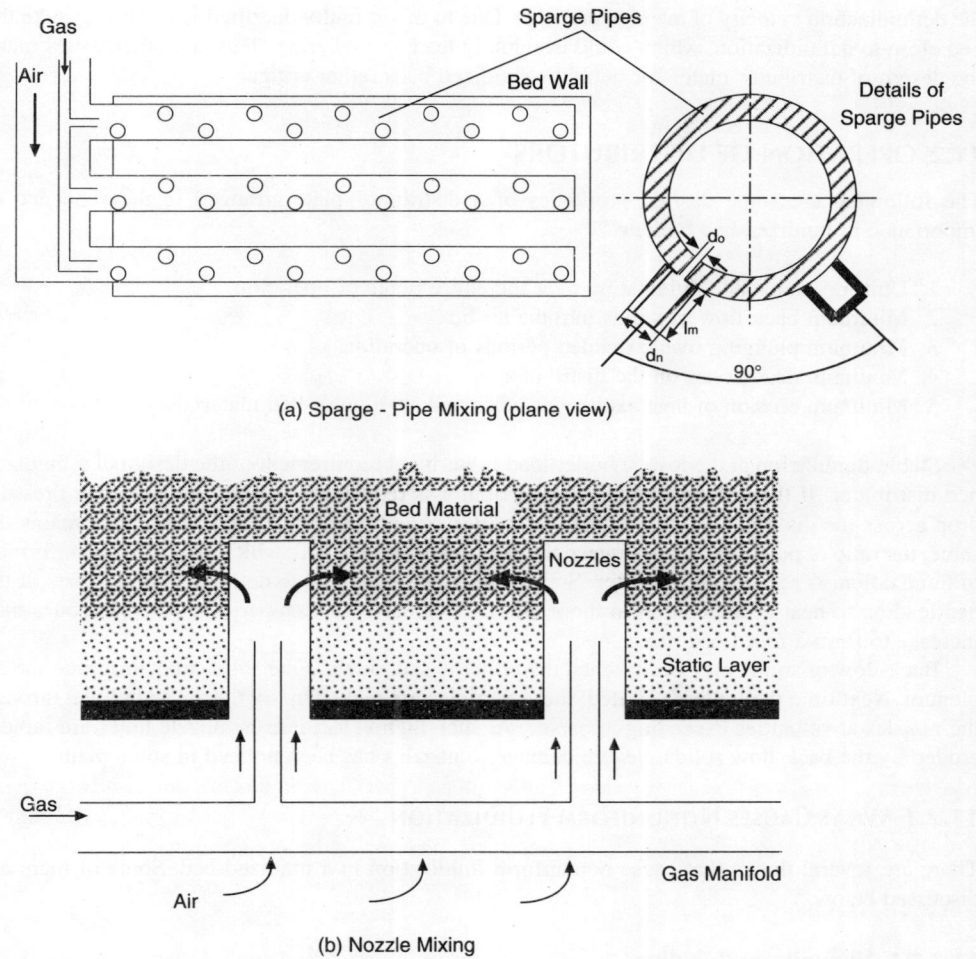

FIGURE 11.7 Sparge pipe distributor.

This accounts for a considerable part (~60%) of the auxiliary power requirement for a CFB boiler island.

The design of the air distributor grate or grid of the CFB is less critical than that for bubbling fluidized bed boilers, as the furnace of a CFB boiler operates at a much higher velocity (4 to 6 m/sec) and uses finer (200 to 300 μm) particles. Thus, even if one part of the bed receives 20% of the fluidizing velocity it is still sufficient to keep the solids fluidized and mobile.

CFB boilers use a bubbling fluidized bed in some of its components outside the furnace. For example, the loop-seal and the external heat exchanger operate in bubbling fluidization regimes. The distributor problem does exist here but may not be significant because these components are free from combustion and use particles finer than those in most bubbling bed boilers.

11.1.5 Distributor Grids for Bubbling Beds

In a bubbling bed the operating velocity is low (0.5 to 1.7 m/sec) but particle size is large (>1 mm). Thus, the fluidization number is low and, therefore, its superficial velocity is not far from the defluidization velocity of average particles. Due to this, a faulty distribution of air may take the bed close to defluidization, which could eventually lead to clinkering. This and other issues make the design of distributor plates for bubbling fluidized beds rather critical.

11.2 OPERATION OF DISTRIBUTORS

The following are some desired properties of a distributor plate arranged in the sequence of importance in fluidized bed boilers:

1. Uniform and stable fluidization over the entire range of operation
2. Minimum back-flow of solids into the air box
3. Minimum plugging over extended periods of operation
4. Minimum dead zones on the distributor
5. Minimum erosion of heat exchanger tubes and attrition of bed materials

Stable fluidization at the lowest boiler load is a critical requirement of the design of a fluidized bed distributor. If the airflow through the distributor is reduced by half at low load, the pressure drop across the distributor will reduce by a quarter. Since the mass of bed materials remains the same, the ratio of pressure drops between the distributor and the bed, which controls the uniformity of fluidization, is reduced to a quarter. Such a situation could lead to defluidization in parts of the bed leading to heat accumulation in those areas. Eventually, the temperature of those zones may increase to form a fused bed mass.

Back-flow of solids may occur when the boiler is shut down. Fine solids may drop into the air plenum. Next time the boiler is started; these particles are picked up by the air and blown through the nozzles at velocities exceeding 30 m/sec. At such high velocities, the nozzle holes are rapidly eroded by the back-flow solids. Severe damage to nozzles has been noticed in some plants.

11.2.1 WHAT CAUSES NONUNIFORM FLUIDIZATION

There are several things that cause nonuniform fluidization in a fluidized bed. Some of them are discussed below.

11.2.1.1 Nonuniform Bubbling

The bursting bubbles and general flow conditions can lead to substantial variations in pressure drop across a fluidized bed. The static pressure difference between the air box and freeboard of a bed generally remains constant across the bed. Therefore, the flow across a particular section of a fluidized bed is governed by the pressure drop across the grid and the fluidized solids in that section. When the bed density above the grid changes from one point to another due to varying concentration of bubbles across the bed (Figure 11.8a), the total resistance across the bed and the grid would naturally vary from point to point. This, in turn, leads to nonuniform airflow distribution across the bed. By making the resistance of the grid appreciably larger than the fluctuation in the bed pressure drop as shown in Figure 11.8b, one dampens the variation of total pressure drop across the bed and the grid. This way, nearly uniform air flow across the bubbling fluidized bed is assured.

11.2.1.2 Nonuniform Air Distribution in Air Box

Nonuniform fluidization is also caused by a variation in the static pressure in the air box. In large fluidized bed boilers, air generally enters the air box or plenum from the side. If the entry section is very small compared to the cross section of the plenum, an air jet is formed underneath the bed.

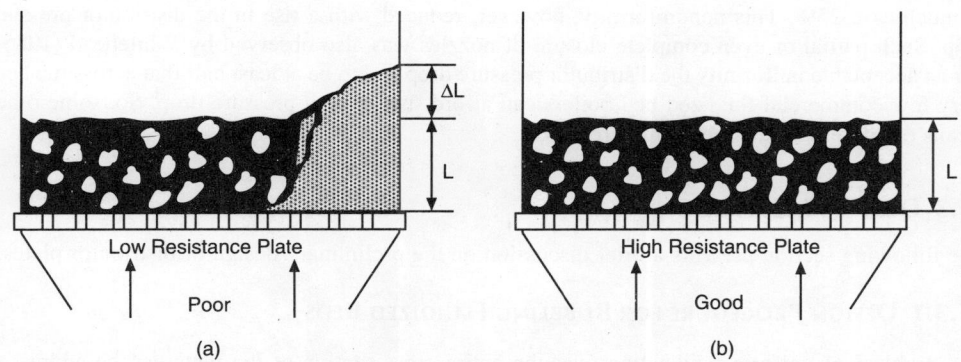

FIGURE 11.8 Nonuniform fluidization could occur due to nonuniform bubbling.

This leads to a variation in static pressure across the air box. In such cases, one would observe nonuniform flow across the bed owing to the variation in pressure differences between the air box and the freeboard from one point to the other. If airflow through one section of the bed is high, the bed density in that section drops below that of the rest of the bed due to a greater presence of bubbles. This further reduces the resistance across the vertical section, allowing more air to pass through this region. If the distributor resistance is very high, increased airflow would increase the pressure drop through the grid sufficiently to offset the lowering of resistance across the bed. This will help bring uniform fluidization across the bed.

11.2.1.3 Blockage of Orifices

Solid particles sometimes drop into the orifices and are trapped for some reason. This could happen while the bed is shut down or even while in operation. The latter situation arises preferentially in orifices receiving less air than others for whatever reason. Such particles could either drop into the air box or become trapped if their size is comparable to smallest dimension of the orifice. When this happens, the fluid resistance through this orifice increases which further reduces air flow through it.

The orifice-velocity is proportional to the square root of the pressure drop across the orifice. Thus, if the distributor is designed for a high pressure drop, design velocity through the orifice would also be higher. Because of this, any solid trapped in the orifice will be subject to much higher upward aerodynamic drag. Those particles are naturally more likely to be dislodged and carried into the bed than in a low-pressure drop distributor.

Downward angled orifices also help prevent solid from entering the nozzles. For reliable operation the length of orifices should be more than 5 to10 cm (Werther, 2005).

11.2.1.4 Inoperative Orifice

In a fluidized bed, not all orifices operate all the time. Sometimes air does not flow through some orifices, making them nonoperative. The exact reason for this occurrence is beyond the scope of this book, however, it occurs especially in low-pressure drop distributors. When air does not flow through an orifice there is local defluidization. It is interesting to note that the position of nonoperative orifices could shift during operation. In any case such an occurrence gives rise to nonuniform fluidization.

Guo and Werther (2004), noticed that the airflow through a set of nozzles in an empty bed was uniform, but, when solids were added to the bed, local air velocity through the nozzles varied by as much as ±25%. This nonuniformity, however, reduced with a rise in the distributor pressure drop. Such partial or even complete closure of nozzles was also observed by Whitehead (1985). For an acceptable uniformity the distributor pressure drop had to be at least half that across the bed. Very few commercial fluidized bed boilers can afford such a high pressure drop. So, some other means of addressing this problem is required.

11.3 DESIGN METHODS

The following section presents a brief discussion on the preliminary design of distributor plates.

11.3.1 DESIGN PROCEDURE FOR BUBBLING FLUIDIZED BEDS

The problem of uniform fluidization over the entire cross section of the plate can be addressed with a distributor pressure drop considerably in excess of the probable nonhomogeneities in the local pressure. Design experience suggests that in bubbling beds, a distributor pressure drop of 0.1 to 0.3 times that across the bed can meet this requirement. No such criterion is available for a circulating fluidized bed. Kunii and Levenspiel (1991) suggest that for stable fluidization of Group A particles, the ratio of pressure drop through the grid ΔP_d, and that across the bed ΔP_b, is a function of bed depths at operation H, and that at minimum fluidized condition H_{mf}.

$$\frac{\Delta P_d}{\Delta P_b} > \frac{H}{H_{mf}} - 1 \qquad (11.1)$$

when $U/U_{mf} > 10$ for porous plates and $U/U_{mf} > 3$ for perforated plates.

For Group B particles, Whitehead (1985) noted that while the gas velocity was increased through a large bed, only 50% of the tuyeres were operational at $U/U_{mf} = 2.5$. All tuyeres were not operational until the velocity reached $U/U_{mf} > 6$ and this value increased with the depth of the bed. For even distribution of fluidizing gas to a bed where U is close to U_{mf}, one could choose $(\Delta P_d/\Delta P_b) > 0.15$.

The key consideration is to ensure that all orifices are open under all operating conditions because under low load some nozzles remain closed. Most practical designs of bubbling beds keep the distributor pressure drop within 15 to 30% of the bed pressure drop.

$$\frac{\Delta P_d}{\Delta P_b} = 0.15 \rightarrow 0.3 \qquad (11.2)$$

In a CFB, there is no such guideline for distributor pressure drop. Still, prevention of back-shifting and keeping all nozzles operational at all loads are important considerations. The bed pressure drop in a CFB boiler being higher than that in bubbling beds, some manufacturers specify the grid pressure drop to be about 15 to 25% of the total bed pressure drop. This high pressure drop is more for prevention of back-sifting of solids rather than for ensuring uniform fluidization.

11.3.1.1 Orifices

The gas velocity through an orifice U_0 is related to the pressure drop through the grid as:

$$U_0 = C_D \left[\frac{2\Delta P_d}{\rho_{gor}} \right]^{0.5} \qquad (11.3)$$

where ΔP_d distributor pressure drop and ρ_{gor} is the density of the gas (kg/m^3) passing through the orifice.

Air Distribution Grate

The orifice coefficient C_D, is given by Zenz (1981) as 0.8. For a thick plate ($t/d_{or} > 0.09$) one may use the relation $C_D = 0.82(t/d_{or})^{0.13}$ (Quereshi and Creasy, 1979).

To achieve uniform and stable fluidization, the resistance offered by the distributor to the air should be significant compared to that offered by the bed solids. Thus, should there be a small difference in the resistance between one section of the bed and another, the combined resistance of the bed and distributor to the air passing will not vary appreciably between sections. In this way, a uniform airflow through all sections of the bed is ensured. Experience suggests that for bubbling fluidized beds the ratio of distributor to bed pressure drops should be around 0.15 to 0.3.

The pressure drop through a bubbling bed of height H, may be calculated from:

$$\Delta P_b = \rho_p(1 - \varepsilon)Hg = \rho_p(1 - \varepsilon_{mf})H_{mf}g \tag{11.4}$$

The bed voidage ε, depends on a large number of factors, such as terminal velocity, type of particles, vessel diameter, fluidization regime, and fluidization velocity. The last factor is the one most affecting the voidage. An approximate relation to find the bed voidage for up to 5 m/sec for Group A particles is given by King (1989). It may be used for a rough estimation of voidage in the lower section of the CFB riser, and other bubbling fluidized beds.

$$\varepsilon = \frac{U + 1}{U + 2} \tag{11.5}$$

where U is the superficial gas velocity in m/s.

If N is the number of orifices (diameter, d_0) per unit area of the distributor, the orifice velocity U_0, and the superficial gas velocity U, may be related to the fraction of the grid plate opened for gas by using the mass balance as follows:

$$N\frac{\pi}{4}d_0^2 U_0 \rho_{gor} = U\rho_g$$

$$\text{Fractional opening area of the orifices} = N\frac{\pi}{4}d_0^2 = \frac{U\rho_g}{U_0\rho_{gor}} \tag{11.6}$$

Here, N depends on the arrangement of orifices on the plate and the pitch P:

$$N = \frac{2}{\sqrt{3}P^2} \text{ for triangular pitch}$$

$$N = \frac{1}{P^2} \text{ for square pitch} \tag{11.7}$$

From the choice of suitable pressure drop ΔP_b, across the bed one can calculate the orifice velocity using Equation 11.2 and Equation 11.3.

An excessively high orifice velocity may lead to the attrition of bed particles. Orifice velocities less than 30 m/sec are generally considered safe (Pell, 1990), while any value above 90 m/sec is considered to be risky (Geldart and Baeyens, 1985). Means of reducing the risk of attrition of particles are discussed later.

Smaller diameter orifices may have several advantages, such as more uniform fluidization, smaller bubbles, and less chance of particles dropping into the plenum during shut-down (especially for straight-hole orifice plates). From Equation 11.6 we note that a smaller diameter of orifice implies a greater number of orifices per unit grate area and hence a shorter pitch P. Too short a pitch may erode the advantages of small bubble diameters because bubbles formed above the orifice jets may coalesce to form larger bubbles. This coalescence may occur if the pitch is less than 1.5 times the initial diameter of bubbles (Zenz, 1968).

Experience suggests that the fractional opening of the distributor ($\pi/4Nd_0^2$) should be in the neighbourhood of 0.01 to 0.02 for a bubbling fluidized bed, while for some laboratory or pilot-plant fast fluidized beds it may be much higher.

11.3.1.2 Jets

The erosion of bed internals and in-bed heat exchanger tubes is often a major problem in a bubbling fluidized bed. Bed solids entrained in the jets formed on the distributor orifices impinge on tubes if they are too close to the jets causing serious erosion. For a safe design, the tube bundles or internals must be located above the jet height, l_j, which may be calculated using the correlation of Merry (1975).

$$\frac{l_j}{d_0} = 5.2 \left[\frac{\rho_g d_0}{\rho_p d_p} \right]^{0.3} \left[1.3 \left(\frac{U_0^2}{g d_0} \right)^{0.2} - 1 \right] \quad \text{for a vertical jet} \tag{11.8}$$

where U_0 is the velocity through an orifice having a diameter d_0.

11.3.2 DESIGN PROCEDURE FOR CIRCULATING FLUIDIZED BEDS

In bubbling fluidized beds, the superficial velocity through the grids, U, is about 0.5 to 2.0 m/sec, while in fast beds it is of the order of 4 to 6 m/sec. If the fractional opening ratio in the distributor of a CFB is maintained at the same level as that in bubbling bed, the orifice velocity through the grid will be very high resulting in an exceptionally high pressure drop across the distributor plate. As discussed earlier, the pressure drop across the grid of a CFB riser does not have to be as high a fraction of the bed pressure drop as that in a bubbling bed. Therefore, the fractional opening in a CFB grid could be much larger than that stipulated above for bubbling bed boilers.

However, in commercial units, only a fraction of the total combustion air is passed through the grid as the primary air, so the superficial velocity through the grid in a fast bed may not be very high.

Considering the above, the orifice velocity through a CFB distributor or grid could be same as that in a BFB, which is in the range of 30 to 90 m/sec.

To ensure uniform fluidization, designers often recommend a high pressure drop across the distributor. A high-resistance distributor with its high orifice velocity easily helps clean a temporary blockage in an orifice and overrides any local difference in pressure drop through the bed. For a bubbling bed Zenz (1981) recommends the ratio of pressure drops across the distributor and the bubbling bed to be 0.3. For the risers of CFB boilers, owing to the high velocity and highly expanded bed, the probability of nonuniform fluidization is less. So, it is guided by a net pressure drop ΔP_d across the plate which varies in the range of 1.0 to 4.0 kPa depending upon a large number of parameters including the size of the unit, criticality of auxiliary power consumption.

The multi-orifice nozzle is a popular type of air distributor (Figure 11.9). In this system, n number of orifices having diameter d_0, are drilled around a nozzle of much large diameter d_n. The size of nozzle is chosen such that its cross section is sufficiently larger than the combined cross-sectional areas of the orifices. This is done to ensure that the fluid resistance across the orifices dominates rather than that across the nozzle:

$$n \frac{\pi}{4} d_0^2 < \frac{\pi}{4} d_n^2 \tag{11.9}$$

It may be noted that some modern nozzles designs do not follow the above norm. These nozzles deliberately use smaller entry size d_n, into the nozzle and relatively larger size of holes d_0, on

Air Distribution Grate

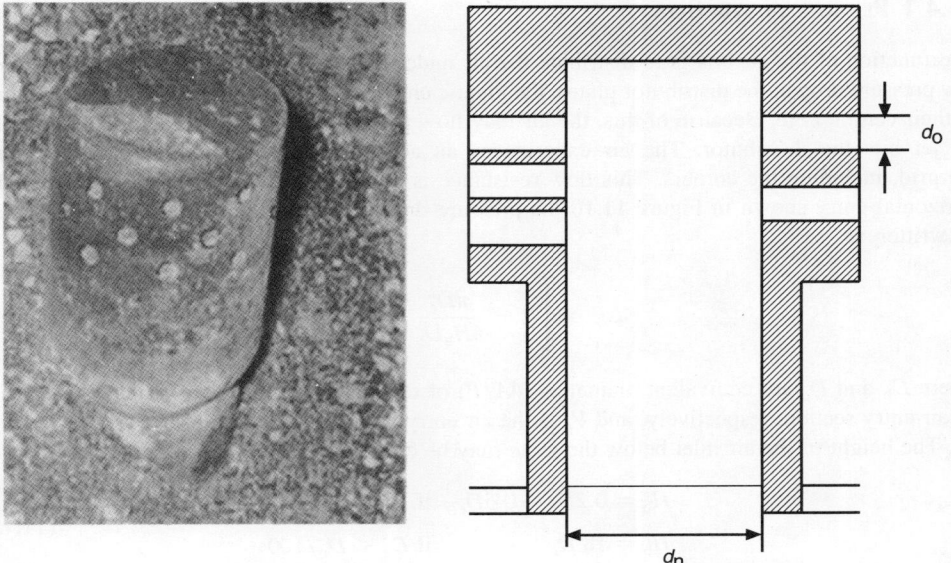

FIGURE 11.9 A multi-orifice nozzle on the grid plate of a bubbling fluidized bed boiler.

the nozzles. In this case velocity at the entrance of the nozzle is higher than that at the exit of the nozzles. This helps reduce erosion at the exit of the nozzles.

If the number of orifices works out to be n_t, the total number of nozzles is found to be:

$$N_n = \frac{n_t}{n} \tag{11.10}$$

The number of nozzles N_n, must conform to a geometrical pattern with a fixed pitch P_n:

$$N_n = \frac{2}{\sqrt{3}P_n^2} \quad \text{for triangular pitch}$$
$$N_n = \frac{1}{P_n^2} \quad \text{for square pitch} \tag{11.11}$$

For this to happen, one may have to adjust the number of orifices, their diameter and even the ratio of orifices to nozzles. Furthermore, once the pitch of nozzles, P_n is found from the above equation it must be large enough to accommodate the physical dimensions of the nozzles leaving adequate space in between as shown in Figure 11.4. If the nozzles are too close together, the air jet from one might erode the neighbouring nozzles.

In the case of a CFB boiler, the bed depth is much greater than that of a bubbling bed boiler. Hence, a choice of a pressure drop across the distributor equal to 30% that across the bed would lead to a very high pressure drop distributor plate. Since the superficial gas velocity through the bed is very high in the CFB furnace, the pressure drop through the distributor is chosen to be much smaller than 30% of the bed pressure drop. This design is guided by a minimum pressure drop under the lowest load, which will prevent the orifices from being plugged.

11.4 PRACTICAL CONSIDERATIONS

Several practical problems need special attention during the design process. Some of them are discussed below.

11.4.1 PLENUM OR AIR BOX

The function of the air box is to distribute the air under the grid as uniformly as possible. For a low pressure drop at the distributor plates, the kinetic energy of the incoming air may be very high in their central core. Because of this, the air may flow preferentially through the point where the air jet hits the distributor. The air experiences an additional resistance in spreading around the grid, including the corners. This flow resistance is called *rearrangement resistance*. For the horizontal entry shown in Figure 11.10 the pressure drop or rearrangement resistance, ΔP_r may be written as:

$$\Delta P_r = \left[1 - \frac{\pi D_e^2}{4 H_b D_b} \right] \left[\frac{\rho V_e^2}{2g} \right] \quad (11.12)$$

where D_b and D_e are equivalent diameters $(4A/P)$ of the air box exit (i.e., distributor plate) and its air entry section, respectively, and V_e is the air entry velocity into the air box.

The height of the air inlet below the grate may be chosen as below:

$$\begin{aligned} H_b &= 0.2 D_b + 0.5 D_e & \text{if } D_e > D_b/100 \\ H_b &= 18 D_e & \text{if } D_e < D_b/100 \end{aligned} \quad (11.13)$$

Zenz (1989) showed that if one uses the above criteria for air box design the rearrangement pressure drop ΔP_r, will be equal to (0.644 to 2.0) ρV_e^2, which is in most cases much smaller than the resistance of the grid plate.

Kunii and Levenspiel (1991) suggest that the distributor pressure drop should be at least 100 times the loss ΔP_r due to sudden expansion into the plenum. If adequate room is available, one should avoid side entry of air and use alternative arrangements. In the arrangement shown in Figure 11.11, air enters from the bottom of the plenum through a central pipe. To avoid high-velocity jets hitting the nozzles at the centre, this arrangement uses a baffle plate. More on such arrangement is available in Burdett (1988) and Zenz (1989).

In a commercial fluidized bed gasifier or boiler, air generally enters the air box from its side. A practical means for ensuring uniform air distribution under the grid is to use flow-straightening vanes as shown in Figure 11.12. These vanes should have a depth of twice the spacing and a curvature at the upstream end of the vanes of a circular arc which is a tangent to the discharge angle of a slot without vanes and perpendicular at the downstream end of the vanes. The vane angle θ, is given in terms of the discharge coefficient of the slot C_d, slot area A_s, and duct cross-sectional area at the upstream end A_d (Perry et al., 1997).

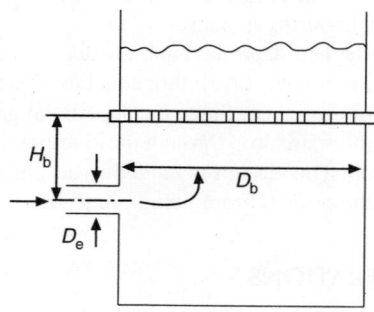

FIGURE 11.10 An air plenum with side-entry air.

Air Distribution Grate

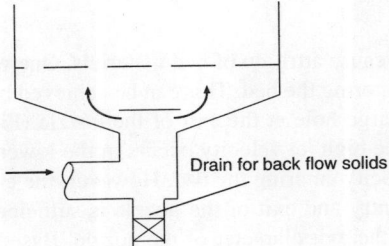

FIGURE 11.11 Bottom-entry air plenum with a baffle plate at entry.

$$\cot \theta = C_d \frac{A_s}{A_d} \tag{11.14}$$

Such vanes are particularly useful where limited space is available for the air box. Additionally, it also helps reduce flow resistance.

11.4.2 Sealing of Distributor

The problem of leakage around grid plates occurs in many fluidized beds. A number of methods of sealing were presented for avoiding the leakage problems in bubbling beds (Basu, 1984). Cooling during shut-off and heating during start-up develop a great thermal stress on the distributor plate, causing it to warp, which could destroy even the best seal. Some suggested means of reducing this problem are:

1. Welding a concave-shaped distributor to the vessel wall so it can accept the expansion and have a positive seal
2. Using fluidizing nozzles extending 10 to 15 cm above the plate and covering the plate with insulating refractory to prevent direct heating (Figure 11.6)
3. Locating the flange between the grid and the plenum, far from the hot zone

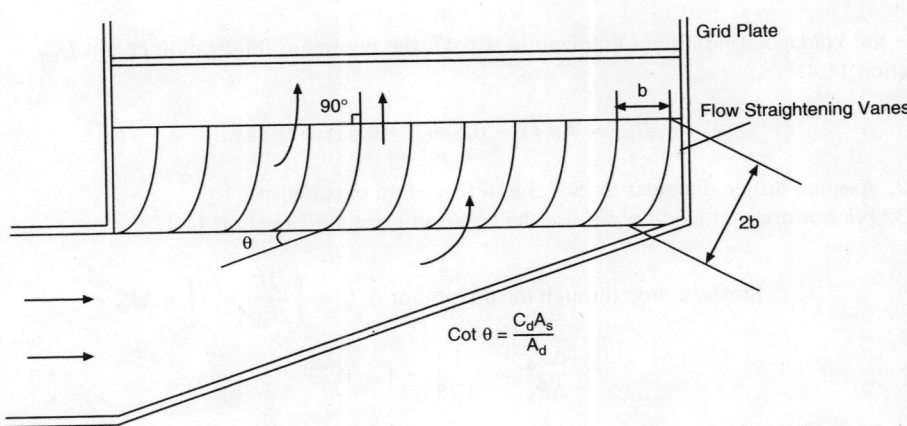

FIGURE 11.12 Flow-straightening vanes can help uniform distribution of air in the plenum chamber.

11.4.3 Attrition

Gas jets from the orifices may cause attrition of bed materials. One way of reducing this attrition is to reduce the velocity of jets entering the bed. This can be achieved by using the main orifice below the nozzle and a sufficiently large hole at the exit of the nozzle (Figure 11.6) such that the main pressure drop and therefore, the high jet velocity occurs in the lower orifice. The larger hole at the nozzle exit reduces the jet velocity entering the bed. However, the condition that must be met here is that the distance between entry and exit of the nozzle is sufficiently long for the gas jet at the entry to expand to the size of the exit diameter of the nozzle. Based on this, the minimum nozzle height l_m, is given as (Zenz, 1981):

$$l_m = \frac{(d_n - d_0)}{0.193} \tag{11.15}$$

where d_n is the inner diameter of the nozzle. Dimensions refer to Figure 11.9.

Example 11.1

> Find the pitch, diameter and desirable pressure drop through a straight orifice-type distributor plate for bubbling fluidized bed (4.5×2.5 m^2) with fluidizing solids of mean size 220 μm at 0.5 m/sec at 850°C. The depth of the fluidized bed is 1.00 m and the air entering the plenum is at 27°C. Voidage at minimum fluidization is 0.47.

Solution

1. We find the bed voidage using Equation 11.5:

$$\varepsilon = \frac{(0.5 + 1)}{(0.5 + 2)} = 0.6$$

From Equation 11.4, the pressure drop through the bed

$$\Delta P_b = (1 - \varepsilon)\rho_p H g \text{ N/m}^2$$
$$= (1 - 0.6) \times 2500 \times 1 \times 9.81 = 9810 \text{ Pa}$$

Since the voidage at minimum fluidization is 0.47, the minimum fluidization height H_{mf}, is from Equation 11.4:

$$H_{mf} = 1 \times (1 - 0.6)/(1 - 0.47) = 0.75 \text{ m}$$

2. Assume orifice diameter to be 1.5 mm (based on experience)
3. For uniform fluidization, we use the criterion given by Equation 11.1.

Pressure drop through the distributor $\Delta P_d = \left\{\dfrac{H}{H_{mf}} - 1\right\} \times \Delta P_b$

$$\frac{\Delta P_d}{\Delta P_b} = \frac{1}{0.75} - 1 = 0.33$$

$$\Delta P_d = 0.33 \times 9810 = 3267 \text{ Pa}$$

4. From Equation 11.3:

$$\text{Orifice velocity} = U_0 = C_D \left[\frac{2\Delta P_d}{\rho_{gor}} \right]^{0.5}$$

The orifice coefficient is taken as 0.8 (from Zenz, 1981).
The gas through the orifice is likely to be slightly heated by conduction from the distributor plate. Assuming the temperature to be raised by 300°C, the average gas temperature is therefore

$$27 + \frac{300}{2} = 177°C$$

From the table for properties of air:

$$\rho_g = 0.7774 \text{ kg/m}^3 \text{ at } 177°C$$

$$U_0 = 0.8 \left[\frac{2 \times 3267}{0.7774} \right]^{0.5} = 73 \text{ m/sec}$$

This is higher than the optimum value, but less than the maximum permissible value of 90 m/sec
5. From Equation 11.6, one can find the number of orifices:
The density of gas at 850°C, $\rho_g = 0.31$:

$$N = \frac{U \rho_g}{U_0 \rho_{gor} \frac{\pi}{4} d_0^2} = \frac{0.5 \times 0.31}{73 \times 0.7774 \frac{\pi}{4} 0.0015^2}$$

$$= 1546 \text{ orifices/m}^2$$

6. Assume the orifices to be in square pitch p:

$$\rho = 1/\sqrt{N} = 1/\sqrt{1546} = 0.025 \text{ m} = 25 \text{ mm}$$

11.4.4 Back-Flow of Solid

The back-flow of solids through the nozzles or orifices is a major problem in the grids of fluidized beds. Bed solids drop into the air plenum due to this back-sifting and are picked up again by the fluidizing air passing through the grid. The velocity of air through the grid orifices is in the range of 30 to 90 m/sec; solids picked up by the air hit the orifice wall and the nozzles around it at that exceptionally high velocity. This gives a severe sand-blasting effect causing the nozzles to erode.

Figure 11.13 shows the pressure fluctuation between the air plenum and a point 0.3 m above the distributor (Werther, 2005). The pressure drop occasionally goes below zero against its time-averaged value of 40 mbar. A similar observation was also made by Mirek et al. (2005) during start-up and load change in their 235-MWe CFB boiler. The reasons are not known yet, but it clearly explains why particles flow back into the plenum.

To avoid the back-flow one could do one of two things:

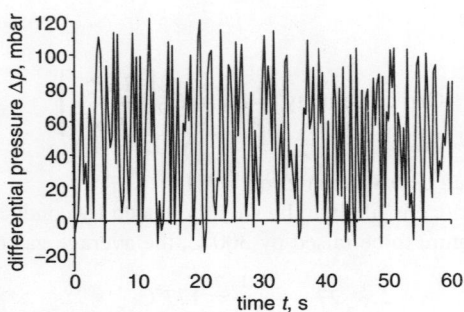

FIGURE 11.13 Time series of the pressure drop over the distributor. (From Werther, J., in *Circulating Fluidized Bed Technology VIII*, International Academic Publishers, Beijing, 2005, 1–25. With permission.)

1. Increase the distributor pressure drop well in excess of the amplitude of pressure fluctuation
2. Use nozzles inclined downward and long enough such that the time interval of pressure reversal is not long enough to transport the solids into the central tube from where they could fall down into the plenum. Werther (2005) found that this distance to be between 50 and 100 mm

Figure 11.14 shows photographs of one nozzle eroded due to the back-shifting and one with long vertical length.

Ideally the solids should not drop through the orifices while air is flowing through them at such high velocities as indicated above. In practice, the solids do drop into the plenum. This is probably due to underperforming or nonperforming orifices. To study this phenomenon, the author opened a small access hole in the plenum chamber of a 6-MWt CFB boiler while the boiler was in operation. It was found that the leakage of air through the hole blew a constant rate of solids as well. This suggested that the back-flow of fresh solids was continuously occurring from the bed through the nozzle orifices.

A possible explanation for this is that in a large grid plate not all nozzles operate simultaneously. For example, the author found in the above boiler (6 MWt) that at low velocity in the loop seal, one nozzle suddenly started spouting, leaving others dead. A small disturbance caused this nozzle to die, opening up another nozzle elsewhere to spout. Thus, there could always be some inoperative nozzles through which solids will back-flow into the plenum. Another possible explanation is the presence of a low velocity boundary layer zone around the orifice wall, which may not prevent the back-flow of the solids.

Several nonweeping grid designs are available (Zenz, 1989), but even they will allow back-sifting (weeping) of solids when the solids have a long deaeration time (time for a fluidized bed to settle down after air is turned off) as in Group A solids or when the gas flow is reduced abruptly. Thus, it is advisable to provide means for draining the solids from the air plenum as shown in Figure 11.11 and access door in for cleaning.

The best means to avoid the weeping or back-sifting of solids is to use porous plate grid (Fig. 11.2b). No solids can flow through this grid. These are not used in commercial boilers as they lack the mechanical strength to handle thermal shock or mechanical impact and there is concern about potential clogging of pores by moisture or dust in the fluidizing air entering it. One way to avoid this is to use a chequered grid above and below the porous plate and use small tiles instead of one large piece of ceramic or sintered plate.

Air Distribution Grate

(a) Eroded nozzle

(b) Improved bubble cap type nozzle

FIGURE 11.14 An eroded opening in the out cap of a nozzle and an improved bubble cap design. (From Werther, J., in *Circulating Fluidized Bed Technology VIII*, International Academic Publishers, Beijing, 2005, 1–25. With permission.)

In a special design, the nozzle is a tube bent in the form of an arc (Figure 11.6). Here, the solids cannot rise into the nozzle to drop on the floor during shut-down. The nozzle diameter is often too large to give a good distribution of air. An orifice as shown in Figure 11.6 may be inserted at its entrance to provide the required pressure drop for flow equalization. This would also avoid high jet velocity at the nozzle exit, which could damage the base of the grate.

11.4.5 Opening and Closing of Nozzles

The interesting phenomena of nonoperative nozzles occur in industrial-sized grids especially while working with coarse Group D particles. Even after supplying the bed with air above that required for minimum fluidization, not all nozzles contribute to fluidization. Air passing through some nozzles creates an empty air pocket in front of the nozzle, as was noticed by the author. Once the air pocket in front of a nozzle is disturbed by some mechanical means the nozzle starts operating and that portion of the bed operates properly.

A possible explanation of this is the formation of a particle arch offering high resistance, which reduces air flow through that particular nozzle. Once that arch is broken, the solids start fluidizing. This is similar to the spouting phenomenon. In a spouted bed, when the airflow through the bed is increased, the pressure drop increases, reaching a maximum. Once the spout punches its passage through the fixed bed, the bed starts spouting and the pressure drop across the bed drops.

11.4.6 EROSION AND CORROSION OF NOZZLES

Nozzles can wear due to erosion or corrosion. Corrosion is less common but has occurred in some cases due to intense localized combustion or some chemical constituents in the fuel. Corrosion weakens the surface of the metal, which is then worn away by strong ash flows.

Erosion typically originates from large pieces of ash or refractory wedging between the nozzles and redirecting the air jet from one nozzle into another. Over a long period of time, the high velocity jet and its entrained ash will erode the adjacent nozzle. In this situation, the riser pipe of the nozzle is more frequently eroded than the head itself, although the head can also be worn out.

Mirek et al. (2005) observed in his experiments with arrowhead nozzles for a 235-MWe CFB boiler that the velocity distribution of air at the orifice outlet was such that solids were entering into the nozzle through the inner edge while air was coming out through the outer. Furthermore, in some cases air issuing out of a nozzle was directly entering through the exit of a neighboring nozzle carrying solids with it. In both cases severe erosion could occur.

Example 11.2

> The loop seal of a CFB boiler operates at 0.5 m/sec and 850°C. The mean size of solids is 220 μm. The bed depth is 1.0 m and the air is entering the plenum at 27°C. Estimated bed voidage at fluidizing condition is 0.6.
>
> Design a distributor grate for the loop seal.

1. The pressure drop through the bed is found from Equation 11.4:

$$\Delta P_b = \rho_p(1-\varepsilon)Hg = 2500(1-0.6)1.00 \times 9.81 = 9810 \text{ Pa}$$

2. Take the distributor pressure drop to be 30% that of the bed:

$$\Delta P_d = 0.3 \, \Delta P_b = 0.3 \times 9810 = 2963 \text{ Pa}$$

3. From Equation 11.3 we find the orifice velocity. Here we take the orifice coefficient C_d, as 0.8. The cold air will be heated while passing through the orifice due to back-radiation from the bed. If we take a temperature rise of 300°C, the average air temperature inside the orifice is $27 + 300/2 = 177$°C. From Table 4.3, $\rho_{gor} = 0.774$ kg/m^3

$$U_0 = C_d \left[\frac{2\Delta P_d}{\rho_{gor}} \right]^{0.5} = 0.8 \left[\frac{2 \times 2963}{0.774} \right]^{0.5} = 70 \text{ m/sec}$$

4. We choose the diameter of the orifice to be 2 mm. Now we use Equation 11.6 to calculate the fractional opening and the number of orifices per unit area of the grid.

$$\text{Area fraction} = \frac{U\rho_g}{U_0 \rho_{gor}} = N\left(\frac{\pi}{4}\right)d_{0r}^2 = \frac{0.5 \times 0.31}{70 \times 0.774} = 0.00286$$

From here we calculate N as 910 orifices per m^2

Note: An orifice velocity of 70 m/sec is quite high though it is less than the upper limit of 90 m/sec. This can be reduced by taking the ΔP_d to be a smaller fraction, i.e., 0.186 of the ΔP_b. After repeating the above calculation we get a more reasonable orifice velocity of 55 m/sec, while the number of orifices works out to be 1159 per m^2.

NOMENCLATURE

A_s: slot area, m^2
C_D: orifice discharge coefficient
D_b: equivalent diameter ($4A/P$) of the air box exit (i.e., distributor plate), m
D_e: equivalent diameters of air entry to the plenum, m
d_0: diameter of orifice, m
d_p: diameter of particle, m
d_n: diameter of nozzle, m
g: acceleration due to gravity, 9.81 m/sec^2
H: expanded bed height, m
H_b: height of air entry duct below the distributor, m
H_{mf}: minimum fluidization height, m
l_j: height of gas jet, m
l_m: minimum height of nozzle, m
N: number of orifices per unit cross section of distributor, m^2
N_n: number of nozzles per unit cross section of distributor, m^2
n: number of orifices per nozzle
n_t: total number of orifices per unit cross section of distributor, m^2
P: pitch of orifice, m
P_n: pitch of nozzles, m
ΔP_b: pressure drop through bed, N/m^2
ΔP_d: pressure drop through distributor, N/m^2
ΔP_r: rearrangement resistance, N/m^2
t: thickness of distributor plate, m
U: fluidizing velocity at bed temperature, m/sec
U_0: orifice velocity, m/sec
U_{mf}: minimum fluidization velocity, m/sec
V_e: air entry velocity into the air box, m/sec

GREEK SYMBOLS

ρ_a: density of air at air box entry temperature, kg/m^3
ρ_g: density of gas bed temperature, kg/m^3
ρ_{gor}: density of air at orifice temperature, kg/m^3
ρ_p: density of solid, kg/m^3
ε: bulk voidage of the bed
ε_{mf}: voidage at minimum fluidization
θ: vane angle

REFERENCES

Basu, P., *Fluidized Bed Boiler: Design and Applications*, Pergamon Press, Toronto, 1984.
Burdett, I. D., *The Union Carbide Unipol Process: Polymerisation of Olefins in a Gas-Phase Fluidized Bed*, Presented at the AIChE Meeting, Washington D.C., December, 1988.
Geldart, D. and Baeyens, J., The design of distributor for gas fluidized bed, *Powder Technol.*, 42, 67–78, 1985.
Guo, Q. Q. and Werther, J., Flow behaviours in a circulating fluidized bed with various bubble cap distributors, *Ind. Eng. Chem. Res.*, 43, 1756–1764, 2004.
King, D., Estimation of dense bed voidage in fast and slow fluidized beds of FFC catalysts, In *Fluidization VI*, Grace, J. R., Shemilt, L. W., and Bergougnou, M. A., Eds., Engineering Foundation, New York, pp. 1–8, 1989.
Kunii, D. and Levenspiel, O., *Fluidization Engineering*, Butterworth Heinemam, Stoneham, pp. 102–112, 1991.
Merry, J. M. D., Penetration of vertical jets into fluidized beds, *AIChE J.*, 21, 507–510, 1975.
Mirek, P., Mirek, J. and Nowak, W., The experimental investigation of arrowhead nozzles operating in a 235 MWe CFB boiler, In *Circulating Fluidized Bed Technology VIII*, Cen, K., Ed., International Academic Publishers, Beijing, pp. 885–890, 2005.
Pell, M., *Gas Fluidization*, Elsevier, New York, pp. 21–30, 1990.
Perry, R. H., Green, D. W. and Maloney, J. O., *Perry's Chemical Engineering Handbook*, McGraw Hill, New York, pp. 6–33, 1997.
Quereshi, A. E. and Creasy, D. E., *Powder Technol.*, 22, 113–119, 1979.
Werther, J., Fluid dynamics, temperature and concentration fields in large-scale CFB combustors, In *Circulating Fluidized Bed Technology VIII*, Cen, K., Ed., International Academic Publishers, Beijing, pp. 1–25, 2005.
Whitehead, A. B., Chapter 5 in *Fluidization*, 2nd ed., Davidson, J. F., Clift, R., and Harrison, D., Eds., Academic Press, New York, pp. 173–198, 1985.
Zenz, F. A., Bubble formation and grid design, *Proceedings of the Tripartite Chemical Engineering Conference, Symposium on Fluidization II*, Montreal, pp. 36–39, 1968.
Zenz, F. A., *Elements of Grid Design, Gas Particle Industrial Symposium*, Engineering Society, Western PA, Pittsburgh, 1981.
Zenz, F. A., *Fluidization and Fluid Particle Systems*, Pemm-Corp. Publication, Nelsonville, pp. 101–109, 1989.

12 Gas–Solid Separators

Gas–solid separation plays an important role in the performance of fluidized bed boilers. These are used in both bubbling (BFB) and circulating fluidized beds (CFB). Gas–solid separators are used in two different locations of a power plant for two distinct purposes. Bag-houses or electrostatic precipitators are used at the relatively cold downstream end of the boiler to reduce the boiler's particulate emission, while cyclone and other types of separators are used within the CFB boiler loop to aid in the circulation of hot solids through the furnace and, in a BFB or gasifier, to improve carbon conversion. The present chapter discusses the principles of gas–solid separation, with special emphasis on cyclone, inertial, and impingement separators.

Solids can be separated from the carrier gas by means of one the following forces:

1. External forces (gravity, electrostatic, magnetic, etc.)
2. Internal forces (inertial, centrifugal, diffusion, impingement, etc.)

Equipment used for gas–solid separation in a power plant can be broadly classified as shown in Figure 12.1.

The separation efficiency of gas–solid separators depends on the size of the particles separated (Figure 12.2). Fine particles are more difficult to separate, thus have a characteristically lower separation efficiency, while coarser particles have higher separation efficiency. Some separators are used for process needs while some are used for air pollution control. Separators like filter bags and electrostatic precipitators, used for the control of particulate emissions, are usually more efficient in collecting fine particulates than those used in the CFB loop, such as cyclone separators.

Bag-house or electrostatic precipitators are commonly used for particulate-emission control and therefore placed in the cooler section of the boiler plant. In the hot loop of a CFB boiler, solids leaving the top of the furnace are separated from the flue gas by some type of mechanical separator and are recirculated into the bottom of the furnace. The cyclone is the most common type of mechanical separators used in the primary loop of CFB boilers, but several other types are emerging.

A cyclone is a simple device providing a high degree of separation for a minimum pressure drop (50 to 100 mm of water gauge). The cyclone, though well proven in CFB boilers, suffers from a number of inherent disadvantages. For one, its cylindrical-shaped body is not compatible with the rectangular shape of the boiler. Therefore, the mechanical integration of cyclones into the boiler structure is relatively difficult. Thus, other types, more compatible with the geometric configuration, are emerging. Such separators also allow easier scale-up of the designs. Several types of mechanical gas–solid separators are discussed below.

12.1 CYCLONES

The absence of moving parts, simple construction, and high efficiency make a cyclone especially suitable for CFB boilers. Proper operation of the cyclone is critical for the desired gas–solid flow structure of a CFB riser. For example, the positive effects from improving the cyclone efficiency of a CFB boiler are:

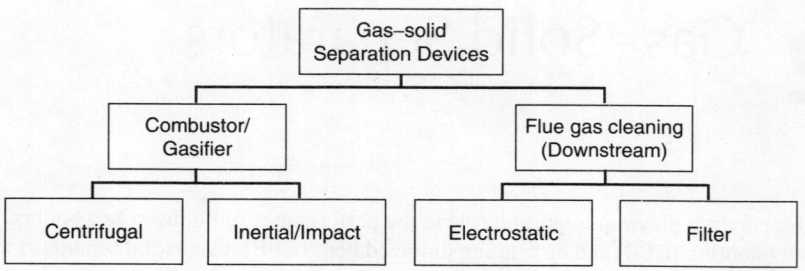

FIGURE 12.1 Types of gas–solid separators used in a power plant.

- Improved limestone consumption
- Lower bed temperatures
- Lower NO_x production
- Reduced fouling or corrosion of back-pass tubes

12.1.1 Types of Cyclone

There are several types of cyclones used in different applications:

1. Vertical-axis, cylindrical horizontal inlet, and axial discharge (reverse-flow) (Figure 12.3)
2. Vertical-axis, noncylindrical cyclone with horizontal inlet and vertical discharge (Figure 12.4a)
3. Vertical-axis, multi-entry, and vertical discharge (Figure 12.4b)
4. Vertical-axis, horizontal entry, and downward discharge (Figure 12.4c)
5. Axial entry and axial discharge (Figure 12.4d)

12.1.1.1 Vertical-Axis Cylindrical Cyclone

The vertical-axis cyclone with tangential inlet and axial discharge is the most common design used in CFB boilers and gasifiers. A cross section of a typical cyclone is shown in Figure 12.3. The upper

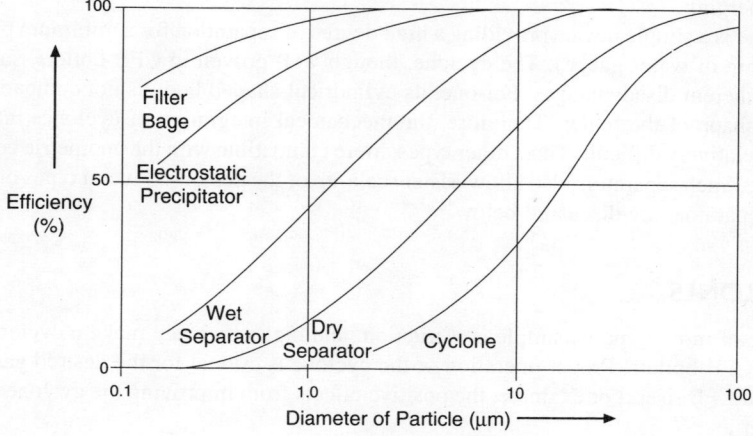

FIGURE 12.2 Comparison of separation characteristics of different gas–solid separators. (Adapted from Geldart, D., *Gas/Solid Fluidization*, John Wiley & Sons, p. 199, 1986.)

Gas–Solid Separators

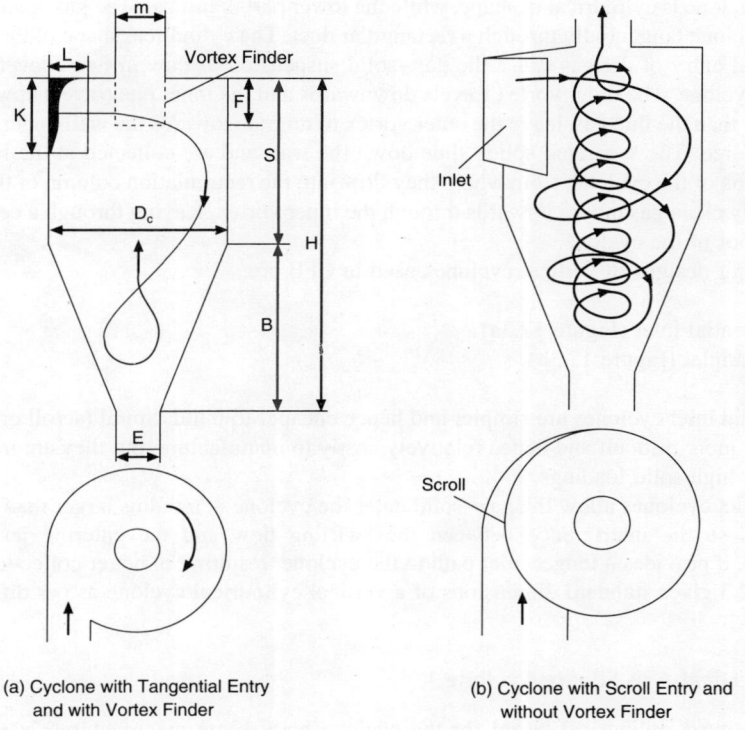

FIGURE 12.3 Typical dimensions of a vertical axis tangential entry reverse-flow cyclone.

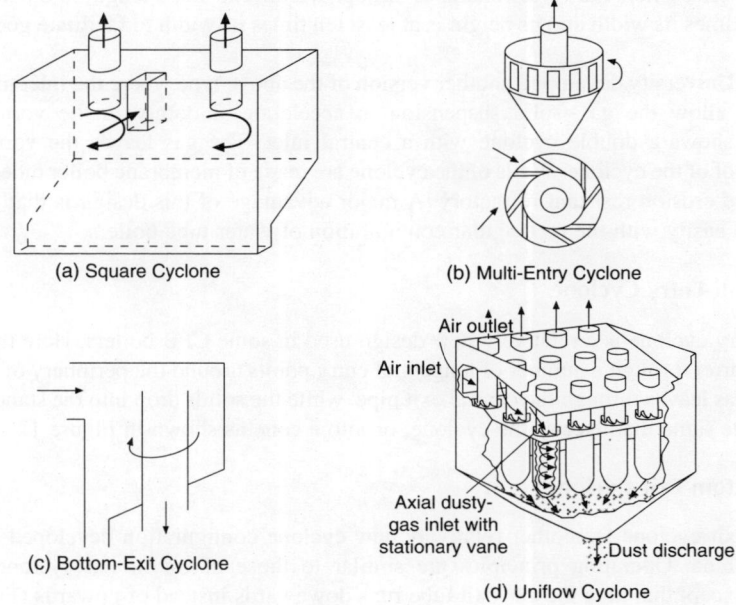

FIGURE 12.4 Different types of vertical-axis cyclones.

part of the cyclone is cylindrical in shape, while the lower part is conical. The gas–solid suspension enters the cyclone tangentially through a rectangular duct. The cylindrical shape of the cyclone and the tangential entry of the gas cause the gas–solid suspension to flow in two concentric vortices around the cyclone. The outer vortex travels downwards and the inner one travels upwards. Solids, being denser than the flue gas, leave the outer vortex to migrate towards the wall under the action of centrifugal force. The separated solids slide down the wall and are collected at the bottom of the conical section of the cyclone, from where they flow into the recirculation column of the CFB loop. The relatively clean gas moves upwards through the inner vortex, leaving through a central vertical exit at the roof of the cyclone.

Two major designs of vertical cyclones used in CFB are:

1. Tangential inlet (Figure 12.3a)
2. Spiral inlet (Figure 12.3b)

Tangential inlet cyclones are simpler and hence cheaper to build. Spiral (scroll or volute) inlet cyclones are more difficult and hence relatively costly to manufacture, but they are more efficient, especially at high solid loadings.

Spiral inlet cyclones allow the gas–solid enter the cyclone at a radius larger than the radius of the cyclone, so the interference between the swirling flow and the entering jet is minimal. Furthermore, it provides a longer inlet path to the cyclone, resulting in better collection efficiency.

Table 12.3 gives standard dimensions of a vertical cylindrical cyclone as per different design models.

12.1.1.2 Vertical-Axis Square Cyclone

Instead of using a cylindrical barrel for the cyclone body some manufacturers use a square or octagonal-shaped barrel (Figure 12.4a) to make it better suited to the rectangular geometry of a CFB boiler. A Canadian patent (Hypannen and Ristola, 1998) defines such a cyclone in which the gas from the boiler enters the vortex chamber through a slot. The slot's length in the flow direction is at least 0.8 times its width and its height is at least ten times its width to facilitate good collection efficiency.

Tsinghua University developed another version of the above type where the inlet to the cyclone is tapered to allow the gas–solid suspension to accelerate smoothly in the vortex chamber. Figure 12.4a shows a double cyclone with a central inlet. The gas leaves the vortex chamber through the roof of the cyclone. Walls of the cyclone are made of membrane boiler tubes covered by a thin heat and erosion resistant refractory. A major advantage of this design is that it makes the cyclone blend easily with the rectangular configuration of water tube boilers.

12.1.1.3 Multi-Entry Cyclone

The multi-entry cyclone is a relatively new design used in some CFB boilers. Here the gas enters the cyclone barrel through a number of tangential entry points around the periphery of the cyclone. The cleaned gas leaves through the central exit pipe, while the solids drop into the standpipe, which could be of the same diameter as the cyclone, or into a cone as shown in Figure 12.4b.

12.1.1.4 Bottom-Exit Cyclone

The bottom-exit cyclone is another relatively new cyclone configuration developed by Zhejiang University, China. Operating principles are similar to those of traditional cyclones shown in Figure 12.3, except that here the gas exit tube runs downwards instead of upwards (Figure 12.4c). This makes the boiler design more appropriate for a two-pass boiler. The cyclone body can be either cylindrical or square.

Gas–Solid Separators

12.1.1.5 Axial Inlet–Axial Discharge Cyclone

The axial inlet and axial discharge-type of cyclone (Figure 12.4d) is generally of small diameter. A battery of such cyclones, popularly known as a multi-clone, is often used for cleaning the flue gas leaving a boiler. The small diameter results in high efficiency. The flue gas enters the cyclones swirling downwards. Solids separated drop into a common collection chamber. The relatively clean gas leaves the cyclone moving upwards through the central tube.

12.1.2 Theory

To explain the operation of the cyclone and the effect of different design and operating parameters, a simplified treatment is presented below. A more rigorous treatment, which is especially useful for the design of CFB cyclones, is given in the VDI Heat Atlas (1993).

When a particle enters a cyclone tangentially, it travels in a vortex and is subject to a centrifugal force towards the wall and a viscous drag in the direction of the flow. For a particle of diameter d_p, and density ρ_p, moving in a cyclone of radius r, the centrifugal force F_c, on the particle entering with a tangential velocity V_e, may be written as:

$$F_c = \frac{\pi}{6} d_p^3 (\rho_p - \rho_g) \frac{V_e^2}{r} \tag{12.1}$$

If the centrifugal force causes it to migrate towards the wall with a radial velocity V_r, then using Stokes law the drag force F_d, on it can be written as:

$$F_d = \frac{6\pi\mu V_r}{2} \tag{12.2}$$

where μ is the viscosity of flue gas.

Under steady state, $F_c = F_d$. Solving Equation 12.1 and Equation 12.2:

$$V_r = \frac{d_p^2 \rho_p V_e^2}{18 \mu r}. \tag{12.3}$$

In an ideal situation, solids reaching the wall are fully captured. The efficiency of the cyclone, thus, will be directly proportional to the solid migration velocity V_r. The faster the particle moves towards the wall, greater is its probability of being captured before it is entrained. Thus, V_r may be taken as an index of cyclone efficiency, where a higher migration velocity implies higher efficiency. It is clear from Equation 12.3 that the cyclone efficiency will increase for:

- Higher entry velocity, V_e
- Larger size of solid, d_p
- Higher density of particles, ρ_p
- Smaller radius of the cyclone, r
- Higher value of viscosity of the gas, μ

12.1.3 Critical Size of Particles

The solid particles entering the cyclone move in a double-helix spiral (Figure 12.3b) while approaching the wall with a radial velocity given by Equation 12.3. The spiral takes the particles to within the boundary layer near the wall, where the fluid drag drops considerably, allowing the particle to fall under gravity. If the particle is not collected on the wall during the several spirals it goes through, the particle leaves the cyclone through the vertical exit on the roof of the cyclone.

The longer the particle's trajectory, the greater is the probability of reaching the wall region. The theoretical cut-size of a cyclone is found from Equation 12.3:

$$d_p = \sqrt{\frac{18 V_r \mu r}{V_e^2 (\rho_p - \rho_g)}} \quad (12.4)$$

From Equation 12.3 one can speculate that ideally, any particle with a diameter larger than d_p will be collected while those smaller will be entrained. Collection efficiency for such an ideal cyclone is shown by the dotted line in Figure 12.5. However, the actual operation of cyclones is much different from this ideal situation, owing to particle-particle interaction, nonuniform velocity distribution, secondary flow and other factors. It is very difficult to define a specific cut-off size for which all smaller particles will be entrained and larger ones will be collected. However, for the sake of design convenience, designers often define a practical *cut-off* size like d_{50}. Particles larger than this have a 50% probability of being captured. Therefore, one could expect a collection probability distribution as shown by the solid line in Figure 12.5.

A particle turns around a cyclone several times as shown in Figure 12.3. The number of such turns or spirals N_c, is dictated by either exit or entry gas velocity, whichever is higher. Zenz (1989) plotted this as a function of the gas velocity (Figure 12.6).

Pell and Dunson (1997) define a 50% cut-off size d_{50}, considering the number of spirals in the cyclone. This gives the size of particles that are likely to be collected with 50% efficiency by a cyclone of given geometry for the given particulate properties and operating conditions. It is given as:

$$d_{50} = \sqrt{\frac{18 \mu L}{2 \pi N_c V_e (\rho_p - \rho_g)}} \quad (12.5)$$

where μ is the gas viscosity, L is the width of the rectangular inlet duct, V_e is the inlet gas velocity, ρ_p is the particle density, ρ_g is the gas density, and N_c is the effective number of turns made by the gas stream in the separator. This is generally taken as five (Frisch and Halow, 1987). It can be found from Figure 12.6, which shows that the number of spirals increases with inlet velocity.

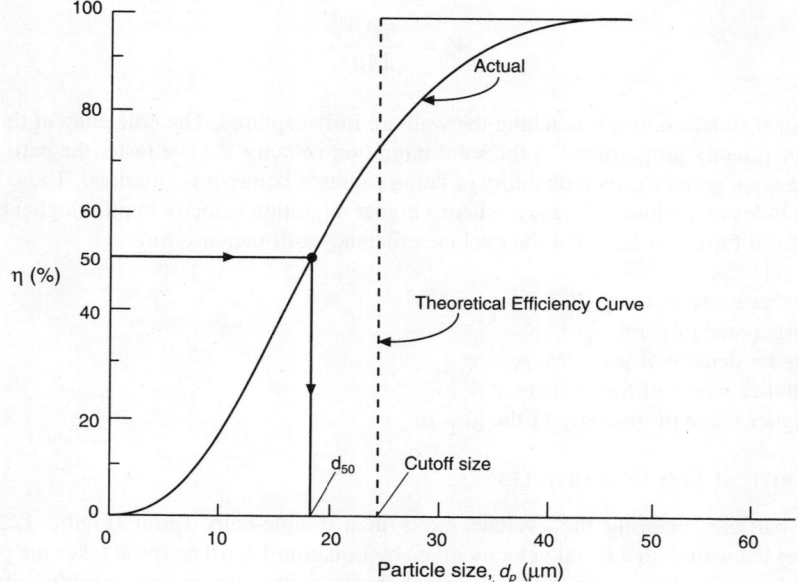

FIGURE 12.5 An ideal and practical grade efficiency curve of a typical cyclone.

Gas–Solid Separators

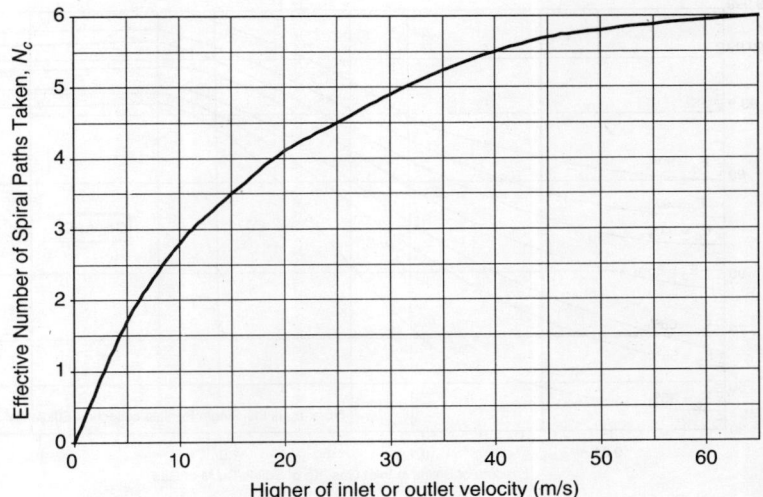

FIGURE 12.6 Number of spirals in cyclone increases with inlet/exit velocity. (From Zenz, F. A., *Perry's Chemical Engineering Handbook*, PSRI, Chicago, 1997. With permission.)

Cut-off size is a characteristic of the performance of the cyclone. For a cyclone of a given proportion, one can relate its performance in terms of its critical or cut-off size. Zenz (1989) gave the efficiency of the collection of single particles (Figure 12.7) in a cyclone of standard dimensions as a function of the ratio of particle size to the cut-off size of the cyclone. The effect of the amount

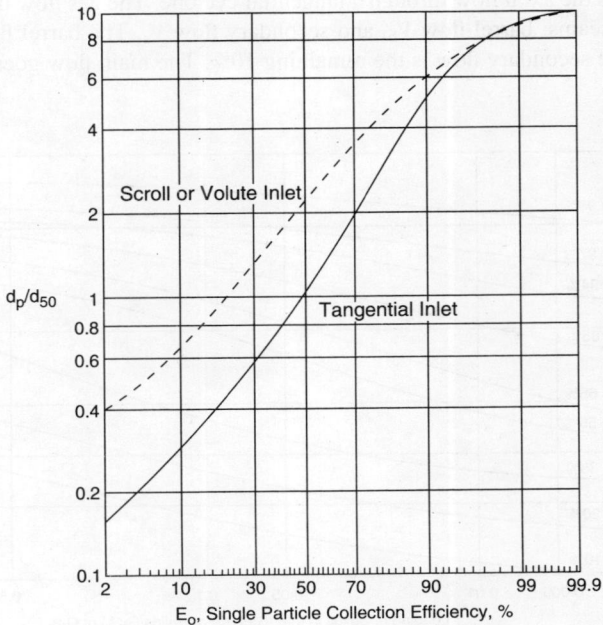

FIGURE 12.7 Single particle collection efficiency curve. (From Zenz, F. A., *Perry's Chemical Engineering Handbook*, PSRI, Chicago, 1997. With permission.)

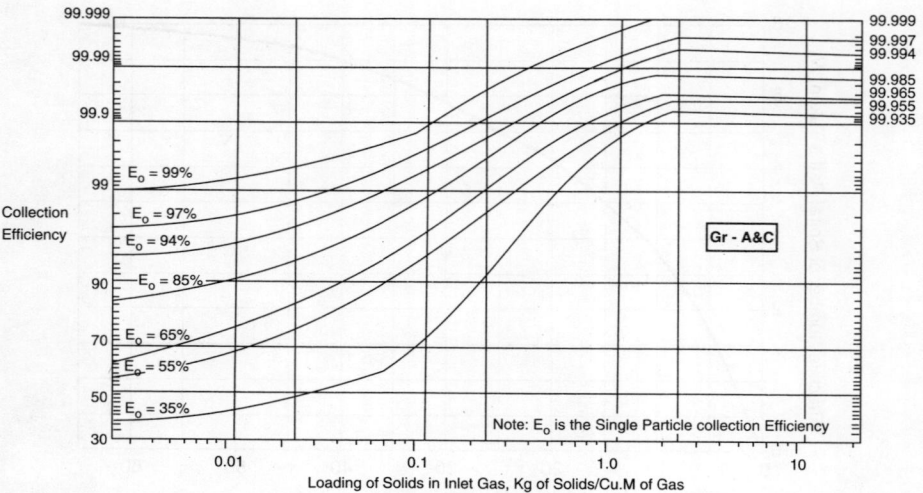

FIGURE 12.8 Effect of inlet loading on collection efficiency for Geldart Group A and Group C particles. (From Zenz, F. A., *Perry's Chemical Engineering Handbook*, PSRI, Chicago, 1997. With permission.)

of solids handled on the collection efficiency is shown in Figure 12.8 and Figure 12.9. This is discussed in later in more detail.

12.1.3.1 VDI Approach

Figure 12.10 shows the axial flow through a tangential cyclone. The gas flow into the cyclone V_{gas}, is split into two streams: barrel flow V_b, and secondary flow V_s. The barrel flow is about 90% of the total flow while secondary flow is the remaining 10%. The main flow goes round the barrel in

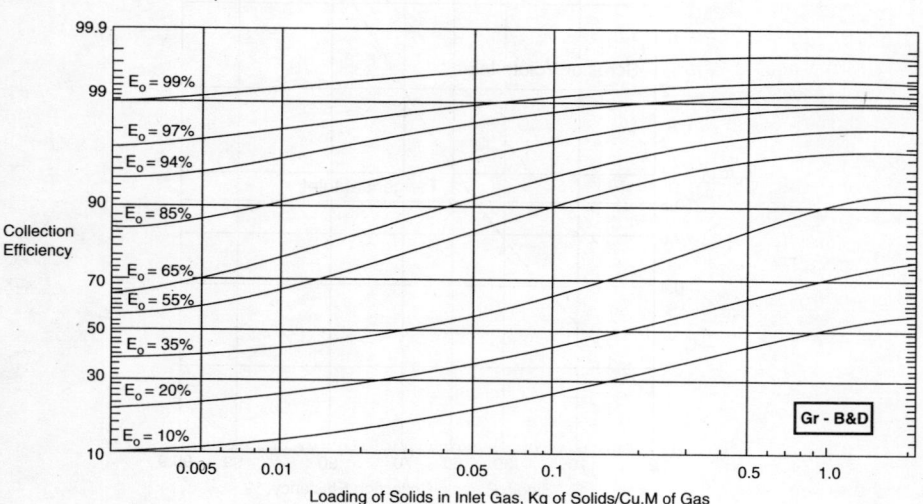

FIGURE 12.9 Effect of inlet loading on collection efficiency for Geldart Group B and Group D Particles. (From Zenz, F. A., *Perry's Chemical Engineering Handbook*, PSRI, Chicago, 1997. With permission.)

Gas–Solid Separators

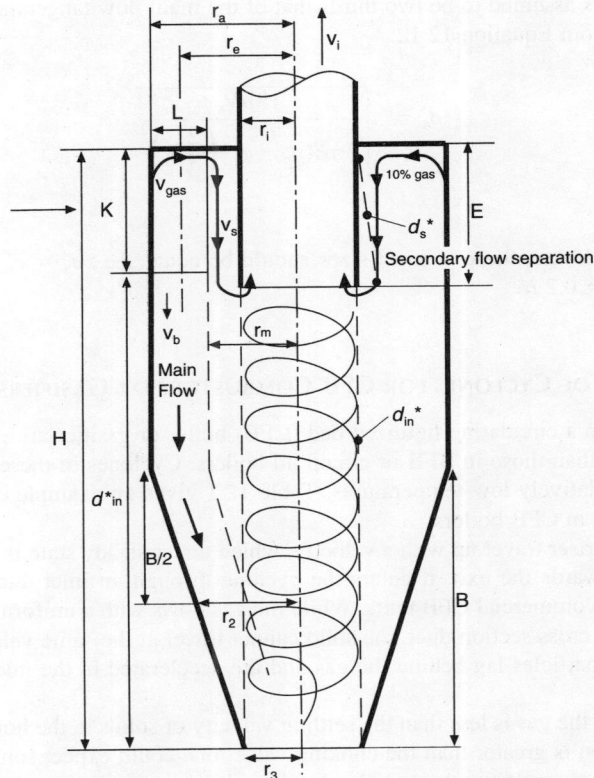

FIGURE 12.10 Gas flow through a cyclone is split into two parts: main stream V_b, and secondary flow V_s.

a spiral, turning upwards into the inner vortex then into the exit pipe as shown in Figure 12.3. The secondary flow takes a short-cut by hugging the cyclone roof and moving down the outer wall of the exit pipe as shown in Figure 12.10. These two flow streams would have separate cut sizes (d_{in}^*, d_s^*).

Muschelknautz and Greif (1997) considered that the particles lying on the boundary of downward and upward spirals are least likely to be separated. They assumed this boundary to lie on a radius equal to the exit pipe radius r_i. Equation 12.6 is developed by expressing in equation 12.4 the radial velocity V_r, in terms of gas flow rate through the cyclone barrel V_b, and the tangential velocity u_i, at the exit radius. The radial velocity is equal to the barrel flow rate (it is about 90% of the total flow V_{gas}) divided by the cylindrical area of the cyclone for collection $[2\pi(H - F)r_i]$. The characteristic cut size d_{in}^*, for this centrifugal separation is expressed as:

$$d_{in}^* = \sqrt{\frac{18\mu V_b}{(\rho_p - \rho_g)u_i^2 2\pi(H - F)}} \qquad (12.6)$$

where $V_b \sim 0.9 V_{gas}$, $(H - F)$ is the height of the separation chamber below the vortex (exit) tube and u_i is the tangential velocity at radius r_i.

The cut size, d_s^* for the secondary air V_s, which occurs within the boundary layer flowing down the exit pipe until it reaches the edge of the tube, is found by calculating the radial velocity on the basis of flow rate V_s, and the settling area is equal to $(2\pi F r_i)$. The tangential velocity in the secondary flow is assumed to be two thirds that of the main flow tangential velocity u_i, which may be computed from Equation 12.12:

$$d_s^* = \sqrt{\frac{18\mu V_s}{2\pi E(\rho_p - \rho_g)\left(\frac{2}{3}u_i\right)^2}} \qquad (12.7)$$

where $V_s \sim 0.1\, V_{gas}$.

For an optimal design, these two cut sizes should be equal, i.e., $d_s^* = d_{in}^*$ (VDI, 1993). This stipulation gives $E = 0.2\, H$.

12.1.4 Working of Cyclones for CFB Combustors or Gasifiers

The cyclones used in a circulating fluidized bed (CFB) boiler or gasifier are subjected to a much higher dust loading than those in BFB or circofluid boilers. Cyclones in these latter two types of boilers operate at relatively low temperatures. Table 12.1 gives an example of design data for a typical cyclone used in CFB boilers.

Solids in a CFB riser travel up with a velocity, which under steady state is equal to $(U_f - U_t)$. As the gas turns towards the exit, it enters the cyclone through an inlet duct, which is several meters long in most commercial CFB units. While the gas flows with a uniform steady velocity V_e through the uniform cross section duct, the solid cannot travel at the same velocity because of its greater inertia. The particles lag behind the gas and are accelerated in the inlet duct as shown in Figure 12.11.

If the velocity of the gas is less than the settling velocity of solids in the horizontal duct and/or its solid concentration is greater than the choking value, one could expect some settling of solids in the inlet duct. The gas velocity through the duct remains constant if its cross section is uniform, but the particles continue to accelerate until the velocity reaches their equilibrium value (Figure 12.11). On entering the barrel of the cyclone the gas velocity drops due to expansion of the flow area. The exchange of momentum between gas and solids occurs immediately after they enter the barrel, and the solid velocity catches up with the gas velocity (Figure 12.11).

Figure 12.12 shows the plan view of the flow of gas and solids through a vertical, tangential-entry cyclone. The gas flow stream enters the cyclone with an entry velocity V_e. The radial pressure field forces the stream against the wall forming a vena contracta. The resulting flow field is shown in Figure 12.12. The outer tangential velocity u_a, is obtained from the momentum of the inlet jet

TABLE 12.1
Characteristic Operating Conditions for a Typical Hot Cyclone in a CFB Boiler

Particle Size at Entrance	90% Fly Ash (5–500 μm)
	5–10% fuel (5–500 μm)
	1% lime (1–15 μm)
	Average size (75–150 μm)
Dust loading kg solid/kg gas	1–10
Gas velocity at cyclone inlet	15–30 m/sec
Pressure drop in cyclone barrel	400–600 Pa
Pressured drop in entry section	200–300 Pa
Diameter of cyclone barrel	0.5–8 m
d_{so} (for 50% efficiency)	≈ 10 μm

Gas–Solid Separators

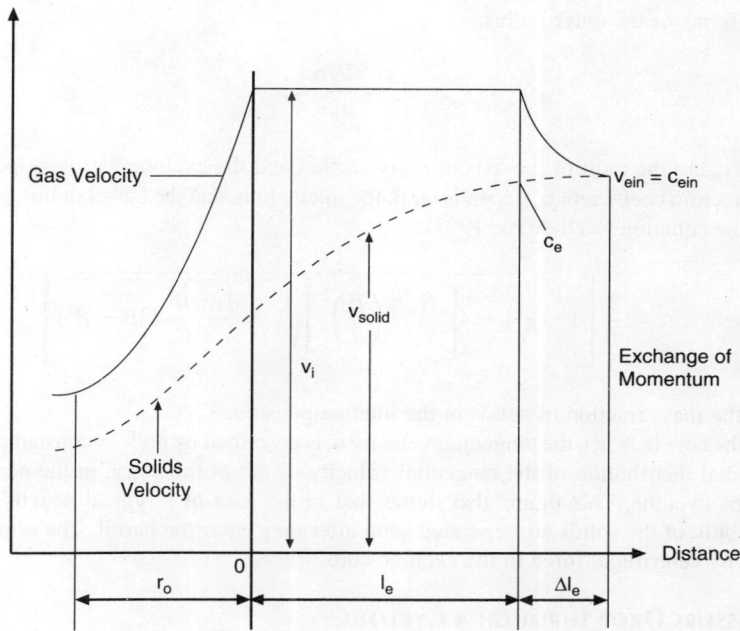

FIGURE 12.11 Acceleration of solids along the entrance duct, and exchange of momentum immediately after the inlet.

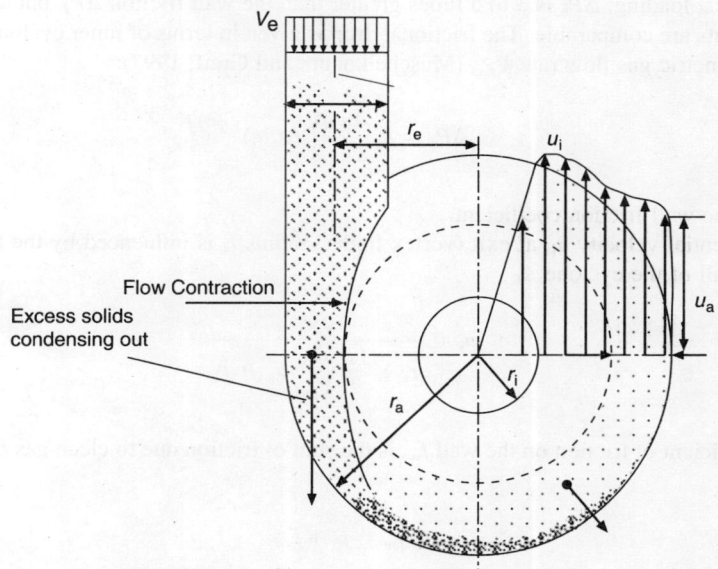

FIGURE 12.12 Solid separation after the cyclone inlet.

expressed in terms of the outer radius:

$$u_a = \frac{V_e r_e}{r_a \alpha} \quad (12.8)$$

where r_e and r_a are the radii of the axis of entry section and the cyclone barrel, respectively.

The contraction coefficient α, depends on β, the inlet width, and the barrel radius ratio L/r_a. It is found from the equation (VDI Atlas, 1993):

$$\alpha = \frac{1}{\beta}\left[1 - \sqrt{1 - 4\left[\frac{\beta}{2} - \left(\frac{\beta}{2}\right)^2\right]\sqrt{1 - \frac{1-\beta^2}{1+C_e}(2\beta - \beta^2)}}\right] \quad (12.9)$$

where C_e is the mass fraction of solids in the inlet suspension.

Outside the core ($r > r_i$), the tangential velocity u, is described by $u.r^n$ = constant. Figure 12.12 shows the radial distribution of the tangential velocity of gas at the entry, in the core, and in the annulus of the cyclone. This figure also shows that in the case of a typical heavily-loaded CFB cyclone, the bulk of the solids are separated soon after they enter the barrel. The remaining solids are captured by centrifugal force in the cyclone core.

12.1.5 Pressure Drop through a Cyclone

The pressure drop through the cyclone is an important parameter in the design of cyclones. Muschelknautz and Greif (1997) presented a method for the calculation of the total pressure drop ΔP, in a cyclone used especially in CFB, by taking it as a sum of pressure drops due to friction on the cyclone wall ΔP_f, and the hydrodynamic loss in the inner vortex ΔP_e:

$$\Delta P_c = \Delta P_f + \Delta P_e \quad (12.10)$$

At low dust loading, ΔP_e is 3 to 5 times greater than the wall friction ΔP_f, but at high loading the components are comparable. The frictional drop is given in terms of inner cyclone surface area A_w, and volumetric gas flow rate V_{gas} (Muschelknautz and Greif, 1997):

$$\Delta P_f = f_w \frac{A_R}{V_b} \frac{\rho_g}{2} (u_a u_i)^{1.5} \quad (12.11)$$

where f_w is the wall friction coefficient.

The tangential velocity u_i, at exit (vortex finder) radius r_i is influenced by the friction on the total inner wall of the cyclone A_R.

$$u_i = \frac{u_a r_i}{r_i + \dfrac{f_w}{2} \dfrac{A_A}{V_{gas}} u_a \sqrt{r_a r_i}} \quad (12.12)$$

The coefficient of friction on the wall f_w, is the sum of friction due to clean gas f_0, and that due to solids f_s:

$$f_w = f_0 + f_s. \quad (12.13)$$

For solid ratio C_e, in the range of 0.001 to 10, the friction coefficient f_w, may be approximated as (VDI Atlas, 1993) below:

Gas–Solid Separators

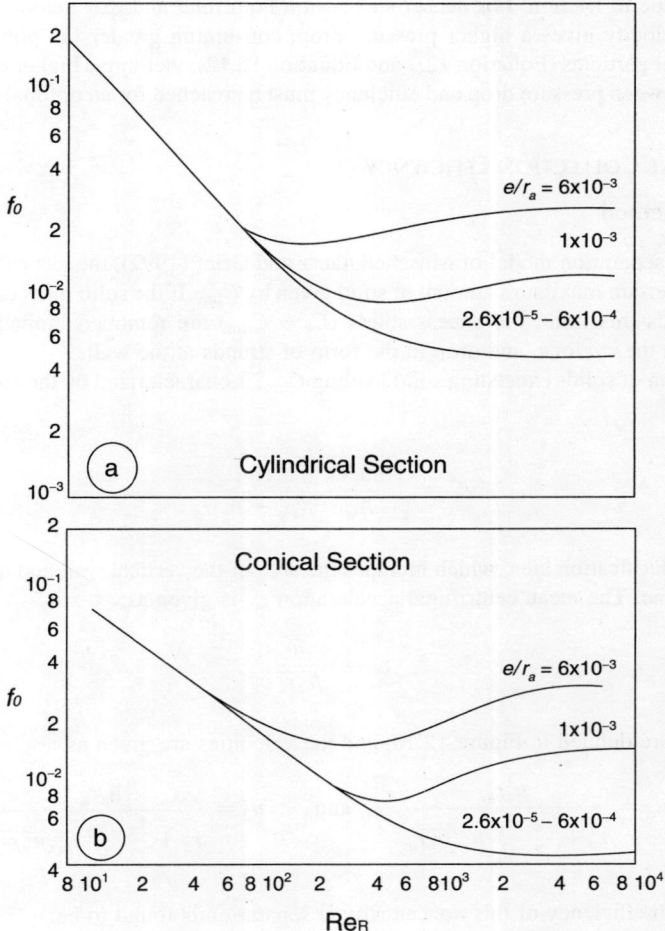

FIGURE 12.13 Friction coefficient for cylindrical and conical cyclones.

The friction coefficient in clean gas f_0, may be taken from graphs similar to the Moody chart based on relative roughnesss (e/r_a) (Figure 12.13):

$$2f_0\sqrt{C_e} \qquad (12.14)$$

The pressure drop in the vortex tube ΔP_e, is calculated from (Muschelknautz and Greif, 1997):

$$\Delta P_e = \left[2 + 3\left(\frac{u_i}{v_i}\right)^{4/3} + \left(\frac{u_i}{v_i}\right)^2\right]\frac{\rho_g v_i^2}{2} \qquad (12.15)$$

The mean velocity in the exit tube v_i, is found by dividing the gas flow by the cross-sectional area of the tube $v_i = \dfrac{V_{gas}}{\pi r_i^2}$.

It is clear from the above relations that a higher exit velocity v_i, or higher tangential inlet velocity u_a, results in a higher pressure drop (Equation 12.11 and Equation 12.15). For low solid concentration, the exit velocity predominates and the pressure drop in the exit tubes is several times larger than that due to friction. For denser suspension both inlet and exit velocity are important. A higher inlet velocity gives a higher pressure drop, consuming greater fan power, but it helps capture much finer particles (Equation 12.7 and Equation 12.12), yielding a higher efficiency. Thus, a compromise between pressure drop and efficiency must be reached for an optimal cyclone design.

12.1.6 CYCLONE COLLECTION EFFICIENCY

12.1.6.1 VDI Method

According to the separation model of Muschelknautz and Grief (1997), the gas entering a cyclone can carry only a certain maximum amount of solid given by C_{lim}. If the solid mass concentration C_e, in the gas exceeds this limit, the excess solids ($C_e - C_{\text{lim}}$) are removed immediately after the suspension enters the cyclone, and drop in the form of strands at the wall.

This separation of solids exceeding solid loading C_{lim}, is characterized by the cut size d_e^*, which is given as:

$$d_e^* = \sqrt{\frac{0.5\, V_b}{A_w} \cdot \frac{18\,\mu}{(\rho_p - \rho_g)z_e}} \qquad (12.16)$$

where A_w is the clarification area, which includes the area of the vertical wall and upper half of the cone of the cyclone. The mean centrifugal acceleration z_e, is given as:

$$z_e = \frac{u_e u_2}{\sqrt{r_e r_2}} \qquad (12.17)$$

where, r_e and r_2 are defined in Figure 12.10, and the velocities are given as:

$$u_e = \frac{u_a r_a}{r_e + \dfrac{f_w}{2}\dfrac{A_w}{V_b} u_a \sqrt{r_e r_a}} \quad \text{and} \quad u_2 = \frac{u_a r_a}{r_2 + \dfrac{f_w}{2}\dfrac{A_w}{V_b} u_a \sqrt{r_a r_2}}$$

The separation efficiency of this noncentrifugal separation is found to be:

$$\eta_{\text{lim}} = 1 - \frac{C_{\text{lim}}}{C_e} \qquad (12.18)$$

Remaining solids, which are generally small, undergo centrifugal separation in the vortex of the cyclone.

The limiting solid concentration C_{lim}, is generally in the vicinity of 0.25, but it depends on inlet solid concentration C_e, characteristic particle cut size d_e^* (Equation 12.16), and mean particle size d_{50}, at the inlet. It is given as (Muschelknautz and Greif, 1993):

$$C_{\text{lim}} = 0.025\, \frac{d_e^*}{d_{50}} (10 C_e)^k \qquad (12.19)$$

where d_{50} is the mean particle size of feed particles entering the cyclone, or 50% of the particles in the solids entering the cyclone are below this size. The exponent k is equal to 0.15 for inlet concentrations exceeding 0.1. For lower concentrations it can go up to 0.8.

Now the remaining solids, whose concentration is C_{lim}, enter the vortex of the cyclone where they undergo centrifugal separation. Muschelknautz and Greif (1997) present the size distribution of particles in the inner vortex of the cyclone. If one assumes that particles larger than d_{in}^* will be

Gas–Solid Separators

captured then the capture efficiency due to centrifugal separation may be written as:

$$\eta_{in} \approx \frac{C_{lim}}{C_e} R_{Ai}(d_i^*) = \frac{C_{lim}}{C_e} \exp\left[-\left(\frac{d_{in}^* 0.7^{1/na}}{d_e^*}\right)^{na}\right] \quad (12.20)$$

The cut size of the inner vortex d_{in}^*, is given by Equation 12.6, and that of the other separation d_e^*, is given by Equation 12.16. Typical value of the exponent na, is 1.2.

The total separation efficiency is the sum of the two separation efficiencies:

$$\eta_{tot} = \eta_{lim} + \eta_{in} \quad (12.21)$$

So, the total efficiency when $C_e > C_{lim}$ (Muschelknautz and Greif, 1997) is:

$$\eta_{tot} = \left(1 - \frac{C_{lim}}{C_e}\right) + \frac{C_{lim}}{C_e} R_{Ai}(d_i^*) \quad (12.22)$$

Example 12.1

Determine the separation efficiency and the pressure drop for the hot gas cyclone given below.

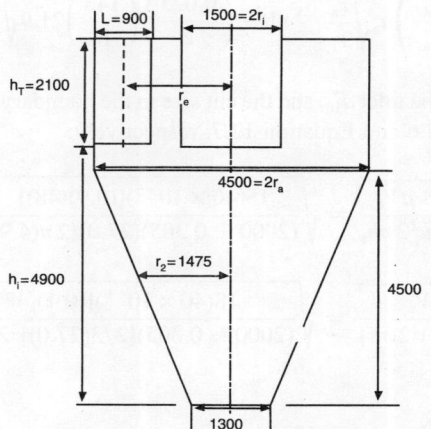

$T_{gas} = 855°C$; $V_{gas} = 48$ m³/sec; $\rho_g = 0.305$ kg/m³; $\rho_p = 2000$ kg/m³; $\mu = 40 \times 10^{-6}$ Pa sec; $A_w = 75$ m²; $A_R = 145$ m²; Effective gas flow rate $= 55$ m³/sec; $C_e = 4$ kg/kg; $d_{50} = 100$ μm; Assume $Re = 1 \times 10^3$

Solution

1. Determine inlet velocity V_e:

$$V_e = \frac{V_{gas}}{A_e} = \frac{48}{(0.9)(2.1)} = 25.4 \text{ m/sec}$$

2. Determine coefficient of contraction, α:

$$\beta = \frac{L}{r_a} = \frac{0.9}{2.25} = 0.4 \rightarrow \quad \text{using Equation 12.9 } \alpha = 0.93$$

3. Find the tangential velocity at r_a using Equation 12.8:

$$u_a = \frac{V_e\left(\dfrac{r_e}{r_a}\right)}{\alpha} = \frac{25.4\left(\dfrac{1.8}{2.25}\right)}{0.93} = 21.9 \text{ m/sec}$$

4. Find the coefficient of friction for gas carrying the solids f_w, using Equation 12.14:

$$f_w = f_0(1 + 2\sqrt{C_e}) = 5 \times 10^{-3}(1 + 2\sqrt{4}) = 0.025$$

5. Solve for the tangential velocity at r_2 using Equation 12.12:

$$u_2 = \frac{u_a\left(\dfrac{r_a}{r_2}\right)}{1 + \left(\dfrac{f_w}{2}\right)\left(\dfrac{A_w}{V_b}\right)u_a\sqrt{\dfrac{r_a}{r_2}}} = \frac{21.9\left(\dfrac{2.25}{1.475}\right)}{1 + \left(\dfrac{0.025}{2}\right)\left(\dfrac{75}{(0.9)(48)}\right)21.9\sqrt{\dfrac{2.25}{1.475}}} = 21.0 \text{ m/sec}$$

6. Solve for the tangential velocity at r_i, using Equation 12.12:

$$u_i = \frac{u_a\left(\dfrac{r_a}{r_i}\right)}{1 + \left(\dfrac{f_w}{2}\right)\left(\dfrac{A_R}{V_{gas}}\right)u_a\sqrt{\dfrac{r_a}{r_i}}} = \frac{21.9\left(\dfrac{2.25}{0.75}\right)}{1 + \left(\dfrac{0.025}{2}\right)\left(\dfrac{145}{48}\right)21.9\sqrt{\dfrac{2.25}{0.75}}} = 27.0 \text{ m/sec}$$

7. Determine the cut size for the inlet d_{in}^*, and the cut size in the boundary layer around the gas outlet pipe d_s^*, using Equation 12.6 and Equation 12.7, respectively:

$$d_{in}^* = \sqrt{\frac{18\,\mu V_b}{\Delta \rho u_i^2 2\pi h_i}} = \sqrt{\frac{18(40 \times 10^{-6})[(0.9)(48)]}{(2000 - 0.305)(27.0)^2 2\pi(4.9)}} = 26 \text{ } \mu m$$

$$d_s^* = \sqrt{\frac{18\mu V_s}{\Delta \rho (2/3u_i)^2 2\pi h_T}} = \sqrt{\frac{18(40 \times 10^{-6})[(0.1)(48)]}{(2000 - 0.305)[2/3(27.0)]^2 2\pi(2.1)}} = 20 \text{ } \mu m$$

8. Calculate u_e:

$$r_e = \frac{4500}{2} - \frac{900}{2} = 1800 \text{ mm}$$

$$u_e = \frac{u_a\left(\dfrac{r_a}{r_e}\right)}{1 + \left(\dfrac{f_w}{2}\right)\left(\dfrac{A_w}{V_b}\right)u_a\sqrt{\dfrac{r_a}{r_e}}} = \frac{21.9\left(\dfrac{2.25}{1.8}\right)}{1 + \left(\dfrac{0.025}{2}\right)\left(\dfrac{75}{[(0.9)(48)]}\right)21.9\sqrt{\dfrac{2.25}{1.8}}} = 17.9 \text{ m/sec}$$

9. Determine the centrifugal acceleration z_e, using Equation 12.17:

$$z_e = \frac{u_e u_2}{\sqrt{r_e r_2}} = \frac{(17.9)(21.0)}{\sqrt{(1.475)(1.8)}} = 231 \text{ m/sec}^2$$

Gas–Solid Separators

10. Determine the cut size d_e^*, using Equation 12.16:

$$d_e^* = \sqrt{\left(\frac{0.5\,V_b}{A_w}\right)\frac{18\,\mu}{\Delta\rho\overline{z}_e}} = \sqrt{\left(\frac{0.5(0.9)(48)}{75}\right)\frac{18(40\times 10^{-6})}{(2000-0.305)231}} = 21\ \mu m$$

11. Determine the separation efficiency $\eta_{\lim}$, using Equation 12.18 and Equation 12.19

$$C_{\lim} = 0.025\left(\frac{d_e^*}{d_{50}}\right)(10 C_e)^{0.15} = 0.025\left(\frac{21}{100}\right)[10(4)]^{0.15} = 0.009130$$

$$\eta_{\lim} = 1 - \frac{C_{\lim}}{C_e} = 1 - \frac{0.009130}{4} = 0.9977$$

12. Determine the capture efficiency η_{in}, using Equation 12.20:

$$\eta_{in} \approx R_{Ai}(d_i^*) = \left(\frac{C_{\lim}}{C_e}\right)\exp\left[-\left(\frac{d_{in}^* 0.7^{1/na}}{d_e^*}\right)^{na}\right]$$

$$= \left(\frac{0.009130}{4}\right)\exp\left[-\left(\frac{(26)0.7^{1/1.2}}{21}\right)^{1.2}\right] = 0.0009238$$

13. Determine the overall separation efficiency using Equation 12.21:

$$\eta_{tot} = \eta_{\lim} + \eta_{in} = 0.9977 + 0.0009238 = 0.9986 = 99.86\%$$

14. Determine the total solids leaving cyclone:

$$S \doteq (1 - \eta_{tot})C_e\rho_g = (1 - 0.9986)(4)(0.305) = 0.001679$$

15. Determine the pressure drop in separation compartment using Equation 12.11:

$$\Delta P_f = f_w\left(\frac{A_R}{V_b}\right)\left(\frac{\rho_g}{2}\right)(u_a u_i)^{3/2}$$

$$= 0.025\left(\frac{145}{(0.9)(48)}\right)\left(\frac{0.305}{2}\right)[(21.9)(27.0)]^{3/2} = 184\ Pa$$

16. Determine the pressure drop in gas outlet pipe using Equation 12.15:

$$v_i = \frac{V_{gas}}{\pi r_i^2} = \frac{48}{\pi(0.75)^2} = 27.16\ m/sec$$

$$\Delta P_e = \left[2 + 3\left(\frac{u_i}{v_i}\right)^{4/3} + \left(\frac{u_i}{v_i}\right)^2\right]\frac{\rho_g}{2}v_i^2$$

$$= \left[2 + 3\left(\frac{27.0}{27.16}\right)^{4/3} + \left(\frac{27.0}{27.16}\right)^2\right]\frac{0.305}{2}(27.16)^2 = 671\ Pa$$

17. Determine the total pressure drop using Equation 12.10:

$$\Delta P = \Delta P_f + \Delta P_e = 184 + 671 = 855\ Pa$$

12.1.6.2 PSRI Method

Figure 12.7 is a design graph developed by the Particulate Solid Research Institute to calculate the collection efficiency of particular cut sizes. The size is nondimensional with respect to d_{50}, which is the size at which 50% of solids are collected. Figure 12.7 presents data for very dilute suspensions where the solid concentration is less than 2.3 g/m^3.

The fractional efficiency of a cyclone increases with volume concentration of solids in the gas–solid suspension entering it. This is due to coarse particles colliding with fine ones as they settle, taking the fine particles to the wall faster. Furthermore, solids have a lower drag coefficient in multiparticle environments. At a very high solid loading the gas is supersaturated, i.e., cannot hold any more solid in the suspension. The bulk of the solids simply *condense* out of the gas stream (Pell and Dunson, 1997). This has been observed in very large diameter (4 to 8 m) cyclones, which should theoretically have low separation efficiency according to Equation 12.4, yet registered much higher efficiency than that observed in much smaller diameter cyclones.

CFB boilers use solids of wide size distribution and with concentrations far in excess of that shown in Figure 12.7, which gives the collection efficiency of an isolated particle. To find the collection efficiency one should also use Figure 12.8 and Figure 12.9, which give the improvement in collection efficiency as a function of the solid concentration for different single particle collection efficiencies.

To use this graph one computes the single particle collection efficiency E_0, using Figure 12.7. Then using the graph (Figure 12.8 or Figure 12.9) corresponding to that single particle efficiency E_0, one reads the efficiency for the given solid loading (kg solid in cu. m gas). Figure 12.8 is for particles belonging to Groups A and C, while Figure 12.9 is for particles belonging to Groups B and D.

12.1.6.3 Design Steps

Most manufacturers have proprietary designs for their cyclones based on the operating experience of commercial units. However, a tentative approach to the design of a cyclone for a CFB gasifier or boiler may be as follows:

1. Choose an inlet velocity from Table 12.2, which should preferably be in the range of 20 to 30 m/sec. Since the inlet area ($K.L$) is given by:

$$K.L = \frac{\text{Gas flow rate}}{V_e},$$

   One can then calculate an acceptable diameter of the cyclone using the geometric proportion given in Table 12.3.
2. Calculate other dimensions of the cyclone from Table 12.3 and calculate the saltation velocity from Equation 12.23 to check for reentrainment potential.

TABLE 12.2
Typical Dimensions and Inlet Velocities Used in Cyclones Used in Commercial CFB Boilers

Heat input (MW)	67	75	109	124	124	207	211	234	327	394	396	422
Number of cyclones	2	2	2	2	1	2	2	2	2	3	2	4
Body diameter (m)	3	4.1	3.9	4.1	7.2	7.0	6.7	6.7	7.0	5.9	7.3	7.1
Estimated gas throughput per cyclone at 850°C ($\times 10^6$ m^3/h)	0.175	0.19	0.285	0.325	0.54	0.55	0.55	0.61	0.85	0.68	1.03	0.6
Inlet gas velocity[a] (m/sec)	43	25	41	43	23	25	27	30	38	43	43	24

[a] The area of the cyclone inlet is assumed to be (body diameter 2/8).

Gas–Solid Separators

TABLE 12.3
Standard Proportion of Vertical Reverse-Flow Cyclones (refer to Figure 12.3)

Recommended Duty	$\dfrac{K}{D_c}$	$\dfrac{L}{D_c}$	$\dfrac{M}{D_c}$	$\dfrac{F}{D_c}$	$\dfrac{S}{D_c}$	$\dfrac{H}{D_c}$	$\dfrac{E}{D_c}$	N_h	$\dfrac{Q}{D_c^2}$	References
a. High throughput	0.75	0.375	0.75	0.875	1.5	4.0	0.375	7.2	4.58	Stairmand (1951)
b. High throughput	0.8	0.35	0.75	0.85	1.7	3.7	0.4	7.0	3.47	Swift (1969)
c. General purpose	0.5	0.25	0.5	0.6	1.75	3.75	0.4	7.6	1.86	Swift (1969)
d. General purpose	0.5	0.25	0.5	0.625	2.0	4.0	0.25	8.0	1.91	Lapple (1951)
e. High efficiency	0.44	0.21	0.4	0.5	1.4	3.9	0.4	9.2	1.37	Swift (1969)
f. High efficiency	0.5	0.22	0.5	0.5	1.5	4.0	0.375	5.4	1.53	Stairmand (1951)
General CFB	0.8	0.35	0.45	0.5–1.0	1.0	2.0	—	—	—	—

3. Find the cut-off size of the particles using Equation 12.5, taking N_c from Figure 12.6.
4. Estimate single particle collection efficiency E_T, for all diameters using Figure 12.7.
5. Using Figure 12.8 for Group A and C particles and Figure 12.9 for Group B and D particles read off the particle collection efficiency after accounting for the solid concentration.
6. Multiply the collection efficiency of each particle fraction by its grade efficiency.

12.1.7 Reentrainment of Solids

The efficiency of the cyclone does not increase continuously with an increase in the entry gas velocity V_i. The process of the collection of particles under centrifugal force in a cyclone is similar to that of saltation under gravity in a horizontal pipe. If the gas velocity at the inlet of the cyclone is higher than the saltation velocity, particles will not settle on the wall of the cyclone. On the contrary, at a higher velocity the particles, already collected on the wall, may be picked by the gas and reentrained. This severely limits the efficiency of the cyclone. The velocity for reentrainment is proportional to the saltation velocity. For the cyclone, the saltation velocity V_{salt}, is given in imperial units as (Frisch and Halow, 1987):

$$V_{\text{salt}} = 2.055 W \left[\frac{\left(\dfrac{L}{D_c}\right)^{0.4}}{\left(1 - \dfrac{L}{D_c}\right)^{1/3}} \right] D_c^{0.067} V_e^{2/3}, \text{ft/sec} \qquad (12.23)$$

where

$$W = \left[\frac{4g\mu(\rho_p - \rho_g)}{3\rho_g^2} \right]^{\frac{1}{3}} \text{ ft/sec}$$

L = width of the cyclone inlet in ft
D_c = cyclone barrel diameter in ft
μ = gas viscosity in lb. mass/(ft/sec)
ρ_p, ρ_g = density of particle and gas, respectively in lb. mass/ft^3
and V_e = cyclone inlet velocity in ft/sec

The collection efficiency of a cyclone increases with increased inlet velocity V_i, until it reaches the velocity $1.25 V_s$, beyond which reentrainment reduces the collection efficiency. There is also another operational problem with high inlet velocity; a higher V_i increases the rate of erosion of the target and other areas of the cyclone.

12.1.8 CYCLONE GEOMETRY

Vertical reverse-flow cyclones (Figure 12.3), used in the process industry, generally follow a certain geometric proportion. Cyclones of this geometry have similar characteristics. Table 12.3 gives the proportions of several types of cyclones, which are not necessarily rigid. This table also compares the dimensions of cyclones used in some large commercial CFB boilers.

As has been mentioned earlier, scroll- or volute-entry cyclones (Figure 12.3b) generally have greater efficiency than the tangential-entry cyclones shown in Figure 12.3a.

The performance characteristics of the cyclone in Figure 12.7 relate more to the general-purpose cyclone (type *d* in Table 12.3).

12.1.8.1 Cyclone Inlet

The shape and size of the entry section of the cyclone have an important influence on its performance. The larger the cross-sectional area of the entry section, the lower the cyclone inlet velocity (for given flow rate) and hence poorer the collection efficiency. On the other hand, a smaller entry section could improve the collection efficiency, but it would increase the pressure drop and the risk of erosion of the cyclone interior. The entry section is usually rectangular (the height of the entry section is twice its width). For vertical cyclones width of the entry is smaller than the height. Equation 12.5 also suggests that, the lower the width L, the smaller is the cut-off size.

Spiral inlet cyclones, shown in Figure 12.3b, offer higher collection efficiency than that given by tangential-entry shown in Figure 12.3a. Muschelknautz and Muschelknautz (1997) suggested some improvement of the inlet duct by inclining it at 30°. The cross section could change from square to rectangular—width height to width ratio 3:1 to 4:1 (Figure 12.14). Here, the gas inlet is spiral to the cyclone for better efficiency.

12.1.8.2 Asymmetric Exit

The inlet jet in a tangential-entry cyclone expands at high dust loadings, making the velocity distribution around the vortex tube asymmetrical (Muschelknautz and Muschelknautz, 1997) as shown in Figure 12.15a. This results in reduced collection efficiency. However, in a spiral inlet (Figure 12.3b), the expanding jet is further away from the vortex tube and therefore does not disturb the secondary (vortex) flow. The velocity distribution therefore remains symmetrical resulting in a higher separation efficiency of the secondary flow. By moving the exit tube to the centre of the vortex, as shown in Figure 12.15b, a definite improvement in cyclone efficiency has been observed (Muschelknautz and Muschelknautz, 1997).

12.1.9 EFFECT OF THE VORTEX FINDER

The *vortex finder* is the extension of the central exit tube inside the cyclone barrel. It prevents the entering gas from escaping through the outlet directly. Figure 12.3a shows a cyclone with a vortex finder while Figure 12.3b shows one without. Ideally, gas–solids entering the barrel tangentially should form a reverse vortex and have good collection efficiency without the need of a vortex finder. In practice, a secondary gas flow created at the roof of the cyclone adversely affects its collection efficiency. Thus, the collection efficiency improves with the installation of a vortex finder, although this also increases the pressure drop across the cyclone. Both measured and calculated data (Trefz, 1992) suggest that the improvement increases with increasing penetration of the vortex finder in the cyclone. Table 12.4 lists this improvement. It shows that the exit dust concentrations are lower after the installation of vortex finder, i.e., $(C'_{exit}/C_{exit} < 1.0)$. This ratio,

Gas–Solid Separators

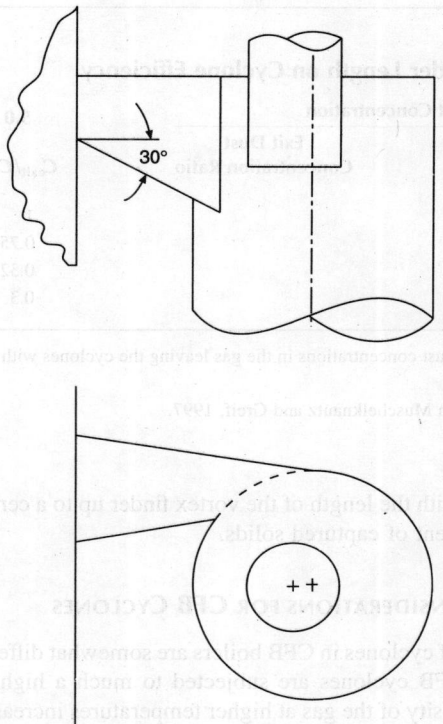

FIGURE 12.14 Modified entrance duct to cyclone for a CFB boiler.

a measure of the improvement in cyclone efficiency, is greater for a leaner suspension. The improvement increases with length of the vortex finder, but it tapers off beyond a length roughly equal to the exit diameter.

Data from several sources (Zenz, 1989) show that the collection efficiency improves with length of the vortex tube, but the improvement tapers off beyond a length equal to the height of the gas inlet section. Qiang (2004) also observed that in a vertical square cyclone the separation

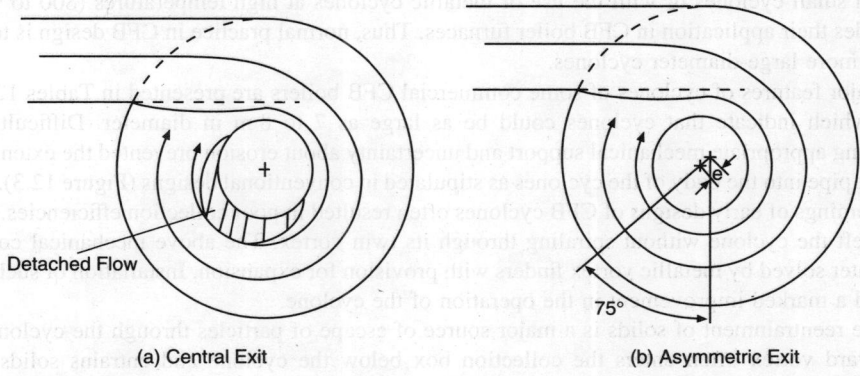

(a) Central Exit   (b) Asymmetric Exit

FIGURE 12.15 Asymmetric vortex in a tangential-entry cyclone at high dust levels.

TABLE 12.4
Effect of Vortex Finder Length on Cyclone Efficiency

Vortex Finder Depth/ Exit Diameter	Inlet Dust Concentration Exit Dust Concentration Ratio	5.0 C_{exit}/C_{exit}	0.5 C_{exit}/C_{exit}
0		1	1
0.5		0.75	0.4
1.0		0.32	0.65
1.5		0.3	0.65

Where C and C_0 are the dust concentrations in the gas leaving the cyclones with and without vortex finders, respectively.
Source: Recalculated from Muschelknautz and Greif, 1997.

efficiency also improved with the length of the vortex finder up to a certain length, beyond which it reduced due to reentrainment of captured solids.

12.1.10 PRACTICAL CONSIDERATIONS FOR CFB CYCLONES

The operating conditions of cyclones in CFB boilers are somewhat different from those in BFB or in process industries. The CFB cyclones are subjected to much a higher solid concentrations and temperatures. Higher viscosity of the gas at higher temperatures increases the drag on the particles, which decreases their inertial separation. The gas density, being several orders of magnitude smaller than the solid density, does not influence the cyclone performance directly. So, owing to its effect on gas viscosity, the net effect of temperature in principle will be to decrease the cyclone efficiency. Patterson and Munz (1989) observed this effect in experiments in the range of 30 to 1700°C. If a CFB unit designed to operate at 850°C operates at 950°C for some reason, the gas inlet velocity will rise due to reduced gas density because the furnace or gasifier throughput does not generally change to accommodate furnace temperature. Such a rise in gas volume may increase the collection efficiency and increase the furnace pressure. This could also happen if the moisture content of the fuel fired increases for some reason.

Cyclones in CFB boilers handle large volumes of gas at high temperatures. Small-diameter cyclones have higher collection efficiencies, but the difficulty associated with the refractory lining of such small cyclones or with the use of metallic cyclones at high temperatures (800 to 900°C) precludes their application in CFB boiler furnaces. Thus, normal practice in CFB design is to build one or more large-diameter cyclones.

Major features of cyclones of some commercial CFB boilers are presented in Tables 12.2 and 12.3, which indicate that cyclones could be as large as 7 to 8 m in diameter. Difficulty with providing appropriate mechanical support and uncertainty about erosion prevented the extension of the exit pipe into the body of the cyclones as stipulated in conventional designs (Figure 12.3). These shortcomings of early designs of CFB cyclones often resulted in poor collection efficiencies. Solids often left the cyclone without spiraling through its twin vortex. The above mechanical concerns were later solved by metallic vortex finders with provision for expansion. Installation of such items showed a marked improvement in the operation of the cyclone.

The reentrainment of solids is a major source of escape of particles through the cyclone. The downward vortex often enters the collection box below the cyclone and entrains solids. Thus special devices, such as baffles and vanes, are used to reduce the reentrainment. Bleeding a small volume of gas from the collection will help to create a slight negative pressure, which

Gas–Solid Separators

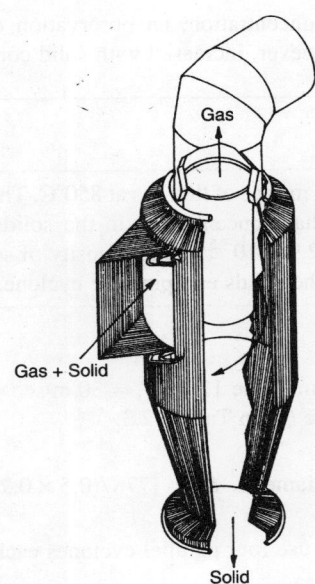

FIGURE 12.16 Water- or steam-cooled cyclones require less refractory.

may also reduce reentrainment. Particles separated sometimes bounce back into the exiting stream, a phenomenon that adversely affects collection efficiency, particularly for coarse particles.

Large CFB units need multiple cyclones. So, parallel operation of cyclones is common. It allows the use of multiple numbers of smaller size, and hence more efficient cyclones instead of one large unit. However, one should ensure that the gas flow is uniformly distributed amongst parallel cyclones. Poor distribution will lead to poor overall efficiency. Cyclones operating in series, however, do not have this problem, but their use is not economical, unless there is a need for very high collection efficiency.

Cyclones used in first-generation CFB boilers are uncooled and contain a large mass of refractory, which takes a long time to heat up and cool down. Therefore, these CFB boilers take a much longer time to start-up, and in the case of a breakdown, take a long time to cool down before repairs to the interior of the cyclone can be carried out. Such a long down-time is very costly for commercial boiler plants. Furthermore, refractories are costly in terms of both capital investment and operating costs.

The outer skin temperature of the cyclone is relatively high since the wall cannot be made excessively thick to bring its skin temperature close to the ambient temperature. Thus, much heat is lost through natural convection and radiation from the cyclone surface. This is responsible for higher levels of surface heat loss from CFB boilers with refractory lined cyclones. To get around this problem, some cyclones are water- or steam-cooled (Figure 12.16), use relatively thin refractory, and have lower heat losses. These cyclones are expected to reduce the mass of refractory used and to reduce the radiation loss from the external surface from the cyclone. The reduced refractory mass has two additional benefits: shorter start-up and cool-down time and reduced structural load.

12.1.10.1 Effect of Particle Concentration

As is seen in the graph in Figure 12.9, collection efficiency increases with solids concentration in the inlet gas stream. Karri and Knowlton (2005) found that in their tangential cyclones the collection efficiency decreases with solid concentration, an observation different from other works. The efficiency of a volute cyclone, however, increased with solid concentration.

Example 12.1

Design a cyclone to process 778 m³/sec of flue gas at 850°C. The bulk density of the gas–solid suspension is 10 kg/m³. The char concentration in the solids entering the cyclone is 2%. Viscosity of air, μ at 850°C = 2.9×10^{-5} lb/s.ft. Density of sorbent is 2500 kg/m³
Find the char concentration in the solids escaping the cyclone.

Solution

1. Gas inlet velocity (assumed from Table 12.2), $V_i = 30$ m/sec $= 984$ ft/sec
2. For general-purpose Lapple type (d) in Table 12.3:

$$\text{Cyclone diameter, } D_c = [778/(0.5 \times 0.25 \times 30)]^{1/2}$$

This figure is too large. So, we use four parallel cyclones each having diameter:

$$D_c = 14 \times \frac{1}{\sqrt{4}} = 7 \text{ m}$$

3. From standard cyclone proportions:
   (a) $L = 7 \times 0.25 = 1.75$ m; $K = 7 \times 0.5 = 3.5$ m; $m = 7 \times 0.5 = 3.5$ m
   $H = 4 \times 7 = 28$ m; $S = 2 \times 7 = 14$ m
   $L = 1.75/0.3048 = 5.7$ ft
   $\rho_p = 2500 \times 0.0625 = 156$ lb/ft³; $\rho_g = 0.305 \times 0.0625 = 0.0191$ lb/ft³
   (b) Check saltation velocity Equation 12.6

$$V_{salt} = 2.055 \, W \left[\frac{(L/D_c)^{0.4}}{(1 - L/D_c)^{1/3}} \right] D_c^{0.067} V_e^{2/3},$$

$V_e = 98.4$ ft/sec
$W = [4 \times 32.2 \times 2.9 \times 10^{-5} \times 156/(3 \times 0.0191 \times 0.0191)]^{1/3} = 8.1$ ft/sec
$D_c = 7/0.3048 = 23$ ft
$V_{salt} = 2.055 \times 8.1 \times [(1/4)^{0.4}/(1 - 1/4)^{0.333}] \times 23^{0.067} \times 98.4^{0.666} = 172$ ft/sec

The inlet velocity (98 ft/sec) is less than the saltation velocity, which rules out the possibility of reentrainment. Now, to reduce the number of cyclones one can increase the throughput to each cyclone by increasing V_e to a maximum 172 ft/sec.

4. The number of loops in the cyclone is estimated from Figure 12.6 as suggested by Zenz (1989) to be $N_c = 5$,
5. Theoretical cut size is determined next:
Viscosity, μ for air at 850°C is taken from Table A4.2 as 449×10^{-7} N s/m².
From Equation 12.5: $d'_{pth} = [9\mu L/(\pi N_c V_i (\rho_p - \rho_g))]^{1/2}$.
For sorbents:

$$d'_{pth} \sqrt{9 \times 449 \times 10^{-7} \times 1.75/(3.14 \times 5 \times 30 \times (2500 - 0.305)} = 24 \, \mu\text{m}$$

Gas–Solid Separators

For carbon particles:

$$d'_{pth} = 24 \times (2500/1200)^{1/2} = 38 \ \mu m$$

The single particle collection efficiency E_0, is found from Figure 12.7 (Zenz, 1989). To account for solid concentration, the collection efficiency is found from Figure 12.8 for Group A particles.
6. Here we calculate the collection efficiency of each size fraction

X Weight Fraction Sorbent	d_p μm Given	d_p/d_{th} Calculated	E_0 (%) From Figure 12.7	E_L Correction for Solid Concentration	$E_L \times X/100$ Collected Fraction
0.02	7.5	0.31	10	0.10	0.002
0.06	12	0.5	22	0.70	0.041
0.05	17	0.7	34	0.93	0.046
0.10	26	1.08	51	0.96	0.096
0.10	47	1.95	70	0.985	0.0985
0.05	72	3.0	82	0.99	0.0495
0.5	112	4.6	87	0.992	0.497
0.12	170	7.1	95	0.995	0.12
				Total = 0.95	

X Weight Fraction Char	d_p μm Given	d_p/d_{th} Calculated	E_0 (%) From Figure 12.7	E_L Correction for Solid Concentration	$E_L \times X/100$ Collected Fraction
0.2	12	0.31	16	0.16	0.08
0.3	22	0.57	29	0.80	0.24
0.05	35	0.92	48	0.95	0.0475
0.10	66	1.68	67	0.985	0.00986
0.10	112	2.94	89	0.993	0.00993
0.10	380	31.5	100	0.100	0.1
0.15	1200	31.5	100	0.100	0.15
				Total = 0.637	

Sorbents entering the cyclone = $(1 - 0.02) \times 778 \times 10 = 7624.4$ kg/sec.
Total sorbent escaping = $(1 - 0.95) \times 7624.4 = 381.22$ kg/sec.
Char entering the bed $0.02 \times 776 \times 10 = 155.2$ kg/sec.
Total char escaped = $(1 - 0.637) \times 155.2 = 56.3$ kg/sec.
Char concentration in the mixture of char and sorbent leaving the cyclone = $56.5 \times 100/(56.3 + 381.22) = 12.9\%$.

12.2 IMPACT SEPARATOR

Low thermal inertia, simpler construction, and lower cost are major attractions of impact separators, but these find application only in CFB boilers because of their low collection efficiency for fine particles. The U-beam separator, originally developed by Studsvik and extensively used by Babcock Wilcox in their CFB boilers, is a good example of an impact separator.

12.2.1 FEATURES AND TYPES

Impact separators separate solids from the gas through impingement on collecting bodies arranged across the path of the gas. In general, impingement separators are designed for collection of coarser particles (>10 to 20 μm) while offering low resistance (0.25 to 0.40 kPa).

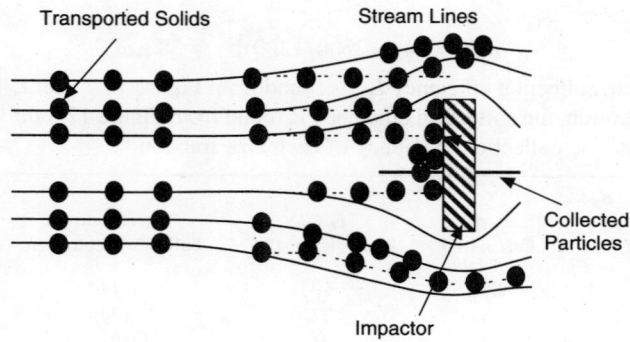

FIGURE 12.17 The working principle of an impingement separator.

In an impingement separator, the gas–solid suspension passes through a series of obstacles and undergoes sudden and repeated changes in flow direction. Due to their inertia, the particles, which have a momentum higher than that of the gas, tend to retain their original direction of movement. Therefore, instead of following the flow stream around the collecting object, the solid particles impact on the collecting object and are separated from the gas stream (Figure 12.17).

The following features make impingement separators especially suitable for large commercial CFB boilers:

- Simple construction and low building cost
- Stable operation at high temperatures
- Low pressure drop
- Easy scale-up

The reverse nozzle is another version of the impingement separator. Here, dust-laden gases impinge on a curved, slotted surface that reverses the gas flow but allows the dust particles to pass through the slots to an enclosed channel, where they can fall into the hopper. Some of the different shapes of targets for impingement separators are given in Figure 12.18.

The efficiency of an impact separator is dependent on the degree of impaction that occurs on the target. The separation efficiency and the pressure drop are functions of four variables:

- Number of targets in series
- Length of separator
- Spacing of separators
- Configuration of targets

Impaction and interception are the two major mechanisms for the collection of solids by impingement separators. A brief discussion is presented below.

12.2.2 SEPARATION THROUGH IMPACTION

Streamlines bend around the collection body, but solid particles, which carry a momentum greater than that of the gas, tend to continue in their original direction, deviating from the main gas stream and finally impacting on the collector. Figure 12.20 shows that there is a limiting trajectory of solids beyond which they are not collected. The distance of the parent streamline from the axis Y_{lim}, is an important parameter influencing the efficiency of collection through impaction. If all particles stick to the target surface on impaction, one can say that all particles flowing within a distance Y_{lim}, from

Gas–Solid Separators

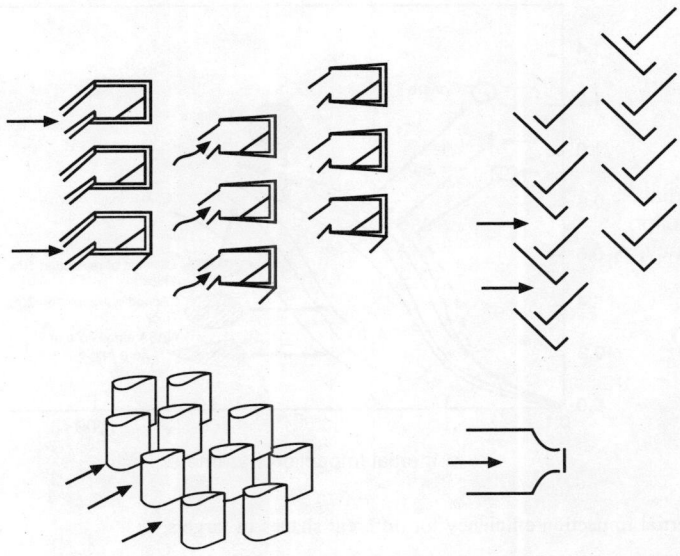

FIGURE 12.18 Different types of targets in an impingement separator.

the axis of the separator will be collected. For a two-dimensional collector, the fractional collection efficiency will be $2Y_{\lim}/D$, where D is the width of the target.

Some investigators (Golovina and Putnam, 1962) used a dimensionless parameter K_t, known as the inertial impaction parameter, to characterize the impaction efficiency of a target. The parameter K_t, is defined by:

$$K_t = \frac{\rho_p d_p^2 V_0}{9\mu D} \qquad (12.24)$$

The Stokes number Stk, is also used to characterize impact separators. It is defined in the present context as:

$$Stk = \frac{\rho_p C' V_0 d_p^2}{9\mu W_n} \qquad (12.25)$$

where C' is a constant, V_0 is the nozzle velocity, d_p is the particle size, and W_n is the width of the nozzle. The Stokes number becomes equal to K_t in case of $C' = 1$, and $W_n = D$. The inertial impaction efficiencies of a number of different shapes of target are shown in Figure 12.19.

12.2.3 SEPARATION THROUGH INTERCEPTION

A particle following a stream around an impactor may be captured if its dimensions are such that the impactor intercepts it. As shown in Figure 12.20, the particle moving around the object is stopped because the particle radius is larger than the distance of the flow stream from the surface of the object. If the streamline on which the center of the particle lies gets to the target any closer than half the particle diameter, the particle will touch the target and be intercepted (Figure 12.20). The analysis assumes that particles have no mass but have finite sizes. These are collected on contact with the target without reentrainment.

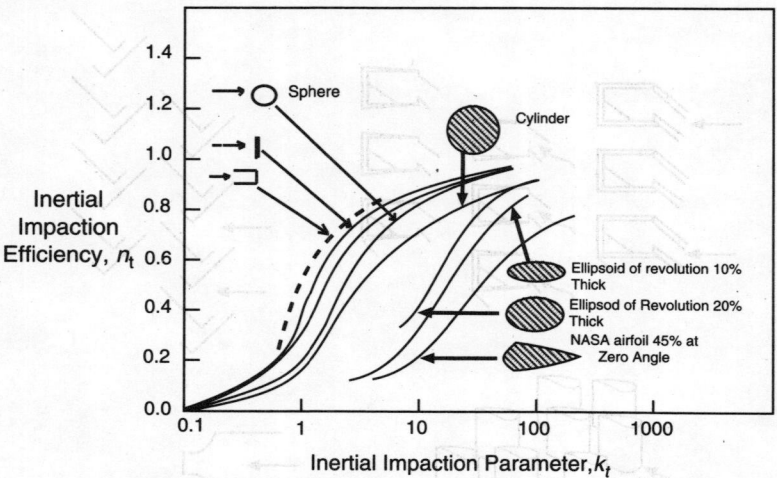

FIGURE 12.19 Inertial impaction efficiency for different shapes of targets.

Interception is characterized by a parameter:

$$R = \frac{d_p}{D}$$

where d_p is the particle diameter and D is the intercepting diameter.

Assuming the particle to have size but no mass, a simple analysis can be made. Assuming a potential flow, Strauss (1975) gave the following equation for calculating the efficiency of collection by interception $\eta_c(d)$.

$$\eta_c(d) = 1 + R - \frac{1}{1 + R} \text{(cylindrical collector)} \qquad (12.26)$$

In the case of impingement separators used in a CFB, the characteristic dimension of the target is of the order of a few centimeters, and the diameter of the particles is of the order of microns.

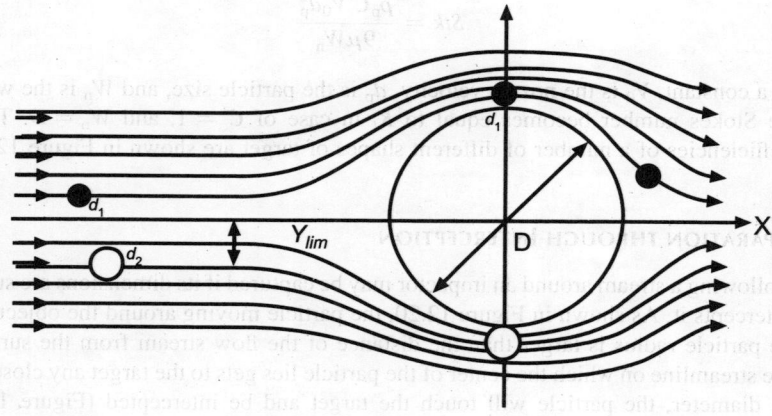

FIGURE 12.20 Collection of particles by interception of a target.

Gas–Solid Separators

Therefore, the contribution of interception between particles and the target, as seen in Equation 12.26, to the total collection is negligible.

12.2.4 Grade Efficiency of Impingement Collectors

An impingement collector would normally have several rows of collecting bodies, each row consisting of a number of targets. The targets may have complicated shapes forming jets or traps. These irregular targets make the flow field within the separator extremely complex. Thus, an exact mathematical description or solution is rather difficult, so, a simplified treatment of the flow pattern is needed. Presently, results with regard to impaction efficiency are available only for certain shapes of targets, such as cylinders, spheres, ribbons, rectangular half-bodies, and airfoils.

12.2.5 Some Types of Impact Separators

12.2.5.1 U-Beam

The U-beam, as shown in Figure 12.21, is the most common type of impact separator. It is made of a simple U-shaped steel section. Simplicity of the design makes this separator very attractive, and it is successfully used in CFB boilers as shown in Figure 8.4. High temperature and direct impact of particles on the separator wall require the use of special, expensive steel.

Gas–solid flow through a series of U-beam separators is shown in Figure 12.21. The solids, having greater momentum, are detached from the gas stream carrying them through the passage between two separators, and impinge on the rear wall of the U-beam, where in the absence of drag force from the gas, they are collected.

12.2.5.2 Lamella Separator

Another type of impact separator is shown in Figure 12.22. The solids pass through zig-zag passages as shown. These passages form a multiple-stage collection device with each stage being a separate inertial separator. The average size of the collected particles d_{50}, in a stage may be calculated from (Muschelknautz and Grief, 1997):

$$\frac{\rho_p z_e d_{50}^2}{18\mu} = \frac{V_{gas}}{\phi r h} \qquad (12.27)$$

where h is the height of the flow passage, r is the flow radius, and ϕ is the angle of curvature as shown in Figure 12.22.

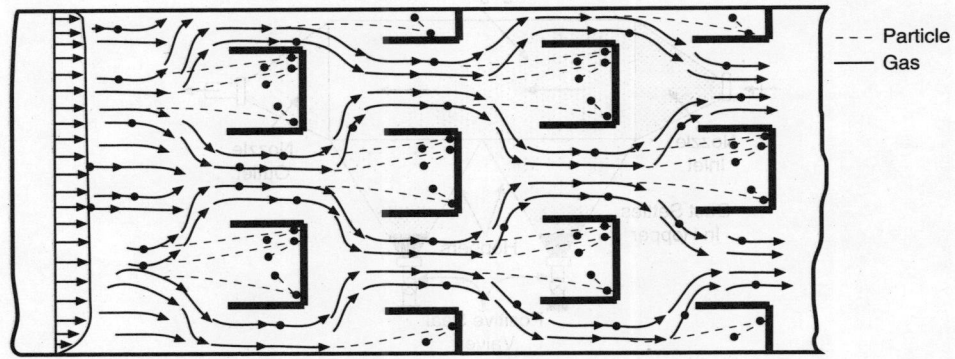

FIGURE 12.21 Gas–solid flow through an array of U-beam separators.

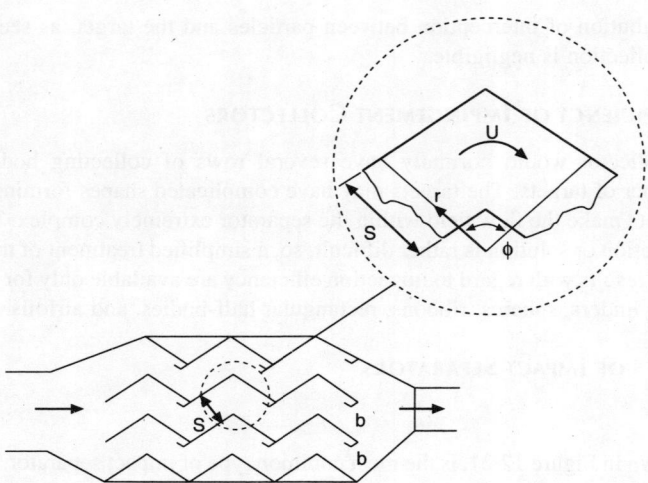

FIGURE 12.22 Lamella separator.

This equation is similar to Equation 12.32 for gravity settling where the acceleration due to gravity g, is replaced here with average centrifugal acceleration z_e, in each section, which is given as:

$$z_e = \frac{u^2}{r} \tag{12.28}$$

where u is the velocity through the passage and r is the radius of flow as shown in Figure 12.22.

The pressure drop through each section is:

$$\Delta P = (0.5 \sim 0.7)\frac{\rho_g u^2}{2} \tag{12.29}$$

12.3 INERTIAL SEPARATORS

The gravity settling chamber (Figure 12.23) is another type of inertial separator. Low separation efficiency and a large space requirement are its major limitations. These shortcomings, which limit

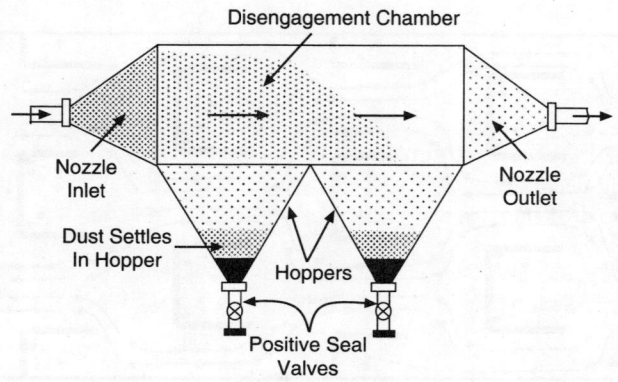

FIGURE 12.23 Gravity settling chamber.

the use of settling chambers in industry, can be improved by introducing baffles in the chamber to give particles a downward momentum in addition to the gravity settling effect. The settling chamber often is very effective as a precleaner for a cyclone. The author observed in a CFB pilot plant that more than 80% of the solids separated in a gravity settling chamber located between the riser and the cyclone. Large CFB boilers use very large diameter cyclones. According to the cyclone theory (Equation 12.4), these cyclones are supposed to be very inefficient, but practical observation suggests just the opposite. From this, one could speculate that these large cyclones perhaps work also as a gravity settling chamber.

12.3.1 Separation through Gravity Settling

In the absence of collecting bodies, an impingement separator is a settling chamber (Figure 12.23). Owing to the disturbance of the collecting bodies in the chamber, the flow can be considered turbulent, with complete mixing in the direction perpendicular to the flow. The grade efficiency of gravity settling for this case is given by (Svarovsky, 1981)

$$\eta_g(d_p) = 1 - \exp\left(-\frac{U_{tdp}A}{Q}\right), \tag{12.30}$$

where U_{tdp} is the terminal velocity of particles of diameter d_p, A is the cross-sectional area of the chamber, and Q is the volumetric flow rate of the gas.

If the velocity is sufficiently small ($V_e \sim 3$ m/sec) and there is no turbulence, the particles will drop at their terminal velocity u_t. The efficiency of collection is the ratio of solid flux in the downward direction to that of the flow direction.

$$\eta_g = \frac{u_t L_s B_s}{V_s H_s B_s} = \frac{u_t L_s B_s}{Q} \tag{12.31}$$

From here one can see that for a given volumetric gas flow rate Q, the separation efficiency is independent of the height of the settling chamber, but increases with the floor area of the vessel ($L_s B_s$), and the terminal or settling velocity, which depends on the particle size.

By using the Stokes law the settling velocity u_t, for a particle of diameter d_p, in a gravity settling chamber is:

$$u_t = \frac{\rho_p d_p^2 g}{18\mu} \tag{12.32}$$

12.3.2 Cavity Separator

Such separators appear similar to the impingement separator described in Section 12.2, but their working principle is different. These separators (Dutta and Basu, 2004) are in the form of a deep cavity, whose depth is much longer than its width. Such separators are supported vertically in the gas–solid path as shown in Figure 12.24. The height of the separator is sufficient to cover the entire height of the flow stream. When a gas–solid suspension approaches the cavity, heavier solids enter the cavity due to their greater momentum and the gas goes around it.

The cavity being deep, the particles lose their momentum before hitting the back wall. This causes the particle to drop into the cavity, thus saving these separator elements from the impact of the erosive particles that occurs in impact separators. For this reason this type of inertial separators are relatively free from erosion. This feature allows cavity separators to be made of heat-absorbing boiler or superheater tubes, reducing the cost of boiler.

12.3.3 Design Steps

To obtain a first estimate of the efficiency of separation, or to design a separator for a given efficiency, one must resort to a number of simplifications and extrapolations from the existing data.

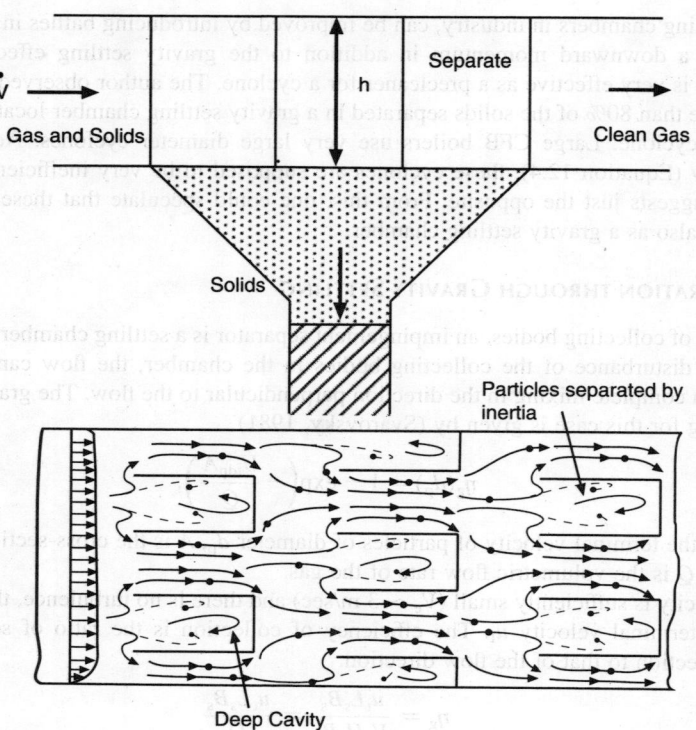

FIGURE 12.24 Deep inertial-type separators.

The following is an example of the design of a separator unit comprising a series of impact separators followed by a gravity settling chamber.

The inertial impaction efficiency η_t, for one target gives the fraction of solids approaching a target that will be collected. It implies that if the collection efficiency is 100%, all particles within the shadow of a target will be collected. Thus the collection efficiency of one row of targets will be the ratio of target area nA_t, and field area A, where n is the number of targets within the area.

Step 1. Simplify the target as one for which data on impaction efficiency are available (Figure 12.19). Then determine the characteristic dimension of the target D. Calculate the impaction parameter K_t, from Equation 12.24.

Step 2. Take the inertial impaction efficiency η_t, for one target from Figure 12.19 for the above value of K_t. Particle collection by one target is $A_t \, \eta_t$:

Step 3. Calculate the separation efficiency η_r, of one row of targets from:

$$\eta_r = \frac{nA_t}{A}\eta_t \qquad (12.33)$$

where n is the number of targets placed across the flow area. Each has projected area A_t, covering a total flow area of A.

Step 4. The amount of solid entering the next row is $(1 - \eta_r)$. If efficiency of this row is η_2, the amount of solid collected here is $(1 - \eta_r)\eta_2$. Thus, the total amount of solids collected in these two rows is $\eta_r + (1 - \eta_r)\eta_2$. Similarly, we can find the solid collected in the ith row by:

$$\eta_i(1 - \eta_{i-1}) + \eta_{i-1} \qquad (12.34)$$

Gas–Solid Separators 413

Calculate the separation efficiency of a target of p rows deep (in series): Here, $\eta_1 = \eta_i$, where the number of rows; $i = 2,\ldots p$:

$$\eta_{\text{total}} = \eta_1 + \eta_2 + \cdots + \eta_p \quad (12.35)$$

If M is the mass of solids entering the separator, the mass of solids leaving it will be $M(1 - \eta_{\text{total}})$. This solid now enters the gravity settling chamber for further separation.

Step 5. Now calculate the efficiency of gravity settling, Equation 12.30:

$$\eta_g(d_p) = 1 - \exp\left(-\frac{U_{\text{tdp}}A}{Q}\right),$$

We can integrate it over the entire size distribution to get the overall collection efficiency of the gravity settling chamber as η_g.

Step 6. The solid collected by the gravity-settling chamber is:

$$\eta_g M(1 - \eta_{\text{total}}).$$

The total solid collected is $[M\eta_{\text{total}} + \eta_g M(1 - \eta_{\text{total}})]$. Thus, the overall efficiency is

$$\eta = \eta_{\text{total}} + \eta_g - \eta_g \eta_{\text{total}} \quad (12.36)$$

An alternative approach has been suggested by Marple and Liu (1974), in which the designer makes use of graphs of Re vs. jet width and efficiency vs. $\sqrt{Stk}$ with the ratio of jet-to-plate distance and jet width as parameters in the latter case.

NOMENCLATURE

A: flow cross section of separator, m^2
A_t: projected area of one target, m^2
A_R: area of cyclone wall including roof and outer surface of exit tube, m^2
A_w: clarification area, m^2
B: width of nozzle of impact separator, m
B_s: width of the separator chamber, m
$K, L, F,$
E, H, S: dimensions of cyclone (Figure 12.3a)
C: constant in Equation 12.5
C': constant in Stokes number
C_e: mass fraction in inlet gas–solid suspension
C_{exit}: mass fraction of solids in gas exiting cyclone (no vortex finder)
C'_{exit}: mass fraction of solids in gas exiting cyclone (after installation of vortex finder)
C_{lim}: limiting solid concentration
D: characteristic dimension of target, m
D_c: cyclone barrel diameter, m
d, d_p-: particle diameter, m
d_e^*: cut size of solids corresponding to solid separation exceeding critical concentration, C_{lim} m
d'_{pth}: theoretical cut-size, m
d_{th} or d_{50}: cut-off diameter for 50% efficiency, m
d_{in}^*, d_s^*: cut-off sizes for the barrel-flow and secondary flow, respectively, m
E_L: collection efficiency after correction for solid concentration

E_0: theoretical collection efficiency
E_T: total efficiency of separator
E: Length of the vortex finder tube, m
F_c: centrifugal force on a particle, N
F_d: drag force on particle, N
$f(d)\,dd$: weight fraction of entering solids in size range d and $d + \Delta d$
f_0: coefficient of friction due to clean gas
$f_c(d)\,dd$: weight fraction of collected solids in size range d and $d + \Delta d$
f_s: coefficient of friction due to solids
f_w: total wall friction coefficient
g: acceleration due to gravity, 9.81 m/sec^2
h: height of flow section in separator chamber, m
H: total height (barrel + cone) of cyclone, m
H_s: height of inertial separator, m
k: exponent in Equation 12.19
K: height of inlet, m
K_t: inertial impaction parameter
L: width of the rectangular inlet duct, m
L_s: length of inertial separator, m
M: mass flow rate of dust entering collector, kg/sec
M_c: mass flow rate of escaped fine particles, kg/sec
m: exit diameter of the cyclone, m
m_p: mass of a particle, kg
n: number of targets in one row across flow area
na: exponent in Equation 12.20
N_c: effective number of turns
N_h: number of inlet velocity head
P: static pressure drop through cyclone, Pa
ΔP: pressure drop through each section, Pa
ΔP_e: hydrodynamic loss in inner vortex, Pa
ΔP_f: pressure drop due to friction on cyclone wall, Pa
ΔP_c: pressure drop through cyclone, Pa
p: number of targets in series
Q: gas flow rate entering cyclone, m^3/sec
R: interception parameter (d_p/D)
R_{Ai}: RRSB distribution
r: radius of cyclone or flow path, m
r_2: average radius of the cyclone cone, m
r_a: radius of cyclone barrel
r_e: radius of axis of entry section
r_i: exit pipe radius, m
S: total solids leaving cyclone, kg/sec
Stk: Stokes number, $\dfrac{\rho_p C' V_0 d_p^2}{9\mu W}$
u_2: average tangential velocity halfway in the cone, i.e., r_2, m/sec
u_a: tangential velocity at cyclone barrel radius, m/sec
u_i: tangential velocity at exit radius, m/sec
u_t: terminal velocity of particles, m/sec
T_{gas}: flue gas temperature, °C
U: particle velocity, m/sec

U_{tdp}:	terminal velocity of size d_p, m/sec
v_i:	mean velocity in exit tube, m/sec
V:	gas velocity, m/sec
V_b:	gas flow through cyclone barrel, m³/sec
V_e:	entry velocity to cyclone, m/sec
V_{gas}:	total gas flow into the cyclone, m³/sec
V_s:	flow rate of the secondary flow in cyclone, m³/sec
V_0:	nozzle velocity, m/sec
V_r:	radial velocity of solids, m/sec
V_{salt}:	saltation velocity, m/sec
W:	parameter as defined in Equation 12.6
W_n:	width of nozzle, m
X, Y:	dimensionless form of x, y
Y_{lim}:	distance of the collector surface axis within which all particles are collected, m
z_e:	mean centrifugal acceleration, m/sec²

GREEK SYMBOLS

α:	contraction coefficient for gas flow into the cyclone
β:	ratio of inlet width to barrel radius, r
$_{str}$:	voidage of strand
ϕ:	angle of curvature of flow path as shown in Figure 12.22.
η_g:	net efficiency of collection including all forms of separation
η_{in}:	efficiency of centrifugal separation
η_{lim}:	separation efficiency of noncentrifugal separation
η_{tot}:	gross efficiency of p rows of collectors
$\eta_c(d)$:	efficiency of collection through interception for size d
$\eta_g(d_p)$:	efficiency of collection of size d_p through gravity settling
$\eta_g(d_p)$:	grade efficiency or collection efficiency of size d
η_r, η_i:	efficiency of collection of one row and ith row, respectively
η_t:	target impaction efficiency
$\eta_x(d)$:	efficiency of collection by interception
ρ_g:	gas density, kg/m³
ρ_p:	particle density
μ:	gas viscosity, kg/msec

REFERENCES

Dutta, A. and Basu, P., Experimental investigation into heat transfer and hydrodynamics to cavity-type inertial separators—a novel technique for development of subcompact circulating fluidized bed boilers, *J. Energy Res.*, 2004, August.

Frisch, N. A. and Halow, J. S., in Atmospheric Fluidized Bed Combustion, *S.E. Tung and G.C. Williams, Technical Source Book*, DOE/MC//14536-2544, 51–536, 1987.

Geldart, D., *Gas/Solid Fluidization*, John Wiley & Sons, New York, 1986.

Golovina, M. N. and Putnam, A. A., Inertial impaction on single elements, *I&EC Fund.*, 1, 4, 264–273, 1962.

Hyppanen, T., Ristola, J., An apparatus and a method for separating particles from hot gases, Canadian Patent no. CA 2,271,158, May 28, 1998.

Karri, R. S. B. and Knowlton, T. M., The Effect of Riser exit geometry on the performance of Close coupled cyclones, In *Circulating Fluidized Bed Technology III*, Cen, K., Ed., International Academic Publishers, Beijing, pp. 899–912, 2005.

Lapple, C. E., Process Use Many Collectors, *Chem. Eng.*, 58, 144, 1951.
Marple, A. A. and Liu, B. Y. U., *Environ. Sci. Technol.*, 8(7), 648–654, 1974.
Muschelknautz, E. and Greif, V., Fundamental and practical aspects of cyclones, In *Circulating Fluidized Bed Technology IV*, Avidan, A., Ed., AIChE, New York, pp. 20–27, 1993.
Muschelknautz, E. and Greif, V., Cyclone and other gas-solids separators, In *Circulating Fluidized Beds*, Grace, J. R., Avidan, A. A. and Knowlton, T., Eds., Blackie Academic, London, pp. 181–213, 1997.
Muschelknautz, U. and Muschelknautz, E., Special design of inserts and short entrance ducts, In *Circulating Fluidized Bed Technology V*, Kwauk, M. and Li, J., Eds., p. 597–602, 1997.
Patterson, P. A. and Munz, R. J., Cyclone collection efficiencies at very high temperatures, *Can. J. Chem. Eng.*, 5, 321–328, 1989.
Pell, M. and Dunson, B. J., *Gas-solid separations, Perry's Chemical Engineering Handbook*, 7th ed., McGraw Hill, New York, Chapter 17, pp. 17-26–17-31, 1997.
Qiang, Z., Research on horizontal square cyclone and its application in a CFB boiler Ph.D. Thesis, Zhejiang University, China, 2004.
Stairmand, C. J., The design and performance of cyclone separators, *Trans. Instn. Chem. Engrs.*, 29, 356–383, 1951.
Strauss, W., *Industrial Gas Cleaning*, 2nd ed., Pergamon Press, New York, 1975.
Svarovsky, L., *Solid-Gas Separation*, Elsevier, Amsterdam, 1981.
Swift, P., Dust Control in Industry, 2, *Steam Heat Engineer*, 38, 453–456, 1969.
Trefz, M., *Boundary layer flows in cyclones and their influence on the solid separation*, Second European Symposium on Separation of Particles from gases. Nurnberg, Germany, 1992.
Verein Deutscher Ingenieure Heat Atlas. *Association of German Engineers/VDI Society for Chemical and Process Engineering*, VDI-Cwerlag GnbH, Dusseldorf, pp. Lj1–Lj8, transla. J.W. Fullarton, 1993.
Zenz, F. A., *Fluidization and Fluid Particle Systems*, Pemm-Corporation Publication, Nelsonville, pp. 287–291, 1989.
Zenz, F. A., *Perry's Chemical Engineering Handbook*, PSRI, Chicago, 1997.

13 Solid Recycle Systems

Circulating fluidized bed boilers, as the name suggests, circulate solids around an endless loop. The recycle system moves the solids from the low-pressure cyclone to the high-pressure bottom of the furnace or riser. The amount of solid to be moved is so large that it is nearly impossible to find an inexpensive, motorized mechanical device to transfer it from the standpipe to the furnace. For example, A 165-MWe CFB boiler would need to transfer nearly 3600 tons/hr of hot solids. It is very difficult for any mechanical feeder to move such a large amount of hot solid from the low-pressure cyclone to the high-pressure end of the riser of a CFB gasifier or boiler. A simple nonmechanical device called *loop-seal* with no moving parts is used for this purpose. It is also sometimes called a *J-valve*, *fluoseal*, *siphon seal*, or simply an *overflow standpipe*. Several other types of nonmechanical devices are also used in other CFB reactors. As the loop-seal transfers the solids across a pressure barrier which allows the solids to move from the standpipe to the furnace (riser) but does not allow the gas to move from furnace to the standpipe it is also called a *nonmechanical valve*. It may not be an exaggeration to call these valves the heart of a CFB plant. Even a short interruption of this device will fill the cyclone with solids and stop the entire circulating fluidized bed. Thus a good understanding of its operating principle and design is very important. This chapter describes the operating principles of nonmechanical valve, and the design methods of some commonly used valves.

13.1 TYPES OF NONMECHANICAL VALVES

Nonmechanical valves are devices that facilitate the flow of solids between the return leg (standpipe) and the furnace without any external mechanical force. Air assists the movement of solids through these valves. The circulation of solids between the furnace and cyclone is vital to the CFB boiler. The absence of moving parts makes nonmechanical valves robust, inexpensive, simple in construction, and easy to maintain.

These valves, however, work best with Group B ($>150\mu$) particles. They may experience some difficulties with Group A particles, but they will have considerable difficulties in moving Group D particles as this would require a large diameter pipe and a greater amount of aeration. Particles with small shape factors like biomass particles may pose a considerable challenge to the smooth operation of nonmechanical valves. The addition of some fine particles, however, may facilitate their operation. Group C particles cannot be fluidized, so, they may not work in conventional loop-seals.

The flow rate of solids through these systems is governed by the overall system including the aeration rate through the valve. Several types of nonmechanical valves, as shown in Figure 13.1, are available for moving solids in a CFB loop, but circulating fluidized bed boilers and gasifiers generally use one of the following two types of valves:

1. Loop-seal (Figure 13.1b and Figure 13.2) (also called *seal-pot*)
2. L-valve (Figure 13.1a and Figure 13.3)

The V-valve (Figure 13.1c), a special purpose valve available for CFB reactors, is not normally used in CFB boilers or gasifiers. The L-valve is also used very rarely in such applications. However, a brief description of these valves is given below:

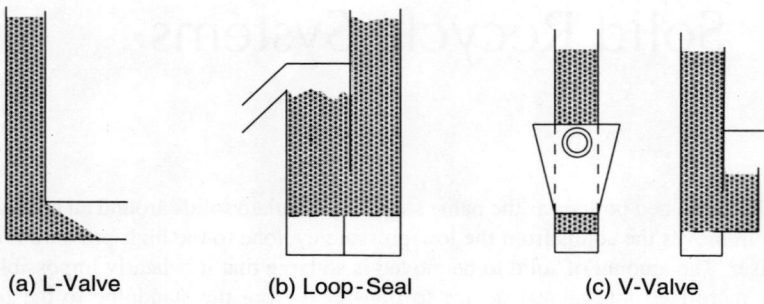

FIGURE 13.1 Schematic diagrams of nonmechanical valves.

13.1.1 Loop-Seal

Loop-seals are extensively used in CFB boilers and gasifiers. A typical loop-seal (Figure 13.4) consists of two sections: *supply chamber* and *recycle chamber*. The supply chamber (also called *down-leg chamber, standpipe*) is connected directly to the standpipe or in some cases, just the bottom section of the standpipe (Figure 13.2). This chamber supplies solids to the recycle chamber, which transfers the solids to the riser through a recycle pipe.

The solids from the supply chamber (also called *return chamber*) move horizontally through a partition wall separating the supply and recycle chambers. The recycle chamber then fluidizes these solids to allow them to move up like a liquid. There is one weir (the lower wall shown in

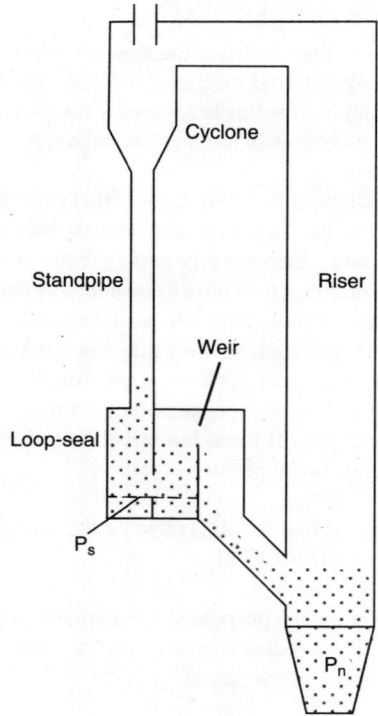

FIGURE 13.2 A typical loop-seal-based solid recycle system for a CFB boiler or gasifier.

Solid Recycle Systems

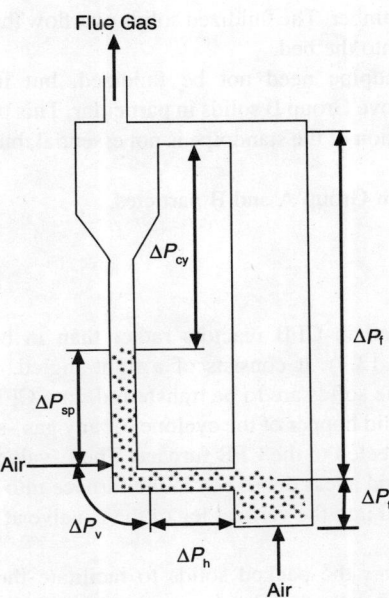

FIGURE 13.3 An L-valve-based solid recycle system for a CFB reactor showing the pressure balance around its loop.

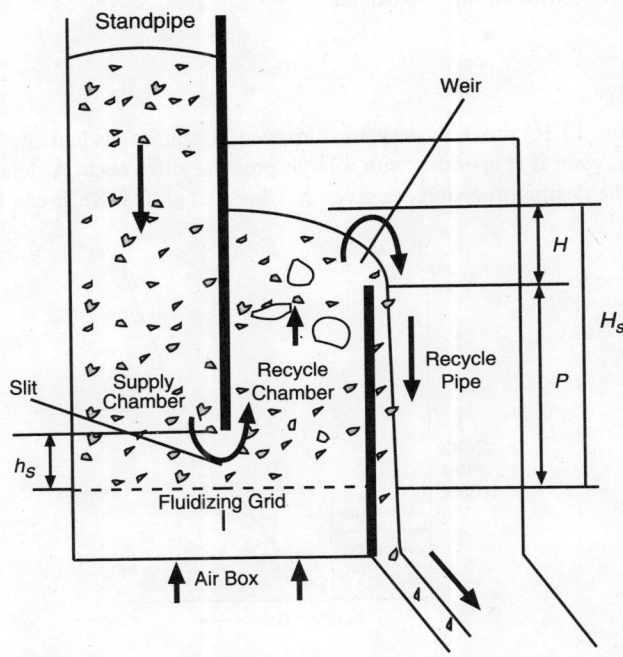

FIGURE 13.4 Flow of solids through loop-seal is similar to the water flow over a weir.

Figure 13.4) in the recycle chamber. The fluidized solids overflow this weir to fall under gravity into the recycle pipe or directly into the bed.

The bottom of the standpipe need not be fluidized, but it often is necessary in large commercially units to help move Group B solids in particular. This point is elaborated in the section on operating principles. Aeration of the standpipe is not essential, but is used sometimes to aid solid flow down the standpipe.

Loop-seals work best with Group A and B particles.

13.1.2 L-Valve

The L-valve is generally used in CFB reactors rather than in boilers. It is, however, one of the simplest designs (Figure 13.3). It consists of a right-angled, L-shaped pipe connecting the two vessels between which the solids are to be transferred. In a CFB boiler, the vertical leg of the L-valve is connected to the solid hopper of the cyclone (or any gas–solid separator). The horizontal section of the L-valve is connected to the CFB furnace. The L-valve would usually have a moving packed bed in the standpipe and a less dense bed in the furnace into which the solid is fed. A small amount of air or gas is injected into the vertical leg of the L-valve at a short distance above the axis of its horizontal arm.

This air (or gas) *lubricates* the packed solids to facilitate their movement from the dense vertical leg to the relatively dilute bed. In large commercial units this air may have to be added in multiple points around the vertical leg and underneath the horizontal section of the L-valve to facilitate smooth solid flow. The aeration air moves downwards through the particles and then through the constricting bend of the L-valve. Solid flow commences only beyond a certain minimum value of air rate (Figure 13.5), a property also noticed in the case of a loop-seal.

Unlike loop-seals, L-valves do not work well with Group A. They work best with Group B and experience difficulties with Group C particles.

13.1.3 V-Valve

The V-valve (Figure 13.1c) provides very good protection against gas leakage between the furnace and the return leg, even if it operates with a large pressure difference. A detailed account of this valve, including the design procedure, is given in Chong et al. (1988). It can be used in gasifiers.

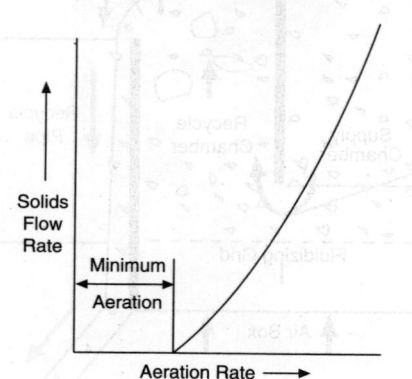

FIGURE 13.5 A typical characteristic of a nonmechanical valve shows that solid flow begins only above a certain aeration rate.

Solid Recycle Systems

13.2 PRINCIPLES OF OPERATION

Figure 13.2 shows a schematic of a CFB riser with a loop-seal-type solid recycle system, while Figure 13.3 shows a CFB plant with an L-valve-type device. To explain the working of the CFB loop we use a simple experiment as shown in Figure 13.6. The glass vessel with two columns (AB and CD) is filled with water. Under normal conditions, the water will remain stationary in the column, as static pressures at B and C are both equal to the hydrostatic head of water column. Now, when air is injected at the bottom of column CD, streams of bubbles will rise through it and escape out of the top. This will reduce the bulk density of column CD due to the presence of lighter bubbles, which will in turn reduce the static pressure at C to below that of B. This will cause water to flow from B to C creating a continuous circulation of water around the vessel. It is analogous to the natural circulation of water and steam in boilers.

This analogy can be applied directly to gas–solid systems in CFB provided the solids in the standpipe behave as a liquid, meaning they deform continuously under the action of shear stress. The function of a loop-seal is to give, albeit to a limited extent, fluid properties to the solids in the loop-seal so they can move from the standpipe to the riser when subjected to a static pressure difference. This is done by the addition of appropriate amounts of air in the nonmechanical valve, which works best with Geldart Group B and D particles (typical sizes between 100 and 5000 μm) (Knowlton, 1997).

The pressure balance around a CFB loop-seal, described in Section 2.4.1, is illustrated here by a pressure profile measured by the author in a CFB pilot plant (Figure 13.7). In this figure, the loop-seal stands at a height of 0.6 m. The static pressure in the standpipe at this height is higher than that in the riser at the same height. The function of the loop-seal is to facilitate movement of solids from the high-pressure standpipe to the low-pressure riser.

At high gas velocities in the CFB riser, the limit of solid mass flux in the riser is set by the height of solids in the standpipe. This is because the low height of solids in the standpipe cannot develop adequate head to push the solids into the riser at a rate high enough to equal the maximum solid-carrying capacity of the gas flowing through the riser. The solid flow rate in the riser naturally cannot reach its choking value for the particular gas velocity.

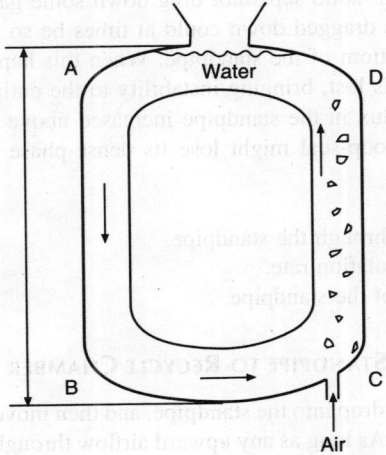

FIGURE 13.6 The principle of working a nonmechanical valve can be explained by the example of moving water in a loop by aeration.

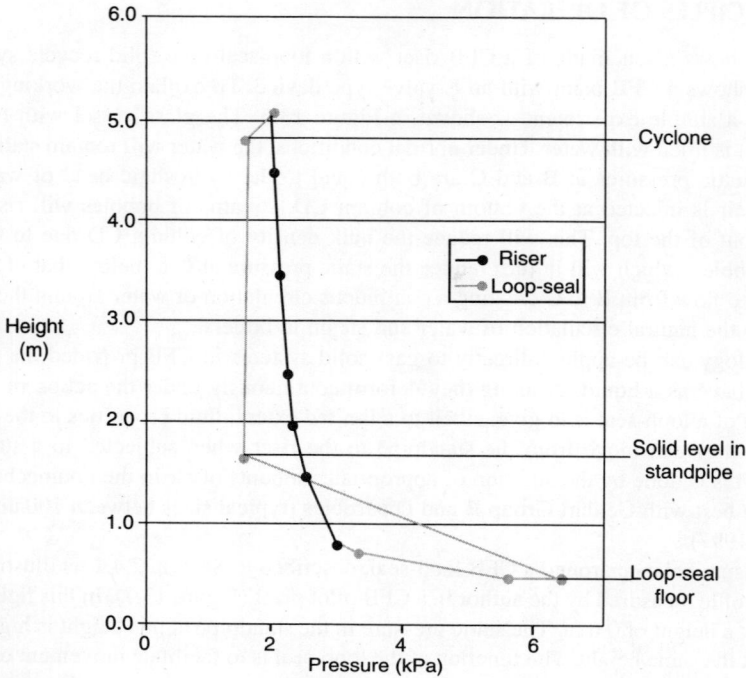

FIGURE 13.7 Pressure drop around a CFB loop.

When the fluidizing velocity in the riser is low, but the level of solids in the standpipe is high, the limiting mass flux in the riser would depend on the solids inventory in the riser. In this case, the height of solids in the standpipe can push solids into the riser, exceeding its solid flow rate at that gas velocity, which creates a dense-phase at the bottom of the riser (Knowlton, 1997).

(i) Solid flow through a standpipe.

Solids collected in the gas–solid separator drag down some gas as they descend through the standpipe. The amount of gas dragged down could at times be so great that it prevents a dense-phase from forming at the bottom of the standpipe. When this happens, the head that drives the solids through the CFB loop is lost, bringing instability to the entire system. Also, Wirth (1995) found that if the solid mass flux in the standpipe increased above a certain value, the standpipe connecting the cyclone and loop-seal might lose its dense-phase seal at the bottom. To avoid this one can:

1. Reduce the mass flux through the standpipe.
2. Decrease the solid circulation rate.
3. Increase the diameter of the standpipe.

13.2.1 SOLID FLOW FROM STANDPIPE TO RECYCLE CHAMBER

Solids collected in the cyclone drop into the standpipe, and then move under gravity into the supply chamber located at its bottom. As long as any upward airflow through standpipe remains below the minimum fluidizing velocity the solids will move through it as a moving packed bed. If no air enters the supply chamber the solids in it will move out only to form a slope whose angle (α) will be dictated by the angle of repose of the solids (Figure 13.8).

Solid Recycle Systems

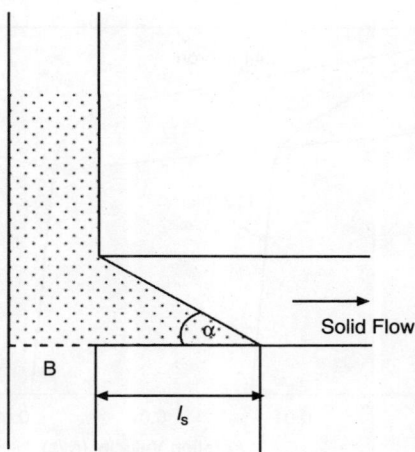

FIGURE 13.8 Solids in a standpipe without airflow forms a slope equal to the angle of repose (α) of the solids.

Solids at the open end of the supply chamber are subjected to gravity force, which is balanced by reaction forces at their points of contact and that between the particles and walls below it. The horizontal component of the reaction forces will cause particles to move sideways. Unlike water, these particles cannot move out freely due to interparticle friction, which is a function of particle shape and size. Interparticle friction is smallest for smooth spherical particles. For this reason different particles will move out to different extents depending on their physical characteristics. Table 13.1 lists the angle of repose of several particles. The angle is related to the coefficient of friction as follows

$$f = \tan \alpha \tag{13.1}$$

Air percolating through the particles acts like a lubricant as its drag reduces the interparticle friction. Higher airflow rates reduce the friction and therefore the angle of repose α. Figure 13.9 shows how α decreases with airflow rate through the standpipe. The angle reduces until the airflow rate reaches the minimum fluidizing value of the particles, at which point the particles move like fluidized solids or liquids. In laboratory-scale units the thickness of the partition between recycle and supply chambers is negligible. Thus the solid flow through it is similar to fluid flow through an aperture. In large industrial units, the partition will have a finite thickness and therefore, solid flow through it must be examined as a flow through a horizontal duct.

TABLE 13.1
Angle of Repose of Some Type and Sizes of Particles

Sand Type and Size	Silica Sand	Silica Sand	Glass Beads
Size (μm)	244	113	262
Angle (α) degree	35	34	27
Sphericity comparison	Less spherical	Less spherical	More spherical
U_{mf} (m/s)	0.059	0.013	0.067

Source: From Botsio, E.P., 2004.

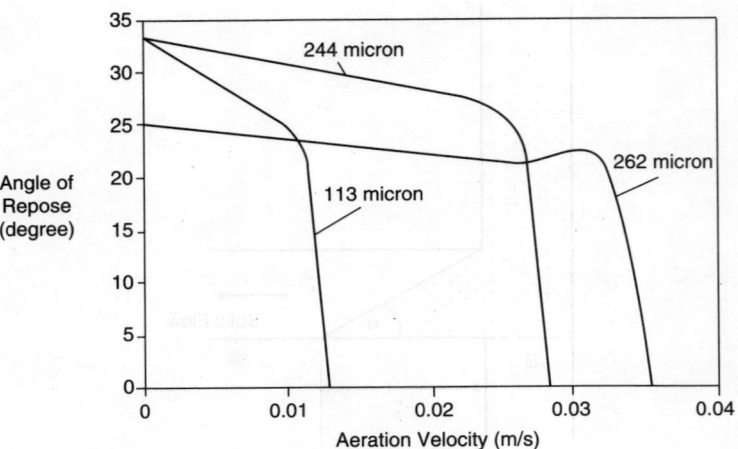

FIGURE 13.9 The angle of repose of different particles decreases with aeration rates.

Figure 13.4 explains the working of a loop-seal in greater detail. The loop-seal is made of two chambers; the *supply chamber* and *recycle chamber*. The supply chamber is at the base of the standpipe or downcomer of the cyclone. A free-flowing liquid will easily flow from the standpipe to the riser through a loop-seal like shown here, but granular solids would not flow that easily due to their inter-particle frictional forces. It is especially difficult in cases of large units where the horizontal passage in the loop-seal is long. For proper function of this type of loop-seal some air must be added into the supply chamber or standpipe. It may, however be noted that this air does not carry the solids. It simply reduces the inter-particle force allowing the solids to move under gravitational force. The flow of solids G_s, through the loop-seal is affected by the ratio of slit height to slit length (H_s/L_s) rather than only the slit height H_s, or slit length L_s.

13.2.2 Flow of Solids through Recycle Chamber

We see from above how solids from the standpipe and supply chamber flow into the recycle chamber. One of the walls of the recycle chamber is considered to be a *weir* as its height is below that of all other walls of the chamber (Figure 13.4). The recycle chamber is kept in a fluidized state by feeding air through its bottom. As expected from the definition of fluidization, the solids in the chamber behave like a liquid. So when the level of fluidized solids in the chamber rises above the weir, they overflow as a liquid would. The fluidized bed expands above the weir height P (Figure 13.4), due to the presence of bubbles, so the fluidized solids overflow into the recycle pipe or directly into the riser. The inclined recycle pipe, if used to convey solids from the loop-seal to the riser, is generally not filled with solids.

The flow of solids over the weir in the recycle chamber is similar to that over a regular sharp-crested weir, due to its shape and geometry. Therefore, one should be able to predict all hydrodynamic parameters using the sharp-crested weir theory. Basu and Botsio (2005) found the following empirical relation relating flow rate through the loop-seal W_s, and with the height of the expanded bed above the weir H, as:

$$W_s = \frac{2}{3}(1-\varepsilon)\rho_p CW\sqrt{2g} \times H^{3/2} \tag{13.2}$$

where C is in the range of 0.065 to 0.08 for 150 to 250 μm particles, while W and H are the width and height of the weir, respectively.

Solid Recycle Systems

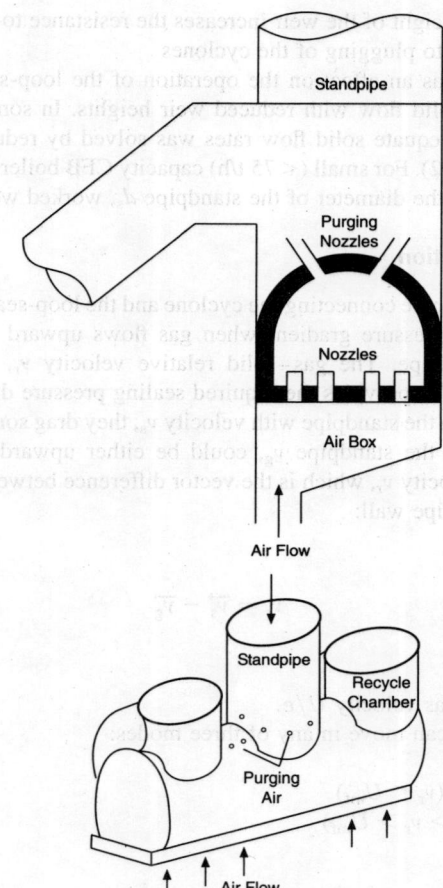

FIGURE 13.10 Sketch of the loop-seal of a commercial CFB boiler.

For proper operation of nonmechanical valves, the bulk of the aeration air must flow in the desired direction of solid flow. The pressure balance around the valve governs this. The air produces a drag on the solids, which upon exceeding the resistance to the solid flow in the bend, causes the solids to flow. This principle is applicable to all nonmechanical valves described.

When the recycle chamber is fluidized and bubbling, the solids in the chamber flow like a liquid. The fluidized bed expands above the weir height P (Figure 13.4), and the solids flow into the riser through the inclined recycle pipe.

A small amount of aeration at selected points as shown in Figure 13.10 could help reduce the friction between the wall and the particles, which in turn would help move the solids better. Excessive aeration may increase the operating cost as loop-seal air comes at high head.

13.2.2.1 Height of Solids in the Recycle Chamber

The height $(P + H)$ of fluidized solids in the recycle chamber provides the seal against any fluctuation in riser pressure and guards against any sudden blow-up of solids from the loop-seal in the event of an accidental pressure surge. It also provides the pressure seal preventing air from the riser to shortcut through the standpipe. Thus, a larger value of the weir height P, could provide better protection. A higher height of the weir increases the resistance to the solid flow through the loop-seal, which could lead to plugging of the cyclones.

This height, however, has an effect on the operation of the loop-seal. Some operating units (Johnk, 2002), had better solid flow with reduced weir heights. In some large commercial CFB boilers, the problem of inadequate solid flow rates was solved by reducing the height H, of the recycle chamber (Johnk, 2002). For small (<75 t/h) capacity CFB boilers, Cheng (2005) found that a loop-seal with H equal to the diameter of the standpipe d_s, worked well.

13.2.2.2 Standpipe Operation

The standpipe is the vertical pipe connecting the cyclone and the loop-seal. Solids are transferred by gravity against an adverse-pressure gradient when gas flows upward relative to the downward moving solids in the standpipe. The gas–solid relative velocity v_r, is important because the frictional pressure drop $\Delta P/L$, provides the required sealing pressure drop.

When solids travel down the standpipe with velocity v_s, they drag some air with them. Thus, the absolute velocity of gas in the standpipe v_g, could be either upwards or downwards. What is important is the relative velocity v_r, which is the vector difference between solid and gas velocities with reference to the standpipe wall:

$$v_r = \vec{v_s} - \vec{v_g} \tag{13.3}$$

where v_g is the interstitial gas velocity U/ε.

Solids in the standpipe can move in any of three modes:

1. Moving packed bed ($v_r < U_{mf}$)
2. Fluidized bed ($U_{mb} > v_r > U_{mf}$)
3. Dilute phase

The dilute phase occurs when the solid concentration is too low compared to the gas flow rate though the standpipe.

13.3 DESIGN OF LOOP-SEAL

Figure 13.10 presents a sketch of the loop-seal of a typical commercial CFB boiler. Although this construction seems much different from the simplified version shown in Figure 13.2, they are essentially the same in design. In this unit, the standpipe joins a half-cylindrical vessel into which two vertical cylindrical pipes are connected on two ends forming two cylindrical chambers which serve as recycle chambers. The fluidized solids overflow into two inclined pipes connected to these two recycle vessels. The cylindrical vessels are used here primarily for construction convenience.

Sizing of the loop-seal is very important because a small loop-seal may not allow adequate solid flow if the system demands a little extra. Too large a loop-seal would require a larger volume of loop-seal air at a great cost because this air needs a very high head. For example, if a typical CFB boiler needs 16,000 Pa for the primary air fan, it may need about 50,000 Pa for the loop-seal air fan.

13.3.1 PRESSURE BALANCE

The loop-seal abides by a pressure balance as shown in Figure 13.7. In CFB boilers, the loop-seal is designed so that the pressure at the exit of the loop-seal is greater than the pressure at the lowest elevation of the furnace, where solids are recycled. Neglecting the pressure drop through the exit

Solid Recycle Systems

pipe of the loop-seal, the pressure balance can be written as (Figure 13.2):

$$P_s - \Delta P_{dist,s} - \Delta P_{sb} > P_0 - \Delta P_{dist,b} - \Delta P_b \quad (13.4)$$

where:

P_s: air pressure in the plenum of the loop-seal
P_0: air pressure in the plenum of the main bed
$\Delta P_{dist,s}$: pressure drops across the distributor of the loop-seal
$\Delta P_{dist,b}$: pressure drops across the distributor of the main bed
ΔP_b: pressure drops across the riser bed below the level where recycle solids enter
ΔP_{sb}: pressure drops across the fluidized bed in the loop-seal

The pressure drop across the fluidized bed in the recycle chamber ΔP_{sb}, can be conservatively estimated as:

$$\Delta P_{sb} = (1 - \varepsilon_s)\rho_p H_s g \quad (13.5)$$

where the voidage of the loop-seal ε_s, is taken as 0.5. H_s is the height of solids in the seal pot above its distributor plate.

The lower section of the furnace where solids enter from the loop-seal generally remains in a turbulent fluidized condition. For design purposes, ΔP_b, the pressure drop through this section is found from:

$$\Delta P_b = (1 - \varepsilon_t)\rho_p H_t g \quad (13.6)$$

where H_t is the height of the turbulent bed above the lowest point of the solid recycle port. The voidage of the turbulent bed ε_t, may be taken as 0.8.

13.3.2 Size of Loop-Seal

The diameter of the return leg or standpipe d_s, should ensure a smooth flow of solids through the standpipe without any choking. Based on their experience, Luo et al. (1989) recommended an empirical guide for the solid velocity V_s:

$$V_s = 1.6\, d_s^{0.5} \quad \text{in SI units for } d_s < 0.2 \text{ m} \quad (13.7)$$

Here, V_s should be less than 0.20 m/s for $d_s < 0.5$ m. Interestingly, this value is very close to that used for the design of standpipes in FCC reactors. Based on his experience with the design of a large number of CFB boilers in the range of 35 to 230 t/h capacities, Cheng (2005) suggests a solid velocity of about 0.025 m/s, or use of a diameter close to that of the bottom of the cyclone arrived at from the design of the cyclone.

Wirth (1995) studied the flow of solids through a standpipe connecting a cyclone and a loop-seal. He developed a correlation between the pressure gradient $\Delta P/\Delta h$, in the standpipe and the solid mass flow W_s, through it:

$$\frac{\Delta P}{(\rho_p - \rho_g)g\Delta h} = 0.22 Tr^{0.7} \quad (13.8)$$

where

$$Tr = \frac{W_s}{\rho_p(1-\varepsilon_{mf})}\left[\frac{\rho_g}{(\rho_p-\rho_g)gd_p}\right]^{0.5} < 1.0;$$

and

$$\frac{\rho_g(\rho_p-\rho_g)gd_p^3}{\mu^2} > 200$$

The pressure drop across the standpipe ΔP, is fixed by the pressure balance across the rest of the CFB loop. Thus, once this and the external circulation rate are known, the standpipe inner diameter d_s, can be found.

Other dimensions of the loop-seal should be chosen to accommodate the standpipe while minimizing the size of the seal. In the absence of any other constraints, the following guidelines may be used:

Length of the loop-seal, $L = 2.5\, d_s$ Width of the loop-seal, $W = 1.25\, d_s$ (13.9)

The length of the loop-seal may be divided into two halves by vertical partitions, with an opening at its bottom for solid flow between the chambers. The opening may cover the width of the loop-seal. Its height can be designed on the basis of the horizontal solid velocity V_h (in the range of 0.05 to 0.25 m/s), subject to a minimum height of 10 times the diameter of the coarsest particle recycled. Cheng (2005) suggests this velocity to be double the solid descending velocity in the standpipe. The deeper the loop-seal and the greater its fluidizing velocity, the higher is the allowable horizontal velocity. Some commercial units do not, however, use this partition. The flow through the loop-seal is set by a one-time adjustment of the fluidizing air through its nozzles. This adjustment is not usually changed:

$$\text{Height of the opening } h_s = \frac{W_s}{(1-\varepsilon_s)\rho_p V_h W} \quad (13.10)$$

where W_s is the solid circulation rate in kg/s.

13.3.3 AIR-FLOW THROUGH LOOP-SEAL

13.3.3.1 Supply Chamber or Standpipe

Theoretically, no air is required for fluidizing the supply chamber at the feet of the standpipe or the cyclone downcomer. In large industrial units the diameter of the standpipe, and hence the supply chamber, is so large that particles may not easily flow on their own even if the entire recycle chamber is empty. Some air is required through the base and from around the lower periphery of the standpipe. A certain amount of air from the recycle chamber invariably leaks into the supply chamber allowing the solids to move down and sideways in this chamber. Air to the supply chamber should not exceed the minimum fluidization velocity for the average particle size of the circulating solids. If that happens the pressure seal around the loop-seal will break down and the normal operation of the loop-seal will be interrupted. Such an occurrence may lead to an occasional surge of a large mass of solids dropping into the riser.

13.3.3.2 Recycle Chamber

The recycle chamber of the loop-seal is kept fluidized at a velocity V_{seal}, greater than 1.25 times the U_{mf} of the larger particles in the loop-seal. Muschelknautz and Greif (1997) suggest that the fluidizing velocity in the recycle chamber be based on the terminal velocity of the average particles according to the following equation:

$$V_{seal} = (0.033 \text{ to } 0.2) U_t^* \tag{13.11}$$

The terminal (settling velocity) U_t^*, of 50% cut sized particle of the cyclone d_{50}, is given as $\rho_p g d_{50}^2/18\mu$. It may be noted that for fine particles, minimum fluidization velocity $U_{mf} \sim 0.025\, U_t^*$.

The pressure drop between the supply chamber and the exit to the riser must exceed probable pressure fluctuations P_b', in the lower bed. The discharge from a normal centrifugal fan drops if the flow resistance increases for any reason. So, to guard against any possible loss in fluidization in the recycle chamber, many commercial CFB boilers use high pressure positive displacement blowers.

Experience suggests that the amount of airflow should be sufficient to fluidize all parts of the recycle chamber, but inadequate for vigorous bubbling (Cheng, 2005).

The exit pipe of the loop-seal carries solids at a much higher velocity but at lower density than that in the standpipe. Its cross section should thus be at least equal to that of the standpipe. Also, its inclination must exceed the angle of repose of the bed materials.

13.3.3.3 Effect of Loop-Seal Air on Recirculation Rate

The solid circulation (mass flux) through a loop-seal is affected by several operating and design parameters. It increases with (Basu and Cheng, 2000):

1. Increasing fluidizing velocity through the recycle chamber
2. Increasing aeration on the vertical wall of the supply chamber
3. Higher solid inventory in system
4. Larger standpipe size
5. Finer particle size
6. Higher gas velocity in the riser
7. System pressure (Cheng and Basu, 1999)

Figure 13.11 shows the variation of solids circulation rate with loop-seal fluidizing air as well as riser gas velocity.

It is apparent from the figure that solids do not flow until the loop-seal airflow rate exceeds a certain value. Johansson et al. (2003) noted that the solid recirculation rate increases with increased fluidization velocity in the recycle chamber and with the gas velocity in the riser.

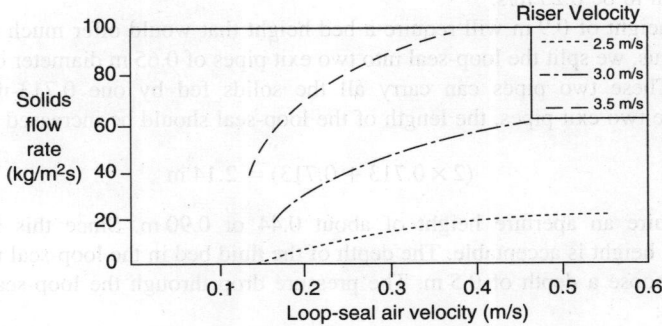

FIGURE 13.11 Experimental data on the effect of loop-seal air on the solid mass flow of different sized particles through it.

Example 13.1

> Design a loop-seal to circulate 250 kg/s solids, the largest particle size of which is 0.5 mm, to be fed to a point 2.5 m above the grid of the furnace.

Solution

From Equation 13.7, we can find the solid velocity in the standpipe for smooth flow. By multiplying that with the cross-sectional area and solid density in the standpipe, we find the circulation rate:

$$W_s = (\pi/4) d_s^2 (1 - \varepsilon_s) \rho_p 1.6 d_s^{0.5}$$

$$d_s^{2.5} = \frac{250 \times 4}{3.14 \times (1 - 0.5) \times 2500 \times 1.6} = 0.159$$

Solving this we get:

$$d_s = 0.479$$

$$V_s = 1.6(0.479)^{0.5} = 1.107 \text{ m}$$

This exceeds the allowable limit of solid velocity 0.2 m/s (Section 13.3.2) for $d_s < 0.5$ m. For a larger bed the limit could be higher. For illustration we recalculated the diameter of the standpipe on the basis of the maximum value of solid velocity, 0.5 m/s:

$$d_s = \left[\frac{250 \times 4}{0.5 \times (1 - 0.5) \times 2500 \times 3.14} \right]^{0.5} = 0.713 \text{ m}$$

Using the proportion suggested in Equation 13.9:

Length of the seal: $L = 2.5 \times 0.713 = 1.78$ m
Width of the seal: $W = 1.25 \times 0.713 = 0.89$ m

The height of the aperture between the two sections of the seal using Equation 13.10:

$$h_s = \frac{250}{(1 - 0.5) \times 2500 \times 0.25 \times 0.89} = 0.90 \text{ m}$$

where V_h is taken to be 0.25 m/s.

An aperture height of 0.9 m will require a bed height that would offer much resistance to the fluidizing air. Thus, we split the loop-seal into two exit pipes of 0.65 m diameter on either sides of the standpipe. These two pipes can carry all the solids fed by one 0.713-m standpipe. To accommodate the two exit pipes, the length of the loop-seal should be increased to

$$(2 \times 0.713 + 0.713) = 2.14 \text{ m}$$

This would require an aperture height of about 0.44 or 0.90 m. Since this is much greater than $10 \times d_p$ this height is acceptable. The depth of the fluid bed in the loop-seal must exceed this height. So we choose a depth of 0.5 m. The pressure drop through the loop-seal is found from Equation 13.5 as:

$$\Delta P_{sb} = (1 - 0.5) \times 2500 \times 0.5 \times 9.81 = 6131 \text{ Pa}$$

where the bed voidage is assumed to be 0.5.

Solid Recycle Systems

The pressure drop through the main bed below the recycle point is determined from Equation 13.6:

$$\Delta P_{tb} = (1 - 0.8) \times 2500 \times 2.5 \times 9.81 = 12,263 \text{ Pa}$$

where the voidage of the turbulent bed is assumed to be 0.8.

Assuming the distributor pressure drops through the main bed and the loop-seal are identical, the air pressure at the plenum of the loop-seal is:

$$P_s = P_0 - 12,250 + 6131 = P_0 - 6118$$

The blower and ducting for the loop-seal should thus be designed such that air pressure in the loop-seal air box will be no less than 6118 Pa below that in the air box of the main bed. In fact, in order to overcome the possible pressure surge in the main bed, the loop-seal fan should have a much higher head than this.

The minimum fluidizing velocity for a 0.5 mm particle at 825°C is about 0.08 m/s (estimated from Chapter 2). The fluidizing velocity may thus be chosen as $3 \times U_{mf}$, which is 0.55 m/s.

The total air requirement for the loop-seal is found to be $0.55 \times 1.78 \times 0.89 = 0.87$ m³/s at the operating temperature of the loop-seal.

13.4 DESIGN OF L-VALVE

Design methods of L-valve can be developed from a pressure balance across the CFB loop. From Figure 13.3 we note that the pressure drop across the standpipe must be balanced by the pressure drop across rest of the CFB loop. So, we can write:

Pressure drop through standpipe

= pressure drop through (L-valve bend + horizontal arm of L-valve + furnace + cyclone)

$$L_{min} = \frac{(\Delta P_{lv} + \Delta P_h + \Delta P_{fb} + \Delta P_{cy})}{\left(\dfrac{\Delta P_{sp}}{L_{sl}}\right)_{mf}} \qquad (13.12)$$

where $L_{min}(\Delta P_{sp}/L_{sl})_{mf}$ is the maximum pressure drop one can have in the standpipe. This could occur when the standpipe is at minimum fluidization. ΔP_{lv}, ΔP_h, ΔP_{fb}, and ΔP_{cy} are pressure drops through L-valve, horizontal section and fast bed, and cyclone, respectively. This balance gives L_{min}, the minimum height of solids in the standpipe. The height of solids in the vertical column is generally 1.5 to 2 times L_{min}.

Solids in the standpipe are in a moving packed-bed condition. So the pressure drop through the standpipe will rise if the relative velocity between the solids and the gas is increased. The pressure drop will rise until it is equal to the weight of solids in the standpipe, that is, when a minimum fluidizing condition in the standpipe is reached. If U_s is the superficial velocity of the downward flow of solids through the standpipe and U_g is the superficial velocity of the air in the standpipe, the relative velocity U_r, can be written as:

$$U_r = \frac{U_g}{\varepsilon_{sp}} - \frac{U_s}{(1 - \varepsilon_{sp})} \qquad (13.13)$$

The maximum allowable U_r can reach is U_{mf}. The maximum resistance offered by the standpipe will be that under minimum fluidizing conditions, i.e., $\varepsilon_{sp} = \varepsilon_{mf}$, and is found by:

$$\Delta P_{sp}|_{max} = L_{sl}(1 - \varepsilon_{mf})\rho_p g \qquad (13.14)$$

where L_{sl} is the height of the solid column in the standpipe and ε_{mf} is the voidage in the minimum fluidization condition.

The pressure drop through the bend of the L-valve was correlated by Luo et al. (1989) as:

$$\Delta P_{lv} = K\rho_{bs}\left(\frac{D_L}{d_p}\right)^{-0.25} V_s^{0.45} \quad \text{Pa} \qquad (13.15)$$

where ρ_{bs} is the bulk density of solid carried in the standpipe in kg/m³, D_L is the diameter of the L-valve in m, and d_p is the average diameter of solids transported in m. The constant, K was determined by Luo et al. (1989) for sand (Group B) particles to be 13.8 in SI units. One should exercise caution in using this equation especially for L-valves larger than 0.25 m in diameter as the above expression was developed based on laboratory-scale units.

The upper limit of the superficial velocity of solids V_{sl} through the L-valve may be calculated from the empirical equation (Luo et al., 1989) in SI units:

$$V_{sl} = 0.95\sqrt{D_L} \quad \text{m/s} \qquad (13.16)$$

The pressure drop between the aeration point and the horizontal section ΔP_{lv} was also correlated by Wen and Simons (1959) from their experiments as:

$$\Delta P_{lv} = 4.266 L\rho_{bs}g\left(\frac{D_L}{d_p}\right)^{-0.25} V_s^{0.45} \quad \text{in SI units} \qquad (13.17)$$

where L is the horizontal length of the L-valve.

An alternative expression for a Group B particle is (Geldart and Jones, 1991):

$$\Delta P_{lv} = 216 \frac{G_s^{0.17}}{D_L^{0.63} d_p^{0.15}} \quad \text{N/m}^2 \qquad (13.18)$$

The pressure drop through the bed below the solid entry point ΔP_b, is known from the solid inventory in the bed. The pressure drop through the cyclone ΔP_{cy}, is found from (Tung and Williams, 1987):

$$\Delta P_{cy} = 26 \frac{\rho_g Q^2}{D_c^4} \qquad (13.19)$$

where Q is the volumetric flow rate of flue gas passing through the cyclone, D_c is the diameter of the cyclone body, and ρ_g is the density of flue gas at the cyclone inlet temperature.

13.4.1 MAXIMUM SOLID FLOW RATE

In a CFB, the maximum flow rate through the L-valve is reached when the sum of pressure drops through the bed above the recirculation pipe, across the L-valve, and cyclone approaches the pressure drop through the standpipe under the minimum fluidized condition. The linear velocity of solids down the standpipe is another limiting factor. A superficial solid velocity as high as 0.5 m/s has been attained in laboratory L-valves. However, for most practical applications it may not exceed 0.3 m/s (Knowlton, 1988).

Solid Recycle Systems

13.4.2 Additional Design Considerations

In addition to the above, the following practical considerations are important to the design of an L-valve:

1. The solid velocity through the L-valve calculated by Equation 13.7 may give too high a value for large-diameter L-valves, but it should not generally exceed 0.2 m/s.
2. The length of the horizontal section of the L-valve should preferably lie within the range of (1.5 to 5) D_L and (8 to 10) D_L. Luo et al. (1989) suggested the minimum length of the horizontal section be $2D_L \cot \alpha$, where α is the angle of repose of the particle conveyed. The minimum length ensures that solids do not flow when the aeration stops. If the cyclone and the furnace are located such that a longer horizontal length of the L-valve is required, additional fluidizing air through the base of the horizontal arm as well as aeration air around the standpipe may help the flow of the solid.
3. A single aeration point may serve L-valves up to 0.3 m in diameter, but for larger valves at least four aeration points around its circumference may be needed (Knowlton, 1988).
4. The aeration point should be at a height of about $2D_L$ above the center of the horizontal axis of the L-valve.

13.5 PRACTICAL CONSIDERATIONS

The loop-seal or seal pot is the heart of a circulating fluidized bed gasifier or boiler. If it stops or malfunctions, solids accumulate in the standpipe and cyclone, choking the entire system. Under typical malfunction conditions, there may be an interruption in the flow of solids from the cyclone to the riser. Then, a large slug of solids may suddenly flow into the bed creating a massive fluctuation in bed pressure. Sometimes solid inventory in the loop-seal may be too low to maintain a backpressure in the riser, causing riser gas to take short-cut into the cyclone through the standpipe. Such a short circuit invariably reduces the efficiency of the cyclone.

13.5.1 Plugging of Loop-Seal

Plugging of the loop-seal has been observed in the following cases:

1. Bed solids are too flaky to flow freely through the loop-seal.
2. Boiler burns fuels that makes the solids sticky, affecting their free flow through the loop-seal.

In addition the plugging has been observed due to the formation of small agglomerates in the bed. This could happen if there are pockets of less fluidized solids and the temperature of the solids is above 900°C. At such high temperatures solids at the base of the standpipe or the supply chamber could undergo sintering owing to the high static pressure to which they are exposed. Solids move in the standpipe as a moving fixed bed. Heat dissipation from the particle is minimal in this section, and the particles remain in contact with each other for a long time. These conditions create favourable conditions for sintering.

Agglomerations could also occur in the bubbling bed external heat exchangers of large CFB units, where the partition walls may be sufficiently wide to allow formation of defluidized zones on top of the partition walls.

Many plants experienced plugging of loop-seals when for some reason the plant was required to feed a higher than designed amount of limestone. Reducing the weir height in the supply chamber helped in some cases. Even that was not adequate in other cases.

13.5.2 Pressure Surge

If the external solid circulation rate increases due to changes such as improved cyclone efficiency, higher solid inventory in the collective size range and pressure surge may be noticed in loop-seals that are not designed for high solid rates. This surge may in the worst case lead to a complete stoppage of solid flow through the loop-seal, causing the cyclone to fill up with solids.

The exact cause for this is not studied in detail. One can speculate that as solid flux through the standpipe or cyclone downcomer increases, the density of solids in it decreases due to more gas being brought in by the solids. This reduces the solid flow from the loop-seal to the riser, which in turn reduces the suspension density at the entry of the cyclone, reducing the amount of solids collected. This increases the packing density of solids in the standpipe, increasing solid flow into the standpipe, which results in higher solid flow into the riser. Thus the flow from the loop-seal rises and falls creating the surge.

This can be avoided by using a more liberal size of loop-seal with a larger standpipe because for the given flow rate, a larger area will give lower mass flux. Alternately one can reduce the height of loop-seal weir or the fluidized solids in the recycle chamber. This will reduce the resistance of the loop-seal and hence increase its solid-carrying capacity. However, this makes it vulnerable to pressure surge in the riser.

13.5.3 Avoiding Loop-Seal Plugging

Two possible means for solving the above problem are discussed below. They could be used either individually or in combination.

13.5.3.1 Mechanical Means

It was mentioned earlier in Section 13.2.1 that the free flow of solids is the key to the smooth operation of a loop-seal, which can be ensured by reduced inter-particle friction. A wider and larger loop-seal may help smooth the flow of solids. Increased aeration in strategic points around the loop-seal also helps.

13.5.3.2 Chemical Means

A number of CFB boilers have experienced malfunction of their loop-seal while attempting to burn 100% PetCoke. While this could be due to the increased solid recycling necessary for greater Petcoke firing, the formation of agglomerates or sintering is also a distinct possibility. PetCoke contains a high level of vanadium (0.15%), which could either form a low-temperature sintering mixture with free CaO, or form V_2O_5 which melts at 673°C (Zierold and Voyles, 1993). Irbane et al. (2003) argued against the vanadium-driven agglomeration.

Calcium sulfate ($CaSO_4$) is expected to exhibit sintering in the temperature region of 600 to 1100°C when the ratio of absolute sintering temperature to absolute melting temperature lies between 0.5 and 0.8 (German, 1996). This could affect the free flow of solids through the loop-seal. Besides this, sintering associated with the conversion of CaO to $CaSO_4$ could also start in the loop-seal until all CaO is depleted at which time the surface sintering takes over (Irbane et al., 2003).

Spent sorbent moves in the standpipe as a moving packed bed, where particles remain in touch with one another for a finite time. During this period spent limestone particles are subjected to hydrostatic pressure due to solids inventory in the standpipe. This could enhance the sintering potential in the loop-seal while working with excessive amounts of sorbents. This could also reduce the flow of solids through the loop-seal leading to plugging of the cyclone.

NOMENCLATURE

C: coefficient in Equation 13.2
D_c: diameter of cyclone body, m
D_L: diameter of L-valve, m
d_{50}: diameter of particles with 50% collection efficiency, m
d_p: diameter of particle, m
d_s: diameter of the standpipe
f: coefficient of inter-particle friction
G_s: solid recycle rate, kg/m² s
g: acceleration due to gravity, 9.81 m/s²
H: expanded bed height above the weir, m
H_s: height of fluidized solids in the loop-seal, m
H_t: height of turbulent bed above the exit pipe of loop-seal, m
h: height of solids in the standpipe, m
K: constant in Equation 13.13
L: length of the loop-seal, m
L_{min}: minimum height of solid in L-valve, m
L_s: horizontal length of the loop-seal, m
L_{sl}: height of solids in the standpipe, m
P: weir height, m
P'_b: pressure fluctuations in lower bed
P_0, P_s: pressure in the plenums of main bed and loop-seal, respectively, N/m²
ΔP: pressure drop across standpipe, N/m²
ΔP_{cy}: pressure drop across cyclone, N/m²
ΔP_b: pressure drop across bubbling bed in riser below the solid entry point, N/m²
$\Delta P_{dist,b}$: pressure drop through distributor of fast bed, N/m²
$\Delta P_{dist,s}$: pressure drop through distributor of loop-seal, N/m²
ΔP_{lv}: pressure drop through L-valve, N/m²
ΔP_h: pressure drop through the horizontal section of L-valve, N/m²
ΔP_{fb}: pressure drop through the fast bed above the solid entry, N/m²
ΔP_{sb}: pressure drop across bubbling bed in loop-seal, N/m²
ΔP_{sp}: pressure drop per unit height of solid in L-valve, N/m²
ΔP_{tb}: pressure drop across the turbulent bed, N/m²
Q: volumetric flow rate of flue gas passing through cyclone, m³/s
U_{mf}: minimum fluidization velocity, m/s
U_r: gas–solid velocity through the standpipe m/s
U_g: downward superficial gas velocity in the standpipe, m/s
U_{mb}: minimum bubbling velocity, m/s
U_s: downward superficial solid velocity, m/s
U_t^*: terminal velocity of average particles, d_{50}^*, m/s
Q: gas flow rate through the cyclone, m³/s
v_g: absolute velocity of gas in standpipe, m/s
V_h: solid transfer velocity in loop-seal, m/s
V_s: solid velocity through standpipe, m/s
V_{seal}: fluidization velocity in recycle chamber, m/s
V_{sl}: solid velocity through L-valve, m/s
v_r: difference between solid and gas velocities with reference to standpipe wall, m/s
W: width of loop-seal or weir, m
W_s: solid circulation rate through loop-seal, kg/s

GREEK SYMBOLS

α:	angle of repose
ρ_{bs}:	bulk density of solid conveyed, kg/m^3
ρ_b:	bed density, kg/m^3
ρ_g:	density of gas at bed temperature, kg/m^3
ρ_{gor}:	density of air at orifice temperature, kg/m^3
ρ_p, ρ_s:	density of solid, kg/m^3
ε:	voidage of the bed
ε_{mf}:	voidage at minimum fluidization
ε_s:	voidage in the loop-seal
ε_{sp}:	voidage in the standpipe
ε_t:	voidage in the turbulent bed
μ:	dynamic viscosity of gas, kg m s

REFERENCES

Basu, P. and Botsio, E., Experimental investigation into the hydrodynamics of flow of solids through a loop-seal recycle chamber, *CSChE J.*, 83(3), 554–558, 2005.

Basu, P. and Cheng, L., An analysis of loop-seal operating in a circulating fluidized bed, *Trans. IChemE*, 78(part A), 991–997, October, 2000.

Botsio, E. P., *Hydrodynamic Studies of the Loop-seal of a Circulating Fluidized Bed Boiler*, Master of Applied Science Thesis, Dalhousie University, 2004.

Cheng, L. and Basu, P., Effect of pressure on loop-Seal operation for a pressurized circulating fluidized bed, *Powder Technol.*, 103, 203–211, 1999.

Cheng, L., Personal communications, 2005.

Chong, Y. O., O'Dea, D. P., Leung, L. S., and Nicklin, D. C., Design of stand pipe and non-mechanical v-valves for CFB, In *Curculating Fluidized Bed Technology—II*, Basu, P. and Large, J. F., Eds., Pergamon Press, Oxford, pp. 493–499, 1988.

German, R. M., *Sintering Theory and Practice*, Wiley, New York, 1996.

Irbane, J. V., Anthony, E. J., and Irbane, A., A scanning electron-microspcope study on agglomeration in petroleum coke fired FBC boiler, In *Proceedings of the 17th International FBC Conference*, Pisupati, S., Ed., ASME, New York, FBC2003-082, 2003.

Geldart, D. and Jones, P., The behavior of L-valves with granular solids, *Powder Technol.*, 67, 163, 1991.

Johnk Engineering, Discussion group, www.johnkengineering.com/discus/, 2002.

Johansson, A., Johnsson, F., and Andersson, B., The performance of a scaled down fluidized loop seal, In *Proceedings of the 17th International Conference on Fluidized Bed Combustion*, Pisupati, S., Ed., ASME, New York, FBC2003-082, 2003.

Knowlton, T. M., Non-mechanical solid feed and recycle devices for CFB, In *Circulating Fluidized Bed Technology II*, Basu, P. and Large, J. F., Eds., Pergamon Press, Oxford, pp. 31–42, 1988.

Knowlton, T. M., Standpipes and return systems, In *Circulating Fluidized Beds*, Grace, J. R., Avidan, A. A. and Knowlton, T. M., Eds., Blackie Academic and Professionals, London, pp. 214–260, 1997.

Luo, Z., Ni, M., Zhou, J., Cheng, L., Chang, Z., and Cen, K., Solid recycle system for CFB, In *Proceedings of the 10th International Conference on Fluidized Bed Combustion*, Manaker, A., Ed., ASME, New York, pp. 557–562, 1989.

Muschelknautz, E. and Greif, V., Cyclone and other gas-solids separators, In *Circulating Fluidized Beds*, Grace, J. R., Avidan, A. A. and Knowlton, T., Eds., Blackie Academic, New York, pp. 181–213, 1997.

Tung, S. E. and Williams, G. C., *Atmospheric Fluidized Bed Combustion — A Technical Source Book*, US DOE, DOE/MC/14536-2544, pp. 5–10, 1987.

Wen, C. Y. and Simons, H. P., Flow Characteristic in Horizontal Fluidized Bed Transport, *AIChE J.*, 2, 263, 1959.

Wirth, K. E., Fluid mechanics of the downcomer in circulating fluidized beds, In *Eighth International Conference on Fluidization*, Laguerie, C. and Large, J. F., Eds., AIChE, New York, pp. 567–576, 1995.

Zierold, D. M. and Voyles, R. W., NISCO cogenerating facility, In *Proceedings of the 12th International Conference on Fluidized Bed Combustion*, Rubow, L., Ed., ASME, New York, pp. 501–510, 1993.

Appendix 1
Characteristics of Solid Particles

A particle may be defined as a small object having a precise physical boundary in all directions. The particle is characterized by its volume and interfacial surface in contact with the environment.

A1.1 SOLID PARTICLES

Solid particles are rigid and have a definite shape. A sphere is a natural choice to define a particle, though most natural particles are not spherical. Hence, natural particles are characterized by their degree of deviation from spherical shape, sphericity, and an equivalent diameter.

A1.1.1 Equivalent Diameters

Let us take a nonspherical particle having a surface area S, and a volume V. Several types of equivalent diameter of the particle can be defined to describe the particle, as shown in Figure A1.1. Four more frequently used definitions are:

A1.1.1.1 Volume Diameter (d_v)

Volume diameter is the diameter of a sphere that has the same volume as the particle:

$$d_v = \left[\left(\frac{6}{\pi}\right) \times \text{volume of particle}\right]^{\frac{1}{3}} = \left[\left(\frac{6V}{\pi}\right)\right]^{\frac{1}{3}} \quad (A1.1)$$

A1.1.1.2 Surface Diameter (d_s)

Surface diameter is the diameter of a sphere that has the same external surface area as the particle. Thus,

$$d_s = \left[\frac{\text{surface area of particle}}{\pi}\right]^{\frac{1}{2}} = \left[\frac{S}{\pi}\right]^{\frac{1}{2}}. \quad (A1.2)$$

A1.1.1.3 Sieve Size (d_p)

Sieve size is the width of the minimum square aperture of the sieve through which the particle will pass.

A1.1.1.4 Surface-Volume Diameter (d_{sv})

Surface-volume diameter is the diameter of a sphere having the same surface to volume ratio as that of the particle:

$$\frac{6\pi d_{sv}^2}{\pi d_{sv}^3} = \frac{S}{V}$$

$$d_{sv} = 6\frac{V}{S} \quad (A1.3)$$

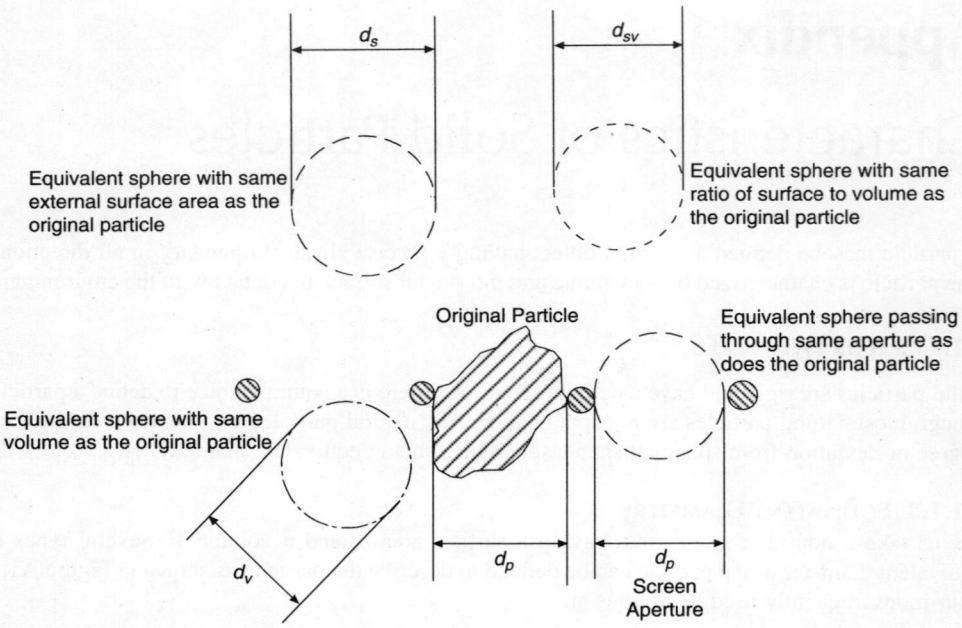

FIGURE A1.1 Different representations of a nonregular shaped particle.

A1.1.2 Sphericity (ϕ)

Sphericity describes the departure of the particle from a spherical shape. For example, a spherical particle has a sphericity of 1.0:

$$\text{Sphericity }(\phi) = \frac{\text{Surface area of a sphere with the volume same as the particle}}{\text{Actual surface area of the particle}} = \frac{\pi d_v^2}{S}. \quad \text{(A1.4)}$$

Eliminating S and V from Equation A1.1, Equation A1.3, and Equation A1.4, one gets:

$$d_{sv} = \phi d_v. \quad \text{(A1.5)}$$

The relationship between the above sizes and the sieve size d_p can be derived through experiments for irregular particles and through calculations for geometrically shaped particles. An approximate relation for crushed quartz of sphericity 0.8 was given as (Abrahamsen and Geldart, 1980):

$$d_v \approx 1.13\, d_p, \quad d_{sv} \approx 0.773\, d_v, \quad d_{sv} \approx 0.87\, d_p, \quad d_s \approx 1.28\, d_p$$

The sphericity is usually measured. Typical values of some commonly used particles are given in Table A1.1. Biomass particles often have very low sphericity. Characteristics of some typical particles are shown in Table A1.2.

A1.1.3 Mean Particle Size and Its Measurement

Millions of small particles are simultaneously handled in an industry for the purposes of reactions, heat and mass transfer, or homogeneity. In such a particulate mass, generally particles are not uniform in size and are characterized by particle size distribution.

Appendix 1: Characteristics of Solid Particles

TABLE A1.1
Sphericity of Some Granular Solids

Particle	Sphericity
Sand (Ottawa)	0.95
Sand (flint, jagged)	0.65
Sand (average of all types)	0.75
Limestone	0.45
Gypsum	0.40
Coal (crushed)	0.65
Coal (pulverized)	0.73
Alumina	0.3–0.8
Catalysts	0.4–0.9
Crushed glass	0.65
FCC catalyst	0.58

Source: *Coal Conversion System Data Handbook*. Table IVB10.1, DOE/FE/05157-2, 1982.

There are several characteristic properties that define a particulate mass:

- Number of particles
- Total surface area
- Total volume

It is difficult to provide individual attention to these properties, and hence it is necessary to define some average properties. Therefore, one finds it convenient to imagine an equivalent particulate mass of particles of uniform size that matches the properties of the actual particulate mass. However, it is possible to match only two properties between the actual and the equivalent. In fluidization and in most chemical engineering applications, total volume and surface area are the two chosen properties. These represent the material content and interfacial area across which transfer processes occur. For a pressure drop through the bed, the surface area is most important. The mean particle size is thus defined in such a way that it equals the average surface area of particles of sizes in the bed.

TABLE A1.2
Sphericity and Density of Some Biomass Fuels and Ash Produced from Them

Particle	Particle Density (kg/m^3)	Voidage at Minimum Fluidization	Sphericity
Biomass			
Saw dust	430	0.586	0.95
Rice husk	500	0.795	0.65
Ash			
Saw dust ash	380	0.603	0.75
Rice hull ash	410	0.678	0.45

Source: Chen et al., *Circulating Fluidized Bed Technology V*, Kefa, C. ed., International Academic Publishers, Beijing, p. 508, 2005.

Sieving is the most commonly used technique for the measurement of the surface area of granular solid particles. Particles of size greater than 44 microns are measured by using a set of standard test sieves with square aperture openings. The test sieves are stacked with the one with the largest aperture on the top. The lower sieves are selected such that the apertures are smaller. After vibrating and shaking the stack using a sieve shaker for a period of 20 to 30 minutes, the particles collected on each sieve are weighed and assigned a size by taking the arithmetical average of the aperture size of the sieve through which the material just passed and the sieve on which it is retained:

$$d_m = \frac{1}{\Sigma \left(\frac{x_i}{d_i}\right)}, \qquad (A1.6)$$

where d_i is the arithmetic mean of the aperture (opening) of two adjacent sieves, and x_i is the weight fraction of samples collected between these two sieves. The above equation will match the surface/volume ratio of the actual poly-size particles. For nonspherical particles, all having the same sphericity, ϕ, the mean size d_m, would then be ϕd_m. Equation A1.6 is, however, not valid for a discontinuous particle size.

In industries the particle size distribution is sometimes described by d_{50}, which is a size below which lies 50% of the sample by weight. The *relative size* range R, is another characteristic used to describe the spread of the size distribution. It is defined as

$$R = \frac{d_{84} - d_{16}}{2d_m}. \qquad (A1.7)$$

where d_{84} and d_{16} are the diameter corresponding to size below which particles constitute 84% and 16%, respectively, by weight.

A1.2 PACKING CHARACTERISTICS

In a particulate mass, particles rest on each other due to the force of gravity to form a *packed bed*. Depending on the shape of particles and packing characteristics, a certain volume of space in between the particles remains unoccupied. Such space is called a *void volume* and is specified as voidage or porosity, defined as

$$\text{Voidage, } \varepsilon = \text{porosity} = \frac{\text{void volume}}{\text{volume of (particles + voids)}}. \qquad (A1.8)$$

The measurement of particle volume is simple, but the precise measurement of its surface area is very difficult. This problem compounds when one attempts to define the sphericity of a mass of a large number of dissimilar particles. The packing characteristics of particles are important parameters that depend on the particle's shape and mode of packing. In some special situations, such as in the vicinity of a sphere or a plane wall, the distribution of local voidage becomes important. Unlike bulk voidage, it is not uniform or monotonically varying. It follows a damped oscillatory pattern.

A1.3 PARTICLE CLASSIFICATION

In the light of fluidization experience, Geldart (1972) classified solids broadly under four groups, **A**, **B**, **C**, and **D** as shown in Figure A1.2. The particle's classification is plotted against the density difference between the solid and the fluidizing gas. This classification is important in understanding the fluidization behaviour of solid particles, because under similar operating conditions particles of different groups may behave entirely differently.

Appendix 1: Characteristics of Solid Particles

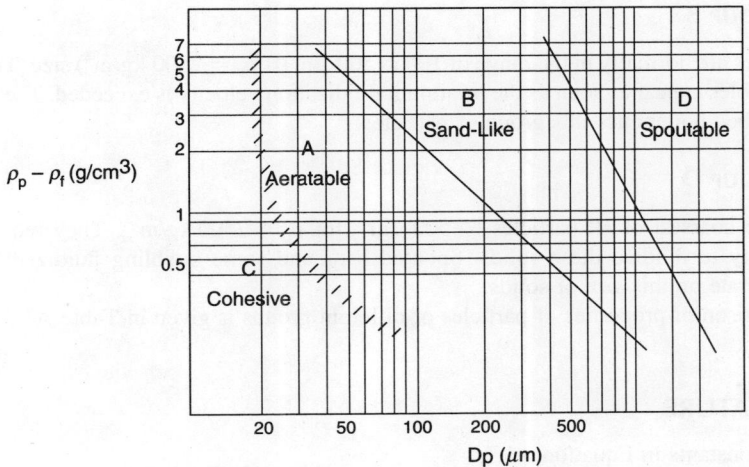

FIGURE A1.2 Powder classification developed by Geldart.

A1.3.1 GROUP C

These particles are very fine and are typically smaller than 30 μm (ρ_p = 2500 kg/m³). The interparticle forces are comparable to the gravitational force on these particles. So, these particles are very difficult to fluidize. An attempt at fluidization often results in channelling. Special techniques are required to fluidize these particles.

A1.3.2 GROUP A

These particles are typically in the range of 30 to 100 μm (ρ_p = 2500 kg/m³). These particles fluidize well, but expand considerably after exceeding the minimum fluidization velocity and before bubbles start appearing. Many circulating fluidized bed systems use Group A particles.

TABLE A1.3
Distinguishing Feature of Four Groups of Particles

Group	C	A	B	D
Particle size for ρ_p = 2500 kg/m³	<20 μm	20–90 μm	90–650 μm	>650 μm
Channeling	Severe	Little	Negligible	Negligible
Spoutability	None	None	Shallow bed	Readily
Expansion	Low	High	Medium	Medium
Minimum bubbling velocity, U_{mb}	No bubble	$>U_{mf}$	$=U_{mf}$	$=U_{mf}$
Bubble shape	Only channel	Flat base spherical cap	Rounded with small indentation	Rounded
Solid mixing	Very low	High	Medium	Low
Gas back-mixing	Very low	High	Medium	Low
Slugging mode	Flat raining plugs	Axisymmetric	Mostly axisymmetric	Mostly wall slugs
Effect of particle size on hydrodynamics	Unknown	Appreciable	Minor	Unknown

A1.3.3 GROUP B

These particles are normally in the range of 100 to 500 μm (if $\rho_p = 2500$ kg/m^3) size. They fluidize well, and bubbles appear as soon as the minimum fluidization velocity is exceeded. The majority of the fluidized bed boilers use this group of particles.

A1.3.4 GROUP D

These are the coarsest of all particles (>500 μm) (for $\rho_p = 2500$ kg/m^3). They require a much higher velocity to fluidize these solids. Spouted beds and some bubbling fluidized bed boilers generally operate on this size of solids.

A comparison of properties of particles of different groups is given in Table A1.3.

NOMENCLATURE

a, b:	constants in Equation A1.9
C_D:	coefficient of drag in a particle
d_i:	mean opening of successive sieves, $(d_{pi} + d_{pi+1})/2$
d_m:	mean diameter of a particulate mass with varying sizes, m
d_p:	sieve size (diameter), m
d_s:	surface diameter, m
d_v:	volume diameter, m
d_{sv}:	surface volume diameter, m
$d_{84, 50, 16}$:	diameters corresponding to cumulative weights of 84%, 50%, and 16%, respectively
F_D:	drag force in a particle, N
m_p:	mass of particle, kg
R:	relative size range defined in Equation A1.6
S:	actual surface of the particle, m^2
U_m:	minimum fluidization velocity, m/sec
U:	superficial gas velocity, m/sec
V:	actual volume of the particle, m^3
x_i:	weight fraction of particles collected between sieve i and $i + 1$
ϕ:	sphericity
ϵ:	voidage
ρ_g:	density of gas, kg/m^3
ρ_p:	density of solids, kg/m^3
μ:	viscosity of gas, kg/sq.m
Ar:	Archimedes number, $(gd_p^3(\rho_p - \rho_g))/\mu^2$
Re:	Reynolds number, $(Ud_v\rho_g)/\mu$

REFERENCES

Abrahamsen, A. R. and Geldart, D., *Powder Technol.*, 26, p. 35, 1980.

Chen et al., *Circulating Fluidized Bed Technology V*, Kefa C., ed., International Academic Publishers, Beijing, p. 508, 2005.

Geldart, D., The effect of particle size and size distribution on the behaviour of gas-fluidized beds, *Powder Technol.*, 6, 201–215, 1972.

Institute of Gas Technology, *Coal Conversion System Data Handbook*, DOE/FE/05157-2, Table IVB 10.1, 1982.

Appendix 2
Stoichiometric Calculations

Stoichiometric calculations (also known as *combustion calculations*) provide much of the basic information necessary for the design of a boiler plant. They help find the amount of fuel to be fed for the required thermal output of the plant. The specifications of fans and blowers are based on the air required for burning or gasifying that quantity of fuel. Combustion calculations also give the amount of limestone required to achieve a certain amount of sulfur capture. Finally the solid and gaseous pollutants produced from the combustion are computed from this. Most of the calculations are based on overall chemical reactions.

A2.1 CHEMICAL REACTIONS

Some boiler furnaces burning high-sulfur coal are required to retain the sulfur released from the coal during combustion in solid form such that it is not emitted into the atmosphere. Thus, stoichiometric calculations of these boilers require special considerations. The overall combustion reactions for this type of furnace can be written as follows:

$$C + O_2 = CO_2 + 32,790 \text{ kJ/kg of carbon} \quad (A2.1)$$

$$C_nH_m + \left(n + \frac{m}{4}\right)O_2 = nCO_2 + \frac{m}{2}H_2O + \text{heat} \quad (A2.2)$$

$$S + O_2 = SO_2 + 9260 \text{ kJ/kg of sulfur} \quad (A2.3)$$

where m and n are stoichiometric coefficients of Equation A2.2.

For absorption of the SO_2, limestone is fed into the furnace. Limestone is first calcined to CaO through the following reaction:

$$CaCO_3 = CaO + CO_2 - 1830 \text{ kJ/kg of } CaCO_3 \quad (A2.4)$$

If the sorbent contains magnesium carbonate, an additional reaction occurs:

$$MgCO_3 = MgO + CO_2 - 1183 \text{ kJ/kg of } MgCO_3 \quad (A2.5)$$

Calcium oxide, from either limestone or coal ash, absorbs a fraction of the sulfur dioxide released from the coal during combustion. The reaction is:

$$CaO + SO_2 + \tfrac{1}{2}O_2 = CaSO_4 + 15,141 \text{ kJ/kg of sulfur} \quad (A2.6)$$

The above equations show that oxygen is required for both the combustion and the sulfation reactions. Since in any gas–solid process the contact between the gas and solid is less than perfect, an excess amount of oxygen is needed for complete combustion. The extra air that provides this oxygen is called *excess air*. This excess air is about 20% for the combustion and sulfation reactions combined.

A2.2 AIR REQUIRED

Noting that dry air contains 23.16% oxygen, 76.8% nitrogen and 0.04% inert gases by weight, the dry air required for complete combustion of a unit weight of coal M_{da}, is given by:

$$M_{da} = [11.53C + 34.34(H - O/8) + 4.34S + A \cdot S] \text{ kg/kg coal} \tag{A2.7}$$

where C, H, O, and S are weight fractions of fuel constituents known from the ultimate analysis. For each unit mass of sulfur converted to calcium sulfate, an additional amount of dry air A, is required (see Reaction A2.6). So the extra air for a unit weight of coal is $A \cdot S$, where A is 2.16 for sulfur capture and is zero when no sulfur is captured as calcium sulfate.

For efficient combustion, a certain amount of air, in excess of what is required theoretically, is provided. To get the total air one must multiply the theoretical air by the excess air coefficient, EAC. The total dry air T_a, is the sum of the theoretical requirement and whatever excess air is allowed to complete the combustion.

$$T_{da} = \text{EAC} \cdot M_{da} \text{ kg/kg burned} \tag{A2.8}$$

The excess air coefficient EAC, is defined in such a way that EAC = 1.2 would mean 20% excess air. Air usually contains some moisture. In standard air this weight fraction of moisture X_m, is about 0.013 kg/kg air, and X_m is the weight fraction of moisture in the air. Thus, M_{wa}, the total wet air is:

$$M_{wa} = T_{da}(1 + X_m) \tag{A2.9}$$

A2.3 SORBENT REQUIREMENT

If the coal ash contains a negligible amount of calcium oxide, the sorbent required L_q, to retain the sulfur in a unit weight of fuel is found from the following equation:

$$L_q = \frac{100S}{32X_{CaCO_3}} X R \tag{A2.10}$$

where S is the weight fraction of sulfur in coal, and X_{CaCO_3} is the weight fraction of $CaCO_3$ in the sorbent. R is defined as the calcium to sulfur molar ratio in the feed of sorbent and coal respectively.

Sometimes the coal ash contains an appreciable amount of calcium oxide, which removes a part of the sulfur released from the coal. If X_{CaO} is the weight of calcium oxide per unit weight of fuel fed, the inherent Ca/S ratio is $32X_{CaO}/56S$. Therefore the limestone required for removal of same amount of sulfur ($E_{sor}S$) will be reduced by the above amount. Thus, R is to be replaced by R' in Equation A2.10 and elsewhere as below:

$$R' = \left\{ R - \frac{32X_{CaO}}{56S} \right\} \tag{A2.11}$$

A2.4 SOLID WASTE PRODUCED

From Reaction A2.4 to Reaction A2.5 we find that the sorbent decomposes into MgO and CaO. Out of this a part of the CaO is converted into $CaSO_4$. The spent sorbent would thus contain $CaSO_4$, unconverted CaO, unconverted MgO, and inert components of the sorbent. The weight of spent sorbent produced per unit weight of coal burned L_w, is the sum of $CaSO_4$, CaO, MgO, and inerts.

Spent sorbents = calcium sulfate + calcium oxide + magnesium oxide + inert

$$L_w = 136 \frac{S}{32} E_{sor} + 56\left(\frac{L_q X_{CaCO_3}}{100} - \frac{SE_{sor}}{32} \right) + \frac{40 L_q X_{MgCO_3}}{84} + L_q X_{inert} \tag{A2.12}$$

where L_q is the sorbent fed per unit weight of coal burned and is given by Equation A2.10.

Appendix 2: Stoichiometric Calculations

The total solid waste contains, in addition to the spent sorbent L_w, coal ash ASH, and unburned carbon $(1 - E_c)$, less the CaO content X_{CaO}, of coal converted to $CaSO_4$ and included in L_w. The solid waste produced per unit weight of fuel burned is thus:

$$W_a = [L_w + ASH + (1 - E_c) - X_{CaO}] \quad (A2.13)$$

where E_c is the combustion efficiency expressed as a fraction.

A2.5 GASEOUS WASTE PRODUCTS

The weight of flue gas due to the combustion reaction, W_c is the sum of carbon dioxide, water vapor, nitrogen, oxygen, sulfur dioxide, and fly ash. These individual constituents of a flue gas can be found as follows.

A2.5.1 CARBON DIOXIDE

$$\text{Carbon dioxide produced from fixed carbon in coal} = 3.66C \quad (A2.14)$$

In addition to the CO_2 produced from the fixed carbon, an extra amount of carbon dioxide is generated due to calcination of $CaCO_3$ and $MgCO_3$ in the sorbent material, see Equation A2.4 and Equation A2.5. This amount W_{CO_2}, may be calculated as:

$$W_{CO_2} = \frac{44SR}{32}\left(1 + \frac{100X_{MgCO_3}}{84X_{CaCO_3}}\right) = 1.375SR\left(1 + \frac{1.19X_{MgCO_3}}{X_{CaCO_3}}\right) \quad (A2.15)$$

where R is the calcium/sulfur molar ratio.

Example A2.1

The following table gives an analysis of a coal. Calculate the amount of CO_2 produced per kilogram of the coal burned. Assume that a calcium/sulfur molar ratio of 2.5 is used for sulfur capture. The limestone contains 88% $CaCO_3$ by weight and 10% $MgCO_3$ by weight.

	C	H	O	N	S	ASH	Moisture
Ultimate analysis of coal (%)	72.8	4.8	6.2	1.5	2.2	9.0	3.5

Solution

CO_2 produced from coal combustion Equation A2.14 = 3.66×0.728.

$$= 2.66 \text{ kg/kg coal due to combustion.}$$

CO_2 produced from calcination Equation A2.15 = $1.375 \times 0.022 \times 2.5 \, (1 + 1.19 \times 0.1/0.88)$

$$= 0.0858 \text{ kg/kg fuel.}$$

Total carbon dioxide produced = $2.66 + 0.0858$.

$$= 2.745 \text{ kg/kg of coal burned.}$$

A2.5.2 Water Vapor

Water in the flue gas comes from the combustion of hydrogen in the coal and the moisture present in the combustion air, coal, and limestone. The water in the flue gas per unit weight of coal burned is:

$$9H + \text{EAC} \cdot M_{da}X_m + M_f + L_qX_{ml} \tag{A2.16}$$

A2.5.3 Nitrogen

Nitrogen in the flue gas comes from the coal as well as the combustion air

$$\text{Nitrogen from fuel and air} = N + 0.768\, M_{da}\text{EAC} \tag{A2.17}$$

A2.5.4 Oxygen

The oxygen in the flue gas comes from oxygen in the coal, excess oxygen in the combustion air, and the oxygen left in the flue gas for incomplete capture of sulfur. We recall from Equation A2.6 that for each mole of unconverted sulfur, 1/2 mol of oxygen is saved. Thus:

$$\text{Oxygen in flue gas} = O + 0.2315\, M_{da}(\text{EAC} - 1) + (1 - E_{sor})S/2 \tag{A2.18}$$

A2.5.5 Sulfur Dioxide

If only E_{sor} fraction of the sulfur is converted to $CaSO_4$, the SO_2 present in the flue gas is:

$$2S(1 - E_{sor}) \tag{A2.19}$$

A2.5.6 Fly Ash

The flue gas may carry a part of the coal ash or sorbents. Though it is very small in amount and is eventually collected in the dust collector, it carries through the convective section of the boiler a fraction of the sensible heat that may not be negligible. The flue gas is approximated on the basis of:

$$\text{Unit weight of coal burned} = a_c\text{ASH} \tag{A2.20}$$

where a_c is the fraction of the ash in coal as it appears as fly ash ($\approx 0.1 - 0.5$).

The total weight of the flue gas can be found by adding up the above components. Simplifying them, the total weight of the flue gas per unit weight of coal burned is:

$$W_c = M_{wa} - 0.2315 M_{da} + 3.66C + 9H + L_qX_{ml} + N + O + 2.5S(1 - E_{sor})$$

$$+ 1.375 SR\left(1 + \frac{1.19 X_{MgCO_3}}{X_{CaCO_3}}\right) + a_c\text{ASH kg/kg burned} \tag{A2.21}$$

A2.6 HEATING VALUE OF FUELS

The approximate higher heating value, HHV, of a solid fuel may be calculated from the Dulong and Petit formula:

$$\text{HHV} = 33,823C + 144,249(H - O/8) + 9418S \text{ kJ/kg} \tag{A2.22}$$

Appendix 2: Stoichiometric Calculations

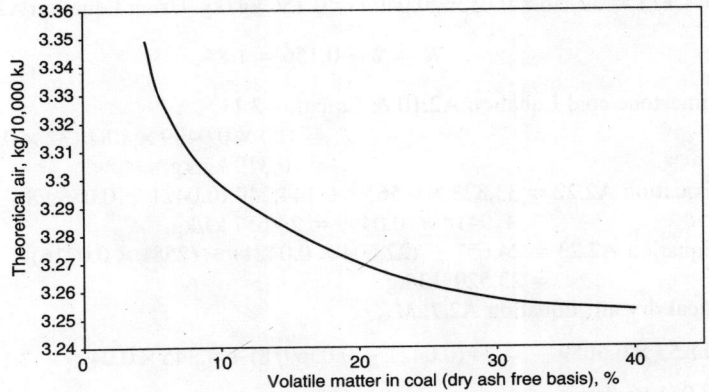

FIGURE A2.1 Theoretical dry air required to release 10,000 kJ of heat as a function of volatile matter in the coal on dry-ash-free basis.

The lower or net heating value, LHV, can be calculated by subtracting the heat of vaporization of the moisture in the fuel from the higher heating value (Ganapathy, 1994):

$$\text{LHV} = \text{HHV} - 22{,}604H - 2581 M_f \text{ kJ/kg} \tag{A2.23}$$

If the ultimate analysis of the coal is not known but the proximate analysis and heating value of the fuel are available, one can use Figure A2.1 to get a rough estimate of the theoretical air requirement. This figure plots the theoretical dry air required in kg/10,000 kJ fired against the volatile matter of coal on a dry ash-free basis.

For example, the theoretical air requirement for 10,000 kJ from a coal with 30% volatile matter on a dry ash-free basis will be (from Figure A2.1) 3.25 kg. If the heating value of the coal is 23,255 kJ/kg, the theoretical dry air requirement is $3.259 \times 23{,}255/10{,}000 = 7.578$ kg/kg. The weight of wet air (1.3% moisture and 20% excess air) is $7.578 \times 1.013 \times 1.2 = 9.21$ kg/kg.

Example A2.2

A coal is required to achieve 90% sulfur retention using a limestone that contains 90% calcium carbonate, 9% magnesium carbonate, and 1% inert material. It uses a sulfur molar ratio of 2. The fly ash is 10% of the total ash. The surface moistures of both the coal and limestone are 7.1%. Find the stoichiometric quantities for the design of the boiler worked out in Example 8.1.

Solution

1. The ultimate analysis of the coal is given as
   $X_m = 1.3\%$, $X_{ml} = M_f = 7.1\%$, $C = 56.59\%$, $H = 4.21\%$,
   $N = 0.9\%$, $Cl = 0.4\%$, $S = 4.99\%$, $O = 5.69\%$,
   ASH $= 20.06\%$, $X_{CaO} = 0.0136$.
2. $E_{sor} = 90\%$, $R = 2$, $X_{CaCO_3} = 0.9$, $X_{MgCO_3} = 0.09$, $E_c = 98\%$.

3. Inherent $Ca/S = 32/56 \times 0.0136/0.0499 = 0.156$ kg/kg. Using Equation A2.11

$$R' = 2 - 0.156 = 1.84$$

Using limestone/coal Equation A2.10 & Equation 2.11
$$= 100 \times 0.0499 \times 1.84/(32 \times 0.9)$$
$$= 0.319 \text{ kg/kg}.$$

4. HHV, Equation A2.22 $= 33,823 \times 0.5659 + 144,249 (0.0421 - 0.0569/8)$
$+ 9418 \times 0.0499 = 24,657$ kJ/kg.

5. LHV, Equation A2.23 $= 24,657 - (22,604 \times 0.0421) - (2581 \times 0.0716)$
$= 23,520$ kJ/kg.

6. Theoretical dry air, Equation A2.7, M_{da}

$$M_{da} = 11.53 \times 0.5659 + 34.34 (0.0421 - 0.0569/8) + 4.345 \times 0.0499 + 2.17 \times 0.0499$$
$$= 8.051 \text{ kg/kg}$$

7. Total dry air, Equation A2.8 $= 8.05 \times 1.2 = 9.66$ kg/kg.
Total wet air $= 9.66 (1 + 0.013) = 9.78$ kg/kg.

8. Total flue gas weight:
N_2 Equation A2.17 $= 0.7685 \times 8.051 \times 1.2 + 0.009 = 7.43$ kg/kg.
H_2O Equation A2.16 $= 9 \times 0.0421 + 0.013 \times 8.051 \times 1.2 + 0.319 \times 0.071 + 0.071$
$= 0.600$ kg/kg.
CO_2 Equation A2.14 and Equation A2.15 $= 3.66 \times 0.5659 + 1.375 \times 0.0499$
$\times 2[1 + (1.19 \times 0.09/ 0.9)] = 2.22$ kg/kg.
SO_2 Equation A2.19 $= 2 \times 0.05 (1 - 0.9) = 0.01$ kg/kg.
O_2 Equation A2.18 $= 0.0569 + 0.2315 \times 8.05 (1.2 - 1) + 0.5 \times 0.0499 (1 - 0.9)$
$= 0.432$ kg/kg.
Fly ash Equation A2.20 $= 0.2 \times 0.1 = 0.02$ kg/kg.
(Ash fraction in fly ash, $a_c = 0.1$)
W_c Equation A2.21 $= 9.78 - 0.2315 \times 8.05 + 3.66 \times 0.5659 + 9 \times 0.0421 + 0.071 +$
$0.343 \times 0.071 + 0.009 + 0.0569 + 2.5 \times 0.0499 (1 - 0.9)$
$+ 1.375 \times 0.0499 (1 + 0.09/0.9) + 0.1 \times 0.20 = 10.7$ kg/kg.

9. Weight of spent sorbent, Equation A2.12 $= (0.0499 \times 136 \times 0.9/32)$
$+ 56[(0.319 \times 0.9/100)$
$- (0.0499 \times 0.9/32)]$
$+ (0.319 \times 0.09 \times 40/84)$
$+ 0.319 \times 0.01 = 0.290$ kg/kg.

NOMENCLATURE

A: factor in Equation A2.7
a_c: fly ash leaving the furnace as a fraction of the total ash in coal
ASH: weight fraction of ash in coal
$[Ca/S]$: calcium to sulfur molar ratio
E_c: fractional efficiency of combustion
E_{sor}:   fraction of sulfur captured in the bed
EAC: excess air coefficient
H, O, S: weight fraction of hydrogen, oxygen, sulfur, carbon and nitrogen, respectively, in C, N, coal

Appendix 2: Stoichiometric Calculations

HHV: higher heat value of coal (kJ/kg)
L_q: limestone required per unit weight of coal fired (kg/kg fuel)
L_w: weight of spent sorbent per unit weight of coal burnt (kg/kg fuel)
LHV: lower heating value (kJ/kg)
m, n: stoichiometric coefficients in Equation A2.2
M_{da}: weight of dry air required per unit of coal (kg/kg)
M_f: moisture content of coal (kg/kg)
M_{wa}: weight of wet air required per unit of coal (kg/kg)
R: calcium to sulfur molar ratio in the feed to the boiler
R': reduced calcium to sulfur molar ratio due to the CaO in the sorbent
T_{da}: total dry air to be supplied to fire unit weight of coal (kg/kg)
W_a: weight of ash produced per unit weight of coal burned
W_c: weight of flue gas due to combustion of unit weight of fuel (kg/kg fuel)
W_{CO_2}: weight of carbon dioxide produced due to calcination per unit weight A1 burned (kg/kg)
X_{inert}: weight fraction of inert in limestone
X_m, X_{ml}: weight fraction of moisture in dry air and limestone respectively
X_{CaO}: weight fraction of CaO inherent in coal
X_{CaCO_3}: weight fraction of calcium carbonate in limestone
X_{MgCO_3}: weight fraction of magnesium carbonate in limestone

REFERENCE

Ganapathy, V., *Steam Plant Calculations Manual*, Marcel Dekker, NY, p. 56, 1994.

Appendix 3
Simplified Model for Sulfur Capture

The mass balance of sulfur dioxide over a horizontal section of the combustor is written below from Equation 5.25 as:

$$U \frac{dC_{SO_2}}{dx} = m - K C_{SO_2} C_c \qquad (A3.1)$$

The molar concentration of calcium (or $CaCO_3$) in the bed is:

$$C_c = \frac{X_{CaCO_3} f_c \rho_b(x)}{M_{CaCO_3}}$$

The axial distribution of bed density is given in Equation 5.24 as:

$$\rho_b(x) = \rho_b(\infty) + [\rho_b(0) - \rho_b(\infty)] e^{-ax} \qquad (A3.2)$$

The average bed density ρ_{bav}, can be found by integrating the above expression:

$$\rho_{bav} = \frac{1}{H} \int_0^H \rho_b(x) dx = \frac{1}{H} \left[\rho_b(\infty) x - \left(\frac{1}{a}\right) [\rho_b(\infty) - \rho_b(0)] \exp(-ax) \right]_0^H$$

$$= \rho_b(\infty) + \left[\frac{\{\rho_b(0) - \rho_b(\infty)\}}{aH} \right] (1 - e^{-aH}) \qquad (A3.3)$$

Assuming that the entire sulfur dioxide is released below the secondary zone, we have $m = 0$ for $x > 0$. So, Equation A3.1 can be rearranged as:

$$\frac{dC_{SO_2}}{C_{SO_2}} = \frac{-K f_c X_{CaCO_3}}{U M_{CaCO_3}} [\rho_b(\infty) + \{\rho_b(0) - \rho_b(\infty)\} \exp(-ax)] dx$$

Integrating it between $x = 0$ and $x = H$ and using Equation A3.3, one gets:

$$\ln\left(\frac{C_{SO_2}}{C_{s0}}\right) = -C_0 \rho_{bav} H \qquad (A3.4)$$

where:

$$x = 0; \quad C_{SO_2} = C_{s0} \text{ and } C_0 = \frac{K f_c X_{CaCO_3}}{U M_{CaCO_3}}$$

The sulfur capture efficiency as defined in Equation 5.26 can be written as:

$$E_{sor} = 1 - \frac{C_{SO_2}(H)}{C_{s0}} = 1 - \exp\left(\frac{-K f_c X_{CaCO_3}}{U M_{CaCO_3}} \rho_{bav} H \right)$$

This can be rearranged as:

$$Kf_c = -\left[\frac{\ln(1 - E_{sor})}{C_1}\right]$$

where

$$C_1 = \frac{X_{CaCO_3}\rho_{bav}H}{100U}$$

and the molecular weight of calcium carbonate, M_{CaCO_3} is 100.

Using the expression of reactivity K from Equation 5.21:

$$Kf_c = \frac{1}{P^*}[\delta(\infty) - \delta]f_c = -\frac{[\ln(1 - E_{sor})]}{C_1} \qquad (A3.5)$$

The average level of sulfation δ, is related to the fraction of sulfur retained in the combustor (or sulfur capture efficiency E_{sor}) and the calcium to sulfur mole ratio (Ca/S) by Equation 5.33. So we write:

$$\delta = \frac{E_{sor}}{E_c(Ca/S)} = \frac{S100E_{sor}}{32X_{CaCO_3}E_c\psi} \qquad (A3.6)$$

where ψ (or F_s/F_c) is the ratio of the feed rates of sorbent and fuel. E_c is the cyclone collection efficiency, X_{CaCO_3} is the weight fraction of calcium carbonate in the limestone, and 100 is its molecular weight. S is the weight fraction of sulfur in coal and 32 is its molecular weight.

Also, the weight fraction of reacting sorbents f_c, in bed materials can be found from the ratio of feed rates of fresh sorbents and that of total solids:

$$f_c = \frac{F_{sor}}{F_c[ASH] + F_{sor}} = \frac{1}{\frac{[ASH]}{\psi} + 1} \qquad (A3.7)$$

where F_c, F_{sor} are feed rates of coal and sorbents, respectively, and [ASH] is the ash fraction in the coal.

Substituting the values of f_c and E_{sor} from the above equations, one can write Equation A3.5 as:

$$\delta(\infty) - \frac{E_{sor}S100}{E_cX_{CaCO_3}32\psi} = \frac{-P^*\ln(1 - E_{sor})}{C_1}\left(\frac{ASH}{\psi} + 1\right)\ln$$

After substitution of the value of C_1 and some algebraic manipulations, the above equation reduces to:

$$\psi = \frac{F_{sor}}{F_c} = \frac{3.12E_{sor}S\rho_{bav}H - 100P^*UASHE_c\ln(1 - E_{sor})}{E_c[\delta(\infty)X_{CaCO_3}\rho_{bav}H + 100P^*U\ln(1 - E_{sor})]} \qquad (A3.8)$$

NOMENCLATURE

Refer to the nomenclature list in Chapter 5.

Appendix 4
Tables of Design Data

TABLE A4.1
Physical Constants

Atmospheric pressure = 1.013 bar
Avogadro's number = 6.024×10^{23} molecules/mol
Boltzmann's constant = 1.380×10^{-23} J/K molecule
Blackbody radiation constants: $C_1 = 3.7420 \times 10^8$ W $\mu m^4/m^2$
$C_2 = 1.4388 \times 10^4$ μm K
$C_3 = 2897.7$ μm K
Gravitational acceleration (sea level), $g = 9.807$ m/s^2
Planck's constant = 6.625×10^{-34} J s/molecule
Stefan–Boltzmann constant, $\sigma = 5.670 \times 10^{-8}$ W/m^2 K^4
Universal Gas Constant, $R = 8.205 \times 10^{-2}$ m^3 atm/kmol K
 = 8.314×10^{-2} m^3 bar/kmol K
 = 8.314 kJ/kmol K
 = 282 N m/kg K
 = 8.314 kPa m^3/kmol K
 = 1.98 kCal/kmol K

TABLE A4.2
Properties of Standard Air at Atmospheric and at Different Temperatures

T (K)	ρ (kg/m^3)	μ ($10^7 \times$ N s/m^2)	τ ($10^6 \times$ m^2/s)	K_g ($10^3 \times$ W/m K)	α ($10^6 \times$ m^2/s)	Pr
100	3.5562	71.1	2.00	9.34	2.54	0.786
150	2.3364	103.4	4.426	13.8	5.84	0.758
200	1.7458	132.5	7.590	18.1	10.3	0.737
250	1.3947	159.6	11.44	22.3	15.9	0.720
300	1.1614	184.6	15.89	26.3	22.5	0.707
350	0.9950	208.2	20.92	30.0	29.9	0.700
400	0.8711	230.1	26.41	33.8	38.3	0.690
450	0.7740	250.7	32.39	37.3	47.2	0.686
500	0.6964	270.1	38.79	40.7	56.7	0.684
550	0.6329	288.4	45.57	43.9	66.7	0.683
600	0.5804	305.8	52.69	46.9	76.9	0.685
650	0.5356	322.5	60.21	49.7	87.3	0.690
700	0.4975	338.8	68.10	52.4	98.0	0.695

Continued

TABLE A4.2 (Continued)

T (K)	ρ (kg/m^3)	μ (10^7 × N s/m^2)	τ (10^6 × m^2/s)	K_g (10^3 × W/m K)	α (10^6 × m^2/s)	Pr
750	0.4643	354.6	.796.37	54.9	109	0.702
800	0.4354	369.8	84.93	57.3	120	0.709
850	0.4097	384.3	93.80	59.6	131	0.716
900	0.3868	398.1	102.9	62.0	143	0.720
950	0.3666	411.3	112.2	64.3	155	0.723
1000	0.3482	424.4	121.9	66.7	168	0.726
1100	0.3166	449.0	141.8	71.5	195	0.728
1200	0.2902	473.0	162.9	76.3	224	0.728
1300	0.2679	496.0	185.1	82	238	0.719
1400	0.2488	530	213	91	303	0.703
1500	0.2322	557	240	100	350	0.685
1600	0.2177	584	268	106	390	0.688

(Density ρ, dynamic viscosity μ, kinematic viscosity τ, thermal conductivity K_g, thermal diffusivity α, and Prandtl Number Pr)

TABLE A4.3
Properties of Coal

	Bituminous	Brown Coal	Anthracite
Particle density (kg/m^3)	2200	1400	2400
Bulk density (kg/m^3) ($d_p \sim 1.5$ mm)	1100	820	1200

The bulk density of coal is for the mean diameter of 1.5 mm.

Appendix 4: Tables of Design Data

TABLE A4.4
Regime Transition Values

d_p (μm)	27°C (in air) $\rho_g = 1.16$ kg/m³; $\mu = 1.84 \times 10^{-5}$ N.s/m²							800°C (in a typical flue gas from coal firing) (CO_2 — 13.7%, N_2 — 76.4%, O_2 — 3.3%, SO_2 — 0.23%, H_2O — 6.3%), $\rho_g = 0.338$ kg/m³; $\mu = 5.68 \times 10^{-5}$ N.s/m²							Powder Group[g]
	U_{mf}^a (m/s)	U_{mb}^b (cm/s)	U_t^c (m/s)	U_c^d (m/s)	U_{tr}^e (m/s)	U_{ch}^f (m/s)	ε_{ch}	U_{mf}^a (cm/s)	U_{mb}^b (cm/s)	U_t^c (m/s)	U_c^d (m/s)	U_{tr}^e (m/s)	U_{ch}^f (m/s)	
2 > 0	0.04	0.369	0.029	0.615	0.948	1.953	0.9936	0.013	0.23	0.0096	1.257	1.858	4.725	Group A
40	0.159	0.72	0.317	0.821	1.297	2.239	0.9936	0.051	0.45	0.038	1.677	2.542	4.754	Group A
60	0.359	0.53	0.476	0.971	1.558	2.392	0.9937	0.116	0.33	0.086	1.985	3.053	4.802	Group A
80	0.639	0.71	0.634	1.095	1.775	2.550	0.9937	0.207	0.44	0.153	2.237	3.477	4.869	Group A
100	0.998	—	0.792	1.202	1.963	2.707	0.9936	0.323	0.055	3.239	2.455	3.847	4.955	Group A
200	3.960	—	1.583	1.603	2.686	3.495	0.9936	1.290	—	1.644	3.276	5.262	6.357	Group B
300	8.727	—	2.372	1.989	3.226	4.281	0.9936	1.290	—	2.464	3.877	6.321	7.175	Group B
400	14.95	—	3.162	2.139	3.674	5.067	0.9936	5.16	—	3.284	4.370	7.198	7.993	Group B
600	29.79	—	4.739	2.532	4.413	6.637	0.9936	11.56	—	4.922	5.174	8.646	9.628	Group B
800	45.09	—	6.315	2.854	5.026	8.207	0.9936	20.37	—	6.558	5.138	9.847	11.26	Group D
1000	59.24	—	7.890	3.132	5.559	9.774	0.9936	31.30	—	8.190	6.399	10.89	13.41	Group D
2000	111.5	—	11.31	4.179	7.605	13.18	0.9935	105.2	—	16.36	8.537	14.90	21.57	Group D
3000	147.0	—	13.86	4.947	9.135	15.72	.9934	182.3	—	24.33	10.10	17.89	29.71	Group D
4000	175.2	—	16.06	5.576	10.40	17.85	.9934	248.3	—	29.65	11.39	20.38	34.83	Group D
5000	199.1	—	17.96	6.118	11.50	19.92	—	304.0	—	33.15	12.49	22.54	38.32	Group D

Calculated values of transition velocities for a spherical solid particle with a density of 2500 kg/m³ in air at 27°C and flue gas at 800°C.

[a] Minimum fluidization velocity. Calculated from Equation 2.3.
[b] Minimum bubbling velocity. Calculated from Equation 2.4 and used for Group A particles.
[c] Terminal velocity. Calculated from Equation 2.23 to Equation 2.25.
[d] Onset velocity for turbulent fluidization. Calculated from Equation 2.15 to Equation 2.16.
[e] Transport velocity for fast fluidization. Calculated from Equation 2.31.
[f] Choking velocity and voidage calculated using Equation 2.29 and Equation 2.30 and based on bed diameter 0.2 m, circulation rate, 30 kg/m² s.
[g] From Figure A1.2.

TABLE A4.5
Specific Heat of Gases

Substance	Formula	Molecular Weight (kg/kmol)	a	$b (\times 10^{-2})$	$c (\times 10^{-5})$	$d (\times 10^{-9})$	Temperature Range (K)
Nitrogen	N_2	28	28.9	−0.1571	0.8081	−2.873	273–1800
Oxygen	O_2	32	25.48	1.520	−0.7155	1.312	273–1800
Air		28.97	28.11	0.1967	0.4802	−1.966	273–1800
Hydrogen	H_2	2	29.11	−0.1916	0.4003	−0.8704	273–1800
Carbon monoxide	CO	28	28.16	0.1675	0.5372	−3.595	273–1800
Carbon dioxide	CO_2	44	22.26	5.981	−3.501	7.469	273–1800
Water vapor	H_2O	18	32.24	0.1923	1.055	−3595	273–1800
Nitric oxide	NO	30	29.34	−0.09395	0.9747	−4.187	273–1500
Nitrous oxide	N_2O	44	24.11	5.8632	−3.562	10.58	273–1500
Nitrogen dioxide	NO_2	46	22.9	5.715	−3.52	7.87	273–1500
Ammonia	NH_3	17	27.56	2.5630	0.99072	−6.6909	273–1500
Sulfur dioxide	SO_2	46	25.78	5.795	−3.812	8.612	243–1800
Acetylene	C_2H_2	26	21.8	9.2143	−6.527	18.21	273–1500
Benzene	C_6H_6	78	36.22	48.475	−32.57	77.62	273–1500
Methanol	CH_3OH	32	19.0	9.152	−1.22	−8.039	273–1000
Ethanol	C_2H_6	30	19.9	20.96	−10.38	20.05	273–1500
Methane	CH_4	16	19.89	5.024	1.269	−11.01	273–1500
Ethane	C_2H_6	30	6.900	17.27	−6.406	7.285	273–1500
Propane	C_3H_8	44	−4.04	30.48	−15.72	31.74	273–1500
Ethane	C_2H_4	28	3.95	15.64	−8.344	17.67	273–1500

Molar specific heats C_p, in kJ/(kmol K) at atmospheric pressure for various common gases as a function of temperature $C_p = a + bT + cT^2 + dT^3$ (T in K).

TABLE A4.6
Design Velocities: Range of Velocities Generally Used in Boilers.

Nature of Services	Range of Design Velocity (m/s)
Air	
Air heater	5–25
Coal and air lines, pulverized coal	15–23
Compressed air lines	7.5–10
Forced draft ducts	7.5–18
Forced draft ducts, entrance to burners	7.5–10
Register grills	1.5–3
Ventilating ducts	5–15
Pneumatic transport of sorbent (100–400 μm)	22–36
Flue gas	
Air heater	5–25
Boiler back-pass	12–25
Induced draft flues and breeching	10–18
Stacks and chimneys	10–25
Steam lines	
High pressure	40–60
Low pressure	60–75
Vacuum	100–200
Superheater	10–25
Water	
Boiler circulation	0.35–3.5
Economizer tubes	0.75–1.5
Water lines, general	2.5–40

TABLE A4.7
Some Frequently Used Unit Conversions

Area
1 m^2 = 1550.0 inch.^{2} = 10.7639 ft^2 = 1.19599 yd^2 = 2.47104 × 10^{-4} acre = 1 × 10^{-4} hectare
= 10^{-6} km^2 = 3.8610 × 10^{-7}

Density
1 kg/m^3 = 10^{-3} g/cm^3 = 0.06243 lb$_m$/ft^3 = 0.01002 lbm/U.K. gallons = 8.3454 × 10^{-3} lb$_m$/U.S. gallons
= 1.9403 × 10^{-3} slug/ft^3

Energy
1 kJ = 238.85 cal = 2.7778 × 10^{-4} kW h = 737.56 ft. lb$_f$ = 0.94783 Btu = 3.7251 × 10^{-4} hp hr

Heat transfer co-efficient
1 W/(m^2 K) = 0.8598 kcal/(m^2 h °C) = 10^{-4} W/(cm^2 K) = 0.2388 × 10^{-4} cal/(cm^2 s °C) = 0.1761 Btu/(ft^2 h °F)

Continued

TABLE A4.7 (Continued)

Length
1 m = 10^{-3} km = 10^{10} Angstrom units = 10^6 micron = 39.370 inch = 3.28084 ft. = 4.971 links = 1.0936 yd.
= 0.54681 fathoms = 0.04971 chain = 4.97097 × 10^{-3} furlong = 5.3961 × 10^{-4} U.K. nautical miles
= 5.3996 × 10^{-4} U.S. nautical miles = 6.2137 × 10^{-4} miles

Mass
1 kg = 10^{-3} tonne = 1.1023 × 10^{-3} U.S. ton = 0.98421 × 10^{-3} U.K. ton = 2.20462 lb_m = 0.06852 slug

Mass flow rate
1 kg/s = 2.20462 lb/s = 132.28 lb/min = 7936.64 lb/h = 3.54314 long ton/h = 3.96832 short ton/h

Power
1 W = 1 J/s = 10^{-3} kW = 10^{-6} MW = 0.23885 cal/s = 0.8598 kcal/h = 44.2537 ft. lb_f/min = 3.41214 Btu/h
= 0.73756 ft. lb_f/s

Pressure
1 bar = 10^5 N/m^2 = 10^5 Pa = 0.1 MPa = 1.01972 kg/cm^2 = 750.06 mmHg = 750.06 Torr = 10197 mm H_2O
= 401.47 inch H_2O = 29.530 inch Hg = 14.504 psi = 0.98692 atm = 1.45 kip/inch^2

Specific energy
1 kJ/kg = 0.2388 cal/g = 0.2388 kcal/kg = 334.55 ft. lb_f/lb_m = 0.4299 Btu/lb_m

Specific heat
1 kJ/(kg K) = 0.23885 cal/(g °C) = 0.23885 kcal/(kg °C) = 0.23885 Btu/(lb_m °F)

Surface tension
1 N/m = 5.71015 × 10^{-3} lb_f/inch

Temperature
T (K) = T (°C) + 273.15 = [T(F) + 459.67]/1.8 = T(R)/1.8

Thermal conductivity
1 W/(m K) = 0.8604 kcal/(m h °C) = 0.01 W/(cm K) = 0.01 W/(cm K) = 2.390 × 10^{-3} cal/(cm s °C)
= 0.5782 Btu/(ft. h °F)

Torque
1 N m = 141.61 oz. inch = 8.85073 lb_f inch = 0.73756 lb_f ft = 0.10197 kg_f m

Velocity
1 m/s = 100 cm/s = 196.85 ft/min = 3.28084 ft/s = 2.23694 mile/h = 2.23694 mph = 3.6 km/h = 1.94260 U.K. knot
= 1.94384 Int. knot

Viscosity dynamic
1 kg/(m s) = 1 (N s)/m^2 = 1 Pa s = 10 Poise = 2419.1 lb_m/(ft. h) = 10^3 centipoise = 75.188 slug/(ft. h)
= 0.6720 lb_m/(ft. s) = 0.02089 (lb_f s)/ft^2

Viscosity, kinematic
1 m^2/s = 3600 m^2/h = 38,750 ft^2/h = 10.764 ft^2/s

Volume
1 m^3 = 61,024 inch^3 = 1000 liters = 219.97 U.K. gallons = 264.17 U.S. gallons = 35.3147 ft^3 = 1.30795 yd^3
= 1 stere = 0.81071 × 10^{-3} acre-foot = 6.289 barrel

Volume flow rate
1 m^3/s = 35.3147 ft^3/s = 2118.9 ft^3/min = 13,198 U.K. gallons/min = 791,891 U.K. gallons/h = 15,850 gallons/min
= 951,019 U.S. gallon/h

Index

A

ABFB *see* atmospheric bubbling fluidized beds
acid rain, 136–8
additive usage, 129, 317
ADVACAT, 146
aeration, 421, 433
agglomeration, 119, 128–30, 180–1, 271
agricultural waste, 10–11
Ahlstrom group, Finland, 3
air-blown gasifiers, 60, 76, 80–1
air distribution, 213
air distribution grates, 359–79
    attrition, 374
    back-flow of solids, 375–7
    bubbling fluidized beds, 374–5
    design methods, 368–71
    leakages, 373
    non-uniform bubbling, 366
    operation, 366–8
    practical considerations, 371–9
    pressure drop, 364, 369, 372, 374–8
    sealing, 373
    see also distributor plates
air division, 266–7
air-flow, loop-seals, 428–31
air heater heat transfer, 228
air injection, 121
air inlet heights, 372
air mass flow rates, 226
air moisture, 217
air nozzles, 242–3, 310
air pollution, 135–41, 213
air port location, 276–7
air preheat temperatures, 227
air requirements, 446
air staging, 154, 160–1
air systems, CFB boilers, 253–5
air velocity, 459
alkaline metals, 315
alloy anchors, 324
alloy steels, 318–20
alloys, 317–20, 324
ammonia injection, 161, 163
anchors, 324–5
angle of impingement, 303
angle of repose, 423
annulus, 48–52
anthracite, 123, 267
aperture heights, 430
Archimedes number, 32
area, unit conversions, 459

ash
    CFB boilers, 264–7, 269, 271–2
    classifiers design, 269
    content, 271–2
    coolers, 329
    extraction, 213–15
    feed property effects, 73
    particulate emissions, 166–7
    solid handling systems, 339
    see also fly ash
aspect ratio, 273, 275
asymmetric exit cyclones, 400
atmospheric bubbling fluidized beds (ABFB), 211
atmospheric nitrogen, 159
atmospheric-pressure gasifiers, 76, 80–1
attrition, 115–17, 228, 229, 374
augers, 347
auxiliary power consumption, 287
axial distribution, 453
axial inlet–axial discharge cyclones, 395
axial voidage profiles, 43–8

B

back-flow, 366, 375–7
back-pass, 7, 278, 310
bag-house precipitators, 339, 381
barrel flow, 388
Battelle/Columbus indirect gasifiers, 81
bed cross-sectional area, 235–7
bed density, 152, 453
bed expansion, 28
bed height
    BFB boiler furnaces, 237–8, 240
    combustion efficiency, 121
    distributor grids, 364
    over-bed feed systems, 349–50
    solid recycle systems, 424
bed inventory, 48, 51–2, 267–70
bed materials
    agglomeration, 271
    alternatives, 129
    boilers, 6–7
    gasification, 89–90
    make-up, 267
bed preparation, 242
bed pressure, 241
bed quality, 267–70
bed temperature
    agglomeration reduction, 130
    BFB boilers, 241, 243

coefficients, 201
heat transfer, 186–7, 201
bed voidage, 47, 51–2, 369
bend coefficients, 355
BFB *see* bubbling fluidized beds
biomass
    combustion, 127–30
    fuel composition, 127–8
    fuel property effects, 128
    gasification, 89–90, 99
    handling systems, 341–2
bituminous coal, 137
boilers
    boiler islands, 287
    efficiency, 219–21
    fluidized beds, 6–9
    performance, 128
    see also bubbling fluidized bed boilers;
            circulating fluidized beds boilers
bottom ash, 266–7
bottom exit vertical-axis cyclones, 383–4
Boudouard reaction, 66–70
breadth and width ratio, 273, 275
bubble cap distributor plates, 360, 362–3
bubble hydrodynamics, 25–8
bubble phase heat transfer, 191
bubble size reduction, 247
bubbling bed feed systems, 342–51
bubbling bed heat exchangers, 257–9
bubbling bed heat transfer coefficient, 175–6
bubbling to fast bed transition, 38–9
bubbling fluidized beds (BFB)
    air distribution grates, 365–6, 368–70, 374–5
    boilers, 211–51
        advantages, 215–16
        air distribution, 213
        air heater heat transfer, 228
        air mass flow rates, 226
        air pollution, 213
        ash extraction, 213–15
        boiler efficiency, 219–21
        characteristics, 236
        coal combustion, 228–30
        combustion, 211, 215, 228–34, 239
        control procedures, 242–6
        description, 211–15
        emission control, 213, 215
        energy balance, 221–3
        features, 215–16
        feed system design, 248
        feeding, 211–13
        feedstock preparation, 211–13
        flue gases, 213, 215, 216
        fuel mass flow rates, 226
        furnace design, 234–42
        gas paths, 213
        gas-side heat balance, 226–8
        heat balance, 216–19, 221–8
        heat transfer, 228
        introduction, 7
        limitations, 216

        operation problems and remedies, 246–8
        overall heat balance, 216–19
        start-up procedure, 242–6
        steam generation, 215
        thermal design, 216–28
    combustion, 120–1, 211, 215, 228–34, 239
    erosion, 304–7
    feed systems, 342–51
    freeboard heat transfer, 201–2
    gas–solid contacting modes, 76
    gasification, 62, 65, 74–6
    heat transfer, 173, 189–202
        correlation, 197
        experimental observations, 195–7
    hydrodynamics, 25–8, 54
    introduction, 1, 2
    scale-up, 54
    sulfur capture, 150–1
bubbling to turbulent transitions, 29–31, 33, 34
burn-up cells, 232–3
burning rate, 109–15
burnout time, 124
butt welds, 308

C

calcination, 143–4, 153, 218, 445
calcium molar concentration, 453
calcium oxide, 144–5, 434, 445–7
calcium sulfate, 84, 144–5, 434, 446–7
captive stage, 31
capture efficiency, 395, 397, 453–4
carbon
    balance, 94
    BFB combustion, 120
    burn-up cells, 232–3
    char combustion, 108
    conversion efficiency, 92–5
    oxygen reactions, 108
    porosity, 113
    steels, 318–20
carbon dioxide
    absorption bands, 138–9
    acceptor process, 82–3
    coal combustion, 229
    emissions, 135, 139–40, 165
    operating parameter effects, 69, 70
    stoichiometric calculations, 447
carbon monoxide
    BFB combustion, 120
    coal combustion, 229
    emissions, 137, 141, 165
    operating parameter effects, 69, 70
carbonizers, 75
castable materials, 323–4
catalytic cracking, 84
cavity separators, 411
cement industries, 9–10
centrifugal acceleration, 394, 397
centrifugal forces, 54, 385, 394, 397
ceramic anchors, 325

Index

ceramic fibers, 324
CFB *see* circulating fluidized beds
CFC-11 emissions, 140
char
    burning rate, 109–15
    combustion, 107–9, 123–4, 228–30
    concentration, 404–5
    particle combustion, 123–4
    porosity, 113
    pyrolysis, 65
characteristic lengths, 181
characteristics of solid particles, 439–44
chemical contaminants removal, 83–4
chemical looping, 262
chemical plugging, 434
chemical reactions
    stoichiometric calculations, 445
    sulfur dioxide, 142–9
chemical removal methods, tar, 84
Chinese Boiler Standards, 192, 302
chlorine, 130, 314–16
choking, 37–8
circo-fluid boilers, 259–61
circulating fluidized beds (CFB)
    air distribution grates, 364–5, 370–1
    air pollution emissions, 137
    boilers, 253–96
        air system, 253–5
        auxiliary power consumption, 287
        combustion, 122–7, 263–4
        control, 287–91
        cyclone design, 125
        design, 125–7, 264–87
        efficiency, 125
        erosion reduction, 307–11
        external heat absorption, 289–91
        furnace design, 271–7, 289
        general arrangement, 253–7
        heat absorption, 279–85, 289–91
        heat transfer, 184–7, 202–5
        heating surface design, 277–87
        hydrodynamics, 42, 43
        introduction, 7–8
        load control options, 289–91
        non-CFB circulation boilers, 259–62
        performance modeling, 127
        solid recycle systems, 417–22, 426–7
        specific features, 12–13
        sulfur capture, 157
        supercritical boilers, 291–3
        types, 257–9
    combustion, 16
    combustors, 390–2
    cyclones, 390–2, 402–5
    feed systems, 352–3
    furnace combustion, 263–4
    gas–solid contacting modes, 76
    gas–solid separators, 390–2
    gasification, 62, 65, 74–6, 80–1
    gasifiers, 76, 80–1, 390–2
    heat transfer, 173, 174, 178–89, 202–5

    hydrodynamics, 35, 42–3, 53–4
    introduction, 1, 2–3
    nitric oxide emissions, 161–2
    nitrous oxide formation, 162
    reactor hydrodynamics, 42, 43
    scale-up, 53–4
    solid handling systems, 352–3
    sulfur capture, 150–1
circulating moving bed boilers, 260, 262
circulation rate effects, 45–7
circumferential distribution, 203–5
climate change, 136, 138–40
closing of nozzles, 377–8
cluster phase, 178–9, 182
clusters
    formation, 36–7
    heat transfer, 178–84
co-firing with coal, 129
coal
    co-firing, 129
    combustion, 228–30
    fired boiler emissions, 141
    fuel flexibility, 10–11
    gasification, 64
    moisture, 217, 272–3
    partial size distribution, 268
    particle combustion time, 230
    properties, 272–3, 456
    residue, 10–11
    specific heat, 193
    sulfur capture, 154
coatings, erosion reduction, 306
cold gas efficiency, 92–5
collection efficiency, 387–8, 394–9, 401–5, 408–9, 454
Columbus indirect gasifiers, 81
combined-cycle plants, 16–18
combustible loss minimization, 230–4
combustion, 103–32
    air distribution, 316
    air division, 267
    air staging, 154, 160–1
    biomass combustion, 127–30
    bubbling fluidized beds, 120–1, 211, 215, 228–34, 239
    calculations, 264, 283
    CFB boilers, 122–7, 263–4, 267, 283
    CFB furnaces, 263–4
    chambers, 215, 317, 327
    combustion efficiency, 117–20
    communition phenomena, 115–17
    design, 125–7
    efficiency, 11, 117–21, 125, 262
    fuel distribution, 317
    gasification, 63, 66
    modifications, 14
    regimes, 108–9
    stages, 103–17
    temperature
        BFB boiler furnaces, 239
        combustion design, 126–7

combustion efficiency, 119–20
nitric oxide emissions, 160
nitrous oxide emissions, 163
sulfur capture, 151–2
times, 123
combustors, 325–7, 331–3
common inputs, BFB boilers, 244–5
communition phenomena, 115–17
compact design, CFB boilers, 259
composition, biomass fuels, 127–8
conductivity, 181, 193, 321, 460
contact resistance, 192
contact times, 192
contaminants removal, 83–5
contraction efficiency, 392
control
    BFB boilers, 242–6
    CFB boilers, 287–91
convection, 179, 190–4, 264
conveying velocity, 353–4
core-annulus model, 48–52
correction heat transfer coefficient, 198–9
corrosion
    biomass combustion, 130
    erosion–corrosion, 317–18
    material issues, 311–18
    nozzles, 378–9
    resistant alloy usage, 317
cost factors/implications, 18–19, 293
critical heat flux, 300
critical particle size, 385–90
cross-sectional area, 90–1, 235–7
cut size, 385–90, 397, 404
cyclones
    CFB boilers, 122–3, 257–60
    CFB combustion, 122–3
    CFB combustors, 390–2
    CFB gasifiers, 390–2
    collection efficiency, 394–9, 454
    combustion efficiency, 125
    design, 125
    gas–solid separators, 381–405
    geometry, 400
    inlets, 327, 400
    particle critical size, 385–90
    performance, 153
    pressure drop, 392–4
    refractory materials, 327–9
    return loops, 329
    scale-up, 54
    theory, 385
    vortex finder erosion, 310

D

definitions
    fast fluidized beds, 35
    fluidization, 21
    gasification, 59
dense bubbling beds, 221–3
dense-phase flow, 353

dense suspension up-flow, 42, 43
density
    coefficients, 201
    decay constant, 46
    solid particle characteristics, 441
    unit conversions, 459
departure from nucleate boiling (DNB), 300–1
deposit formation, 318
depth design, 275
design
    air distribution grates, 368–71
    CFB boilers, 264–87
    cyclone gas–solid separators, 398–9
    data tables, 455–60
    graph usage, 197–200
    inertial gas–solid separators, 411–13
    L-valves, 431–3
    loop-seals, 426–31
    parameter effects, 120–1
devolatilization, 64–6, 104–7, 118, 228–9
diameters, 27, 201, 374, 439–40
diffusion, 108–9
directional nozzle distributor plates, 362, 363
dished perforated distributor plates, 361
dispersed phases, 178, 182–3
dispersion, 52–3
distributor grids, 364–6
distributor plates, 309–10, 359–79
    see also air distribution grates
DNB *see* departure from nucleate boiling
dolomite, 142, 144
down-leg chambers, 418
downcomers, 256
downdraft fixed/moving beds, 62, 65
downward nozzle distributor plates, 362, 364
drag, 54
drag forces, 31–2, 385
dry flue gas heat losses, 217
drying, 63, 103–4, 228, 229
dust, 339
dynamic viscosity, 460

E

economizers, 224, 227, 256, 278
effective furnace surface area, 282
electricity generation costs, 293
electro-catalytic oxidation, 164
electrostatic precipitators (ESP), 339, 381
Elliott, Douglas, 1
elutriation rates, 28
emissions, 135–69
    air pollution, 135–41
    carbon dioxide, 165
    carbon monoxide, 165
    control, 213, 215
    mercury, 164–5
    national standards, 141
    nitrogen oxide, 159–62
    nitrous oxide emissions, 162–4
    particulates, 166–7

Index

standards, 141
sulfur dioxide, 141–58
trace organics, 166
empirical relations, 194
emulsion phases, 25, 191–3
energy
    balance, 89, 176, 221–3, 286
    gasification, 94–5
    unit conversions, 459, 460
enhancement of heat absorption, 188–9
enthalpy, 87, 224–5, 281
entrained beds, 61, 64
environmental conditions, 303
equilibrium gas composition, 68–9, 88, 97–8
equipment, fluidized beds, 4–10
equivalence ratio, 86–8
equivalent diameter, 27, 439
equivalent size, 304
equivalent volume diameters, 27, 439
erosion
    bubbling fluidized beds, 304–7
    in-bed components, 246–7
    material issues, 301–11, 317–18, 320, 330–1
    nozzles, 378–9
    reduction, 307–11
    refractory materials, 330–1
    resistance, 320
    types, 304
erosion–corrosion process, 317–18
ESP *see* electrostatic precipitators
evaporation/evaporators, 224, 227, 256–7, 280, 329
excess air
    BFB boiler furnaces, 242
    coefficients, 125–6
    combustion efficiency, 119
    nitric oxide emissions, 161
    nitrous oxide emissions, 163
    stoichiometric calculations, 445–6
    sulfur capture, 154
exothermic oxidation, 103
expansion joints, 333
experimental observations, heat transfer, 184–7, 195–7
external heat absorption, 289–91
external heat exchangers, 8, 187–9, 279–80, 329

F

failure analysis, 329–31
fast beds
    characteristics, 35–7
    definition, 35
    gas-solid contacting processes, 22
    gas-to-particle heat transfer, 176–8
    hydrodynamics, 22, 35–52
    regimes, 7–8
fatigue, 312
FBR *see* fluidized bed reactors
feed effects, 270
feed points, 13, 349–53
    allocation, 350–3
    spacing, 349–50

feed property effects, 71–4
feed rates, 454
feed stock, 117–19, 211–13
feed systems
    bubbling fluidized beds, 342–51
    circulating fluidized beds, 352–3
    design, 248
feeding BFB boilers, 211–13
feeding modes, 347–50
feedstock preparation, 211–13
fins, 187–9, 247–8, 305, 306
fire bricks, 323
fire-side ash corrosion, 313
fire-side corrosion, 312–15
firing, solid fuel power generation, 15–16
fixed beds, 62–4, 175–6
flow conditions, 303
flow pulsation, 196–7
flow rates, 460
flow regime diagrams, 40
flow split, 266–7
flue gases
    BFB boilers, 213, 215, 216
    CFB boilers, 255, 279, 281
    recycling, 317
    specific heat, 193
    stoichiometric calculations, 447–8
    thermal conductivity, 193
    velocity, 459
flue temperature control, 317
fluid bed boilers, 325–9
fluid bed heat exchangers, 279–80
fluidization number, 364–6
fluidization regimes, 21–35
fluidization velocity
    bed cross-sectional area, 236
    BFB boilers, 230, 236, 241–2, 247
    coal combustion, 230
    combustion efficiency, 119
    heat transfer, 185, 196
fluidization volume, 91
fluidized bed boilers, 6–12
fluidized bed equipment, 4–10
fluidized bed features, 10–13
fluidized bed firing, 16
fluidized bed gas-solid contacting processes, 22
fluidized bed gasification, 59–100
    air-blown gasifiers, 60, 76, 80–1
    bed materials, 89–90
    biomass, 89–90, 99
    design, 85–96
    equilibrium gas composition, 97–8
    gas composition, 88, 97–8
    gas yield composition, 67–9
    gasifiers
        characteristics, 65
        efficiency, 85–6
        types, 60–3
    modeling, 96–7
    physicochemical processes, 63–9
    theory, 63–9

fluidized bed gasifiers, 5–6, 75–85
fluidized bed reactors (FBR), 148–9
fluidized bed technology options, 13–19
fluidized bed volume, 91
fluoseals, 417
fly ash
   carbon content, 126
   CFB boilers, 266–7
   erosion, 310
   flue gases, 448
   particulate emissions, 166–7
   recirculation, 120, 121
   rejection, 267
   solid handling systems, 339
force balance, 31–2
formed refractory materials, 323
fossil fuel combustion, 139–40
fossil fuel plants, 337–9
Foster Wheeler atmospheric CFB gasifier, 80–1
fouling, 128, 130, 318
fragmentation, 115, 116, 119, 228, 229
freeboard
   BFB hydrodynamics, 28
   heat transfer, 201–2
   height, 35, 121, 233, 238
friction coefficients, 355, 393, 396, 423
fuel
   augers, 347
   bound nitrogen, 159
   combustion design, 126
   distribution, 317
   feed boilers, 211
   feed ports, 276, 352–3
   feeding, 121
   flexibility, 10–11
   heat input, 283
   heating values, 126, 448–50
   mass flow rates, 226
   nitric oxide emissions, 159
   particle size, 118–19
   ratio, 117
   reactivity, 71–2
   reprocessing, 129
   splitters, 355–7
   switch, 13–14
   types, 10–11, 263, 278–9
furnaces
   cross section, 273, 284
   design, 234–42, 271–7
   dimensions, 281–5
   exit gas temperatures, 226–7
   heat absorption, 280–5, 289
   heat load, 278–9
   heights, 35, 152, 203–5, 273–4, 284–5
   openings, 276–7
   outlet, 317
   surface area, 282
   temperatures, 270–1
   wall fire side combustion, 312
   zones, 122–3

G

gas cleaning, 83–5
gas composition, 67–9, 88, 97–8
gas convection, 179, 194
gas dispersion, 52–3
gas fuel flexibility, 10–11
gas heating, 177–8
gas paths, 213
gas residence times, 152
gas-side heat balance, 226–8
gas-side heat transfer coefficients, 189–90
gas–solid contacting modes, 75–6
gas–solid contacting processes/flow regimes, 22–4
gas–solid mixing, 52–3
gas–solid separators, 381–415
   cyclones, 381–405
   impact separators, 405–10
   inertial separators, 410–13
gas–solid slip velocity, 52
gas specific heats, 458
gas temperatures, 227, 317
gas-to-particle heat transfer, 173–8
gas velocity, 52, 301, 368–9, 429
gas voids, 25–8
gas yield composition, 67–9
gaseous pollutants, 135
gaseous waste products, 447–8
gasification
   definition, 59
   feed property effects, 71–4
   operating parameter effects, 69–71
   theory, 66–9
   see also fluidized bed gasification
gasifiers
   characteristics, 65
   examples, 77–83
   gas cleaning, 83–5
   gasifier efficiency, 85–6
   gasifier height, 91–2
   introduction, 5–6
   sizing, 90–5
   types, 60–3, 75–7
gasifying medium, 76
global warming, 138–40
global warming potential (GWP), 140
grade efficiency, 269, 386–7, 409
grate area, 12
grate bar distributor plates, 361
grate distributors, 359–79
grate release rates, 125–6, 273, 274
gravity
   CFB scale-up, 53–4
   chutes, 343, 345
   cyclone scale-up, 54
   fluidization regimes, 21
   hydrodynamics, 21, 53–4
   settling, 410–11, 413
greenhouse gases, 138–40
grit, 339
group A solids, 54, 442, 443

Index

group B solids, 54, 442–4
group C solids, 442, 443
group D solids, 442–4
GWP *see* global warming potential

H

half-calcination, 144
Halliburton KBR transport reactors, 82
hammer mills, 340–1
harvested fuel feeding, 341
HC air pollution emissions, 136
headers, 256–7
heat absorption, 188–9, 279–85, 289–91
heat balance
    BFB boilers, 216–19, 221–8
    CFB boilers, 264–5, 283, 285–7
heat of combustion of gases, 86, 94
heat duty, 279–80, 282, 300–1
heat exchangers, 279–80
heat flux CFB boilers, 293
heat load, 224–5
heat losses, 216–19, 264–5
heat transfer, 173–207
    air heaters, 228
    bubbling fluidized beds, 173, 189–202, 228
    CFB boilers, 202–5, 270–1, 281–5, 288–9
    circulating fluidized beds, 173, 174, 178–89, 202–5
    coefficients
        bubbling beds, 175–6, 189–90
        CFB boilers, 270–1, 281–5, 288–9
        circumferential distribution, 203–5
        design graph usage, 198–200
        fixed beds, 175–6
        unit conversions, 459
    combustion stages, 103–4
    correlation, 197
    design graph usage, 197–200
    freeboard, 201–2
    furnace heights, 203–5
    gas-to-particle heat transfer, 173–8
    mechanisms, 178–9
    mechanistic models, 190–5
    table based calculations, 200–1
    theory, 179–84
heating
    coal combustion, 228, 229
    combustion stages, 103–4
    gases, 177–8
    modes, 77
    solid particles, 176–7
    surfaces, 277–87, 313–15
    time, 177–8
    values, 279, 448–50
heights
    air inlet, 372
    aperture, 430
    freeboard, 35, 121, 233, 238
    furnaces, 35, 152, 203–5, 273–4, 284–5
    gasifiers, 91–2
    primary, 177–8, 277

solids in recycle chambers, 425–6
transport disengaging, 28, 35
see also bed height
high combustion efficiency, 11
high-temperature corrosion, 316–17
high-temperature Winkler (HTW) gasifiers, 78
higher heating values, 448–9
horizontal tube heat transfer coefficient, 198–9
hot corrosion, 128, 130
hot-face insulating refractory materials, 322–3
hot gas efficiency, 86, 95–6
hot spot avoidance, 317
HTW *see* high-temperature Winkler
hybrid fluidized bed gasifiers, 76
hydration, 146
hydrodynamics, 21–57
    bubbling fluidized beds, 25–8
    circulating fluidized beds, 35, 42–3
    fast beds, 43–52
    fast fluidized beds, 35–42
    fluidization regimes, 21–35
    gas-solid mixing, 52–3
    packed bed, 22, 25
    scale-up, 53–4
    turbulent beds, 29–31
    velocity, 25–6, 30–6, 38, 40–2, 52
hydrogen balance, 93, 97–8
hydrogen burning, 217

I

IGCC *see* integrated gasification combined-cycles
ignition, 119
impact separators, 257, 259, 405–10
impingement collectors, 409
in situ sulfur removal, 12
in-bed component erosion, 246–7
in-bed tubes, 246–8, 306
incipiently fluidized beds, 25
induced draft fans, 255
inertial impact separators, 408
inertial impaction efficiency, 412
inertial separators, 257–9, 408, 410–13
inoperative orifices, 367–8
impingement angles, 303
impingement separators, 405–9
insulation materials, 320–33
    properties, 322
    types, 322–4
integrated gasification combined-cycles (IGCC), 16–18, 59, 73, 77
interception separation, 407–9
intrinsic reactivity, 117–18
inventory control, 267–70

J

J-valves, 417
jets, 370

K

Kellogg–Rust–Westinghouse (KRW) gasifier, 79
kinematic viscosity, 460
kinetics, char combustion, 108–9
KRW *see* Kellogg–Rust–Westinghouse

L

L-valves, 417–21, 431–3
lamella separators, 409–10
lateral distribution, 48–52
leakages, 373, 376
length
    characteristic lengths, 181
    heat transfer, 185–6
    loop-seals, 428
    unit conversions, 460
light gases, 65
lime, 84
limestone
    calcines, 83–4
    capture rate, 158
    demand, 152
    feed systems, 353
    injection, 167
    nitrous oxide emissions, 163
    reactivity, 147, 154
    stoichiometric calculations, 445–50
    sulfur dioxide calcination, 143–4
    sulfur dioxide retention, 142
    sulfur retention, 445–50
limiting bubble size, 27
limiting relative velocity, 32
limiting solid concentration, 394–5
linings, 320, 325, 330
load control options, 289–91
load-following capacity, 13
loopseals, 417–22, 425–31
    air distribution grates, 378–9
    blowers, 255
    design of, 426–31
    plugging, 433–4
    refractory materials, 329
low-cement castables, 323
low heat flux, 293
low-solid circulation boilers, 260–2
low-temperature Winkler gasifiers, 77–8
lower combustors, 331–2
lower furnace design, 275
lower heating values, 448
lower zone combustion, 122

M

magnesite, 89–90
magnesium carbonate, 445
magnesium oxide, 446–7
maintenance, refractory materials, 331–3
masonry, 138

mass, unit conversions, 460
mass balance, 89, 266, 284–6, 369, 453
mass flow rates, 226, 460
mass-transfer rate, 110–11
material issues, 299–334
    anchors, 324–5
    behavior, 299
    corrosion potentials, 311–18
    erosion potential, 301–11
    expansion joints, 333
    fire-side corrosion, 312–15
    fluidized bed boilers, 6–7, 318–20
    gasification, 89–90
    insulation materials, 320–33
    lining design, 325
    plastics, 324
    refractory materials, 320–33
    selection criteria, 299–301
    steels used, 318–20
maximum resistance, 432
maximum solid flow rate, 432
mean bed voidage, 47, 51–2
mean particle size, 440–2
mechanical particle properties, 303–4
mechanical plugging, 434
mechanical properties, 303–4, 321
mechanical requirements, linings, 325
mechanistic models, 190–5
mercury emissions, 135, 141, 164–5
metal cutting theory, 302
metal loss, 307, 309
methanation, 66–70, 72
methane, 135, 139, 140
minimum bubbling velocity, 25–6, 33–4
minimum fluidization velocity, 25, 33–4
minimum velocity, 25–6, 33–4, 40–2
modeling fluidized bed gasification, 96–7
moisture content, 73–4, 127, 271–2
moisture losses, 217
molar specific heats, 458
momentum, 303, 308
moving beds, 62–3
moving-hole feeders, 346–7
MSW *see* municipal solid waste
multi-entry vertical-axis cyclones, 383, 384
multi-orifice nozzles, 370–1
multiple combustion chambers, 239
municipal solid waste (MSW), 339–41
municipal waste, 10–11, 339–41

N

national standards, 141
nitric oxide emissions, 135–7, 141, 159–62
nitrogen
    balance, 98
    flue gases, 448
    gasification, 92
nitrogen dioxide, 137
nitrogen oxide emissions, 12, 159–62
nitrous oxide emissions, 135, 140, 162–4

Index.   469

non-CFB circulation boilers, 259–62
non-circular vertical cyclones, 257–9
non-fossil fuels, 339–41
nonharvested fuel feeding, 341–2
nonmechanical valves, 417–36
    L-valves, 417–21, 431–3
    loop-seals, 417–22, 425–31
    operation principles, 421–6
    practical considerations, 433–4
    types, 417–20
nonmetalic expansion joints, 333
nonspherical particles, 33–5
non-uniform bubbling, 366
non-uniform fluidization, 366–8
nonweeping grids, 376
nozzles
    air distribution grates, 360, 362–3, 370–1, 376–9
    back-flow elimination, 376
    corrosion, 378–9
    distributor plates, 360, 362–3, 370–1, 376–9
    erosion, 378–9
    opening/closing, 377–8
nucleate boiling, 300–1

O

omega superheater, 257
opening of nozzles, 377–8
operating conditions
    BFB boiler furnaces, 238–42
    CFB boilers, 270–1
    CFB cyclones, 402–5
    combustion efficiency, 119–20
operating parameter effects, 69–71, 163
operating pressures, 76
operating regimes, 42, 43
operation
    BFB boilers, 246–8
    distributor plates, 366–8
    solid recycle systems, 421–6
    supercritical CFB boilers, 292
orifices
    air distribution grates, 367–70, 374, 378–9
    blockages, 367
    distributor plates, 367–70, 374, 378–9
    fractional opening, 378–9
    velocity, 369, 370, 374, 379
outer heat transfer coefficient, 198–200
over-bed feed systems, 347–9
overall heat balance, 216–19
overflow standpipes, 417
oxygen
    balance, 93–4, 98
    biomass combustion, 127
    blown gasifiers, 60, 76, 92–3
    combustion prevention, 316
    flow, 92–3
    flue gases, 448
    transportment, 108

P

packed beds, 22, 25
packing characteristics, 442
part-load operation, 238–40, 287–9
partial gasification, 17–18, 64–6
partial gasifiers, 75
particles
    classification, 442–4
    collection efficiency, 387–8
    concentration, 403–5
    convection, 179, 190–4
    critical size, 385–90
    density coefficients, 201
    diameter coefficients, 201
    distances, 180
    erosion, 303–4
    gas-to-particle heat transfer, 173–8
    size
        BFB boiler furnaces, 242
        CFB boilers, 268–9
        combustion efficiency, 118–19
        heat transfer, 187, 195–6
        mean, 440–2
        standpipe to recycle chamber solid flow, 423
        suspension density profiles, 47
    terminal velocity, 31–5
    velocity, 31–5, 303
Particulate Solid Research Institute (PSRI) method, 398
particulates
    emissions, 141, 166–7
    removal, 85
    solid handling systems, 339
percolative fragmentation, 116
perforated plates, 360–1
performance modeling, 127
petroleum products, 10–11
PFBC *see* pressurized fluidized bed combustion
photochemical urban pollution, 137
physical constants, 455
physical particle properties, 303–4
physical properties, 303–4, 321, 322
physical removal methods, 84
physicochemical processes, 63–9
pitch, 374
plants, 138
plastics, 324
plate-type distributor plates, 360–2
plenum (airbox), 372–3
plugging loop-seals, 433–4
pneumatic injection, 346, 355
pneumatic transport, 22, 36–7, 39–40, 353–7
pollutants, 136–40
pollution, 135–69
pore plugging, 145
pore-size distribution, 154
porosity, 113, 442
porous distributor plates, 360, 361
postcombustion scrubbing, 14–15
power, unit conversions, 460
power generation, 13–15

precombustion technologies, 165
pressure
    balance, 421–2, 426–7, 430–2
    calcination, 153
    drop
        air distribution grates, 364, 369, 372, 374–8
        back-flow of solids, 375–7
        bubbling to turbulent transitions, 29–31
        cyclone gas–solid separators, 392–7
        distributor grids, 364
        distributor plates, 364, 369, 372, 374–8
        L-valves, 431–2
        loop-seals, 421–2, 426–7, 430–1
        nonmechanical valves, 421–2, 426–7, 430–2
        pneumatic transport lines, 353–5
    gasifiers, 76
    heat balance, 224–5
    surges, 434
    unit conversions, 460
pressurized bubbling, 16
pressurized fluidized bed boilers, 9
pressurized fluidized bed combined-cycles, 16
pressurized fluidized bed combustion (PFBC), 211
pressurized gasification, 76, 82
primary air, 267
primary air fans, 254
primary air port location, 276–7
primary control loops, 245–6
primary fragmentation, 116, 228, 229
primary heights, 177–8, 277
primary loops, 278
Prince Coal char particles, 113–15
PSRI *see* Particulate Solid Research Institute
pulsed air supply, 196–7
pulverized coal
    fired boilers, 1, 2, 137
    firing improvements, 15
pyrolysis/pyrolyzers, 64–6, 75, 104–7, 118
pyrosulfate melting corrosion, 315

Q

quartz wool matrix (QWM) reactors, 149–50

R

radial velocity, 385
radiation, 179, 194–5, 264
ram feeders, 347
ranking sorbents, 157–8
RDF *see* refuse derived fuel
reacting sorbent weight fraction, 454
reaction products, 107–8
reaction rates, 111–15
reactivity, 111–13, 117–18, 454
rearrangement resistance, 372
recirculation rates, 232, 429–31
recycle chambers, 418–20, 422–6, 428–9
reduction methods, 63, 66, 159–61
redundancy, 351

reentrainment, 399–400, 402–3
refractory materials, 320–33
    failure analysis, 329–31
    fluid bed boilers, 325–9
    maintenance, 331–3
    properties, 320–1
    tube interfaces, 307–9
    types, 321–2
refuse derived fuel (RDF), 340–1, 347
regime transition values, 457
reheaters, 257, 278
relative size, 442
relative velocity, 431–2
residence times, 181
return chambers, 418–20
reverse sulfation, 145
rice husk, 128
riser gas velocity, 429

S

safe density, 353, 354
safe velocity, 353, 354
safety examinations, 242
saltation, 353, 404
saw dust, 128
scale-up, 53–4, 96–7
screw conveyors, 344
scrubbing, 14–15
sealing air distribution grates, 373
secondary air, 267
    fans, 254–5
    injection, 121
    nozzles, 310
    port location, 276–7
secondary control loops, 246
secondary flow, 388–9
secondary fragmentation, 116
secondary loops, 278
selection criteria, materials, 299–301
separation efficiency, 381, 394–7, 411–13
sharp-crested weir theory, 424
shift conversion, 66–8
shredding, 340–1
side draft fixed/moving beds, 62, 65
sieving, 439, 442
silica sand, 89
simple nozzle distributor plates, 362
single-pass residence times, 123–4
single sorbent particles, 146–7
sinter formation, 364
siphon seals, 417
size issues
    bubble size reduction, 247
    equivalent size, 304
    gasifiers, 90–5
    limiting bubble size, 27
    loop-seals, 427–8
    pore-size distribution, 154
    relative size, 442
    sieve size, 439

Index

solid particle characteristics, 439–42
    sorbent size effects, 154
    see also particle size
slip velocity, 32, 36, 52
slit nozzle distributor plates, 362
sludge, 10–11
slugging, 29
smog, 137
smoke, 137
solid concentration, 394–5
solids density, 301
solid dispersion, 52–3
solid drain valve, 277
solid entry, 277
solid flow, 422–6
solid fuel firing boilers, 337–9
solid fuel power generation, 13–15
solid–gas separators, 381–415
solid handling systems, 337–57
    biomass, 341–2
    bubbling bed feed systems, 342–51
    circulating fluidized beds, 352–3
    fossil fuel plants, 337–9
    municipal solid waste, 339–41
    non-fossil fuels, 339–41
    pneumatic transport line design, 353–7
    solid fuel firing boilers, 337–9
solid particle characteristics, 439–44
solid particle heating, 176–7
solid recycle systems, 417–36
    L-valves, 417–21, 431–3
    loop-seals, 417–22, 425–31
    operation principles, 421–6
    practical considerations, 433–4
solid reentrainment, 399–400, 402–3
solid residence times, 153
solid stream, 255
solid waste produced, 446–7
sorbents
    CFB boilers, 255
    feeding systems, 213
    injection, 164
    ranking, 157–8
    reactivity, 148–50
    selection, 157–8
    size effects, 154
    solid handling systems, 339
    stoichiometric calculations, 446
    sulfur dioxide reactions, 145–6
    sulfur dioxide retention, 142, 143
sparge pipe-type distributors, 5–6, 360, 363–5
specific burning rate, 111
specific energy, 460
specific heat, 181, 193, 458
spherical particles, 32–3
sphericity, 304, 440, 441
spiral cyclone gas–solid separators, 383, 384, 386–8
splitters, 355–7
spouted beds, 62, 79–81
spouting phenomenon, 377–8
spreaders-stokers, 344–5

square vertical-axis cyclones, 383, 384
staged combustion, 239–40
staging air, 154, 160–1
standard air properties, 455–6
standard safety examinations, 242
standpipes, 418, 422–4, 426, 428, 431–2
start-up procedure, BFB boilers, 242–6
state variables, 244–5
steam
    generation, 215
    heat balance, 223–5
    parameters, 293
    steam-blown fluidized bed gasifiers, 76
    water–steam flow circuits, 255–7
steels for fluidized bed boilers, 318–20
stoichiometry, 282–5, 445–51
stokers, 344–5
Stokes number, 407
straight hole distributor plates, 360, 361
stream line velocity, 459
structural requirements, 299–300
sulfate corrosion, 314–15
sulfation, 144–5, 218
sulfide corrosion, 315
sulfur
    capture, 150–4, 157, 453–4
    CFB boilers furnaces, 272–3
    corrosion, 313–15
    gasification, 95–6
    oxide emissions, 135–7, 141–58
    removal, 12, 83–4
    retention, 445–50
    stoichiometric calculations, 445–7
sulfur dioxide
    emissions, 135, 141–58
    flue gases, 448
    reactions, 142–9
    retention, 142
    sorbent particle reactions, 146–7
    stoichiometric calculations, 445
supercritical boilers, 9, 15, 291–3
superficial velocity, 125–6, 201, 365–6
superheaters
    BFB boilers, 224, 227
    CFB boilers, 278
    fire-side corrosion, 313
    refractory materials, 329
    water–steam flow circuits, 257
supply chambers, 418–20, 424, 428
surface diameter, 439
surface tension, 460
surface-volume diameter, 439–40
suspension density, 44–5, 47–8, 184–5
swelling, 115, 119, 228, 229
synthetic sorbents, 142
system modifications, 15

T

table based calculations, 200–1
tangential entry cyclones, 383, 384

tangential forces, 385
tangential velocity, 396
tapered beds, 231–2
tar, 65, 84–5
target material nature, 304
TDH *see* transport disengaging height
technology options, 13–19
temperature
    BFB boilers, 224–5
    CFB boilers, 186–7, 310–11
    coefficients, 201
    combustion design, 126–7
    combustion efficiency, 119–20
    erosion, 310–11
    heat balance, 224–5
    heat transfer, 186–7, 201, 224–5
    nitric oxide emissions, 160
    nitrous oxide emissions, 163
    sulfur capture, 151–2
    unit conversions, 460
tension, 460
terminal velocity, 26, 31–5, 230
TGA *see* thermo-gravimetric apparatus
theory
    cyclone gas–solid separators, 385
    gasification, 63–9
    heat transfer, 179–84
thermal conductivity, 181, 193, 321, 460
thermal design, BFB boilers, 216–28
thermal nitric oxide emissions, 159
thermal requirements, linings, 325
thermo-chemical processes, 60
thermo-gravimetric apparatus (TGA), 149, 151
thermocouples, 310–11
three-body erosion, 304, 305
torque, 460
trace organic emissions, 166
transition phases, 29–31, 33, 34, 37–42
transition values, 457
transport disengaging height (TDH), 28, 35
transport velocity, 39
tube wastage, 301
tube-to-steam heat transfer coefficient, 198
turbulent beds, 29–31
turndown, 13
Tuyere distributor plates, 360
two-body erosion, 304, 305

U

U-beam impact separators, 409
unburnt carbon losses, 218, 264–5
unburnt carbon monoxide, 120
unburnt solids, 120, 121
under-bed feeders, 233–4, 347–51
uniflow vertical-axis cyclones, 383
uniform air distribution, 372
unit conversions, 459–60
updraft fixed/moving beds, 62–3, 65
upper bed area, 273, 274

upper combustors, 332–3
upper zone combustion, 122
urban pollution, 136, 137

V

V-valves, 417–18, 420
vanadium, 316, 434
VDI approach, 388–90, 392, 394–7
velocity
    cluster properties, 181
    coefficients, 201
    conveying, 353–4
    cyclone gas–solid separators, 393
    design data tables, 459
    flue gases, 459
    fluidization regimes, 33–4
    gas, 52, 301, 368–9, 429
    hydrodynamics, 25–6, 30–6, 38, 40–2, 52
    limiting relative, 32
    minimum, 25–6, 33–4, 40–2
    orifices, 369, 370, 374, 379
    particles, 31–5, 303
    radial, 385
    relative, 431–2
    safe, 353, 354
    slip, 32, 36, 52
    stream line, 459
    superficial, 125–6, 201, 365–6
    tangential, 396
    terminal, 26, 31–5, 230
    transport, 39
    unit conversions, 460
    water, 459
    see also fluidization velocity
Vereingte Aluminum Werke, Luenen, Germany, 3
vertical-axis cyclones
    bottom exit, 383, 384
    cylindrical, 382–4
    multi-entry, 383, 384
    square, 383, 384
    uniflow, 383
vertical cyclones, 257–9, 382–4
vertical fin effects on walls, 187–9
vertical lengths, 185–6
vertical tubes, 198–9, 247, 292–3
vibration, 196–7
viscosity, 460
viscous force constant, 53–4
VOC *see* volatile organic compounds
void volume, 442
voidage, 43–52, 369, 442
volatile combustion, 106–7, 228, 229
volatile matter, 72–3, 104–5, 118, 126
volatile organic compounds (VOC), 135
volatiles effects, 163
volume, unit conversions, 460
volume diameter, 27, 439–40
volume flow rate, 460
volumetric heat release, 125–6
volumetric release rates, 125–6, 273, 274

vortex creation, 234
vortex finders, 310, 400–2

W

walls
　area, 281–2
　fraction covered by clusters, 183–4
　heat transfer, 202–3
　temperatures, 317
　tubes, 257
　vertical fin effects, 187–9
　water interfaces, 307–9
waste wood, 128
water
　absorption bands, 138–9
　content, 281
　vapor, 448
　velocity, 459
　wall tubes, 257
　water–gas reactions, 66, 68–9, 71
　water-side chemistry, 312
　water–steam flow circuits, 255–7
　water–wall interfaces, 307–9
weeping grids, 376
weirs, 424
whytheat castables, 323, 324
width and breadth ratio, 273, 275
wing walls, 202–3, 329
Winkler, Fritz, 1
Winkler gasifiers, 77–8
wood powder, 90
wood residue, 10–11
wrapped tube arrangements, 292–3